Smart Nanomaterials Technology

Series Editors

Azamal Husen, Wolaita Sodo University, Wolaita, Ethiopia

Mohammad Jawaid, Department of Chemical and Petroleum Engineering, United Arab Emirates University, Al Ain, United Arab Emirates

Nanotechnology is a rapidly growing scientific field and has attracted a great interest over the last few years because of its abundant applications in different fields like biology, physics and chemistry. This science deals with the production of minute particles called nanomaterials having dimensions between 1 and 100 nm which may serve as building blocks for various physical and biological systems. On the other hand, there is the class of smart materials where the material that can stimuli by external factors and results a new kind of functional properties. The combination of these two classes forms a new class of smart nanomaterials, which produces unique functional material properties and a great opportunity to larger span of application. Smart nanomaterials have been employed by researchers to use it effectively in agricultural production, soil improvement, disease management, energy and environment, medical science, pharmaceuticals, engineering, food, animal husbandry and forestry sectors.

This book series in Smart Nanomaterials Technology aims to comprehensively cover topics in the fabrication, synthesis and application of these materials for applications in the following fields:

- Energy Systems—Renewable energy, energy storage (supercapacitors and electrochemical cells), hydrogen storage, photocatalytic water splitting for hydrogen production
- Biomedical—controlled release of drugs, treatment of various diseases, biosensors,
- Agricultural—agricultural production, soil improvement, disease management, animal feed, egg, milk and meat production/processing,
- Forestry—wood preservation, protection, disease management
- Environment—wastewater treatment, separation of hazardous contaminants from wastewater, indoor air filters.

Imran Uddin
Editor

Smart Nanosensors

Springer

Editor
Imran Uddin
Department of Conservative Dentistry
and Endodontics
Saveetha Dental College and Hospitals
Saveetha Institute of Medical and Technical
Sciences (SIMATS)
Saveetha University
Chennai, Tamil Nadu, India

ISSN 3004-8273 ISSN 3004-8281 (electronic)
Smart Nanomaterials Technology
ISBN 978-981-96-3877-2 ISBN 978-981-96-3878-9 (eBook)
https://doi.org/10.1007/978-981-96-3878-9

This Springer imprint is published by the registered company Springer Nature Singapore Pte Ltd.
The registered company address is: 152 Beach Road, #21-01/04 Gateway East, Singapore 189721, Singapore

Foreword by Prof. Dr. Mohmed Isaqali Karobari

The intersection of nanotechnology and sensor development is one of the most exciting frontiers in modern science. The ability to design and implement smart nanosensors has opened unprecedented opportunities across disciplines, ranging from healthcare and agriculture to environmental monitoring and industrial applications. These sensors, powered by the remarkable properties of nanomaterials, enable precision detection and monitoring at scales previously unimaginable, offering transformative solutions to some of humanity's most pressing challenges.

The book *Smart Nanosensors* presents an exceptional compilation of research and advancements in this dynamic field. It brings together a diverse range of topics, showcasing the breadth and versatility of nanosensors while offering a detailed exploration of their fabrication, functionality, and applications. From healthcare innovations like early cancer diagnosis and drug delivery systems to agricultural breakthroughs in crop protection and environmental applications such as pollution detection, this book captures the essence of how nanosensors are revolutionizing industries.

The chapters of this book delve deeply into cutting-edge developments across a range of disciplines. In healthcare and diagnostics, topics such as smart nanoparticles for early cancer diagnosis, theranostic drug delivery systems, and biosensors for antimicrobial resistance surveillance offer hope for addressing critical medical challenges with precision and efficiency. In environmental and agricultural applications, carbon-based sensors for pollution detection, biosensors for monitoring climatic stressors, and nanosensors for crop protection demonstrate how these technologies contribute to sustainability and resource optimization. Insights into food quality, safety, and authenticity monitoring underscore the potential of nanosensors to ensure health and safety standards in an increasingly complex global food system. Additionally, advancements in materials and fabrication, such as substrates for colorimetric sensors, polymer-based nanosensors, and rare-earth nanomaterials, highlight innovative approaches to sensor design and functionality for diverse applications.

Each chapter is authored by experts in their respective fields, and the book embodies a multidisciplinary approach, offering readers a well-rounded understanding of the subject. Moreover, the focus on sustainability, ethical considerations, and emerging challenges ensures that this book is not just a technical resource but also a guide to responsible innovation. As we navigate the era of smart technologies, nanosensors will undoubtedly play a pivotal role in shaping the future. Their ability to provide accurate, real-time data and their adaptability to various applications make them indispensable for addressing the demands of modern society. However, with great potential comes great responsibility. Ensuring that these advancements are developed and deployed with careful consideration of their environmental and societal impacts is paramount. It is my privilege to present this book, which I hope will serve as a source of inspiration and knowledge for researchers, students, and professionals. May it spark new ideas, foster collaborations, and drive innovation in this ever-evolving field. With great pride and enthusiasm, I invite you to explore the world of smart nanosensors through the pages of this book and join us in envisioning a future where nanotechnology is harnessed for the greater good of humanity and the planet.

Prof. Mohmed Isaqali Karobari
BDS, MSC.D (Endo), Ph.D. (Endo),
FDS. RCS (Eng.), FPFA, MFDS.
RCPS(Galsg), MBA (Hospital
Administration), PGD (HO and QM),
PDCR, PCPV
Professor and Head (Admin
and Research) at Department
of Conservative Dentistry
and Endodontics
Saveetha Dental College and Hospitals
Saveetha Institute of Medical
and Technical Sciences (SIMATS)
Saveetha University
Chennai, Tamil Nadu, India

Foreword by Prof. Manish Ranjan

The science of nanotechnology and sensor development represents one of the most dynamic and promising areas of modern scientific research. With the advent of innovative nanosensors, we are witnessing the birth of technologies that not only push the boundaries of detection and monitoring but also have the potential to revolutionize industries and improve quality of life globally. From early diagnosis of diseases in healthcare to real-time environmental monitoring, these sensors, driven by the unique characteristics of nanomaterials, are opening up new possibilities across multiple domains. Their ability to function at the molecular level promises to deliver unparalleled accuracy and sensitivity, making it possible to detect and address issues that were once beyond our reach.

This book sounds like a fascinating and comprehensive resource on the cutting-edge field of nanosensors. It highlights a wide range of applications that span healthcare, agriculture, environmental protection, and even food safety, demonstrating the far-reaching potential of this technology. The emphasis on sustainability and ethical considerations is particularly noteworthy, as they play a crucial role in ensuring the responsible use of such advanced technologies.

Given the broad scope, I'm curious—what aspect of nanosensors in this book do you find most exciting or impactful? Is there a particular chapter or development with the most promise for addressing a pressing global issue?

Experts in their respective fields author every chapter of this book, and the book embodies a multidisciplinary approach, offering readers a well-rounded understanding of the subject. Moreover, the focus on sustainability, ethical considerations, and emerging challenges ensures that this book is a technical resource and a guide to responsible innovation. As we navigate the era of innovative technologies, nanosensors will undoubtedly play a pivotal role in shaping the future. Their ability to provide accurate, real-time data and their adaptability to various applications make them indispensable for addressing the demands of modern society. Ensuring that these advancements are developed and deployed with careful consideration of their environmental and societal impacts is paramount. I am privileged to present this book, which will serve as a source of inspiration and knowledge for researchers, students, and professionals. I invite you to explore the world of innovative nanosensors through the pages of this book and join us in envisioning a future where nanotechnology is harnessed for the greater good of humanity and the planet.

Prof. Manish Ranjan, BDS, MDS,
Ph.D. (Endo)
Professor and Clinical Head at
Department of Conservative Dentistry
and Endodontics
Saveetha Dental College and Hospitals
Saveetha Institute of Medical
and Technical Sciences (SIMATS)
Saveetha University
Chennai, Tamil Nadu, India

Preface

We are immensely pleased to present our book, *Smart Nanosensors*. Indeed, as educators and researchers in the realm of nanotechnology, we frequently encountered challenges in locating a thorough reference text that adequately covers the fundamental topics in the field while also offering sufficient depth for novice researchers. Therefore, we decided to address this vacuum in the literature by methodically preparing this book, which we hope will be a significant resource for both students and experienced researchers in the field of nanosensors.

The book serves as a comprehensive compilation of contemporary research conducted in the domain of nanosensors. The amalgamation of many elements within the emerging field contributes to the development of a useful collection specifically designed for inexperienced researchers in the domain of smart materials and nanosensor technologies. Nanosensors have been utilized by researchers to effectively apply them in various fields such as biosensors, diagnostics, portable smart nanodevices, electrochemical nanosensors, and fluorescence-based nanosensors. An adequate range of subjects has been incorporated into the present book. It includes enzyme-mimetic use of smart nanomaterials for enhanced biosensing applications, theranostic utilization of smart nanomaterials for targeted drug delivery, sensors for pollutant detection, and the utilization of smart nanomaterials in the development of biosensors for studying host-microbe interactions.

Nanosensors have emerged as a promising avenue for various applications, including sensing in the fields of medicine, packaging, and heavy metal ion detection. Recent developments in the field of smart nanomaterials have led to significant advancements in the application of intelligent switches and sensors within the domains of agriculture, food production, and water treatment.

The primary emphasis of this book is the study of the synthesis and fabrication processes involved in the production of smart materials, together with their application within the domain of sensor technology. The existing body of literature has two main categories: introductory textbooks that provide fundamental knowledge about the field, and specialized publications that focus exclusively on certain subtopics within the domain of sensor technology.

The book covers a wide range of topics, including the synthesis and characterization of smart nanomaterials, and their various applications in different fields such as diagnostic, agriculture, and solar cells environmental pollution detection. A sufficient number of topics are included on the smart nanomaterials and various aspects of emerging and existing applications, including diagnosis of cancer, target drug delivery, gas sensors, food packaging, monitoring climatic stressors in agricultural crops, and pest management. This book includes chapters on polymer-based nanosensors, rare-earth-based nanosensors, and the perspective of nanosensors in the industries, using a control chart approach. It offers a thorough summary of the present research landscape, emphasizing recent developments and future opportunities. The book contains case studies and practical examples that demonstrate the real-world uses of smart nanomaterials, rendering it a valuable resource for academics and industry experts alike. This book will be pivotal in enhancing the comprehension and use of nanosensors across many sectors and scientific fields.

Furthermore, this book explores the useful applications of the disciplines in addition to covering the fundamental ideas and theories. Moreover, it is a great tool for novices as well as for professionals looking for a thorough knowledge of the topic as it features the most recent developments in the industry. The current content of the book makes it a comprehensive reference resource, ideal for scholars in the field. This book will serve as a vital resource for researchers in several areas of materials science seeking to further the development of smart materials and sensor technology.

Chennai, India Prof. Imran Uddin

Acknowledgments

I would like to express my gratitude for the assistance provided by all the writers who have contributed to this work. I express my gratitude to all the students who have consistently motivated me to contribute to the profession via their countless thought-provoking inquiries. Their curiosity and dedication have pushed me to continually expand my knowledge and refine my writing skills. Additionally, I am grateful to my colleagues who have provided valuable insights and feedback throughout this project, helping me to enhance the quality and depth of my work.

I would like to express my gratitude to the Department of Conservative Dentistry and Endodontics, Saveetha Dental College and Hospitals, Saveetha Institute of Medical and Technical Sciences University (SIMATS), for their financial assistance.

Finally, I would like to express our heartfelt gratitude to the family members who assumed additional obligations during the compilation of this book. The consistent support and comprehension shown by our peers enabled us to allocate several hours towards the completion of this book.

Contents

Editor and Contributors

About the Editor

Imran Uddin is a Professor in the Department of Conservative Dentistry and Endodontics, Saveetha Dental College and Hospital, Saveetha Institute of Medical and Technical Sciences, (SIMATS), Saveetha University, Chennai, Tamil Nadu, India. He worked as a senior research scientist at the University of Pannonia, Hungary; postdoctoral fellows at Rhodes University, South Africa, and Korea University, Seoul, Republic of Korea. He also worked as an assistant professor at Aligarh Muslim. He completed his Ph.D. in biotechnology from the CSIR-National Chemical Laboratory, Pune, India. His research interest lies in the fabrication of biocompatible nanostructures for nanobiosensors and flexible electronics. Moreover, his research work includes studying the interface of biology with inorganic nanomaterials, understanding the underlying biological process, developing new industrially relevant renewable nanomaterials, as well as eco-toxicological aspects. Additionally, he is dedicated to exploring the potential applications of these nanostructures in the medical field, such as in targeted drug delivery systems and tissue engineering. With a focus on sustainability, he is also actively involved in investigating the environmental impact of nanomaterials, ensuring their safe and responsible use in the future. He is the recipient of a number of awards and fellowships, including the Start-up Research Grant (Young Scientists) Scheme funded by Life Sciences, Science & Engineering Research

Board (SERB), Department of Science & Technology, Government of India; Senior Research Associateship (Scientists' Pool Scheme) and Senior Research Fellowship funded by CSIR (Council of Scientific and Industrial Research), Government of India; and the National Eligibility Test (NET) conducted by the Agriculture Scientist Requirement Board (ASRB) for Assistant Professor/Lectureship in Indian Universities.

Contributors

Shaik M. Abzal Department of Physics, SRM University-AP, Amaravati, India

Nafis Ahmad Faculty of Dentistry, Jamia Millia Islamia, New Delhi, India

Mostafa Gamal Ahmed Faculty of Science, Chemistry Department, Cairo University, Giza, Egypt

Insha Akbar Department of Chemistry, Aligarh Muslim University, Aligarh, India

Md Nadim Akhtar Department of Chemistry, Aligarh Muslim University, Aligarh, U.P., India

Oyesolape Basirat Akinsipo Department of Chemical Sciences, College of Science and Information Technology, Tai Solarin University of Education Ijagun, Ijebu-Ode, Nigeria

Abid Ali Department of Applied Chemistry, ZHCET, Aligarh Muslim University, Aligarh, India

Nizamul Haque Ansari Department of Physical Sciences (Chemistry), Sant Baba Bhag Singh University, Jalandhar, Punjab, India

A. Arshad Ahamed Department of Physics, SRM University-AP, Amaravati, India

Sameer Aryan Department of Chemistry, Khandelwal College of Management Science and Technology, Bareilly, Mahatma Jyotiba Phule Rohilkhand University, Bareilly, U.P., India

Aya Abd El Kareem Abd El Aziz Faculty of Science, Chemistry Department, Cairo University, Giza, Egypt

Arijit Bag Department of Applied Chemistry, Maulana Abul Kalam Azad University of Technology West Bengal, Haringhata, West Bengal, India

Shashi Bala Department of Chemistry, Lucknow University, Lucknow, Uttar Pradesh, India

Tushar Borase Swentek Nordic Private Limited, Bld-G1, Atal Incubation Centre, IISER Campus, Pashan, Pune, Maharashtra, India

Tejansh Chandole Sanguine Transformation Innovatives LLP, Pune, Maharastra, India

Jatis Kumar Dash Department of Physics, SRM University-AP, Amaravati, India

Rajkumar Divya School of Advanced Sciences, Vellore Institute of Technology (VIT), Vellore, India;
Centre for Biomaterials, Cellular and Molecular Theranostics, Vellore Institute of Technology (VIT), Vellore, India

Namarta Dixit Department of Chemistry, Lucknow University, Lucknow, Uttar Pradesh, India

A. M. El-Khawaga Department of Basic Medical Sciences, Faculty of Medicine, Galala University, Galala City, Suez, Egypt

Oluwatoyin Adenike Fabiyi Department of Crop Protection, Faculty of Agriculture, University of Ilorin, Ilorin, Nigeria

Saiyyeda Firdous VIT School of Agricultural Innovations and Advanced Learning, Vellore Institute of Technology (VIT), Vellore, India

Juan Santiago Hidalg Research and Development Center, University of Pannonia, Nagykanizsa, Hungary

Md Nazrul Islam Chemistry Department, Jawahar Lal Nehru College, West Singhbhum, Jharkhand, India

Adhish Kumar Jaiswal Department of Chemistry, Lucknow University, Lucknow, Uttar Pradesh, India

Sai Lakshmi Janga Department of Physics, SRM University-AP, Amaravati, India

Santhoshkumar Jayakodi Department of Biotechnology, Saveetha School of Engineering, Saveetha Institute of Medical and Technical Science (SIMATS), Chennai, TN, India

Kurapati Kalyan Department of Physics, SRM University-AP, Amaravati, India

Mohd Shoeb Khan Department of Applied Sciences and Humanities, Faculty of Engineering and Technology, Jamia Millia Islamia, New Delhi, India

Zahid Khan Department of Quantitative Methods, University of Pannonia, Veszprem, Hungary

Béla Kovács Faculty of Agriculture, Food Science & Environmental Management, Institute of Food Science, University of Debrecen, Debrecen, Hungary

Sathyabama Kowsalya VIT School of Agricultural Innovations and Advanced Learning, Vellore Institute of Technology (VIT), Vellore, India

Jaya Lakkakula Amity Institute of Biotechnology, Amity University Mumbai, Panvel, Maharashtra, India;
Centre for Computational Biology and Translational Research, Amity University Mumbai, Panvel, Maharashtra, India

Massimo Lazzari Center for Research in Biological Chemistry and Molecular Materials (CiQUS), University of Santiago de Compostela, Santiago de Compostela, Spain

Edgar R. Lopez-Mena Tecnologico de Monterrey, Escuela de Ingeniería y Ciencias, Zapopan, Jalisco, Mexico

Adil Mahammad Department of Chemistry, Aligarh Muslim University, Aligarh, India

Geetanjali Mallick Amity Institute of Biotechnology, Amity University Mumbai, Panvel, Maharashtra, India

Abd El Rahman Mamdouh Mohamed Faculty of Science, Chemistry Department, Cairo University, Giza, Egypt

Riham R. Mohamed Faculty of Science, Chemistry Department, Cairo University, Giza, Egypt

Esraa Said Mohammed Faculty of Science, Chemistry Department, Cairo University, Giza, Egypt

Shoaib Mukhtar Research Group of Environmental and Inorganic Photochemistry, Center for Natural Sciences, Faculty of Engineering, University of Pannonia, Veszprém, Hungary

Mangey Ram Nagar Department of Electronics and Communication Engineering, Noida Institute of Engineering and Technology, Noida, Uttar Pradesh, India

Tamilmaran Nagarajan Department of Biotechnology, Saveetha School of Engineering, Saveetha Institute of Medical and Technical Science (SIMATS), Chennai, TN, India

Diego E. Navarro-López Tecnologico de Monterrey, Escuela de Ingeniería y Ciencias, Zapopan, Jalisco, Mexico

Ayorinde Victor Ogundele Department of Chemistry, College of Pure and Applied Sciences, Kwara State University, Malete, Nigeria;
Laboratory of Natural Products & Drug Discovery, Centre CEBIM, Department of Basic Sciences, Faculty of Medicine, Universidad de La Frontera, Temuco, Chile

Vinay Kumar Pandey Research and Development cell, Biotechnology Department, Manav Rachna International Institute of Research and Studies, Faridabad, Haryana, India

Theivasigamani Parthasarathi VIT School of Agricultural Innovations and Advanced Learning, Vellore Institute of Technology (VIT), Vellore, India

Virendra Patil Swentek Nordic Private Limited, Bld-G1, Atal Incubation Centre, IISER Campus, Pashan, Pune, Maharashtra, India

József Prokisch Institute of Animal Science, Biotechnology and Nature Conservation, Faculty of Agricultural and Food Sciences and Environmental Management, University of Debrecen, Debrecen, Hungary

Navaid Zafar Rizvi Department of Applied Sciences and Humanities, Faculty of Engineering and Technology, Jamia Millia Islamia, New Delhi, India

Somia Samir Saied Faculty of Science, Chemistry Department, Cairo University, Giza, Egypt

Maheswaran Saravanakumar Centre for Biomaterials, Cellular and Molecular Theranostics, Vellore Institute of Technology (VIT), Vellore, India;
School of Bio Sciences and Technology, Vellore Institute of Technology (VIT), Vellore, India

Mariam Ahmed Sayed Faculty of Science, Chemistry Department, Cairo University, Giza, Egypt

Shumaila Shahid Division of Plant Pathology, ICAR-Indian Agricultural Research Institute, New Delhi, India

Ayaz Mukarram Shaikh Faculty of Agriculture, Food Science & Environmental Management, Institute of Food Science, University of Debrecen, Debrecen, Hungary

Prashant Kumar Singh Department of Biochemistry, Lucknow University, Lucknow, Uttar Pradesh, India

Radhakrishnan Sreena Centre for Biomaterials, Cellular and Molecular Theranostics, Vellore Institute of Technology (VIT), Vellore, India;
School of Bio Sciences and Technology, Vellore Institute of Technology (VIT), Vellore, India

Shivangi Srivastava Department of Food Technology, Harcourt Butler Technical University, Kanpur, Uttar Pradesh, India

Tejas R. Suthar Chicago, IL, USA

Atirah Tauseef Section of Plant Pathology and Nematology, Botany Department, Aligarh Muslim University, Aligarh, India

Naveen Tiwari Center for Research in Biological Chemistry and Molecular Materials (CiQUS), University of Santiago de Compostela, Santiago de Compostela, Spain

Imran Uddin Department of Conservative Dentistry and Endodontics, Saveetha Dental College and Hospital, Saveetha Institute of Medical and Technical Sciences, Saveetha University, Chennai, Tamil Nadu, India

Mohammad Umar Institute of Genetics and Biotechnology, Hungarian University of Agriculture and Life Sciences, Gödöllő, Hungary

Ibrahim Garba Wawata Department of Pharmaceutics and Pharmaceutical Technology, School of Pharmacy, Kampala International University Western Campus, Ishaka Bushenyi District, Uganda;
Institute of Biomedical Research, Kampala International University Western Campus, Ishaka Bushenyi District, Uganda;
Department of Pure and Applied Chemistry, Kebbi State University of Science and Technology Aliero, Aliero, Nigeria

Akanksha Yadav Department of Chemistry, Lucknow University, Lucknow, Uttar Pradesh, India

Smart Nanoparticles for Early Diagnosis of Cancer

Geetanjali Mallick and Jaya Lakkakula

Abstract Cancer remains a significant global health challenge, with increasing incidence and mortality rates projected for the coming years. Early detection plays a crucial role in improving survival rates, highlighting the importance of innovative approaches in cancer diagnosis and treatment. This book chapter explores recent advancements in nanomedicine, focusing on the utilization of smart nanoparticles and biosensors for early cancer detection and targeted therapy. Smart nanoparticles offer unique advantages, including precise drug delivery and controlled release mechanisms triggered by endogenous or exogenous stimuli. Biosensors, with their ability to detect specific biomolecules indicative of cancer, provide additional tools for accurate and timely diagnosis. Early detection strategies involving imaging techniques, biomarkers, and biosensors are discussed, alongside the promising role of nanoparticle-based therapies in improving outcomes for patients. These advancements represent a paradigm shift in cancer management, offering enhanced precision, efficiency, and therapeutic outcomes in the fight against this formidable disease.

Keywords Diagnosis · Gold nanoparticles · Silver nanoparticles · Carbon nanoparticles · MRI · Breast cancer

1 Introduction

Globally, both cancer incidence and mortality rates are on the rise. According to GLOBOCAN 2018 estimates, there will be approximately 18.1 million new cancer cases and 9.6 million cancer-related deaths. Predictions indicate that by 2030, annual cancer-related deaths could reach 30 million. Early detection is crucial in the fight against cancer, significantly reducing mortality rates. For instance, breast cancer

G. Mallick · J. Lakkakula (✉)
Amity Institute of Biotechnology, Amity University Mumbai, Panvel 410206, Maharashtra, India
e-mail: spencerjaya@gmail.com

J. Lakkakula
Centre for Computational Biology and Translational Research, Amity University Mumbai, Panvel 410206, Maharashtra, India

I. Uddin (ed.), *Smart Nanosensors*, Smart Nanomaterials Technology,
https://doi.org/10.1007/978-981-96-3878-9_1

shows a nearly 90% 5-year survival rate when detected locally, compared to only 27% survival rate when diagnosed with distant metastasis [1].

In the year 2020 cancer has been considered one of the most dangerous conditions affecting most people around the world which caused around 10 million deaths. At its early stage, it is not detectable and if not diagnosed at the early stage it leads to severe condition, some of the different types of cancer which were focused [2].

While cancer surgery is often effective for early-stage tumors, its limitations become evident when tumors are deeply seated or large, potentially impacting surrounding tissues or organs. For instance, brain surgery can affect cognitive functions, while breast cancer surgery faces challenges in accurately determining tumor size and position. Other examples include permanent fertility impairment from prostate, ovarian, and uterine surgeries, as well as vocal cord and breathing issues post-lung surgery. These limitations highlight situations where surgery may not be the preferred treatment option [3].

In recent years, advancements in nanomedicine have revolutionized cancer diagnosis and treatment by introducing innovative nanostructures. Nanomaterials, with their small size and unique properties, can efficiently bind, absorb, and transport various substances like drugs, DNA, and proteins. They offer high carrier potential, durability, and compatibility with different delivery routes, making them highly versatile in oncology. Their customizable scale, structure, and surface properties further enhance their appeal across a wide range of cancer-related applications [4].

Early detection with the help of smart nanoparticles which is the strategy and mechanism of the smart drug, loading and subsequent release of drugs have found a groundbreaking platform in smart nanoparticles (NPs), offering a highly effective approach to cancer diagnosis and treatment. These NPs demonstrate the capability to release drugs specifically at tumor sites, triggered by endogenous factors such as pH, redox gradients, or enzymes, as well as exogenous stimuli like light, ultrasound, temperature changes, and electric and magnetic fields. This innovative technology holds great promise for enhancing the precision and efficiency of cancer therapy [5].

One such example is the mitochondrial-targeted nanoparticle for the identification of the tumor cells in the host, when the host is infected by cancer the mitochondria of the cancer cells display a completely altered physiological and bioenergetics profile which distinguishes the cancer cells from the normal cells. For this condition drugs such as cisplatin, arsenic trioxide, and paclitaxel were targeted in the tumors by altering the IMM and OMM membrane potential for their detection and further leading to their apoptosis, graphene oxide was also used to deliver doxorubicin (DOX) into the mitochondria [6]. Specifically addressing ovarian cancer, it has been noted that this type of cancer can possibly be diagnosed in females at any age. Reportedly, 45% of females diagnosed with ovarian cancer can survive for up to five years after detection. According to information published by the World Health Organization (WHO), early detection has the potential to decrease the incidence of ovarian cancer by up to 30%. Imaging showed an effective diagnosis of this cancer and the nanoparticles were then loaded with the following drugs which were effective against the early detection and further treatment of the cancer cell lines, biomarker was a key factor in the detection of ovarian cancer [7].

This book chapter covers different nanoparticles and nanomaterials like gold, silver, and carbon nanoparticles, which are utilized in cancer detection and imaging due to their diverse biological uses. Their unique characteristics, including chemical, magnetic, optical, and chemical properties, enable the development of imaging probes for diagnosis and drug materials for cancer treatment. The progress in advanced nanomaterials presents a promising approach for early cancer detection and diagnosis [8].

2 Smart Nanoparticles for Early Diagnosis of Cancer

2.1 Gold Nanoparticles

Gold nanoparticles (AuNPs) exhibit several fundamental properties that render them particularly well-suited for applications in both diagnosis and treatment. One of the key attributes of AuNPs is their high surface-to-volume ratio respectively, which enhances their reactivity and interaction with biological entities [9]. The ease of synthesis of AuNPs further contributes to their utility in biomedical contexts. Their facile production enables the development of tailored nanoparticles for specific diagnostic or therapeutic purposes. Furthermore, the stable nature of gold nanoparticles ensures their longevity and functionality in biological environments [10].

This section provides a concise overview of how these distinctive properties of AuNPs have been harnessed for early detection and treatment modalities (Fig. 1), emphasizing their role in advancing biomedical technologies.

Sakore et al. (2024) conducted a study on the synthesis of AuNPs as a therapeutic and diagnostic potential using green nanotechnology in cancer treatment. The study focused on the use of AuNPs in PTT and inducing apoptosis in cancer cells, using the seed growth method Polymyxin E gold nanoflowers were synthesized and it also demonstrated their efficacy in tumor suppression (in vitro and in vivo) following with laser irradiation techniques respectively. It was observed that the flower-shaped PE@AuNFs was highly biocompatible and effective in photothermal therapy due to their excellent near-infrared photothermal treatment effects. The study also explored the impact of the tumor site on the efficacy of Gold Nanorod-Mediated Photothermal Therapy (GNRPTT) for bladder tumors. Also, it was observed that with a laser power of 0.3 W superficial tumors could be effectively ablated. However, located tumors which were deeper and differently showed less reaction because of their tissue features whereas skin cooling and boosting laser power were found to be effective in some cases. They also investigated the use of PEGylated AuNPs (GNPs) combined with cetuximab (GNPPEG-Cet) in inducing apoptosis in KRAS mutant SW-480 (G12V) CRC cells. The GNP-PEG-Cet nanoconjugates were observed to inhibit the proliferation of tumor cells by inducing apoptosis and halting the cell cycle (sub-G1 phase). The nanoconjugates were also found to be responsible for

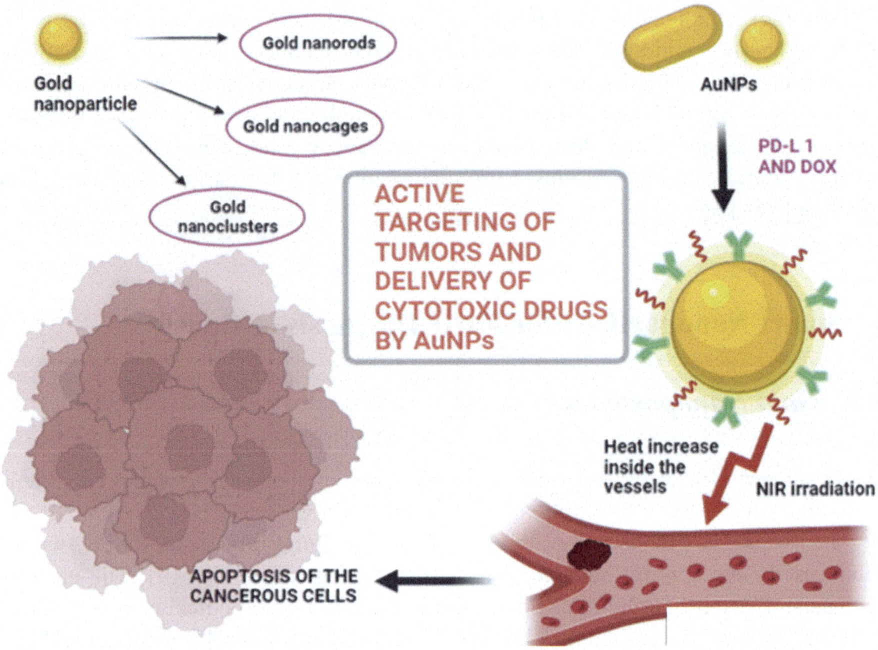

Fig. 1 Gold nanoparticles exhibiting multifaceted effects on tumor cells, promising advancements in cancer treatment

the blockage of the Nrf2-dependent antioxidant pathway, Plasmonic Characteristics, And Simple Functionalization, AuNPs were emerged as an optimistic tool in cancer diagnosis and therapy due to their High Surface Area. The study found that AuNPs KRAS Mutant Colorectal Cancer Cells could be an effective treatment with PEGylated plus cetuximab by inducing apoptosis mechanism and further halting their following proliferation, and anticancer effects of AuNPs were generated from medicinal plants that hence have transformed the treatment of Ovarian Cancer, following this AuNPs have also been successfully used to treat cancers of the Colon, Breast, and Prostate because of their photothermal properties that allowed for the targeted death of the tumor cells with minimal injury to the healthy tissues. Additionally, AuNPs have shown promise as a treatment for leukemia and other malignancies while also improving breast cancer diagnostics [11].

Malekzadeh et al. (2023) conducted a study on the development of targeted AuNPs as a potential contrast agent for molecular CT imaging where they utilized Fe_3O_4@Au NPs in the surrounding atmosphere. To attach the PEG chains to TZ, a bi-functional cross-linker, OPSS-PEG-SVA, was used. The molecular interaction of these compounds was analyzed using Fourier transform infrared (FTIR) spectroscopy, which provided valuable information about their functional groups where the samples' composition and phase purity were identified through X-ray diffraction

(XRD) by recording the peaks obtained from the analysis. The absence of a significant peak related to Fe_2O_3 suggests that the black powder produced is pure Fe_3O_4 and on the other hand, the XRD spectra showed six diffraction peaks corresponding to the following: 220, 311, 400, 422, 440, and 511 plans of Fe_3O_4 where the peaks of Au-Fe_3O_4 describe both Au and Fe_3O_4 NPs. The hydrodynamic sizes were calculated to be 32.6 and 67.4 nm for Fe_3O_4@Au NPs and TZ-PEG-Fe_3O_4@Au NPs, via DLS, which confirmed the appropriate water dispersibility of the NPs. The cell viability of TZ-PEG-Fe_3O_4@Au NPs was tested using MCF-10A and SKBr-3 cells at two incubation time (24 and 48 h) and both MCF-10A and SKBr-3 cells demonstrated a high cell viability at doses up to 100 μg/mL (Fig. 2). In conclusion, Malekzadeh et al. successfully developed targeted AuNPs as a potential contrast agent for molecular CT imaging. The PEG chains were attached to TZ using a bi-functional cross-linker, and Trastuzumab was covalently attached to the Fe_3O_4@Au NPs. The synthesized NPs exhibited appropriate water dispersibility, and their potential use as contrast agents was assessed in in-vitro CT scans at various doses. This study provides valuable insights into the development of targeted gold nanoparticles for molecular CT imaging [12].

Aamri et al. (2023) researched Using Pre-Activated Nylon Membrane and Peroxidase-Mimetic Activity of Cysteamine-Capped AuNPs for Paper-Based Colorimetric Detection of miRNA-21. They aimed to enhance the sensitivity and permeability of the assay and optimize numerous factors that may also affect the signal, which included immobilization situations of the DNA probe and also the nanozyme concentration which were used during the detection process. To achieve maximum

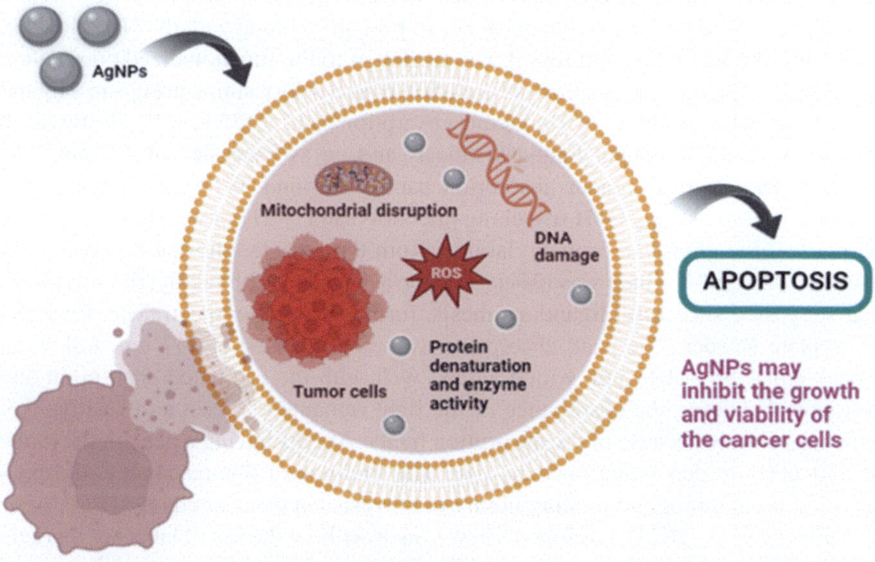

Fig. 2 Silver nanoparticles demonstrating a range of interactions with tumor cells, presenting potential applications in cancer therapeutics

hybridization and stable platform, DNAprobe was immobilized using glutaraldehyde. With a range from 25 to 200 nM, the concentration of immobilized DNAprobe was studied, and the optimum response was gained using 100 nM of DNAprobe, the immobilization time was also optimized and an optimum response was obtained within 60 min. High concentration and long immobilization time of DNAprobe resulted in good immobilization but hindered miRNA-21 hybridization due to steric effect. The immobilization buffer was also tested with different pHs (pH 6, pH 7.4, and pH 8) respectively, and it was observed that the signal was strong with pH 7.4 and 8 which was expected to the nature of the pre-activated membrane. Later different amounts of a prepared Cys/AuNPs suspension to distilled water and found that color intensity increased steadily when the concentration increased from 1 pM to 1 nM of miRNA-21 following a linear range from 1 pM to 1 nM concentrations was also obtained using the 3 σ methods. In FTIR a large band which overlapped with the nylon membrane spectrum from 3000 to 3600 cm^{-1} was observed, which confirmed the existence of the DNA. On the same spectrum, an overlapping band at 1660 cm^{-1} with the nylon membrane band was observed after modification with DNA and glutaraldehyde. Overall, the team's research on optimizing the immobilization conditions of the DNAprobe and the nanozyme concentration has resulted in a sensitive and accurate detection method for miRNA-21 using paper-based colorimetric detection [13].

Vogel et al. (2023) researched the use of fluorescent AuNPs as a theranostic agent for highly radio-resistant cancer cells. During the research, a continuous mass loss of the nanoparticles up to 700 °C was observed, mainly due to the surface desorption of the surface chemical functionalization which was composed of numerous organic constituents. The weight loss of 63% was mainly attributed to water molecules, the AuNPs powder's water of constitution was then eliminated up to approx. 400 °C and was hypothesized because of the existence of porosities along with the Nanopowder. Following the second weight loss it was attributed to the functionalized compound, the IR analysis showed a successful formation of secondary amine groups in a ligand structure with the existence of a moderate absorption band at 3350 cm^{-1}. The research also showed a Triumphant ligand synthesis and an average size of 1.5 nm. The spectrum indicated a broader absorption band of around 3370 cm^{-1}, which was characterised to a strong O–H stretching (alcohols) and C–H (alkyne) rather than any primary amine groups in propargylamine. Both the samples showed an absorption band at 1420 cm^{-1}, which was observed to be associated with methylene scissoring that suggested successful ligand synthesis, further using the SpectraMax Plus 384 Microplate Reader. The peak absorption was then detected at 647 nm, following a peak emission at 673 nm, which agreed well with the estimated absorption and emission. However, the absorbance of the light remained at least at a relative 0.3, which was likely because of the absorption by the AuNPs having a peak in the range of 550–600 nm depending on their size. The findings of this research could be a potential breakthrough in treating highly radio-resistant cancer cells [14].

Yaiwong et al. (2023) developed a new sensor without the use of labels or markers that can detect Alpha-Fetoprotein (AFP). The sensor is based on AuNPs coated with polyethyleneimine (PEI) and an aptamer complexed with toluidine blue (TB). SEM, TEM, and UV-Vis spectroscopy were used to analyze the morphology, and different characteristics of the PEI-AuNPs. The results showed that the PEI-AuNPs were uniformly distributed and had a quasi-spherical shape, with an average diameter of 21.94 ± 2.97 nm. At 524 nm the maximum peak in the UV-Vis absorption spectrum was observed, Cyclic Voltammetry (CV) and Electrochemical Impedance Spectroscopy (EIS) to observe the behavior of the aptasensor during each and every step of the fabrication was used where, the CV responses of the sensor were studied using a redox probe in a 0.01 M Phosphate-Buffered Saline (PBS) solution with a pH of 7.4. The quantitative data, including the Peak-To-Peak Separation (ΔEp), Oxidation Peak current (Ipa), and Reduction Peak Current (Ipc), were collected and analyzed and the results showed that the sensor had good reproducibility, with RSD values of 1.60% and 2.22% for the detection of 0.1 ng/mL and 50 ng/mL AFP, respectively. The researchers also tested the specificity of the sensor by examining its ability to detect AFP in the presence of interfering substances. The results showed that the sensor was able to detect 5.0 ng/mL AFP even in the presence of 100-fold higher concentrations of interfering substances [15].

In another study conducted by Bonamy et al. (2022) the focus was on the impact of green AuNPs coating on the internalization, trafficking, and efficiency of photothermal therapy for skin cancer. The GNPs were obtained by following two different methods: tGNP was synthesized using chemicals like citrate by following Turkevich's protocol, and gGNP was synthesized using extracts of *Hubertia ambavilla* rich in reducer molecules in a reverse protocol. The formulation of spherical nanoparticles was confirmed by the macroscopic color change of the following plant extract after the introduction of the tetrachloroauric acid, confirmed by UV-Vis spectroscopy. The GNPs' size and shape were spherical with an average diameter of around 15 nm, confirmed by TEM images, further to confirm the potential use of gGNPs for PTT, an in vitro study was conducted using laser irradiation to measure the hyperthermia. Experiments showed an increase in temperature of 14 and 20 °C for an irradiation of 0.2 and 0.6 W/cm^2, the nanoparticles were accumulated in cancerous cells since it is known that tumors undergo metabolic reprogramming with an increase in glucose uptake and overexpression of VEGFR1/2 at the membrane of cancerous cells. In conclusion, the study by Bonamy et al. demonstrated the successful synthesis of stable gGNPs using green chemistry rules, which could potentially be used for photothermal therapy of skin cancer. The in vitro study provided promising results, indicating the potential use of gGNPs for hyperthermia induction. The researchers aimed to optimize the endocytosis of the nanoparticles in cancerous cells by targeting the metabolic reprogramming of tumors and overexpression of VEGFR1/2 at the membrane of cancerous cells [16].

Medici et al. (2021) conducted an extensive investigation into the diverse applications of AuNPs in the detection, diagnosis, and treatment of cancer cell lines. The researchers harnessed the versatility of AuNPs in both in vitro and in vivo settings to explore their potential in cancer diagnosis. Various diagnostic techniques, which included MRI, CT scans, Positron Emission Tomography (PET), fluorescence, and X-ray analysis, were employed to detect cancer cells. For in vivo diagnosis of breast cancer cell lines, CT scans were conducted using macrophages and dendrimer-entrapped AuNPs. Simultaneously, PET analysis was utilized for diagnosing HeLa cell lines. In MRI analysis, a novel AuNPs nanoprobe, coupled with Fe_3O_4-NPs, was developed and paired with either antibody FITC or quantum dots for diagnostic purposes. The researchers successfully achieved prostate cancer diagnosis using a linear detection method covering a broad range from 10 pM to 100 nM. The sensitivity limit was remarkable at 10 pM, demonstrating a high level of precision, with a detection sensitivity recorded at 1.65 nF/pM. This multifaceted approach highlighted the potential of AuNPs as highly effective agents in diverse diagnostic modalities, showcasing their adaptability and precision in the realm of cancer detection and diagnosis. The study underlines the importance of AuNPs in offering a wide array of diagnostic tools, allowing for a comprehensive examination of cancerous cell lines through various imaging and analytical techniques. The findings suggested that the unique properties of AuNPs make them promising candidates for advancing cancer diagnostics, potentially leading to more accurate and efficient methods for early detection and characterization of cancer cells [17].

Makableh et al. (2023) conducted research focused on the utilization of selective gold nanotubes to detect HER2 biomarkers in breast cancer cells at an early stage. The study employed atomic force microscopy and AuNP sensors to identify cancer cells, revealing consistent behavior in both anodic and cathodic current during cyclic voltammetry (CV). Incorporating AuNPs enhanced the functionality of the sensors, improving diagnostic accuracy. DPV analysis indicated consistent anodic as well as cathodic current behavior of the CV. Initially, the pristine working electrode (WE) exhibited a peak of approximately 31.4 μA, which decreased to 17.8 μA post-aptamer immobilization. FTIR measurements were conducted to identify crucial binding formations, identifying spectral bands in the range of (3750–3000 cm^{-1}) associated with N–H bonds and C–H vibrations from the aptamer's CH_2 group in the sugar-phosphate structure. Additionally, spectral peaks between 2600–900 cm^{-1} were attributed to the S-CH_2 section in the DNA sequence, while spectra around 1100 cm^{-1} indicated the formation of a covalent bond due to sulfide. Bands around 1200 and 1500 cm^{-1} were linked to the sugar present in the DNA structure. Electrochemical Impedance Spectroscopy (EIS), FTIR, and Cyclic Voltammetry Techniques were utilized, demonstrating the aptasensor's log-linear response over a wide concentration range of 1–25 ng/ml, with a very low detection limit reaching pg/ml levels. These findings signify significant advancements in cancer diagnosis techniques. Overall, the study illustrated the effective utilization of selective gold nanotubes for early detection of HER2 biomarkers in cancer cells. The incorporation of AuNPs and other techniques, including EIS, FTIR, and cyclic volumetry, can substantially enhance the precision and efficacy of cancer diagnosis. The study's outcomes are particularly encouraging, showcasing a remarkably low detection limit and a broad concentration range for identifying cancer cells, with potential implications in cancer diagnosis and treatment [18].

Kumar et al. (2023), conducted a study on the development of an immunosensor for detecting the breast cancer CD44 biomarker. They used gold nanoparticles (AuNPs) combined with graphene quantum dots to create a highly sensitive early detection method for cancer cell lines. The AuNPs were extracted from the compounds and then integrated with the graphene NDs for drug delivery. The UV-Vis spectrum analysis of the sample showed a characteristic peak at a wavelength of 229 nm, indicating the presence of an aromatic group that exhibited $\pi \rightarrow \pi^*$ electron transitions. They then performed XRD analysis, which showed a diffracted pattern captured at 2θ from 10° to 80°. Further analysis was conducted using FTIR, which resulted the presence of various functional groups, such as C–H at 3065 and 2940 cm^{-1}, and C–C at 1577 cm^{-1}. The researchers also conducted TEM analysis that showed the particle size to be around 3 nm, with a d-spacing of 0.23 nm. They also put the sample through XPS analysis, which resulted in two strong readings of 285 and 532 eV. Overall, the study demonstrated that the integration of AuNPs with graphene quantum dots can lead to the development of a highly sensitive immunosensor. The various analyses conducted on the sample provided insights into its physical properties, such as its particle size and functional groups, which are crucial for understanding its potential applications. The findings of this study could pave the way for the development of improved detection methods for breast cancer and other diseases [19].

2.2 *Silver Nanoparticles*

Silver-based nanoparticles have gained widespread prominence in contemporary biomedical applications, particularly in the early detection as well as treatment of cancer. Among the various types of silver nanoparticles (AgNPs), those derived through green synthesis have emerged as the most extensively employed. The utilization of environmentally friendly methods for their synthesis aligns with the growing emphasis on sustainable practices in nanotechnology [20].

This section delves into an analysis of several AgNPs that have been synthesized, encompassing not only biologically driven approaches but also physical and chemical methods (Fig. 2). The focus is on understanding the distinct characteristics and applications of silver nanoparticles produced through diverse synthesis techniques, elucidating their effectiveness in the context of cancer diagnosis and treatment. The exploration of biologically, physically, and chemically synthesized AgNPs provides a comprehensive overview of the evolving landscape of nanomaterials in the field of cancer research and therapy.

Nycz et al. (2021) performed a detailed investigation into the creation of biosensors that could be used for the early detection of cancer cells. The biosensors were created by coating titanium dioxide nanotubes with silver nanoparticles (AgNPs). The process of fabrication involved immersing the substrate in a 1 mM $AgNO_3$ solution and depositing AgNPs onto the titanium dioxide. The potential value ranged from −1.25 to −0.7 V, and the process involved 25 cycles with a scan rate of 50 mV/s within a 3-electrode system. The researchers found that incorporating heat shock protein 70 (HSP70) into biosensors enhanced sensitivity in early cancer detection. In particular, the simultaneous use of HSP70 and prostate-specific antigen (PSA) showed promise for diagnosing and prognosticating cancer cell lines, specifically in the context of prostate cancer. The study found that in the early stages of prostate cancer, where the level of PSA remains relatively unchanged and exhibits limited diagnostic properties, HSP70 showed heightened diagnostic sensitivity. This underscores the importance of incorporating HSP70 in biosensors for early cancer detection. The electrodeposition of AgNPs on titanium nanotubes exhibited a discernible decrease in potential current at 0.3 V, indicating the potential utility of this biosensor configuration in improving the efficacy of early cancer diagnosis, particularly in the context of prostate cancer, by leveraging the complementary diagnostic properties of HSP70 and PSA. In summary, the investigation by Nycz et al. provided insights into the creation of biosensors for early cancer detection. The use of silver nanoparticles coated on titanium dioxide nanotubes, as well as the incorporation of heat shock protein 70, showed promise in enhancing diagnostic sensitivity. Additionally, the simultaneous use of HSP70 and PSA could be a valuable approach in diagnosing and prognosticating cancer cell lines, particularly in the context of prostate cancer. These findings highlight the potential of this biosensor configuration in improving the efficacy of early cancer diagnosis [21].

The study conducted by Miranda et al. (2022) focused on the use of AgNPs for the diagnosis and treatment of cancer. The study conducted found that AgNPs had antitumoral effects and high conductivity due to their catalytic activity and plasmonic properties against cancer cell lines detected during in vitro analysis. They also found that AgNPs were suitable for diagnostic analysis against cancer cell lines, and their toxicity depended on the generation of ROS. According to the study, tumoral pancreatic cells were observed to have higher levels of ROS induced by 2.6 and 18 nm AgNPs as compared to normal cells originating from the same tissues. The study found that there was an increase in harmful substances called ROS, along with a decrease in protective substances, which led to oxidative stress at both protein and mRNA levels. The study also showed that smaller silver nanoparticles (2.6 nm) were more harmful than larger ones (18 nm), with the smaller nanoparticles having an estimated IC50 value of 1.67 μg/mL, and the larger ones having an IC50 value of 26.8 μg/mL. Additionally, when cystamine was added to silver nanoparticles, it caused the particles to aggregate and a new absorption band was observed (560 nm). The effectiveness of AgNPs as a radiosensitizer decreased with increasing particle size. Among 20, 50, and 100 nm AgNPs, the 20 nm particles showed better performance. These modified electrodes, which were functionalized with aptamers, were used to detect furaneol, a commonly used flavoring agent, in a range from 2 pM to 200 nM, and provided good analytical performance. Finally, the biosensors were combined with various techniques, including MRI scanning and CT scan, for early diagnosis of cancer and treatment before the tumor gets worse in severe cases. In summary, the study conducted by Miranda et al. provided insights into the potential of AgNPs for cancer diagnosis and treatment. The findings highlight the importance of size-dependent effects of AgNPs on their radiosensitizing effect and their potential use in biosensors for the early diagnosis of cancer [22].

Jouyandeh et al. (2022) conducted a study to investigate the potential use of various metal nanoparticles in cancer treatment and diagnosis. They prepared the nanoparticles using biological, chemical, and physical methods, including ultrasonication and electrochemical methods. The researchers found that silver-based nanoparticles had unique properties that made them suitable for cancer cell line diagnosis due to their strong extinction bands and electron transport properties. Additionally, they created an electrochemical sensor by combining silver nanoparticles and riboflavin-taurine to diagnose malondialdehyde (MDA) in human samples. The RF-PT-Ag NPs platform showed remarkable sensitivity in MDA detection, bacteriophages and silver nanoparticles were then combined to diagnose promonocytic cancer cells, which showed RS peaks of 862.6, 1132, and 1154 cm^{-1}. They used an AuNP-based SERS biosensor chip LSPR to predict critical complications in neurosurgical diagnoses, which had a LOD of 0.74 pM and enhanced the analytical concept by a factor of 3.29×109 with superior reproducibility. Additionally, they found that five SERS-based biomarkers could provide significant information for the prediction and identification of SAH-induced cerebrovascular complications and hydrocephalus, with a 91.1% confidence and 19.3% repeatability. Lastly, the researchers used fluorescent AgNPs to diagnose HIV (HIV-1 p24 antigen) in a linear detection range of 10–1000 pg mL^{-1}. In summary, the study found that various metal nanoparticles, including silver-based nanoparticles, could be used for cancer diagnosis and treatment through electrochemical sensors and combining with bacteriophages. Additionally, AuNP-based SERS biosensor chips and fluorescent AgNPs showed potential for predicting complications in neurosurgical diagnoses and diagnosing HIV, respectively. These findings could have significant implications for the development of diagnostic and treatment tools for various diseases [23].

The study conducted by Farshchi et al. (2020) focused on the use of nanomaterials in the early diagnosis of prostate cancer through biomedicine. The researchers utilized a photoelectrochemical method for analytical approaches to autosensing, using single-stranded DNA as an interface with AgNPs. In addition, an aptasensor was used, which showed a linear range of 1.0×10^{-11} to 1.0×10^{-7} g mL^{-1}, with the detection range noted at 0.3×10^{-13} g mL^{-1}. The analysis resulted in a substantial reduction of the electrochemical responses, and the linear range was noted to be 5 to 2×10^4 pg ml^{-1}, with the archived LOD at 0.33 pg ml^{-1}. They chose to use this analysis because it is sensitive and accurate, and it can replace traditional antibodies for detecting PSA. Instead, they used a short peptide that has good stability and efficiency. Moreover, the accuracy of GO/AgNPs was also increased. The concentration of PSA ranged from 2 to 150 ng ml^{-1}, with a detection limit of 1.14 ng ml^{-1}. Overall, this study highlighted the potential of nanomaterials and biomedicine in the early diagnosis of prostate cancer. The use of a photoelectrochemical method and an aptasensor proved to be effective in detecting PSA and showed significant improvements in accuracy and sensitivity compared to traditional methods. These findings could lead to the development of more accurate and efficient diagnostic tools for prostate cancer, which could ultimately improve patient outcomes and survival rates [24].

O'Brien et al. (2024) developed a sensing platform to detect breast cancer biomarkers through multiplex electrochemical sensing. The sensor signal is generated by metallic silver deposition on two electrodes, which avoids unwanted signal overlap. The sensor platform was designed to detect two biomarkers using a redox label. The platform was connected to a dual-channel electrochemical reader through an electrical connector that enabled interaction between the working electrodes.

The detection limit of the sensors for CA 15-3 and HER2-ECD biomarkers were found to be 5.0 U/ml and 2.9 ng/ml, respectively, which are comparable to conventional assays measured by ELISA. The amperometric dual platform demonstrated the ability to detect TNF and RANKL with detection limits of 3.0 pg/ml and 2.6 pg/ml, respectively. Another electrochemical biosensor was developed to identify MUC-1 biomarkers and MCF-7 BC cells, which exhibit elevated levels of MUC-1 protein. This duplex sensor utilizes a glassy carbon disk electrode instead of a screen-printed electrode, with each electrode linked to a single-channel electrochemical reader for subsequent biomarker and MUC-1 positive BC cell measurements. Limits of detection (LODs) were also reported for miRNA-21 and CA 15-3 at 1.2 fM and 0.14 U/ml, respectively. In summary, the developed sensing platforms can detect specific biomarkers related to breast cancer with high sensitivity and accuracy. These platforms provide a reliable and efficient diagnosis of breast cancer, which can lead to early detection and better treatment outcomes. The platforms also have the potential for high-throughput screening of biomarkers, which can be useful in large-scale studies for biomarker discovery and validation. Overall, the platforms have great potential for clinical applications in the diagnosis and treatment of breast cancer [25].

In a study by Tunc et al. (2024) AgNPs loaded with carboplatin were synthesized and characterized as a potential therapy for both cancer and antimicrobial purposes. The study found that the average size of AgNPs produced was 2.32 nm, with a nanoparticle size distribution of 6.5 nm. However, the carboplatin drug when bound to AgNPs, the largest particle detected was 43.82 nm, and further the average particle size increased to 28.85 nm. The study also confirmed the DLS results through STEM images and EDX analyses. The STEM images indicated that the size of AgNPs was 10 nm or less. On the other hand, AgNPs-Car particles bound to a drug had particle sizes ranging from 30–40 nm and the drug was attached to AgNPs. Moreover, the nanoparticles remained stable. Although agglomeration was observed in the nanoparticles due to solvent evaporation, STEM images were also sufficient to confirm the DLS results. According to the analysis conducted through EDX, it was observed that the concentration of silver in AgNPs was 1.01%. However, when AgNPs were combined with the drug, the concentration of silver reduced to 0.48%. The drug-bound AgNPs were also found to contain 2.57% platinum, which is a specific element of the drug. Further research also revealed that the drugs attached themselves to the –NH groups in AgNPs-Car, enabling the drug to encase the AgNPs. The results of this study also showed that AgNPs-Car were more effective than using carboplatin alone against all types of cancer cells. The IC50 values and cell viability were lower when AgNPs-Car was used. Overall, the study suggests that loaded AgNPs-Car particles can be used as a potential therapy for both cancer and antimicrobial purposes due to their increased efficacy and low toxicity. The study's findings can pave the way for further research and development in the field of nanotechnology-based drug delivery systems [26].

In their study, Mirshahi et al. (2024) explored the impact of electric pulses and silver nanoparticles on cancerous cells viability, using a mathematical model based on a fuzzy dataset called an adaptive neuro-fuzzy inference system (ANFIS). By correlating pulse specifications with the percentage of cell viability, the ANFIS model predicted the likelihood of cancerous cell viability. The calculated root mean square error (RMSE) was 1.1, and a reliable model for prediction should have an R^2 value closer to 1. The ANFIS model showed an R^2 value of 88%, making it an effective tool for predicting system behavior. In the study, the researchers observed that there was no effective synergism between electric pulses and SNPs at a strength of 500 V/cm. However, synergistic effects were observed when both treatment agents were used in combination at a strength of 700 V/cm and a duration of 100 μs. The study observed the greatest synergistic effect when both treatment agents were administered together at a strength of 700 V/cm and a duration of 100 μs. Different combinations of electric pulses, ranging from 500 V/cm in durations of 300 and 500 μs to 1000 and 1500 V/cm in 100 μs, led to notable cell death compared to a control group. When comparing groups that received both treatment agents with those that only received SNPs, significant cell death occurred at strengths of 500 V/cm in 500 μs duration, 700 V/cm across all durations, and 1000 and 1500 V/cm in 100 μs duration. Overall, the study concluded that the presence of SNPs in most electric pulse-treated groups resulted in a significant difference compared to groups without nanoparticles. The ANFIS model was an effective tool for predicting cancerous cell viability, and the combination of electric pulses and SNPs resulted in significant cell death [27].

One study by Khazaei et al. (2023) evaluated the feasibility of using a nanoprobe to detect Ki67 mRNA and uPA in vivo. The study results showed that the nanoprobe was found to be reliable way to screen the antitumor drugs. The study developed a squaraine dye called SQ1 and created a nanoprobe using the dye that had NIR-II/PA bimodal imaging capabilities with favorable photothermal effects. The SQ1 nanoprobe was found to be suitable for tumor-targeting and imaging angiography in mice which made it a potential candidate for breast cancer and lung metastasis theranostics (Fig. 3). The composites were then tested using Rhodamine 6G (R6G) as a test molecule, it was discovered that exosome samples obtained from cultures of MDA-MB-231 and MCF-10A breast cancer cells can be analyzed on synthetic substrates. By combining PCA with Raman intravariability and intervariability in exosomal samples, a cost-effective, eco-friendly, label-free diagnosis method for breast cancer has been developed. This method shows great potential in clinical diagnosis of breast cancer. Additionally, a fluorescent nanoprobe has been created using a gold–selenium (Au–Se) bond which can be used to evaluate the invasive potential of breast cancer cells. Overall, these studies demonstrated the potential of nanosensors and nanoprobes in the early diagnosis and treatment of cancer. The use of these technologies has the potential to revolutionize cancer diagnosis and treatment by enabling earlier detection and more targeted therapies [28].

Rudrappa et al. (2023) conducted research on the Myco-Nanofabrication of AgNPs by *Penicillium brasilianum* NP5 and their impact on the MDA-MB-231 Cell Line. Ten different fungi were isolated from soil samples and sub-cultured on potato dextrose agar media. NP5 was chosen from the fungi for its superior antimicrobial activity. The genomic DNA from the isolate underwent PCR amplification and sequencing of the 18S rRNA gene, resulting in a DNA fragment of 900 base pairs. The reaction mixture of $AgNO_3$ and NP5 exhibited a maximum absorption peak at

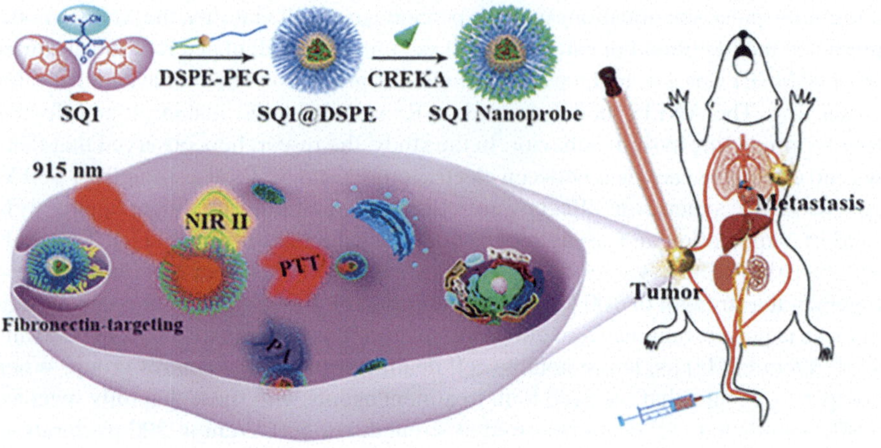

Fig. 3 Nanofunctionalization and molecular engineering of squaraine dye SQ1 for NIR-II/PA bimodal imaging and photothermal ablation of metastatic breast cancer. (Reprinted with permission from Khazaei et al. 2023)

420 nm, while the fungal extract of NP5 showed a peak at 253 nm, which shifted to 420 nm after the synthesis of AgNPs. The synthesized NP5-AgNPs underwent further characterization through FTIR, TEM, EDX, and XRD analyses. The FTIR spectrum analysis of NP5 extract revealed several peaks at different wavelengths ranging from 580 to 3731 cm^{-1}. On the other hand, the synthesized NP5-AgNPs showed peaks at different wavelengths ranging from 1075 to 3713 cm^{-1}. The peaks observed in NP5-AgNPs confirmed the presence of carboxylic acid, primary amine, alcohol, and alkane. Additionally, the TEM images of NP5-AgNPs showed that the nanoparticles were spherical-shaped and had varying sizes ranging from 10 to 60 nm, with an average size of 25.32 nm and a PDI of 0.40. The XRD analysis confirmed the face-centered cubic crystalline nature of the synthesized NP5-AgNPs. The stability and surface charge potential of NP5-AgNPs in an aqueous medium was examined by Zeta potential analysis, which showed that the negative value of 15.3 mV of fabricated NP5-AgNPs exhibited good stability [29].

The study conducted by Félix-Piña et al. (2023) explored the anti-tumor and immunogenic effects of silver nanoparticles synthesized using D-glucose reduction in breast cancer cells. The researchers synthesized silver nanoparticles by reducing them with glucose, resulting a size of 5.991 nm and a nearly spherical shape. These nanoparticles exhibited a specific SPR characteristic of silver in the nanometer size range of 400–480 nm, with a zeta potential value of −7.48 mV, akin to AgNPs prepared via microwave irradiation. Additionally, the study evaluated the cytotoxicity of AgNPs on different breast cancer cell lines, determining the CC_{50} values for each line. The results indicated that treatment with AgNPs-G reduced cell viability in the 4T1, MCF-7, MDA-MB-231, and SKBR-3 cell lines. The study also investigated whether the impact of AgNPs-G on cell viability was associated with cell cycle arrest. Findings revealed variations in cell cycle progression among different cell types.

Furthermore, an in vitro assay was conducted, seeding 5×10^3 cells per well in a 96-well microplate, followed by treatment with varying concentrations of AgNPs-G for 24 h. Subsequently, the wells were washed, and Alamar blue was added, with fluorescence readings taken using a spectrophotometer. Overall, Félix-Piña et al. demonstrated the anti-tumor and immunogenic properties of silver nanoparticles reduced with glucose in breast cancer cells. Treatment with AgNPs-G reduced cell viability across various breast cancer cell lines by inducing cell cycle arrest. The in vitro assay further confirmed the cytotoxic effects of the nanoparticles on the cells. This study offers valuable insights into the potential therapeutic application of silver nanoparticles reduced with glucose in breast cancer treatment [30].

2.3 Carbon Nanoparticles

Research findings indicate widespread utilization of various carbon NPs in the diagnosis and treatment of cancer. Among these, prominent examples include graphene, fullerene, and carbon nanotubes [31]. Although these nanoparticles share a commonality in their carbon composition, they exhibit distinct structural features that set them apart from each other [32]. In this section, our attention is directed toward a comprehensive exploration of the applications of these carbon nanoparticles in the early detection processes associated with cancer [33].

The discussion encompasses the unique structural attributes of graphene, fullerene, and carbon nanotubes, elucidating how these differences play a crucial role in their respective functionalities for early cancer detection [34]. By delving into the specific applications and methodologies employed in utilizing these carbon nanoparticles, the section aims to provide a detailed understanding of their contributions to advancing diagnostic techniques in the realm of cancer research [35].

Feng et al. (2021) conducted a study to evaluate the effectiveness of preoperative carbon nanoparticle (CNP) labeling in patients undergoing laparoscopic radical gastric cancer surgery. The study involves labeling the participants between 24–48 h before surgery and then doing a gastroscopy. The results revealed that 76.3% of the patients had stomach cancer, with 14.2% requiring an initial biopsy to confirm the diagnosis. The research comprised 1199 patients, with 62.2% being male and an average age of 57.7 years. There was no significant change in tumor size, pathological kind, or Lauren categorization between the CNP-labeled and control groups. Additionally, the pathological T stage, which assesses tumor infiltration depth, was identical. The investigation discovered that all of the cases were of a single stomach cancer, with 70.8% being distal and 29.2% proximal. Six instances were diagnosed thanks to the CNP labeling, which would not have been discovered otherwise. However, the study found no significant differences between the groups that got CNP labeling and the control group (Fig. 4). Overall, the study suggested that preoperative CNP labeling can be a useful tool in diagnosing gastric cancer in patients undergoing with laparoscopic radical gastric cancer surgery. However, further research is needed to determine if CNP labeling can improve the overall outcomes and prognosis of gastric cancer patients [36].

Zhu et al. (2021) developed a method for detecting circulating tumor cells (CTCs) using mRNA-wrapped carbon nanotubes. Before detecting the CTCs, the researchers assessed the viability of their SWCTCs colorimetric probe by measuring its absorbance at a peak wavelength of 652 nm, which they labeled A0. They employed MCF-7 cells as the target tumor for detection. Fe_3O_4–SiO_2-Gel MNs were discovered with an absorption peak at 1640 cm^{-1}, corresponding to the stretching vibration of the N–H bond in gelatin. Other peaks emerged at 570 and 1070 cm^{-1}, due to the Fe–O stretching vibration of Fe_3O_4 and Si–O–Si vibration, XRD patterns revealed diffraction peaks at Fe_3O_4, Fe_3O_4–SiO_2, and Fe_3O_4–SiO_2-Gel MNs, which correspond to signals of the Fe_3O_4 cubic structure. The peak locations remained constant, showing that the SiO_2 and gelatin alterations did not damage Fe_3O_4 normal structure. XPS analysis revealed distinct Fe 2p, C 1 s, and O 1 s peaks at Fe_3O_4. In the colorimetric detection, various amounts of mDNA were incubated with SWCTCs. The absorbance change ranged from 0.1–10 μM, with a linear regression of 0.04 CmDNA + 0.11 resulting in an R2 of 0.990. The researchers discovered that the Zeta potential of Fe_3O_4–SiO_2-Gel was -7.28 ± 0.35 mV (n = 3), while the potential of EpCAM aptamer P1 alone was -13.32 ± 0.47 mV. After incubating EpCAM aptamer with Fe_3O_4–SiO_2-Gel, the potential was determined to be -14.70 ± 0.56 mV (n = 3). It was also noted the detection limit for mDNA to be 0.12 μM. When these

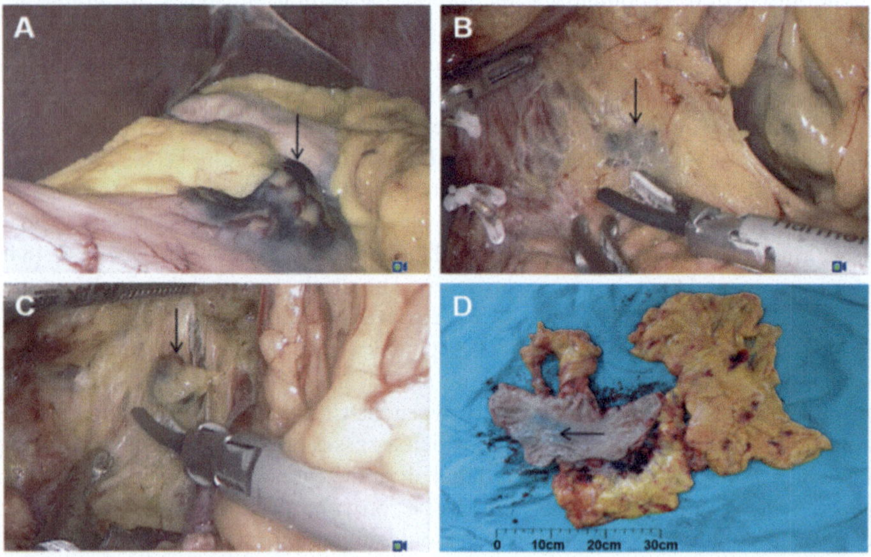

Fig. 4 Carbon nanoparticles labeling helped to find the lesion site (gastric angle) (**a**) and draining lymph nodes (**b**, **c**) in laparoscopic radical gastrectomy. **a** Lesions of gastric angular labeled with carbon nanoparticles; **b** Black staining lymph node No. 8a; **c** Black staining lymph node No. 11p; **d** Gross specimen. (Reprinted with permission from Feng et al. 2021)

complexes were used to detect MCF-7 cell lines, the SWCTCs showed a stronger intensity, enabling the researchers to locate the tumor cells precisely [37].

A research team led by Rashid et al. (2021) conducted a study on the use of dopamine/mucin-1 functionalized electro-active carbon nanotubes as a probe for detecting breast cancer through an electrochemical immunosensing technique. The detection strategy consisted of three main steps, which involved preparing a nanoprobe by modifying the electrode, followed by competitive recognition leading to the formation of MWCNTs. This process was followed by further analysis of the MWCNTs. They used various techniques to analyze the MWCNTs, including FTIR analysis, a spectroscopic technique used to monitor the functional groups present in the MWCNTs. The FTIR analysis showed a spectral peak of 1490 cm^{-1}, indicating C–H bending, 1202 cm^{-1} for C–O–C bending, and 1490 to 1650 cm^{-1} for C==C bending. Furthermore, UV-Vis spectra analysis was performed, which showed a characteristic peak at 250 nm, indicating a red shift due to electronic transition. The SEM (Scanning Electron Microscopy) analysis showed rough surface morphology, and typical XRD (X-ray Diffraction) peaks were obtained at $2\theta = 26.68°$ and 48°. The study has shown promising results for detecting breast cancer using the electrochemical immunosensing technique and the use of MWCNTs functionalized with dopamine/mucin-1 has proved to be an effective nanoprobe for detecting breast cancer. The functionalization of MWCNTs with dopamine/mucin-1 has improved the sensitivity and specificity of the following detection process. The use of FTIR, UV-Vis spectra, SEM, and XRD analysis has provided valuable information on the structural and morphological characteristics of the MWCNTs. In conclusion, Rashid et al.'s research study has demonstrated the potential of using dopamine/mucin-1 functionalized electro-active carbon nanotubes as a probe for detecting breast cancer. The study provides valuable insights into the design and development of effective nanoprobes for cancer detection, which can be further explored for clinical applications [38].

Kim et al. (2022) focused their study on detecting ovarian cancer by using carbon nanotubes and spectral fingerprinting in serum, coupled with machine learning. They employed Single-Walled Carbon Nanotubes (SWCNTs) that were covalently functionalized and were then used to create organic color center-DNA (OCC-DNA) nanosensors. These nanosensors were then used to detect ovarian cancer from patient serum through a nanosensor assay and a computational model. By introducing OCC-DNA to SWCNTs coated with single-stranded DNA, the identification of DNA strands was made possible, forming specific wrapping patterns for morphological analysis and early detection of tumor cells. A study was conducted to measure certain biomarkers in the serum samples of patients, which included HE4, YKL40, creatinine, and bilirubin, using immunoassays. They evaluated the accuracy of these biomarkers in diagnosing high-grade serous ovarian cancer (HGSOC) in these patients. The researchers found that plate number 1-based screening with a 50 U ml^{-1} cutoff resulted in 65.3% sensitivity, 88.3% specificity, and an F-score of 0.621 in their patient sample set. However, logistic regression of additional biomarkers only marginally improved HGSOC prediction. SVM algorithm was used to develop binary classification models to differentiate abnormal levels of HE4 and

YKL40 using the clinical reference values of 150 pM for HE4 and 1,650 pM for YKL40. Overall, the study found that the use of carbon nanotubes and spectral fingerprinting in serum, along with machine learning, can potentially improve the accuracy of ovarian cancer detection [39].

Makableh et al. conducted a study on selective gold nanotubes that demonstrated enhanced response for the early detection of HER2 biomarkers. The researchers used a biosensor composed of carbon and gold nanoparticles to detect HER2 protein against cancerous breast cell lines. The linear range of absorbance was measured using differential pulse voltammetry, which showed a range of 1 Pg mL^{-1}–25 ng mL^{-1}. As a result, the detection range of tumor-containing cells was found to be 4.4 Pg mL^{-1}. Multi-Walled Carbon Nanotubes (MWCNTs) for the detection was used, which led to the early detection of cancer cell lines against breast cancer. For identification important binding formation, FTIR measurements were carried out in the spectral range of 4000–1000 cm^{-1} respectively. After aptamer immobilization, bands with a spectral range between 3750–3000 cm^{-1} were observed, which can be attributed to the N–H bond. The NH_2 bond in the aptamer sequence structure vibrates to produce a specific frequency. Additionally, the CH_2 group of the sugar-phosphate structure in the aptamer produces vibrations in the range of 2800–3000 cm^{-1}. The spectral peaks in the range of 2600–900 cm^{-1} are attributed to the S-CH_2 expansion in the DNA sequence, and the appearance of peaks around 1100 cm^{-1} indicate the bending vibrations of the phosphate (PO^-) group, which indicates the formation of a covalent bond of sulphide. The DNA structure's sugar component can be associated with the bands around 1200 and 1500 cm^{-1} (Fig. 5). Overall, Makableh et al.'s study showed that selective gold nanotubes can lead to enhanced response, making them a promising option for early detection of HER2 biomarkers. The use of carbon nanoparticles in biosensors has also shown potential for detecting cancerous cells against breast cancer and MWCNTs were used for detection, which had a positive impact on early detection. Additionally, FTIR measurements were used to identify important binding formations, which can be attributed to different bonds and vibrations in the aptamer sequence structure, sugar-phosphate structure, and DNA sequence [18].

Tao et al. (2023) conducted a study to investigate the characteristics of systematic lymph node dissection and factors influencing sentinel lymph node biopsy in endometrial carcinoma using carbon nanoparticles. The study group included 54 patients with endometrial carcinoma who underwent surgery with carbon nanoparticle injection. The clinical characteristics of both groups were similar, and patients' vital signs remained stable during the surgery and no adverse events occurred during or after the surgery. The study found that a significantly a greater number of lymph nodes were removed in the study group when compared to the control group. The percentage of removed lymph nodes with an average diameter of $\leq$0.2 cm was also higher in the study group compared to the control group. The detection rate of sentinel lymph node mapping with carbon nanoparticles was 70.4%, and all three cases with lymph node metastases were successfully identified. The sentinel lymph node biopsy demonstrated perfect sensitivity, specificity, and accuracy, with 52.5 years for the study group and 50.5 years for the control group, indicating no notable variance

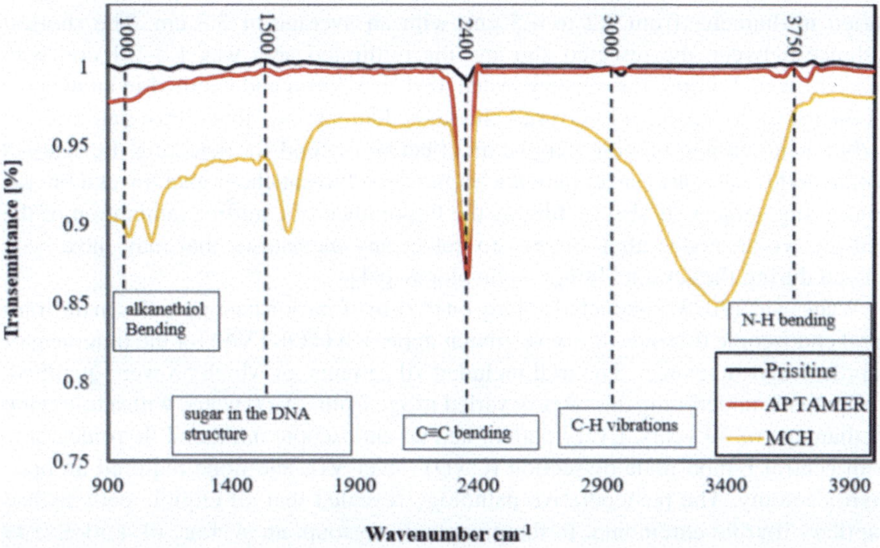

Fig. 5 The FTIR result for each modification step as (pristine CNTs/GNPs, CNTs/GNPs with aptamer, and CNTs/GNPs with aptamer and MCH). (Reprinted with permission from Makableh et al. 2023)

between the two sets of participants. In conclusion, the study suggests that carbon nanoparticles can be used as a safe and effective tracer for sentinel lymph node biopsy in endometrial carcinoma. The procedure can accurately identify lymph node metastases, and its sensitivity, specificity, and accuracy are 100%. The study also indicates that the use of carbon nanoparticles does not cause any adverse effects on patients' health during and after the surgery [40].

A group of researchers led by Lin et al. (2023) conducted a study on the use of ultrasound-assisted carbon nanoparticle labelling for neoadjuvant chemotherapy in breast cancer patients undergoing breast-conserving surgery. The study involved 32 patients, all of whom had following assays: Neoadjuvant Chemotherapy, Carbon Nanoparticle Labeling, and A Sentinel Lymph Node Biopsy. The marking time for the sentinel lymph nodes varied from 10–20 min, with an average of 15.8 min. Out of the 32 patients, 29 showed black sentinel lymph node staining, yielding a sentinel lymph node identification rate of 90.6%, with an average of 1.8 sentinel lymph nodes found in each. Three patients did not exhibit black sentinel lymph node staining, and one of them was diagnosed with axillary lymph node metastases. Following the operation, the axillary lymph nodes were checked, and 14 instances were determined to be negative, while 12 were positive. The sentinel lymph node biopsy found 11 of the 12 positive patients. The study discovered that the sensitivity was 91.7%, specificity was 100%, positive predictive value was 100%, negative predictive value was 93.3%, and the false-negative rate was 3.8%. Out of the 32 patients, 26 had breast-conserving surgery. Five of them required a second operation due to an intraoperative frozen pathology evaluation that revealed a positive incision margin. The removed tumors varied in diameter from 2.2 to 4.5 cm, with an average of 3.3 cm. The shortest distance between the resected site and the infiltrated area was 1.0–2.1 cm, with an average of 1.4 cm. The surgery lasted 48.4 ± 5.3 min and resulted in an average blood loss of 20.8 ± 6.2 mL. Overall, the study demonstrates that ultrasound-assisted carbon nanoparticle tagging may be an effective method for detecting the sentinel lymph nodes in breast cancer patients having neoadjuvant chemotherapy and breast-conserving surgery. It also highlights the importance of careful examination of the axillary lymph nodes after surgery to detect any metastases that may have been missed during the sentinel lymph node biopsy [41].

Wang et al. (2023) conducted a study on the use of carbon nanoparticles in the transoral endoscopic thyroidectomy vestibular approach (TOETVA) for the treatment of papillary thyroid cancer. The trial included 70 patients, of which 53 were qualified. The patients undergoing the surgery varied in age from 20–60 years, with a following median age of 39 years. Every patient had an endoscopic unilateral thyroidectomy with central lymph node dissection (CND) (level VI), and none required an open thyroidectomy. The postoperative pathology revealed that all eligible patients had papillary thyroid carcinoma. In the preoperative group, an average of 4.50 ± 0.49 lymph nodes per patient were dissected, with 20 showing metastases. In the intraoperative group, an average of 5.00 ± 0.56 lymph nodes per patient were dissected, with 15 lymph nodes showing metastases. The total number of central lymph nodes (CLN) and central lymph node metastases (CLNM) did not change significantly between the preoperative and intraoperative groups. The study also assessed the preoperative and postoperative levels of parathyroid hormone (PTH) in both groups. The preoperative and intraoperative groups had similar baseline PTH levels. Although the postoperative PTH level decreased on the first day, there was no meaningful difference between the preoperative and intraoperative groups, and it returned to preoperative levels within the first month. According to the study, there was one episode of unintentional parathyroid gland (PG) removal in the preoperative group, compared to seven in the control group. This indicates that there were more inadvertent PG removals in the intraoperative group. Additionally, the preoperative group had a lower ratio of PG auto transplantation than the intraoperative group. Overall, the study suggests that TOETVA using carbon nanoparticles is a safe and effective method for the treatment of papillary thyroid cancer. The preoperative application of carbon nanoparticles can prevent accidental removal of PG and minimize the need for PG auto transplantation, making the procedure more efficient and less invasive [42].

A study conducted by Wang et al. (2023) explored the application of carbon nanoparticles combined with refined extracapsular anatomy in endoscopic thyroidectomy. The study comprised 108 individuals who had a successful endoscopic thyroidectomy. The majority of the patients were women, with a mean age of 30.98 years. The patient's height, weight, and BMI were all documented. The oral vestibule, axilla-breast, and areolar approaches were employed during surgery. The majority of patients received hemithyroidectomy, with six undergoing complete thyroidectomy and central neck dissection (CND). The study discovered that postoperative transitory hoarseness occurred in 2.77% of patients, whereas transient neuromuscular symptoms occurred in 3.70%. However, no lasting hoarseness or neuromuscular problems were seen. There were no reports of persistent hypoparathyroidism. Patients with transitory hypoparathyroidism healed after three months without long-term calcium therapy. The study discovered that CNs can greatly reduce parathyroid damage. The average operation duration was 156.94 min, and the total postoperative drainage volume was 149.69 mL. The postoperative hospital stay was 3.49 days. The average tumor size was 0.72 cm, with 88.89% of patients having a tumor size of $\leq$1 cm and 11.11% having a tumor size more than 1 cm. The average number of extracted LNs was 5.54, with 57.41% of patients having less than 5 LNs and 42.59% having more than 5 LNs. Of the 108 patients, 37.96% had metastatic lymph nodes, with an average of 0.91. Of individuals with metastatic LNs, 65.85% had $\leq$2 metastatic LNs, whereas 34.15% had more than two. All 108 patients had histological grade I, and the pathological TNM staging was stage I. In conclusion, Wang et al. demonstrated the safety and efficacy of carbon nanoparticles coupled with revised extracapsular architecture in endoscopic thyroidectomy. The study showed that the procedure was associated with low rates of hoarseness, neuromuscular symptoms, and hypoparathyroidism. The study also showed that CNs can significantly prevent parathyroid injuries. These findings suggest that the use of carbon nanoparticles combined with refined extracapsular anatomy in endoscopic thyroidectomy may be a viable option for patients with thyroid disorders [42].

In a study conducted by Koowattanasuchat et al. (2023) a novel approach was developed for detecting cancer DNA in human blood. The researchers employed cysteamine-capped gold nanoparticles (Cyst/AuNPs) and machine learning-enabled cell phones to identify the Methylscape biomarker, which is found in cancer DNA. Cyst/AuNPs were employed as a nanoprobe to detect Methylscapes. The DNA's Methylscape variation can create different self-assembly patterns for DNA fragments. Additionally, the presence of positively charged AuNPs adsorbed onto DNA aggregates might uncover distinctive self-assembly patterns in cancer and normal genomes because of the diverse Methylscapes that exist. The Cyst/AuNPs were produced in accordance with the described literature. The UV-Vis spectra of Cyst/AuNPs showed an absorption peak at 525 nm, which was caused predominantly by the localized surface plasmon resonance (LSPR) phenomena reported in Cyst/AuNPs. Based on the UV-Vis spectroscopy analysis using an extinction coefficient of $3.67 \times 108\ M^{-1}\ cm^{-1}$ at 525 nm, the concentration of the produced Cyst/AuNPs was determined to be 10 nM. Additionally, the TEM image revealed that these nanoparticles had a spherical shape and an average diameter of 33.5 ± 4.8 nm. The Cyst/AuNPs had a zeta potential of 31.7 ± 0.3 mV, suggesting that positively charged cysteamine stabilized the AuNP surfaces. Cyst/AuNPs have outstanding dispersion and anti-aggregation properties due to electrostatic repulsion. Further the hydrodynamic size and zeta potential of DNA clusters were also measured under various circumstances. The size of mDNA aggregate in water is smaller than that of uDNA aggregate because of the stronger hydrophobic force of the methyl group in mDNA. However, when $MgCl_2$ is present, the difference in size between mDNA and uDNA is not as significant as it is in water. Overall, the study offers a promising approach for detecting cancer DNA in human blood using Cyst/AuNPs and machine learning-enabled smartphones [43].

3 Conclusion

In conclusion metal NPs are the future compounds that can be implemented for early diagnosis and also for the treatment of the cancer. On the other hand, it also showed some of the disadvantages such as some of the derivatives of the metal nanoparticles were not effective against the cell lines because of the lack of the property which can be solved in the future by finding some of the combinations that can increase their effectiveness against the cell lines. Also, if not 100% it should be that effective so that the mortality rate can decrease when these NPs are used for early diagnosis before cancer reaches its dangerous stage and the treatment is impossible. The FTIR, X-ray diffraction, and some other analyses showed very positive peaks where the absorbance was excellent and the NPs were used along with the biosensing imaging techniques which caused a positive result when they were implemented for the drug-delivery process.

References

1. Zhang Y, Li M, Gao X, Chen Y, Liu T (2019) Nanotechnology in cancer diagnosis: progress, challenges and opportunities. BioMed Central Ltd. https://doi.org/10.1186/s13045-019-0833-3
2. Sargazi S et al (2022) Application of green gold nanoparticles in cancer therapy and diagnosis. MDPI. https://doi.org/10.3390/nano12071102
3. Gyanani V, Haley JC, Goswami R (2021) Challenges of current anticancer treatment approaches with focus on liposomal drug delivery systems. MDPI. https://doi.org/10.3390/ph14090835
4. Mosleh-Shirazi S et al (2022) Nanotechnology advances in the detection and treatment of cancer: an overview. Ivyspring International Publisher. https://doi.org/10.7150/ntno.74613
5. Wang H et al (2022) UV-irradiating synthesis of cyclodextrin–silver nanocluster decorated TiO_2 nanoparticles for photocatalytic enhanced anticancer effect on HeLa cancer cells. Front Chem 10. https://doi.org/10.3389/fchem.2022.995261
6. Ganji C, Muppala V, Khan M, Purnachandra Nagaraju G, Farran B (2023) Mitochondrial-targeted nanoparticles: Delivery and therapeutic agents in cancer. Elsevier Ltd. https://doi.org/10.1016/j.drudis.2022.103469
7. Hasan MR, Sharma P, Pilloton R, Khanuja M, Narang J (2022) Colorimetric biosensor for the naked-eye detection of ovarian cancer biomarker PDGF using citrate modified gold nanoparticles. Biosens Bioelectron X 11. https://doi.org/10.1016/j.biosx.2022.100142
8. Lan H, Jamil M, Ke G, Dong N (2023) The role of nanoparticles and nanomaterials in cancer diagnosis and treatment: a comprehensive review. [Online]. www.ajcr.us/
9. Liu P, Xie X, Liu M, Hu S, Ding J, Zhou W (2021) A smart MnO2-doped graphene oxide nanosheet for enhanced chemo-photodynamic combinatorial therapy via simultaneous oxygenation and glutathione depletion. Acta Pharm Sin B 11(3):823–834. https://doi.org/10.1016/j.apsb.2020.07.021

10. Yang Y et al (2022) Multifunctional gold nanoparticles in cancer diagnosis and treatment. Dove Medical Press Ltd. https://doi.org/10.2147/IJN.S355142
11. Sakore P, Bhattacharya S, Belemkar S, Prajapati BG, Elossaily GM (2024) The theranostic potential of green nanotechnology-enabled gold nanoparticles in cancer: a paradigm shift on diagnosis and treatment approaches. Results Chem 7. https://doi.org/10.1016/j.rechem.2023.101264
12. Malekzadeh R, Ghorbani M, Faghani P, Abdollahi BB, Mortezazadeh T, Farhood B (2023) Fabrication of targeted gold nanoparticle as potential contrast agent in molecular CT imaging. J Radiat Res Appl Sci 16(1):100490. https://doi.org/10.1016/j.jrras.2022.100490
13. El Aamri M, Mohammadi H, Amine A (2023) Paper-based colorimetric detection of miRNA-21 using pre-activated nylon membrane and peroxidase-mimetic activity of cysteamine-capped gold nanoparticles. Biosensors (Basel) 13(1). https://doi.org/10.3390/bios13010074
14. Vogel S et al (2023) Fluorescent gold nanoparticles in suspension as an efficient theranostic agent for highly radio-resistant cancer cells. J Nanother 4(1):37–54. https://doi.org/10.3390/jnt4010003
15. Yaiwong P, Anuthum S, Sangthong P, Jakmunee J, Bamrungsap S, Ounnunkad K (2023) A new portable toluidine blue/aptamer complex-on-polyethyleneimine-coated gold nanoparticles-based sensor for label-free electrochemical detection of alpha-fetoprotein. Front Bioeng Biotechnol 11. https://doi.org/10.3389/fbioe.2023.1182880
16. Bonamy C et al (2022) Impact of green gold nanoparticle coating on internalization, trafficking, and efficiency for photothermal therapy of skin cancer. ACS Omega. https://doi.org/10.1021/acsomega.2c07054
17. Medici S, Peana M, Coradduzza D, Zoroddu MA (2021) Gold nanoparticles and cancer: detection, diagnosis and therapy. Semin Cancer Biol 76:27–37. https://doi.org/10.1016/j.semcancer.2021.06.017
18. Makableh Y, Athamneh T, Ajlouni M, Hijazi S, Alnaimi A (2023) Enhanced response and selective gold nanoparticles/carbon nanotubes biosensor for the early detection of HER2 biomarker. Sens Actuators Rep 5. https://doi.org/10.1016/j.snr.2023.100158
19. Kumar N et al (2023) Immunosensor for breast cancer CD44 biomarker detection based on exfoliated graphene quantum dots integrated gold nanoparticles. Hybrid Adv 3:100065. https://doi.org/10.1016/j.hybadv.2023.100065
20. Takáč P, Michalková R, Čižmáriková M, Bedlovičová Z, Balážová Ľ, Takáčová G (2023) The role of silver nanoparticles in the diagnosis and treatment of cancer: are there any perspectives for the future? MDPI. https://doi.org/10.3390/life13020466
21. Nycz M, Arkusz K, Pijanowska DG (2021) Fabrication of electrochemical biosensor based on titanium dioxide nanotubes and silver nanoparticles for heat shock protein 70 detection. Materials 14(13). https://doi.org/10.3390/ma14133767
22. Miranda RR, Sampaio I, Zucolotto V (2022) Exploring silver nanoparticles for cancer therapy and diagnosis. Elsevier B.V. https://doi.org/10.1016/j.colsurfb.2021.112254
23. Jouyandeh M et al (2022) Metal nanoparticles-assisted early diagnosis of diseases. Elsevier Inc. https://doi.org/10.1016/j.onano.2022.100104
24. Farshchi F, Hasanzadeh M (2020) Nanomaterial based aptasensing of prostate specific antigen (PSA): recent progress and challenges in efficient diagnosis of prostate cancer using biomedicine. Biomed Pharmacother 132. https://doi.org/10.1016/j.biopha.2020.110878
25. O'Brien C, Khor CK, Ardalan S, Ignaszak A (2024) Multiplex electrochemical sensing platforms for the detection of breast cancer biomarkers. Front Med Technol 6. https://doi.org/10.3389/fmedt.2024.1360510
26. Tunç T (2024) Synthesis and characterization of silver nanoparticles loaded with carboplatin as a potential antimicrobial and cancer therapy. Cancer Nanotechnol 15(1). https://doi.org/10.1186/s12645-023-00243-1
27. Mirshahi S, Vahedi B, Aryana K, Sazgarnia A (2024) Cancerous cell viability affected by synergism between electric pulses and a low dose of silver nanoparticle: an adaptive neuro-fuzzy inference system. Med Nov Technol Devices 21. https://doi.org/10.1016/j.medntd.2023.100280

28. Khazaei M, Hosseini MS, Haghighi AM, Misaghi M (2023) Nanosensors and their applications in early diagnosis of cancer. Elsevier B.V. https://doi.org/10.1016/j.sbsr.2023.100569
29. Rudrappa M et al (2023) Myco-nanofabrication of silver nanoparticles by Penicillium brasilianum NP5 and their antimicrobial, photoprotective and anticancer effect on MDA-MB-231 breast cancer cell line. Antibiotics 12(3). https://doi.org/10.3390/antibiotics12030567
30. Félix-Piña P, Franco Molina MA, Zarate Triviño DG, García Coronado PL, Zapata Benavides P, Rodríguez Padilla C (2023) Antitumoral and immunogenic capacity of β-D-glucose—reduced silver nanoparticles in breast cancer. Int J Mol Sci 24(10). https://doi.org/10.3390/ijms24108485
31. Singh G et al (2021) Carbon based nanodots in early diagnosis of cancer. Frontiers Media S.A. https://doi.org/10.3389/fchem.2021.669169
32. Bao YW, Hua XW, Chen X, Wu FG (2018) Platinum-doped carbon nanoparticles inhibit cancer cell migration under mild laser irradiation: multi-organelle-targeted photothermal therapy. Biomaterials 183:30–42. https://doi.org/10.1016/j.biomaterials.2018.08.031
33. Das S, Saha B, Tiwari M, Tiwari DK (2023) Diagnosis of cancer using carbon nanomaterial-based biosensors. Sens Diagn 2(2):268–289. https://doi.org/10.1039/d2sd00182a
34. Lisik K, Krokosz A (2021) Application of carbon nanoparticles in oncology and regenerative medicine. MDPI. https://doi.org/10.3390/ijms22158341
35. Tang L et al (2021) Insights on functionalized carbon nanotubes for cancer theranostics. BioMed Central Ltd. https://doi.org/10.1186/s12951-021-01174-y
36. Feng Y et al (2021) Value of preoperative gastroscopic carbon nanoparticles labeling in patients undergoing laparoscopic radical gastric cancer surgery. Surg Oncol 38. https://doi.org/10.1016/j.suronc.2021.101628
37. Zhu L et al (2021) Colorimetric detection of immunomagnetically captured rare number CTCs using mDNA-wrapped single-walled carbon nanotubes. Biosens Bioelectron 172. https://doi.org/10.1016/j.bios.2020.112780
38. Rashid S, Nawaz MH, ur Rehman I, Hayat A, Marty JL (2021) Dopamine/mucin-1 functionalized electro-active carbon nanotubes as a probe for direct competitive electrochemical immunosensing of breast cancer biomarker. Sens Actuators B Chem 330. https://doi.org/10.1016/j.snb.2020.129351
39. Kim M et al (2022) Detection of ovarian cancer via the spectral fingerprinting of quantum-defect-modified carbon nanotubes in serum by machine learning. Nat Biomed Eng 6(3):267–275. https://doi.org/10.1038/s41551-022-00860-y
40. Tao S et al (2023) Characteristics of systematic lymph node dissection and influencing factors of sentinel lymph node biopsy using carbon nanoparticles in endometrial carcinoma: a single-center study. World J Surg Oncol 21(1). https://doi.org/10.1186/s12957-023-02922-0
41. Lin N et al (2023) Ultrasound-assisted carbon nanoparticle labeling of neoadjuvant chemotherapy for breast-conserving surgery in breast cancer. Quant Imaging Med Surg 13(3):1825–1837. https://doi.org/10.21037/qims-22-361
42. Wang Z et al (2023) Application of carbon nanoparticles combined with refined extracapsular anatomy in endoscopic thyroidectomy. Front Endocrinol (Lausanne). https://doi.org/10.3389/fendo.2023.1131947
43. Koowattanasuchat S et al (2023) Rapid detection of cancer DNA in human blood using cysteamine-capped AuNPs and a machine learning-enabled smartphone. RSC Adv 13(2):1301–1311. https://doi.org/10.1039/d2ra05725e

Theranostic Application of Smart Nanomaterials in Target Drug Delivery

Adil Mahammad, Insha Akbar, and Shoaib Mukhtar

Abstract Nanotechnology has emerged as a promising field in biomedical research, offering innovative solutions for both diagnosis and therapy. Nanomaterials, characterized by their unique properties at the nanoscale, have garnered significant attention for their potential applications in theranostics and the integration of diagnostics and therapeutics into a single platform. This abstract provides a comprehensive overview of nanomaterials utilized in theranostic applications, highlighting their diverse functionalities and contributions to healthcare. Nanoparticles, such as quantum dots, gold nanoparticles, and magnetic nanoparticles, have demonstrated remarkable imaging and drug-delivery capabilities. Their tunable optical properties enable precise imaging of biological structures, aiding in early disease detection and monitoring of therapeutic responses. Moreover, surface modification techniques enhance their biocompatibility, stability, and targeting efficiency, facilitating selective delivery of therapeutic agents to disease sites while minimizing off-target effects. In addition to imaging and drug delivery, nanomaterials play a pivotal role in multifunctional platforms for theranostics. However, challenges such as biocompatibility, scalability and regulatory considerations remain pertinent in the translation of nanomaterial-based theranostics to clinical applications. Addressing these challenges requires interdisciplinary collaboration, rigorous preclinical evaluations, and adherence to regulatory standards to ensure safety and efficacy. The smart nanomaterials hold immense potential for revolutionizing theranostic strategies, offering precise diagnostics, targeted therapy, and personalized medicine paradigms. Continued research and technological advancements will drive the translation of nanomaterial-based theranostics from bench to bedside, transforming the landscape of healthcare delivery and patient outcomes.

Keywords Smart nanomaterial · Theranostic application · Drug delivery

---

A. Mahammad (✉) · I. Akbar
Department of Chemistry, Aligarh Muslim University, Aligarh, India
e-mail: imdadilchemist@gmail.com

S. Mukhtar
Research Group of Environmental and Inorganic Photochemistry, Center for Natural Sciences, Faculty of Engineering, University of Pannonia, Veszprém, Hungary

I. Uddin (ed.), *Smart Nanosensors*, Smart Nanomaterials Technology,
https://doi.org/10.1007/978-981-96-3878-9_2

1 Introduction

In the twenty-first century, the world is developing at a fast and furious pace toward urbanization and industrialization with technical advancement. In this paced stress-filled lifestyle, there is no time and stare on healthcare and this ignorance introduced several serious challenges like depression, diabetes, hypertension and the most lethal disease cancer. The very recent pandemic COVID-19 which had spread devastatedly and turned into a pervasive and uncontrolled epidemic situation, gave a reason for the enduring concern of public health [1]. Early diagnosis can prevent the bombarding prevalence of diseases. With the advancement of science in the nanotechnology field, advanced nanomaterials can offer early diagnosis as well as determination of disease state, pathological condition, and imaging of causative organisms [2]. Various types of nanomaterials have been developed to implicate as biosensors, microchips, nano diagnostic agents, theranostic agents for extraction and sequencing of genes, detection of blood sugar levels, bioimaging and detection of pathogens, and cancer biomarkers [3]. In the treatment of cancer, diagnosis at an early stage is crucial and an important key step for successful treatment and eliminating the cancer-like dreadful disease from the patient's life while successively reducing the cancer mortality rate. For early detection of cancer and treatment simultaneously, the application of smart nanoparticles as a theranostic agent can make a significant impact on controlling of this life-threatening disease [4, 5].

Currently, biopsy (morphological analysis of the cells) and histopathology (morphological analysis of tissues) with different imaging techniques assist in early cancer detection. These commonly employed imaging techniques like Magnetic Resonance Imaging (MRI), optical imaging, endoscopy, ultrasound, photoacoustic imaging, computing tomography, X-ray, etc., diagnose cancer only when apparent changes occur in the tissue, by that time cancerous cells may have proliferated or even metastasized [6]. Additionally, these imaging techniques cannot discriminate the benign lesions from malignant lesions. Furthermore, morphological analysis of tissues and cells is not efficacious individually for early cancer detection. Once the cancer is detected, the stage at which it's diagnosed figure out the type of treatment it will have to undergo, whether it can be removed surgically or there is a need of combination therapies [7]. The most widely used treatment strategy is chemotherapy but it has some disadvantages like cytotoxicity, poor bioavailability of administered drugs, also it has some deleterious side effects, it lacks specificity which results in low drug concentration at the target site. In order to overdue these limitations theranostic based diagnostic methods and molecular/targeted therapies, photodynamic therapy (PDT), photothermal therapy (PTT), chemo dynamic therapy (CDT), and sonodynamic therapy (SDT), are being used for specific and selective tumor diagnosis and detection [8]. The word theranostic is the combo of two words: therapeutics and diagnostics. It can be both ways therapeutics followed by diagnostics to diagnose the disease at an earlier stage, accordingly treatment can be done, or diagnosis followed by therapeutics, also the codelivery of biomarkers and therapeutic agents is done which is even more effective [9]. Certain proteins are highly overexpressed on cancer cell membranes that act as target for cancer therapeutic agents. As the diagnostic drugs are administered within the patient's body, travel via the bloodstream to all the tissues and organs; as they find their particular target, it will light up on PET scan, thereby the cancer is detected, and then the therapeutic drugs kill out the only cells having that particular protein expressed on their surface membranes. Thus helping the oncologist personalize the treatment based on the output. Nanotechnology-based cancer treatment is a somewhat new area of research involving the study of different optical, magnetic, structural and behavioural properties of materials with less than 100 nm in dimensions [10]. 'Nanos', the literal meaning of the word nano, refers to a very small object or a person of very short height. One billionth of a meter or one millionth of millimeter is nanometer (nm) ("Nanotechnology—Dr. Isaac Manuel 2241—Studocu"). Over the past few decades enormous applications of nanomaterials (NMs) in different fields of life have attracted the interest of researchers over time. Thereby substantial research has been done in Nanotechnology to overcome the drawbacks of conventional chemotherapeutic strategies. Agriculture, food industry, ceramics, green technology to combat rising global warming and climate change damage, telecommunication, material science and other industries use nanomaterials for their effective output [11]. Size, ranging from 1–100 nm is the defining basic/fundamental characteristic property of all nanoparticles with each nanomaterial having at least one dimension in nanoscale. Top-down and bottom-up are the two fundamental techniques for synthesizing NM's.

2 Methods for Synthesis of Nanomaterials

Synthesis of nanomaterials can be carried out using two prominent approaches those are top-down or bottom-up methods. In the top-down method which generally renders the approach from general to specific, the nanoparticle is broken down into fine structures into nano-range dimensions through the conversion of mechanically machined bulk materials. On the other hand, the bottom-up method, which goes from a specific to a general approach, is based on assembling atomic structures into desired nanoparticles via co-precipitation or self-assembly [12].

The nanoparticles prepared by the top-down method go through sputtering, pyrolysis, lithographic techniques, etc. There is a significant limitation of this method is the lack in surface engineering of the synthesized product. In the bottom-up method, the particle grows from simple to desired size in a controllable manner. This advantage makes this method most preferable to synthesize metal-based nanoparticles to the researchers. On the basis of different steps the methods of fabrication for the synthesis of nanoparticles can be categorized into three major parts i.e. Physical, Chemical and Biological methods. Here, the chemical and biological method of synthesis of metal-based nanoparticles is dependent on bottom-up approach.

Physical methods generally recommend UV irradiation, laser ablation, and sonochemistry which provides ease of modification for synthesizing metal-based Nanoparticles. In this method the high purity level can be achieved as this physical

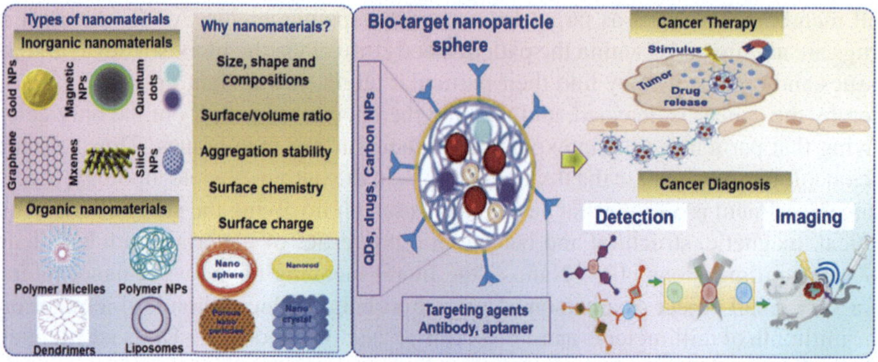

Fig. 1 Synopsis of nanomaterials for cancer therapeutics

process follows the solvation of metal atoms and aggregation into small clusters of NPs. Although this method requires high energy and highly advanced equipment, it can provide thin films with no solvent contamination and uniformity, making the physical synthesis method better than chemicals [13].

For the synthesis of Metal–Organic Framework (MOF), Carbon Organic Framework (COF), POPs, and metal oxide nanomaterials, Chemical synthesis is the most suitable method as it suggests the procedure through thermal decomposition, chemical reduction, micro-emulsion, etching and mechanical procedures for the easiest synthesis. In this method, synthesized nanomaterials can be reduced from their metal salt precursors by chemical reduction method through adding reducing. This method provides a huge range of different reducing agents, such as hydrides for example, $NaBH_4$ (sodium borohydride), formaldehyde, and hydrazine, and a low budget for great-scale industrial production.

The biological method introduces a synthesis process using natural resources like plants, algae, fungi, yeast etc., with the principle of green chemistry to provide green synthesis of various nanoparticles. These synthesized advanced nanoparticles have been considered as theranostic agents for biomedical diagnostics, bioimaging, molecular sensing, and cancer therapy. Figure 1 gives a general review of bio-targeted cancer theranoscopic applications, properties of nanoparticles, and clarifying nanoparticles. This environment-friendly non-toxic method shows its versatile qualities as it can control the size, morphology, nature, and properties of synthesized nanomaterial by varying different stabilizing and reducing agents as well as controlling the parameters such as pH, temperature, and solvent [14].

2.1 Classification of Smart Nanomaterials

Figure 2 illustrates classification of nanomaterials on the basis of dimensions; Zero, One, two and three.

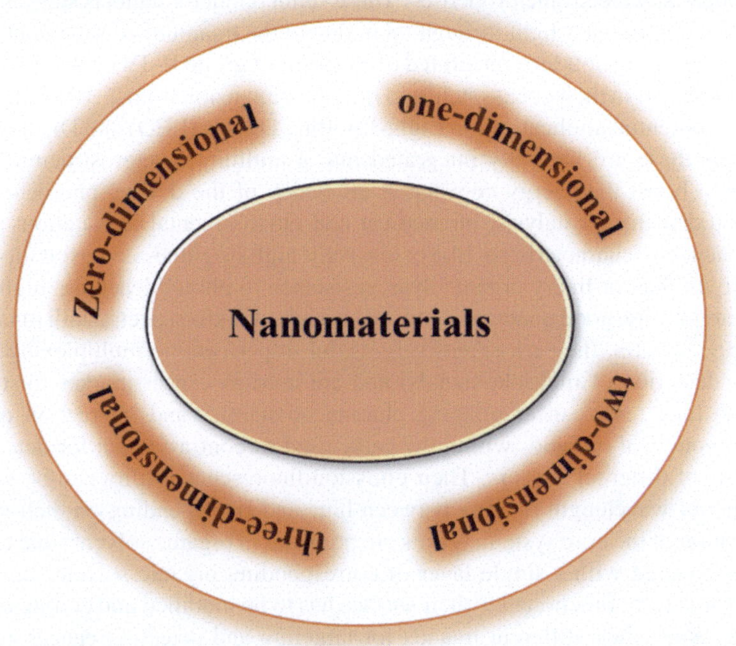

Fig. 2 Classification of nanomaterials on the basis of dimensions

2.1.1 Zero-Dimensional NPs Used in the Theranostic Application

Zero-dimensional nanomaterials have their size confined in all three dimensions (X, Y, Z) falling within Nano range e.g. quantum dot, fullerene, nanoparticles. The term "zero-dimensional" emphasizes that these nanomaterials don't extend in any one or more dimensions beyond the nanoscale, distinguishing them from higher-dimensional nanostructures. It can be composed of various materials and can be inorganic, organic, amorphous, and crystalline or of other shapes. Due to its point size structure, it has applications in photothermal therapy, catalysis, drug delivery, and bioimaging etc.

Quantum Dots (QDs)

Quantum dots are referred to as such small nanoparticles that their point physical size is neither categorized under molecular nor bulk though they have the structure and atomic composition same as bulk materials and exhibit the quantum size effect in their optical and electronic properties. These semiconductor nanocrystals emit fluorescence when passing light through them due to the quantum confinement characteristic property. Q dots are constituted of elements of group II–IV or III–V with 100–1,000,000 atoms in their core system and range in dimension from 2–10 nm. Quantum dots have potential applications in Light-Emitting Diodes (LED) and Display Applications as these are naturally integrated into a multilayer heterostructure, photovoltaics to boost the energy conversion efficiency of the device, photoconductors and photodetectors, catalysis, biomedical and environmental applications. Q dots due to their few characteristics like extensively high brightness on excitation, large absorptivity than ordinary organic dyes, resistance to photo bleaching, high photochemical stability, good quantum yield, and broad excitation spectra with tunable and symmetric spectra, they are extensively useful in performing multiplexing (simultaneous detection of multiple signals) and are used as cancer marker for tracking real-time long term cancer imaging pharmacodynamics and interaction of many labeled biological moieties within the cells, used as contrast agent for detection of prognosis of different cancers. Their emission fluorescent spectra can be varied to any value of wavelength ranging between blue and IR depending on their size and composition of the core system. Q dots are prepared in organic solvents due to which they get covered with a single layer of corresponding organic solvent, in order to make Q dots fully functionalize their surface has to be modified and conjugated with peptides, antibodies, different ligands for targeting and detecting cancer at earlier stages. The toxicity profile of QDs is due to the release of heavy metal ions which they are composed of. The release of metal ions is prohibited by the polymer coating on QDs. Hybrid quantum dots are being synthesized to increase the stability of QDs inside the biological system. Protein-QDs nanohybrids such as Gelatin-QDs has no toxicity up to 5 mg/mL and potentiality in drug delivery applications. There is a minute difference between QDs and C-dots, that is QDs are derived from metal-based materials whereas C-dots are from carbon-based. This fundamental difference brings some changes in photoluminescence properties such as the emission spectra of QDs depends on its size but the C-dots' emits spectrum which depends on laser excitation. Because of such unique properties, C-dots are employed in bioimaging of cell organelles e.g. nucleus, cell membrane, ER, lysosome, golgi body and mitochondria. In spite of the huge success of QDs in theranostics, there are still a few obstacles in enhancing the selectivity, specificity, sensitivity and in reducing cytotoxicity. In the coming few years, the research on conjugated QDs cancer theranostics will lead to breakthroughs in clinical application. The surface derivatization of QDs needs to be further improved in order to reduce its Reticulo Endothelial system (RES) uptake simultaneously enhancing its uptake by cancer cells.

Fullerene

After diamond and graphene, the third allotrope of carbon is buckminsterfullerene. Fullerene (C_{60}) is a nanoscale carbon material having 0.7 nm diameter. This zero-dimensional nanomaterial has the shape of a spherical icosahedron consisting of sixty carbon atoms in the form of pentagons with C_5-C_5 single bonds and Hexagons with C_5-C_6 double bonds. Due to its unique geometrical shape and quantum size, it shows novel photophysical and electrochemical properties, low toxicity and high radical scavenging efficiency. These characteristics attracted interest in biological applications and have shown tremendous promise in the field of inhibition of HIV protease, Specific cleavage of DNA, cancer treatment. In past decades Photodynamic therapy (PDT) has become a promising modality for treating cancer, tumors and various nonmalignant diseases. Fullerene can generate reactive oxygen species (ROS) by absorbing visible light and it has efficient intersystem crossing and delayed triplet state which makes fullerene the best choice to be photosensitizer (PS) that is highly specific to the targeted cancerous cell and resulted cell death by activating photo damage. In theranostic application, fullerene and metallofullerene significant role as it has the empty cavity inside the fullerene which is an ideal site for encapsulation of a metal atom. The most important electronic property exhibited by metallofullerene is intra-fullerene electron transfer i.e. the encaged metal atom can transfer charge to the cage of fullerene, which has been suggested to use for bioimaging in Magnetic Resonance Imaging (MRI). This novel zero-dimensional nanomaterial can absorb and convert the dissipated energy of radio frequency while penetrating the tissue, into heat and made the Radio frequency thermal therapy (RTT) a controllable, minimally invasive and efficient treatment method for cancerous cells. Such interesting photophysical and unique physicochemical properties of C_{60} promoted itself potentially for the applications in theranostics.

2.2 1-D NMs Used in Theranostic Applications

1D nanomaterials are commonly defined as linear structures with two of the three dimensions fall within the nano range with other third dimension outside the nanoscale range e.g. Nanotubes, nanorods, thin films, nanowires etc. Due to its unique wired, tubular, rod-like morphology with large specific surface area, facile strain relaxation, strong electron–hole separation efficiency, very high light absorptivity, plenty of active sites, these materials have attracted researcher's attention toward many advanced and potential applications from photocatalysis, ultrasensitive sensors to nano-opto-electronics, and it is the most promising and significant building block for nano electronic devices. They can be chemically pure or impure, crystalline, amorphous, metallic, ceramic, polymeric etc.

2.2.1 Carbon Nanotubes (CNTs)

CNTs consist of cylindrical tubes composed of sp^2 hybridized carbon atoms arranged hexagonally approximately 1 nm in diameter and 1–100 nm in length. Graphene sheets rollup either in single layer called single walled CNTs (SWCNTs) or in multiple layer called multi-walled CNTs (MWCNTs), both having very high electrical and thermal conductivity with multifunctional uses like biosensing, tumor detection, photoacoustic imaging, targeted drug delivery and photo thermal therapy. Both the ends of SWCNTs and MWCNTs are capped with carbon atoms arranged in hemispherical manner called fullerenes-warped up by graphene sheets. The average separation between different layers in MWCNTs is 0.34 nm with each layer forming a separate tube having outer diameter of 2.5–100 nm while SWCNTs ranges between 0.6–2.4 nm. The prime requirement for CNTs to be used as drug carrier is that CNTs should be soluble in aqueous medium for its biocompatibility, gastrointestinal absorption and blood transportation. Due to the hydrophobic nature of graphene side walls and strong pi–pi interaction between separate tubes, its dispersion in aqueous medium gets hindered, thereby functionalization of CNTs side walls is done, which decreases its cytotoxicity, enhances its biocompatibility, provides ability to appendage drugs, proteins, nucleic acids for targeted drug delivery. Both non-covalent and covalent functionalization is done to CNTs. Many anticancer drugs which are hydrophobic in nature are adsorbed non-covalently on the CNTs surface due to the hydrophobic and pi–pi interactions between different constitutional nanotubes. The charges present on the chains is responsible for the encapsulation of ionic drugs or charged molecules. Non-covalent surface decoration of CNTs is more appealing strategy as it adsorbs chemical substances without influencing the basic electronic structure and keeps the fundamental network of CNTs intact. Non-covalent derivatization is done by exposing/revealing the CNT to vapor stabilization species which react with the decorated layer forming a stabilization covering with conjucted functional groups at the surface. Whereas the hydrophilic groups like –OH, –COOH are conjugated at the surface by covalent decoration at the defect points or on the open ends. Its remarkable strategy for CNT surface functionalization but it's difficult to control over the reactivity of reagents as reaction takes place in extreme conditions for the covalent bond to form and their interaction with the CNTs surface.

2.2.2 Nanorods

Nanorods are also included in the group of 1D nanoparticles having a confined pathway of electron movement in two direction only along its length. Because of this unique physiochemical properties it has been predominantly employed in theranostic applications. Nanorods can be synthesized by surfactant-assisted method using surfactants like CTAB, CTAC etc. Instead of this surfactant, other phospholipids such as Di-palmitoyl phosphatidylcholine (DOPC), Di-stearoyl phosphoryl ethanolamine (DSPE) also can be used. The stability of this nanodiametered nanoparticle in body fluids such as blood, plasma, cell culture media is very high which prevent aggregation. Nanorods can be engineered to carry both diagnostic and therapeutic agents, enabling simultaneous imaging and treatment. This dual functionality allows for real-time monitoring of therapeutic responses and adjustments to treatment regimens. For instance, multifunctional gold nanorods can be used for both photoacoustic imaging and photothermal therapy, providing a comprehensive theranostic platform

[15]. The photothermal efficiency on irradiation with NIR lights of nanorods can generate sufficient heat to kill the cancerous cell. On hitting of photon, the outermost valence electron resonant of gold-nanorods oscillates at a specific frequency which generates heat inside the cancer cell and this working principle of nanorods opens a new avenue in cancer treatment. Scientists are now showing interests to synthesize nanorods with bismuth sulphide iodide for photothermal therapy (PTT). Tang et al. synthesized mesoporous silica nanorods for targeted drug delivery functionalized with targeting ligands in cancer treatment [16]. Nanorods represent a highly versatile and promising platform for theranostic applications, offering unique advantages for integrated diagnosis and therapy. Their high surface area, tunable properties, and multifunctionality make them suitable for a wide range of biomedical applications, from imaging and biosensing to drug delivery and photothermal therapy. Continued advancements in nanorod technology have the potential to significantly enhance personalized medicine, providing more targeted, efficient, and effective treatments for various diseases [17].

2.2.3 Nanowires

Nanorods and nanowires are almost similar but there are subtle differences based on their shape and dimensions. Nanorods have rod-like cylindrical morphology with uniform diameter, whereas nanowires has elongated wire-like shape and may exhibit tapering or uniform diameter along its length. Its versatile structural properties like stretchable, flexible, branched, coiling, wire-like shape have garnered significant attraction for electrical as well as theranostic applications [18]. The high surface area of nanowires enhances their ability to load therapeutic agents and biomolecules, facilitating effective drug delivery and biosensing. Silver nanowires (Ag NWs) and Copper nanowires (Cu NWs) are best choice for the commercial application of soft conductors, sensors and various electronic devices as these both metal has highest electrical conductivity. Nanowires can be used as carriers for drug delivery, ensuring controlled release of therapeutic agents. Their ability to penetrate cells efficiently makes them ideal for targeted drug delivery. For instance, zinc oxide nanowires have been used to deliver anticancer drugs directly to tumor cells, enhancing treatment efficacy and reducing side effects [19]. Nanowires can be employed in photothermal and photodynamic therapies. Gold nanowires, for instance, can absorb near-infrared light and convert it into heat, effectively killing cancer cells through hyperthermia. Similarly, nanowires can be used to generate reactive oxygen species upon light exposure for photodynamic therapy. Nanowires are also used as a potent theranostic agent to treat cancer as well as in the treatment of Alzheimer's disease and Parkinson's. Nanowires can be designed to carry both diagnostic and therapeutic agents, enabling real-time monitoring of treatment. For example, Wu Y. et al. synthesized iron oxide nanowires functionalized with drugs and fluorescent markers that can be used to track drug delivery and release while simultaneously treating the disease [20]. Functionalized nanowires can target specific cells or tissues, enhancing the precision of both diagnosis and therapy. This targeted approach reduces off-target effects and improves the overall efficacy of the treatment.

2.3 2-D NMs

This freestanding sheet-like nanomaterial is an excellent choice for theranostic applications as it possesses the highest possible specific surface area, unique surface chemistry and quantum size effect. It evolved to a more advanced level with these exceptional physiochemical properties and plate-like structure which provides huge drug payload capacity into tumor cells through active or passive targeting pathways. The photoactivity, biodegradability and variable properties by changing the functions such as interlayer forces, diversity of composition, and surface functionalities directed researchers toward the new dimension of a myriad of theranostics. These 2D nanomaterials can be metallic, ceramic, crystalline, and amorphous or made of different chemical compositions like transition-metal dichalcogenide nanosheet (TMD), black phosphorous Nanosheet (BP), layered double hydroxide nanosheet (LDH), transition metal carbides, nitrides and carbonitride nanosheet (MXene nanosheet), hexagonal boron nitride (hBN) nanosheets and Metal–organic framework (MOF).

2.3.1 Graphene Sheet

Graphene, an allotrope of carbon, is a two-dimensional nanomaterial having a lattice structure like honeycomb with a monoatomic layer of sp^2 hybridization [21]. In 2004, Geim et al. has introduced graphene to the world, which brought them a Nobel Prize in physics. The discovery of graphene and its derivatives has shown a new dimension in the field of nanotechnology and targeted drug delivery because of their unique properties like huge surface area, single-layer structure, and electrical and thermal properties. Graphene has a large number of derivatives and all of them have different properties but the basic carbon properties can be found in each derivative. These unique properties offer wide and diverse applications in various biomedical applications like biosensing, gene transfection, drug delivery etc. of these derivatives and have grabbed researchers' attraction. Among all derivatives, the most popular graphene oxide (GO) is a graphene-like nanosheet with a high surface area and uncharged epoxide (O) and hydroxyl (OH) groups. In the GO plane, these epoxide, hydroxyl, and carboxyl groups form H-boding and other interactions to provide stability in aqueous solution, whereas the free unbounded π electrons form π–π and non-covalent bonding. It also gives smooth functionalization of these nanomaterials with various biocompatible polymers like chitosan, PEG, and to conjugate with targeting moieties to develop advanced nanocomposite systems. Unlike hydrophobic graphene, the hydrophilicity nature which provides solubility and stability in colloid and the ultrathin nanosheet, introduces very high surface area for huge drug payload capacity. This multilayered nanoparticle with layer-dependent surface chemistry can be used as therapeutic cargoes to encapsulate drugs, nucleic acids, biomolecules, fluorescent probes etc. [22]. GO can also be used as a potential photothermal agent in photothermal treatment to deliver efficient therapeutic results absorbing its high NIR. Reduced Graphene Oxide (rGO) shows dramatically enhanced absorption of non-infra-red and resulted in significant therapeutic effects in cancer treatment. The applications of GO and its derivatives in biological imaging has been limited due to the negative effects of autofluorescent interference of biological cells on the visible fluorescence of GO. This multifunctional graphene and its derivative have been extensively promoted in theranostics for therapeutic effects combined with imaging.

2.3.2 Metal–Organic Frameworks (MOFs)

Theranostics, a portmanteau of therapeutics and diagnostics, represents a paradigm shift in the field of personalized medicine, integrating diagnostic and therapeutic capabilities into a single platform. This dual functionality aims to provide a more targeted and efficient approach to disease management. Metal–Organic Frameworks (MOFs), a class of crystalline materials composed of metal ions or clusters coordinated to organic ligands, have emerged as promising candidates in the field of theranostics due to their unique properties. The reversibility of bond formation between inorganic clusters and organic linkers of MOFs is a crucial factor for obtaining highly porous and crystalline MOFs [23]. As a hybrid of inorganic and organic linker units, MOFs possess extraordinary properties such as high strength, flexible framework, optical, magnetic and electronic properties, tunable structure, and are highly porous with surface area up to 10,000 m^2/g. The interaction between secondary binding units and organic linkers are strong i.e., of the order of 100 kg/mol. Metal clusters within MOFs are mostly transition metals, with variety of their versatile coordination numbers, these include tetrahedron, cuboctahedron, square planar etc. MOFs can be synthesized at room temperature with the use of microwave, electrochemical, ultrasonic and mechanochemical processing. The post-synthetic treatment of samples produced and the reproducibility of the results of synthesis are of crucial importance. These properties include high surface area, tunable pore sizes, and the ability to incorporate various functional groups, making MOFs highly versatile for a range of biomedical applications.

MOFs possess exceptionally high surface areas and porosity, allowing for the efficient loading of therapeutic agents. This high loading capacity is particularly beneficial for delivering a substantial dose of drugs or imaging agents within a compact carrier. The pore sizes of MOFs can be precisely controlled during synthesis, enabling the encapsulation of various molecules of different sizes. Furthermore, the surface of MOFs can be functionalized with targeting ligands, enhancing their specificity towards certain tissues or cells. Many MOFs are biocompatible and stable under physiological conditions, essential characteristics for any material intended for medical applications. Researchers have developed several biocompatible MOFs that degrade into non-toxic by-products, minimizing potential side effects.

The ability of MOFs to encapsulate and protect therapeutic agents from degradation before reaching the target site is one of their most significant advantages. Controlled release of the drug can be achieved by tuning the pore sizes and surface properties, ensuring a sustained and effective treatment regimen. MOFs can be loaded with photosensitizers or agents that convert light into heat. Upon exposure to specific wavelengths of light, these MOFs can generate reactive oxygen species (ROS) for photodynamic therapy or heat for photothermal therapy, both of which can effectively kill cancer cells [24]. Panaogiota Markopoulou et al. identified differing intracellular cargo release mechanism by monitoring in vitro drug delivery in real time and explored a new dimension in designing MOFs to simultaneously deliver diagnostic and therapeutic agents to the target site [25]. This allows for real-time monitoring of treatment efficacy and adjustments to the therapeutic regimen if necessary. Functionalization with targeting ligands ensures that MOFs accumulate primarily in the diseased tissue, reducing off-target effects and enhancing treatment efficacy with targeting drug delivery process. By encapsulating drugs within the MOF structure, adverse side effects commonly associated with conventional drug delivery methods can be minimized. The controlled release mechanism ensures that the therapeutic agents are released only at the target site.

2.4 3-D NMs

Bulk materials with none of the three dimensions (X, Y, Z) falling within the nano size, all the 3 dimensions are >100 nm in dimension but the individual blocks which constitute the bulk material fall in the nano scale (1–100 nm) e.g. bundles of nano wires, nano particle dispersion, various nano layer in which 0D, 2D, 3D elements are in vicinity and result in formation of interface. e.g. colloids, thin films.

On the basis of length of the pore diameter dimensions NMs are categorized into 3 groups: microporous, mesoporous and macro porous materials. The size of NMs depends on the pore dimension which in turn gives detailed information about diffusion, adsorption and corresponding interaction properties. If the incoming molecules are lesser in size than the pore size, there will be more interaction between the incoming molecules than molecules and wall interaction. Microporous materials: these molecules have very small pore sixe with <2 nm diameter. Only gaseous, molecules or other small molecules can be held within these materials. They have narrow diffusion kinetics with good interaction properties e.g. Clay (naturally occurring material), Na-Y. These materials have applications in gas purification systems, membrane filters, gas storage material. Mesoporous materials: they have pore diameter between 2–50 nm which can encapsulate somewhat large molecules. These have applications in polymerization or adsorption of liquids and gases. Macroporous materials: materials with pore size dimensions >500 nm and can hold very large molecules like poly aromatic substances, biological molecules e.g. porous gel, porous glass, carbon microtubes.

2.4.1 Liposomes

One of the most promising and incompetent drug delivery vesicle, with self-aggregated phospholipid into the bilayer, enveloping a central aqueous chamber forming a sac like structure. Its size varies from 30 nm to few μm with bilayer of 4.5 nm in thickness. Liposomes have some prominent properties in comparison to the conventional drug delivery systems, it prevents the loaded drugs from clearance, degradation, and helps in controlled drug release with on-site targeting, enhances therapeutic effects of payload, increases bioavailability and biocompatibility of encapsulated drugs thereby decreases the side effects. Liposomes discriminately deliver their encapsulated content to target site via both active and passive targeting. The tumour cells have abnormal porous vasculature with pores between endothelial cells unlike other normal tissues having tight neo vasculature. The porosity of solid tumour varies (100 nm–2 μm) depending on the type of tumour tissue. Liposomes along with encapsulated drugs are injected/administered via different routes like oral, parenteral, ophthalmic, nasal, transdermal, they cross the intercellular junction passively via EPR effect. Their active targeting involves the particular interaction between receptor and antibody on tumour cells with that of liposomal surface receptors. Chemotherapeutic drugs injected within the body if unable to overcome the different biological barriers like Blood Brain Barrier (BBR), epithelium, endothelium, cellular membranes will not be able to perform their therapeutic activities, which will affect pharmacokinetics and pharmacodynamics of administered drugs. Targeted drug delivery is needed to overcome such drawbacks. Nanotechnology puts forward the concept of drug targets. NMs enhance the therapeutic effects of drugs in various possible ways: by loading the biological active hydrophilic/hydrophobic molecules in their core systems, decreasing drug clearance time, increasing circulation time more importantly controlling drug release at target site. Loaded NPs are administered within the bloodstream via different routes, after which they interact with the target site, release their payload via both passive and active strategies. Passive targeting harness the leaky vasculature of tumour cells. The endothelium of blood vessels supplying the cancer tissues have pore size large comparative to the healthy tissues, therefore these vessels become more permeable result in preferential enhancement and permeation of NPs inside the tumour tissues, this effect is known as Enhanced Permeability and Retention effect (EPR). Moreover, EPR effect relies on other factors too like hydrophobicity of particles, molecular weight, surface charge of particles and the type of tumour, which results in lower concentration of drugs at target site than optimal value. Active targeting (tuning surface functionality) overcomes the drawbacks of passive targeting, enhances the efficiency of loaded drug with limited side effects. Active targeting significantly increases the therapeutic drug concentration at tumour site by surface modification of NPs with antibodies, biomarker agents or other biological active agents/ligands, which help the NPs to bind with the receptors highly expressed on tumour surface tissues. Proper identification of cancer biomarker as well as precise selection of receptor on tumour tissues assist the NPs to increase the therapeutic efficiency of encapsulated drugs in highly selective and directed manner. Based on their lamellarity and chamber structure, liposomes are classified as unilamellar vesicles (ULVs), Oligo lamellar vesicles (OLVs), Multi lamellar vesicles (MLVs) and Multi vesicular liposomes (MVLs). MVLs have honey comb like structure with >100 nm non-centric aqueous compartment which is bounded with a single phospholipid bilayer, while as MLVs display anionic structure with >5 concentric bilayer and OLVs possess 2–5 bilayer. ULVs is furthermore divided in different size ranges into small unilamellar vesicle (SUVs, 30–100 nm), large unilamellar vesicle (LUVs, >100 nm) and giant unilamellar vesicles (GUVs, >1000 nm).

2.4.2 Dendrimers

Dendrimers are the 3-D hyper branched, compact architect structures with core inside over which Dendron radiate. Size of dendrimers ranges from 1–20 nm while some have diameters 14–15 nm. Ligands or other biological active molecules (drugs, genes, and contrast agents) are conjugated with the outer ends of Dendron both covalently and non-covalently. Dendrimers with amine ($-NH_2$) hydroxyl (–OH) groups at the terminal ends are complete dendrimers while with carboxyl (–COOH) groups are half dendrimers. The conjugation increases the packing density of substance depending on whether dendrimer is of first, second, third generation and so on. This actually retards their further growth henceforth dendrimers are synthesized up to 10th generation only. Dendrimers can be prepared with specific molecular weight as it's possible to control its synthesis at each step of generation. Most widely employed dendrimers include polyamidoamine (PAMAM), polypeptides, polyamides, and polyamines, phosphorous (P-dendrimers), polyesters (PGLSA-OH) and carbosilicates. Sugar moieties and oligonucleotides also discriminate between different types of dendrimers. Convergent and Divergent are the two synthetic method of preparation of dendrimers. Divergent being most employed method of preparation but this results in decreased reactivity of conjugated functional groups while in convergent method firstly branched arms are synthesized followed by adhering them with multifunctional core system. Due to their characteristic structural properties dendrimers act as transporters for pharmacological substances either by conjugation on the surface or by physical loading within the cavities. The hydrophobic pharmaceutical substances are encapsulated and loaded within the hydrophobic cavity of dendrimers thereby increases their solubility in aqueous medium. Dendrimers penetrate the cell membrane, get accumulate within the tumor cells release their payload at controlled and sustained rate consequently increases/enhances biocompatibility, effectiveness, selectivity of chemotherapeutic drugs, decreases drug resistance and off-site effects.

2.4.3 Gold NPs

Gold nanoparticles (AuNPs) have diverse applications in theranostics, leveraging their unique physical, chemical, and optical properties. They can be utilized in both therapeutic and diagnostic aspects such as targeted drug delivery with controlled release and Photodynamic therapy (PDT), Photothermal therapy (PTT), Combined therapy and imaging with personalized medicines. AuNPs can be conjugated with drugs, targeting molecules (like antibodies or peptides), and release drugs specifically at the diseased site, minimizing side effects on healthy tissues. It can convert absorbed light into heat when irradiated with near-infrared light, selectively destroying cancer cells through localized hyperthermia without harming surrounding healthy tissues. these versatile Nps are excellent contrast agents for CT imaging and also exhibit strong surface plasmon resonance (SPR), which enhances contrast in optical imaging techniques like optical coherence tomography (OCT) and photoacoustic imaging. AuNPs can be functionalized with specific ligands to detect biomolecules (e.g., proteins, nucleic acids) associated with diseases. This allows for highly sensitive detection in assays like colorimetric sensing, where the aggregation state of AuNPs changes color upon binding with target molecules. Gold nanoparticles represent a powerful tool in theranostics, offering precise targeting, effective therapy, and enhanced imaging capabilities, leading to improved diagnostic accuracy and therapeutic outcomes.

2.4.4 Metal Derived Nano Particles as Theranostic Agent (NPs)

Elements belonging to 3–12 group of periodic table with incompletely filled d-orbital have the tendency to interact with other elements/substances forming the complex structures. Transition metal derived NPs represent enormous nanosized materials with metal ion combined/conjugated with other anions, molecules, organic, inorganic substances. Fe, Cu, Ag, Au, Zn and few of the f-block elements represent the major constituent of these biomedical NPs. Both magnetic and photothermal properties are shown by different NPs based metal containing compounds (oxides, salts). These materials are well tailored and suited for targeted drug delivery and diagnostics of cancer. These materials have received great attraction due to their myriad of benefits like high efficiency of conversion of photothermal energy, photothermal stability, less cytotoxicity and optimum biocompatibility, biocatalytic property (Nanoenzyme). Magnetic nanoparticles (MNPs) are employed in magnetic resonance imaging (MRI) to improve contrast and provide detailed images of biological tissues. Functionalization of MNPs with specific ligands allows for targeted imaging of specific cells or molecular markers [26]. Nano particle containing iron oxide (IONPs) show strong magnetization with very low to almost no toxicity in vivo. IONPs get accumulated in lysozyme after incorporating in the cells, chemotherapeutic drugs are released there due to low values of pH and protease. Integration of IONPs with gold NPs (Au/Fe NPS) form the composite IONPs, which releases enormous heat than superparamagnetic iron oxide NPs (SPIONPs). These materials have high near IR absorption with good thermal conductivity thereby TM based NPs are commonly used in biosensors, as multifunctional imaging agents, in targeted drug delivery, photothermal ablation (PTA), magnetothermal ablation (MTA) which result in increasing the non-invasive ablation temperature thereby destroying corresponding cancer cells [27]. The bioapplication, therapeutic functionalization of TM based NPs depend on various bio physicochemical properties like stability of NPs, magnetic property, oxidizing potential and intratumoral distribution. Therefore these properties have to be accurately and carefully controlled, modulated via different preparation techniques and surface modification approaches. Currently three separate synthetic ways of preparation; physical, chemical, biological approaches each with their unique characteristic advantages synthesize TM based NPs, furthermore surface modification is done to enhance the functionalization, performance of IONPs e.g. organosulfur, selenium compounds are commonly employed for surface decoration as sulfur containing molecules strongly, coordinatively interact with the metal ions Au, Fe, Cu, Ag etc. also TM based NPs are covered with polymeric materials by absorption for enhancing their overall surface area to volume ratio. Gold nanoparticles (AuNPs) have been widely used in bioimaging due to their biocompatibility and ease of functionalization. AuNPs can enhance contrast in imaging techniques such as optical coherence tomography (OCT) and photoacoustic imaging (PAI), providing detailed images of tissues and organs [28]. Figure 3 illustrates how nanomaterials can be divided into different categories depending on structural configuration.

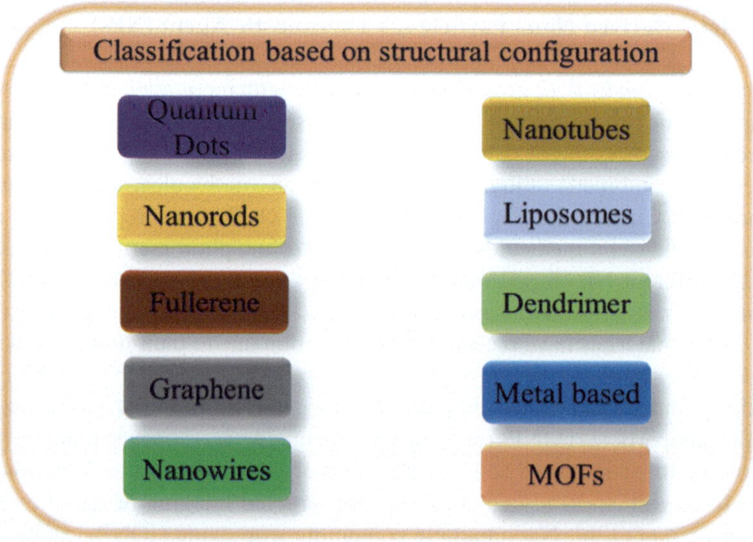

Fig. 3 Classification of smart nanomaterials based on structural configuration

3 Applications of Smart Nanomaterials

Cancer treatments based on smart nanomaterials are more effective and safer. Nanomaterials used in cancer theranostics will soon be accessible to clinics as a result of technological advancements. Nanomaterials, physiological characteristics, targeted areas, and cancer theranostic application utilization are all compared in Table 1.

3.1 Nano-sensors in Bioimaging

In last few decades, nanomaterials have drawn increasing attention in biomedical applications for biosensing and bioimaging due to their optical characteristics, high sensitivity, low toxicity and quick response time. These advanced nanomaterials can be categorized into two major category organic and inorganic nanoparticles. These advanced nanoparticles have versatile application in the field of theranostics where the therapy and diagnostic simultaneously happen. Fluorescent nanomaterials are the most preferable to the scientist for bioimaging because of its biodegradable and biocompatible and low toxicity nature. The inorganic fluorescent nanomaterials are majorly quantum dots of various inorganic materials like carbon quantum dots, silica dots and gold and silver nanoparticles etc. Organic fluorescents are most of organic dyes and various polymers and aggregation induced emission dyes etc. the organic fluorescents are favourable for bioimaging over inorganic one. The only limitation which limits the fluorescent quantum yield is its hydrophobicity and unstability in water. Organic dyes can be utilised as signalling unit i.e. chromophores because of its visible light absorbance capability. Fluorescein can produce the bright signals and its nontoxic nature makes it promising for biological imaging applications. Moreover, coumarins and cyanine dyes are very popular fluorophores because of their high quantum yield and quality to provide brightness and biocompatibility respectively. In recent days, Novel polydopamine-based fluorescent organic Nanoparticles (PDA-FONPs) have been synthesized by Xiaoyong Zhang et al., the oxidation of PDA (polydopamine) with H_2O_2 (Hydrogen peroxide) led to this novel product PDA-FONPs which is extensively used for biological imaging as it possess several exceptional characteristics like high photoluminescence biocompatibility and highly water soluble and well dispersity in cytoplasm of cells [49]. Another talented group Yang et al. are researching on FONPs and successfully designed a water-soluble from ATP (Adenosine 50-triphosphate) with low toxicity and surface functionalization which provides significantly enhanced fluorescence. These highly sensitive novel FNPs show very high selectivity nature for Fe^{3+} sensing with a LOD (limit of detection) of 0.1 nM, which provides the ability to sense Ferric ion in a biological environment through fluorescence quenching. Recently, Wang et al. developed a new nanoparticle combining the nitrogen lone pair e^{-}s as donor part and highly electron deficient aromatic part as receptor for sensing more selectivity and sensitivity towards

Table 1 Comparison of nanomaterials, physiological characteristics, targeted areas, and cancer theranostic application

SN	Nanomaterial	Targeting cancer	Size	Physiological properties	Therapeutic/imaging modality	Refs.
1	Manganese oxide (GC4419)	Squamous cell cancer of the head and neck	e	Intravenous injection; about 2 h half-life with minimal accumulation upon related dosing	Targeting and therapy	[29]
2	Silica NPs (Crgdy-peg-Cy5.5-C dots)	Head and neck melanoma, breast cancer, colorectal cancer	6e7 nm	Superior brightness; enhanced tumor tissue retention with low background signal,	Tracking and imaging (portable and realtime optical detection)	[30]
3	Liposome NPs encapsulating DOX and trastuzumab (for targeting of HER2 receptor protein) (MM-302)	Breast cancer	100 nm	Enhanced antitumor activity, activation of p53 protein levels phosphorylation/activation; decreased intracellular signaling (p-Akt); increased DNA damage signaling	Targeting and therapy	[31]
4	SPION (Ferumoxytol)	Head and neck cancer	17e31 nm	3D quantitative tumor characterization	Targeting and imaging (MRI)	[32]
5	CdS/ZnS core shell quantum dots, carboxylic acid-functionalized, coated with veldoreotide (VELD)	Breast cancer	1e10 nm	Conjugation with VELD at pH 7; somatostatin receptors (SSTR) agonist with anti-cancer activity	Imaging and therapy	[33]
6	Lymphotrophic SPION loaded with drug (Ferumoxytol)	Thyroid cancer	30 nm	Intravenous injection (lyophilized iron oxide injected dose); imaging with high sensitivity in detecting small lymph nodes (even 2 nm)	Imaging (MRI) and therapy	[34]
7	Gd NPs (AgulX)	Brain metastates	4.5 nm	Can be 4 used in combination with fractionated stereotactic radiation in oligo brain metastates	Targeting and imaging (MRI)	[35]

(continued)

Table 1 (continued)

SN	Nanomaterial	Targeting cancer	Size	Physiological properties	Therapeutic/imaging modality	Refs.
8	Spherical gold NPs carrying nucleic acids and drugs (NU-0129)	Gliosarcoma	13 nm	Can cross blood brain barrier; targeting of Bcl2L12 gene to stop cancer cells to grow	Targeting and therapy	[36]
9	Silica NPs (64 Cu-NOTA-PSMAi-PEG-Cy5.5-C0 dots	Prostate cancer	8 nm	Intravenous injection; targeting prostate-specific membrane antigen (PSMA); specific uptake at prostate-specific (PC-3) cells	Targeting and imaging (PET/MRI)	[37]
10	Graphene oxide-polyethyleneimine graphene quantum dots hybrid (GO-PEI-GQDs)	Breast cancer	e	Great photothermal response (44e49 C), oxidative stress responses in breast cancer cells (MDA-MB231)	Targeting and imaging	[38]
11	Gd NPs (AGulX) grafted with additional DOTA chelates, complexation of bismuth	Cervical cancer	4.5 nm	Improved longitudinal relaxivity; contrast of 4.26 Hounsfield unit mM^{-1} (close to clinically used CT contrast agents)	Theranostic agent (imaging, targeting and therapy)	[39]
12	SPION NP iron oxide (Ferumoxytol)	Pancreatic cancer	17e31 nm	Targeting bombesin receptors; relaxivity; identification of small/ undetectable lymph nodes metastatis	Targeting and imaging (MRI)	[40]
13	pA20-36 Chitosan-HA-NPs	Liquid tumor	70 nm	Multimodal imaging depending on the site of tumor, chemotherapy	Imaging (MRI) and therapy	[41]

(continued)

Table 1 (continued)

SN	Nanomaterial	Targeting cancer	Size	Physiological properties	Therapeutic/imaging modality	Refs.
14	Bovine serum albumin-based Gd/Au complex (BSA-Gd/Au)-MSNs e (DOX@bMSN-ss-GABA)	Oral epidermoid carcinoma	e	Receptor is highly expressed in keratin forming KB human oral epidermoid carcinoma cell lines. DOX@bMSN-ss-GABA is redox-responsive, prompting fast drug release in a tumor microenvironment with high glutathione levels	Targeting, imaging, therapy	[42]
15	Polymeric NPs, with hydrophobic polylactic acid polymeric core, encapsulating DOX and hydrophilic PEG	Lung cancer	100 nm	Biocompatible; demonstrated dose linear pharmacokinetic profile: prolonged persistence	Targeting and therapy	[43]
16	PEGylated TiO_2 NPs	Solid tumors	e	Strong photoabsorption, biocompatible deeper penetration, high efficacy for photothermal method	Targeting and imaging	[44]
17	Lip/PEG/Tween 80/SPIONs-PVA/RTX	Liquid tumor	7e10 nm	Blood brain barrier internalization	Imaging (MRI)	[45]
18	SiNc-PNP encapsulated in PEG-PCL	Solid tumor	40 nm	Real-time intraoperative image-guided surgery; process can be tailored via photo irradiation dosage	Imaging and therapy (image guided surgery)	[46]

(continued)

Table 1 (continued)

SN	Nanomaterial	Targeting cancer	Size	Physiological properties	Therapeutic/imaging modality	Refs.
19	FUP cationic nanomicelles	Colorectal cancer	30e150 nm	Antitumor apoptosis mechanism might be related to the regulation of Fas/FasL and activation of caspase-8 and caspase-3	Targeting and therapy	[47]
20	Hyaluronic acid-mediated cell penetrating peptide-modified 10-hydroxycamptothecin-loaded (HA/CPPs-10-HCPT-NPs)	Breast cancer	e	Naturally occurring macromolecular polysaccharide with excellent biocompatibility and biodegradability, HA has anti-immune effects, strong affinity for some overexpressed markers, such as CD44, seen on tumor cell surfaces	Theranostics	[48]

Fe^{3+} ion [50]. The core part made of triazole, and the branch part is of benzothiadiazole derivatives, which has shown exemplary results. Aggregation-induced emission (AIE)—active polymeric nanoparticles are in the limelight for last few years as this anti-aggregation-caused nanoparticle can be applied as light emitting system for bioimaging. Acetylene polymers, and fluorescent diamond particles are the most suitable examples of Nps with AIE properties for bioimaging as it possesses strong fluorescence and high dispersity. The advancement of biological imaging systems requires more precision and high image quality. The advanced version of fluorescence imaging with innovative qualities like rapid acquisition, high resolution of images, and utilisation of non-ionizing radiation is magnetic resonance imaging (MRI). The lower sensitivity in the cell environment is the only limitation of this imaging system. Superparamagnetic iron oxide (SPIO) Nanoparticles are introduced to provide high contrast images for not only cellular but also upto the molecular level. Besides being used as a contrast agent, these advanced nanoparticles are applied in the field of theranostic as it has exhibited higher magnetization in the induced magnetic field. These smart nanomaterial SPIOs can do in-vivo cell tracking and be directed to the targeted site for therapy. With the help of such smart nanoparticles, the theranostic applications can achieve a very practical and futuristic approach for the diagnostic and targeted drug delivery for the treatment.

Nanosensors function by interacting with biological molecules, producing signals that can be detected and measured. The mechanisms include fluorescence resonance energy transfer (FRET), surface plasmon resonance (SPR), and magnetic resonance [51]. FRET-based nanosensors utilize the energy transfer between fluorescent donor and acceptor molecules to detect molecular interactions. This technique provides high sensitivity and spatial resolution, making it suitable for real-time monitoring of dynamic cellular processes. SPR nanosensors detect changes in the refractive index near the sensor surface, which occur when biomolecules bind to the sensor. This method is highly sensitive and can be used for label-free detection of biomolecular interactions.

3.2 Nanosensors as Biosensor

Biosensors are analytical devices that convert a biological response into a measurable signal. They consist of a bioreceptor, transducer, and signal processor. Nanosensors, due to their high surface area-to-volume ratio and unique physicochemical properties, enhance the performance of traditional biosensors, providing greater sensitivity and faster response times. Metal-based sensors have many advantages over conventional biosensors [52]. Usually, microfluidic technique keeps an eye on the liquid at microscale, even volumes which are smaller than microscale. Nano sensors are used in medical diagnostics for the early detection of diseases, monitoring disease progression, and evaluating treatment efficacy. For example, gold nanoparticle-based biosensors can detect cancer biomarkers with high sensitivity, enabling early diagnosis and improved patient outcomes. Range of ZnO nanowires with varying girth and length were synthesized using hydrothermal growth in microdomain and microchannel with different content/amount of PEI. Increasing the amount of PEI corresponding length and diameter of nanometres change to micrometres. Intercalated microelectrodes with microfluidic improved the diagnostic sensitivity. Quantum dots and gold nanoparticles are commonly used in optical biosensors due to their strong optical signals. Mohammed Keshtkar et al. developed aptamer conjugated nanomaterials and explored its application as biosensor in cancer imaging [53]. Literature has evidenced many examples where researchers have synthesized multifunctional and multimodal microfluidic and microdomain chips with some detection zone, mixing zone and micro electrodes to enhance sensitivity of diagnostic using Metal NPs. Huixiang Li et al. developed a gold nanoparticle for coloremetric detection of DNA sequences based on electrostatic interactions [54]. To further improve the diagnostic techniques smartphones-based metal biosensors are used in Nano technological developments. Integrating smartphones in conventional metal-biosensors helps in diagnosis bypassing some sophisticated experimental methods in laboratories. In the food industry, nano sensors are used as biosensor to detect contaminants, pathogens, and spoilage indicators. Carbon nanotube-based biosensors can rapidly detect foodborne pathogens, ensuring the safety and quality of food products.

3.3 Nanosensor in Theranostic

Nanosensors are integral to theranostic platforms, offering a wealth of information through their precise and sensitive detection capabilities at the nanoscale. These miniature sensing devices are designed to detect specific biomolecules, ions, or cellular responses, providing real-time insights into physiological processes and disease states. Nanosensors play a multifaceted role in theranostics by monitoring kay parameters such as drug release kinetics, therapeutic efficacy, and disease progression. Structures, imaging systems, tailored drug delivery systems, and cancer treatments based on nanomaterials are shown in Fig. 4. Quantum dots conjugated with anti-HER2 antibodies have been used to target and image HER2-positive breast cancer cells. This dual functionality allows for both the visualization and targeted treatment of cancer cells, significantly enhancing treatment precision. Au nanosensors have been utilized in photothermal therapy for prostate cancer. When irradiated with near-infrared light, these nanoparticles generate heat, selectively killing cancer cells while sparing healthy tissue. Kalyani Pathak et al. studied application of nanosensors in detection and treatment of diabetes [55].

Nanosensors enable real-time monitoring of disease markers and therapeutic responses. It can provide continuous imaging of tumor cells during and after treatment, offering insights into treatment efficacy. Mark A. Eckert dedicatedly researched on successfully overcoming the challenges of In vivo sensing [56]. Combining different types of nanosensors and therapeutic agents is a growing area of research. For instance, combining verteporfin-photodynamic therapy with 5-aza-2'-deoxycytidine has shown enhanced anti-tumor immune responses in triple-negative

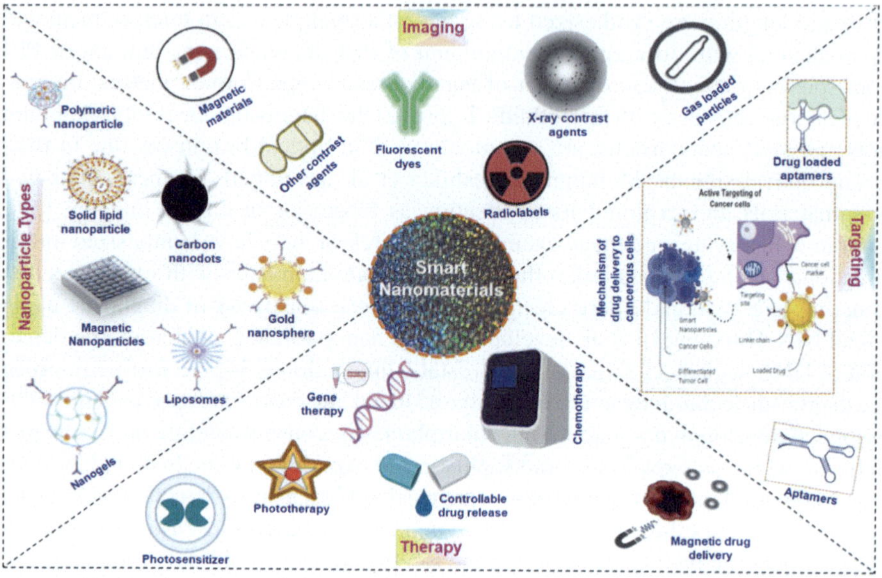

Fig. 4 Phases of smart nanomaterials

breast cancer. This combination therapy leverages the strengths of each component to achieve better therapeutic outcomes. One of the notable advantages of nanosensors is their ability to operate in complex biological environments with high selectivity and sensitivity. For example, nanosensors can detect biomarkers associated with various diseases, including cancer, cardiovascular disorders, and neurological conditions, facilitating early diagnosis and monitoring of treatment outcomes. Additionally, their compatibility with imaging techniques such as fluorescence or surface-enhanced Raman spectroscopy enables multimodal imaging and sensing, allowing for comprehensive characterization of biological systems. Furthermore, nanosensors can be engineered to respond to specific stimuli, such as pH changes, enzyme activity, or biomolecular interactions. This responsiveness enables controlled drug release from nanocarriers, triggered by the microenvironment of target tissues or cells. By Integrating nanosensors with drug delivery systems, theranostic platform achieve synergistic functionalities, enhancing therapeutic precision and minimizing adverse effects on healthy tissues.

4 Economical Aspect of Nanotechnology in the Field of Theranostic Application

NMs produced via myriads of different chemical and physical approaches have various sets of drawbacks such as cytotoxicity, cost, their effectiveness etc. considering these negative effects, bionanotechnology has revolutionized the ordinary nanotechnology, which is a green technology, eco-friendly approach to synthesize cost-effective and non-toxic NMs. Different plants, plant extracts, microorganisms, biological templates, enzymes are utilized to prepare NPs, while harnessing the natural formation methods. Plants which are ubiquitous due to their extensive availability, cost effectiveness, eco-friendliness, non-harmful nature, they are enormously used in NPs preparation/synthesis. The biomass of plants like *Medicago sativa, Azadirachta indica, Aloe vera, Pelargonium grancolens, Coriander sativum* and many more have been used in the synthesis of Au NPs of variety of shapes. Au nano triangles are synthesized from leaf extract of *Aloe vera* plant. Bimetallic Au, Ag, Fe core Ag shade NPs are synthesized from *Azadirachta indica.* Macromolecule templates like nucleic acid matrix, DNA, and proteins are employed to synthesize characteristic and sophisticated NPs. These NPs are used in biosensors, bio NEMs, and bioelectronic systems. Protein molecules form the main component of nanocomposite structures e.g. Ferritin is an iron storage protein found in both prokaryotes and eukaryotes which regulates the iron content within their bodies by releasing Fe in a controlled manner. Similarly, microorganisms like fungi, bacteria, are used to synthesize NPs from metal salts in aqueous solutions such as magnetotactic bacteria found deep inside the seas, under anaerobic conditions synthesize magnetic NPs. *Rhodopseudomonas* synthesize the gold NPs extracellularly which ranges in size between 10–20 nm. *Fungus Fusarium oxysporum* synthesize Ag NPs. Similarly, algae synthesize Ag NPs extracellularly. Researchers are more interested in synthesizing materials from naturally occurring substances for biomedical purposes rather than relying on chemical processes.

5 Challenges and Perspectives in the Field of Theranostic Applications

Personalized cancer treatment results in significant advancement of outcomes, decrease in off-site effects, healthcare costs; by delivering the right medicine to the right cancer patient at right time. Recent advancements in renal-clearable nanoprobes have significantly improved optical imaging for early disease diagnosis. These nanoprobes, which are activated by disease-associated biomarkers, offer high sensitivity and resolution. Once activated, they release signal reporters into urine, facilitating non-invasive and sensitive optical in vitro urinalysis. This technology is particularly useful for detecting superficial diseases and monitoring renal function in real-time. Advances in the use of molecular biomarkers for early diagnosis, detection and the overall prognosis of cancer has remarkably improved at rapid rate. The nanostructure delivery platform is a novel paradigm in cancer theranostics because of its potential biophysiochemical properties for early detection/diagnosis of cancer and effective targeted drug delivery. Preclinical research trials of theragnostic nanomaterials have them as promising techniques in the treatment of different types of cancers like breast, lymph, brain tumors. However, the clinical application of NMs are very limited because the co-administration of therapeutic and diagnostic agents have no prominent additive effects in clinical trials. Additionally, the methods of synthesis and fabrication of NMs for cancer theranostics still has many obstacles like biocompatibility, cost effectiveness, in vivo targeted delivery, therefore further investigation is needed for studying their pharmacokinetics and pharmacodynamics, real-time, long-range toxicity profile of NPs. Another challenge is that it's difficult to manufacture/synthesize the complicated NMs and nanocomposites at large scale. Also some NPs induce toxicity depending on type of NP, their administration route and their functional properties. All these factors have to be considered before the treatment. A better and clear understanding of protocols of NMs synthesis and safety both in vivo, in vitro need to be formulated in order to revolutionize the future of healthcare system and to eradicate the cancer from one's life.

6 Conclusion

In conclusion, the applications of nanomaterials in theranostics represent a paradigm shift in healthcare, offering multifaceted solutions that integrate diagnostics and therapeutics on a nanoscale platform. Through their unique properties and functionalities, nanomaterials have revolutionized medical imaging, drug delivery, and personalized medicine approaches. Nanoparticles, nanosensors, and nanocomposites have demonstrated remarkable capabilities in enabling precise diagnosis, targeted therapy, and real-time monitoring of treatment responses. Their tunable properties, responsiveness to environmental cues, and compatibility with imaging modalities have paved the way for advanced theranostic strategies that enhance patient outcomes and reduce healthcare burdens. However, challenges such as biocompatibility, regulatory considerations, and scalability require ongoing research and collaboration across disciplines. Addressing these challenges will be crucial in harnessing the full potential of nanomaterial-based theranostics and translating these innovations into clinical practice. As we continue to explore and innovate in the field of nanotechnology, the future holds promise for even more sophisticated theranostic platforms that revolutionize disease management, improve patient care, and contribute to the advancement of precision medicine.

Acknowledgements The Figs. 1, 4 and Table 1 shown in this chapter was taken from an article with copyright permission provided in pdf version and Figs. 2 and 3 which is acknowledged by the authors.

Funding Not applicable.

Availability of Information and Resources The corresponding author may provide you with the data used in this research upon request.

Ethics Statements

Competing Interests The writers affirm that they have no financial or other conflicts of interest.

Acceptance of Participation and Ethical Clearance Not applicable

References

1. Mech A et al (2022) Safe- and sustainable-by-design: the case of Smart Nanomaterials. A perspective based on a European workshop. Regul Toxicol Pharmacol. https://doi.org/10.1016/j.yrtph.2021.105093
2. Husen A, Siddiqi KS (2023) Advances in smart nanomaterials and their applications. https://doi.org/10.1016/C2021-0-02202-1
3. Abbasi Kajani A, Haghjooy Javanmard S, Asadnia M, Razmjou A (2021) Recent advances in nanomaterials development for nanomedicine and cancer. ACS Appl Bio Mater 4(8). https://doi.org/10.1021/acsabm.1c00591
4. Fan Z, Fu PP, Yu H, Ray PC (2012) Theranostic nanomedicine for cancer detection and treatment. J Food Drug Anal 22(1). https://doi.org/10.1016/j.jfda.2014.01.001
5. Sumer B, Gao J (2008) Theranostic nanomedicine for cancer. Nanomedicine 3(2). https://doi.org/10.2217/17435889.3.2.137
6. Nicolson F, Kircher MF (2021) Theranostics: agents for diagnosis and therapy. In: Molecular imaging: principles and practice. https://doi.org/10.1016/B978-0-12-816386-3.00040-5
7. Hwang SR, Chakraborty K, An JM, Mondal J, Yoon HY, Lee YK (2021) Pharmaceutical aspects of nanocarriers for smart anticancer therapy. Pharmaceutics 13(11). https://doi.org/10.3390/pharmaceutics13111875
8. Kashyap BK, Singh VV, Solanki MK, Kumar A, Ruokolainen J, Kesari KK (2023) Smart nanomaterials in cancer theranostics: challenges and opportunities. ACS Omega 8(16). https://doi.org/10.1021/acsomega.2c07840
9. Shrivastava S, Jain S, Kumar D, Soni SL, Sharma M (2019) A review on theranostics: an approach to targeted diagnosis and therapy. Asian J Pharmaceut Res Dev 7(2). https://doi.org/10.22270/ajprd.v7i2.463
10. Aflori M (2021) Smart nanomaterials for biomedical applications—a review. Nanomaterials 11(2). https://doi.org/10.3390/nano11020396
11. Amukarimi S, Ramakrishna S, Mozafari M (2021) Smart biomaterials—a proposed definition and overview of the field. Curr Opin Biomed Eng 19. https://doi.org/10.1016/j.cobme.2021.100311
12. Paramasivam G, Palem VV, Sundaram T, Sundaram V, Kishore SC, Bellucci S (2021) Nanomaterials: synthesis and applications in theranostics. Nanomaterials 11(12). https://doi.org/10.3390/nano11123228
13. Chitsazi MR, Korbekandi H, Asghari G, Najafi RB, Badii A, Iravani S (2016) Synthesis of silver nanoparticles using methanol and dichloromethane extracts of Pulicaria gnaphalodes (Vent.) Boiss. aerial parts. Artif Cells Nanomed Biotechnol 44(1). https://doi.org/10.3109/21691401.2014.949726
14. Lee J, Marrocchi A (2024) Advances in green chemistry and engineering. Sci Rep 14(1). https://doi.org/10.1038/s41598-024-53594-z

15. Pan X et al (2023) Gold nanocage-based multifunctional nanosensitizers for programmed photothermal/radiation/chemical coordinated therapy guided by FL/MR/PA multimodal imaging. Int J Nanomed 18. https://doi.org/10.2147/IJN.S436931
16. Ahmadi F et al (2022) A review on the latest developments of mesoporous silica nanoparticles as a promising platform for diagnosis and treatment of cancer. Int J Pharmaceut 625. https://doi.org/10.1016/j.ijpharm.2022.122099
17. Khlebtsov N et al (2013) Analytical and theranostic applications of gold Na-noparticles and multifunctional nanocomposites. Theranostics 3(3). https://doi.org/10.7150/thno.5716
18. Muthu MS, Leong DT, Mei L, Feng SS (2014) Nanotheranostics—application and further development of nanomedicine strategies for advanced theranostics. Theranostics 4(6):660–677, Ivyspring International Publisher. https://doi.org/10.7150/thno.8698
19. Mohammed MH, Hanoon FH (2021) Application of zinc oxide nanosheet in various anticancer drugs delivery: quantum chemical study. Inorg Chem Commun 127. https://doi.org/10.1016/j.inoche.2021.108522
20. Shen Z et al (2017) Multifunctional theranostic nanoparticles based on exceedingly small magnetic iron oxide nanoparticles for T1-weighted magnetic resonance imaging and chemotherapy. ACS Nano 11(11). https://doi.org/10.1021/acsnano.7b04924
21. Mbayachi VB, Ndayiragije E, Sammani T, Taj S, Mbuta ER, Ullah Khan A (2021) Graphene synthesis, characterization and its applications: a review. Results Chem 3. https://doi.org/10.1016/j.rechem.2021.100163
22. Iannazzo D, Celesti C, Giofrè SV, Ettari R, Bitto A (2023) Theranostic applications of 2D graphene-based materials for solid tumors treatment. Nanomaterials 13(16). https://doi.org/10.3390/nano13162380
23. Nasrollahi M, Nabipour H, Valizadeh N, Mozafari M (2020) The role of flexibility in MOFs. In: Metal-organic frameworks for biomedical applications. https://doi.org/10.1016/B978-0-12-816984-1.00006-8
24. Picchi DF, Biglione C, Horcajada P (2023) Nanocomposites based on magnetic nanoparticles and metal-organic frameworks for therapy, diagnosis, and theragnostics. ACS Nanosci Au. https://doi.org/10.1021/acsnanoscienceau.3c00041
25. Markopoulou P et al (2020) Identifying differing intracellular cargo release mechanisms by monitoring in vitro drug delivery from MOFs in real time. Cell Rep Phys Sci 1(11). https://doi.org/10.1016/j.xcrp.2020.100254
26. Gao X, Cui Y, Levenson RM, Chung LWK, Nie S (2004) In vivo cancer targeting and imaging with semiconductor quantum dots. Nat Biotechnol 22(8). https://doi.org/10.1038/nbt994
27. Lee JH et al (2007) Artificially engineered magnetic nanoparticles for ultra-sensitive molecular imaging. Nat Med 13(1). https://doi.org/10.1038/nm1467
28. Russ Algar W, Massey M, Krull UJ (2009) The application of quantum dots, gold nanoparticles and molecular switches to optical nucleic-acid diagnostics. TrAC Trends Anal Chem 28(3). https://doi.org/10.1016/j.trac.2008.11.012
29. Anderson CM et al (2018) Phase 1b/2a trial of the superoxide dismutase mimetic GC4419 to reduce chemoradiotherapy-induced oral mucositis in patients with oral cavity or oropharyngeal carcinoma. Int J Radiat Oncol Biol Phys 100(2). https://doi.org/10.1016/j.ijrobp.2017.10.019
30. Bradbury MS et al (2013) Clinically-translated silica nanoparticles as dual-modality cancer-targeted probes for image-guided surgery and interventions. Integr Biol (UK) 5(1). https://doi.org/10.1039/c2ib20174g
31. Hendriks BS et al (2012) Multiscale kinetic modeling of liposomal doxorubicin delivery quantifies the role of tumor and drug-specific parameters in local delivery to tumors. CPT Pharmacometrics Syst Pharmacol 1(1). https://doi.org/10.1038/psp.2012.16
32. Ng SP et al (2018) Predicting treatment response based on dual assessment of magnetic resonance imaging kinetics and circulating tumor cells in patients with head and neck cancer (PREDICT-HN): matching 'liquid biopsy' and quantitative tumor modelling. BMC Cancer 18(1). https://doi.org/10.1186/s12885-018-4808-5
33. Abdellatif AAH, Tawfeek HM (2018) Development and evaluation of fluorescent gold nanoparticles. Drug Dev Ind Pharm 44(10). https://doi.org/10.1080/03639045.2018.1483400

34. Harisinghani MG et al (2003) Noninvasive detection of clinically occult lymph-node metastases in prostate cancer. New Engl J Med 348(25). https://doi.org/10.1056/nejmoa022749
35. Tanguy R, Métellus P, Mornex F, Mazeron J-J (2013) Place de la radiochirurgie et de la radiothérapie stéréotaxique hypofractionnée dans la prise en charge des métastases cérébrales. Bull Cancer 100(1):75–81. https://doi.org/10.1684/bdc.2012.1683
36. Kumthekar P et al (2019) A phase 0 first-in-human study using NU-0129: a gold base spherical nucleic acid (SNA) nanoconjugate targeting BCL2L12 in recurrent glioblastoma patients. J Clin Oncol 37(15_suppl). https://doi.org/10.1200/jco.2019.37.15_suppl.3012
37. Chen F et al (2019) Ultrasmall renally clearable silica nanoparticles target prostate cancer. ACS Appl Mater Interfaces 11(47). https://doi.org/10.1021/acsami.9b15195
38. Kumawat MK et al (2019) Preparation of graphene oxide-graphene quantum dots hybrid and its application in cancer theranostics. Mater Sci Eng C 103. https://doi.org/10.1016/j.msec.2019.109774
39. Lux F et al (2019) AGuiX ® from bench to bedside-transfer of an ultrasmall theranostic gadolinium-based nanoparticle to clinical medicine. Br J Radiol 92(1093). https://doi.org/10.1259/bjr.20180365
40. Montet X, Weissleder R, Josephson L (2006) Imaging pancreatic cancer with a peptide-nanoparticle conjugate targeted to normal pancreas. Bioconjug Chem 17(4). https://doi.org/10.1021/bc060035
41. Torino E et al (2018) Multimodal imaging for a theranostic approach in a murine model of B-cell lymphoma with engineered nanoparticles. Nanomedicine 14(2). https://doi.org/10.1016/j.nano.2017.11.016
42. Yang S et al (2019) Tumor-targeted biodegradable multifunctional nanoparticles for cancer theranostics. Chem Eng J 378. https://doi.org/10.1016/j.cej.2019.122171
43. Jin M, Jin G, Kang L, Chen L, Gao Z, Huang W (2018) Smart polymeric nanoparticles with pH-responsive and PEG-detachable properties for co-delivering paclitaxel and survivin siRNA to enhance antitumor outcomes. Int J Nanomed 13. https://doi.org/10.2147/IJN.S161426
44. Liu Z et al (2018) 2D magnetic titanium carbide MXene for cancer theranostics. J Mater Chem B 6(21). https://doi.org/10.1039/c8tb00754c
45. Di Venosa G et al (2008) Characterisation of liposomes containing aminolevulinic acid and derived esters. J Photochem Photobiol B 92(1). https://doi.org/10.1016/j.jphotobiol.2008.03.008
46. Taratula O et al (2015) Naphthalocyanine-based biodegradable polymeric nanoparticles for image-guided combinatorial phototherapy. Chem Mater 27(17). https://doi.org/10.1021/acs.chemmater.5b03128
47. Yan Z et al (2018) Cationic nanomicelles derived from Pluronic F127 as delivery vehicles of Chinese herbal medicine active components of ursolic acid for colorectal cancer treatment. RSC Adv 8(29). https://doi.org/10.1039/c8ra01071d
48. Zhao H et al (2018) Cell-penetrating peptide-modified targeted drug-loaded phase-transformation lipid nanoparticles combined with low-intensity focused ultrasound for precision theranostics against hepatocellular carcinoma. Theranostics 8(7). https://doi.org/10.7150/thno.22386
49. Xu X, Ho W, Zhang X, Bertrand N, Farokhzad O (2015) Cancer nanomedicine: from targeted delivery to combination therapy. Trends Mol Med 21(4). https://doi.org/10.1016/j.molmed.2015.01.001
50. Cavanagh K, Strauss C, Forder L, Jones F (2014) Can mindfulness and acceptance be learnt by self-help?: A systematic review and meta-analysis of mindfulness and acceptance-based self-help interventions. Clin Psychol Rev 34(2). https://doi.org/10.1016/j.cpr.2014.01.001
51. Huang X, Jain PK, El-Sayed IH, El-Sayed MA (2007) Gold nanoparticles: interesting optical properties and recent applications in cancer diagnostics and therapy. Nanomedicine 2(5). https://doi.org/10.2217/17435889.2.5.681
52. Michalet X et al (2005) Quantum dots for live cells, in vivo imaging, and diagnostics. Science 307(5709). https://doi.org/10.1126/science.1104274

53. Keshtkar M, Shahbazi-Gahrouei D, Khoshfetrat S, Mehrgardi M, Aghaei M (2016) Aptamer-conjugated magnetic nanoparticles as targeted magnetic resonance imaging contrast agent for breast cancer. J Med Signals Sens6(4). https://doi.org/10.4103/2228-7477.195093
54. Li H, Rothberg L (2004) Colorimetric detection of DNA sequences based on electrostatic interactions with unmodified gold nanoparticles. Proc Natl Acad Sci USA 101(39). https://doi.org/10.1073/pnas.0406115101
55. Gogoi U et al (2022) Recent advances on natural and non-natural Xanthones as potential anticancer agents: a review. Med Chem (Los Angeles) 19(8). https://doi.org/10.2174/1573406419666221226093311
56. Eckert MA et al (2013) Novel molecular and nanosensors for in vivo sensing. Theranostics 3(8). https://doi.org/10.7150/thno.6584

Nanomaterials for Sensing Applications: Present and Future in Medicine, Packaging, and Quality Control

Namarta Dixit, Akanksha Yadav, Prashant Kumar Singh, Shashi Bala, and Adhish Kumar Jaiswal

Abstract Nanotechnology has the potential to address various challenges in engineering, materials science, chemistry, physics, biology, and medicine through sustainable nano-materials. Nanomaterials have been widely utilized to identify the current problem and their solution in various sectors like medical, food, industrial, agricultural, energy, biodefence, pollutants, pesticides, water purification, and waste management. Nowadays, Nanotechnology-based nanosensors are becoming an essential part of our modern lifestyles. Due to their small size and lightweight, nanosensors have been thoroughly investigated and used for sensing applications in many domains. In this chapter, we have covered the present scenario of nanosensors in the medical and packaging sector and discussed its quality control and future prospects. The chapter briefly introduces nanosensors and their application in various fields, the working principle of nano biosensors, and the application of nanosensors in the medical sector, like cancer diagnosis, drug delivery systems, MRIs, and nanobots. Besides this, the potential of edible coating and the role of metal nanoparticles and carbon nanotubes in food packaging are also discussed, as well as their toxicological effect and major challenges in the sector.

Keywords Nanomaterials · Nanosensors · Toxicity · Medical diagnosis · Food packaging

---

N. Dixit · A. Yadav · S. Bala (✉) · A. K. Jaiswal (✉)
Department of Chemistry, Lucknow University, Lucknow, Uttar Pradesh, India
e-mail: shashichem15@gmail.com

A. K. Jaiswal
e-mail: jaiswal_adhish@lkouniv.ac.in

P. K. Singh
Department of Biochemistry, Lucknow University, Lucknow, Uttar Pradesh, India

I. Uddin (ed.), *Smart Nanosensors*, Smart Nanomaterials Technology,
https://doi.org/10.1007/978-981-96-3878-9_3

1 Introduction

In science and engineering, the phrase "nanotechnology" refers to fields where materials, structures, devices, and systems are designed, characterized, and manufactured using phenomena that occur at nanoscale dimensions. A vast range of scientific and technological domains greatly benefits from nanoscale materials because of their distinctive chemical and physical properties [1, 2]. The past decade has witnessed a rise in public financing for the advancement of nanotechnology research and development, indicating that it will bring in a new era of productivity and financial stability [3, 4].

In the healthcare industry, nanotechnology has enabled the development of targeted drug delivery systems and more effective diagnostic tools. In biomedical sensing, metallic nanoparticles have also been used to detect various clinically significant analytes, including metal ions, proteins, and toxins [5]. Electrochemical biosensors with integrated functional nanomaterials have been developed for the quick diagnosis of viral diseases, such as hepatitis, influenza, and Covid-19 [6].

Besides this, nanomaterials-based sensors have also been studied as electrode coatings in bioimpedance sensing to improve the conductive characteristics of bioimpedance sensors [7]. Comparably, the electrode surface in electrochemical sensors has been altered by various nanomaterials, improving cycle stability, selectivity, and sensitivity [8]. Furthermore, sensing techniques based on local surface plasmon resonance (LSPR) excitation have used plasmonic nanomaterials like gold and silver and have applications in data storage, optical imaging, and quantitative detection [9]. In gas sensor networks, nanomaterials with varying shapes can be designed to greatly increase the gas absorbing capacity. Such enhanced nanoparticles can monitor various contaminated gases' concentrations, including SO_2, CO_2, NO_2, N_2O, and NH_3. A carbon allotrope with promise for sensing applications is carbyne [10]. Carbyne nanomaterial has created organically connected surface acoustic wave gas sensors.

2D nanomaterials have revolutionized chemical sensing, enabling efficient detection of gases, volatile compounds, ions, and biomolecules. Gas-sensing nanomaterials offer the potential for modernizing chemical sensing technologies and addressing indoor pollution challenges [11]. These flexible sensing systems enable comprehensive monitoring of personal health and the environment. Graphene-based nanosensors that provide real-time electrical measurements have been created to detect nitrotriazolone (NTO) molecules selectively [12].

Single nanosensor injection and manipulation into cells have been accomplished using fluorescence nanosensors equipped with a glass nanoprobe and optical zeta potential control [13]. Nano-apta-sensors—which combine nanotechnology with aptamer bioreceptors, have been applied for quick detection, quality assurance, and drug-level evaluation [14]. Nanomaterials are used as surface plasmon resonance (SPR) sensors for early-stage disease diagnosis, food quality control, environmental monitoring, and homeland security protection. Electrospun nanomaterials are ideal for ultrasensitive environmental monitoring and medical diagnostics [15]. Nanosensors are being used to detect cancer and other health issues early. Metal-based nanoparticles have demonstrated potential in the treatment and diagnosis of cancer by focusing on metal-dependent illnesses [16]. Nanobarcode particles, for example, can be used to detect DNA and other biomaterials using encoded antibodies. Carbon nanotubes, grown on silicon chips, are being used as a biosensor to attach DNA molecules to the ends of vertical carbon nanotubes [17]. Development of biosensors that can detect asthma attacks up to three weeks before they occur by testing the nitric oxide level in the patient's breath is underway. Regular testing could save lives by alerting patients if their nitric oxide levels were too high or increasing, indicating the risk of an asthma attack [18].

A well-known example of a nanomaterial-based sensor is a hybrid noble nanoparticle/DNAzyme electrochemical biosensor that can detect the presence of heavy metals such as Pb^{2+}, Cd^{2+}, and Cr^{3+} [19]. The sensor takes advantage of a well-designed DNAzyme sequence that selectively binds to the target analytes, allowing the detection of these metals at low concentrations. Another example is a biosensor device that uses a biological or bioinspired receptor unit with unique specificities toward corresponding analytes [20]. These biosensors can detect various analytes, including glucose, cholesterol, and various environmental pollutants [21]. Nanomaterials have been effectively used in diabetes treatment to build enzymatic and non-enzymatic glucose sensors that function in blood, serum samples, urine, tears, sweat, and wound exudates [22]. A glucometer, a sensor that measures glucose concentration in bodily fluids (usually in blood), remains one of the most commonly used biosensors in diagnostics, as it is essential for people suffering from diabetes [23]. Blood glucose monitors use a complex chemical process to measure the sugar content in blood samples. The blood is mixed with glucose oxidase, which reacts with glucose to create gluconic acid and ferrocyanide, which create ferrocyanide. The electrode in the test strip runs a current through the blood sample, influencing the current to measure blood glucose concentration accurately. Nanomaterial-based glucose sensors have been created for continuous, non-invasive monitoring in diabetes management [24]. Nanomaterial-based sensors have been used in gas detection to identify volatile organic compounds (VOCs) in breath, which can act as biomarkers for several different illnesses. The development of highly sensitive and selective sensors for a range of monitoring and diagnostic applications is possible thanks to the versatility and effectiveness of nanomaterials in sensing applications [25].

A new electrochemical sensor, PVP-GR/GCE, has been developed to detect ascorbic acid (AA), dopamine (DA), and uric acid (UA) simultaneously. The sensor, based on a polyvinylpyrrolidone (PVP)-graphene composite film-modified glassy carbon electrode, has excellent electrocatalytic activity for the oxidation of these compounds [26]. The sensor's sensitivity, selectivity, reproducibility, and long-term stability make it a promising strategy for early disease diagnosis. The sensor's unique structure and synergistic effect of GR and PVP have allowed for the simultaneous determination of these small biomolecules. This sensor could revolutionize the field of biosensing due to its ability to detect multiple analytes with high accuracy and efficiency. The sensor has been successfully applied to human urine samples, demonstrating its potential for further development. The sensor's performance was further enhanced by the unique structure of the modified layer and the synergistic effect of GR and PVP [27]. The sensor's potential for early disease diagnosis is promising.

A biosensor is being developed for aeronautics that is capable of detecting early radiation damage or infection in astronauts by sensing signs of biochemical changes [28]. These sensors are created from synthetic polymers called dendrimers and are layer-by-layer into spheres with a diameter of less than five nanometers. These nanosensors can be administered transdermally, eliminating the need for injections during space missions. This would also eliminate the need to draw and test blood samples, greatly improving conditions for astronauts going into space. A deployable nanosensor, a type of sensor not widely researched, is primarily used in military or national security applications. One such sensor is the SnifferSTAR, a lightweight, portable chemical detection system that integrates into micro unmanned aerial vehicles [29]. This sensor uses a nanomaterial for sample collection and concentration, combined with a microelectromechanical (MEM)-based "chemical lab-on-a-chip" detector for efficient chemical detection.

Nanosensors are being utilized in transportation vehicles on land, sea, air, and even in space [30]. They are being used in communications in wired and wireless technologies, optical and RF (radio frequencies) technologies, and even in buildings and facilities like factories, offices, and homes.

Besides medical applications, nanosensors are also being utilized in various fields, such as robotics, transportation, and communication. In Fig. 1, different applications of sensors are shown.

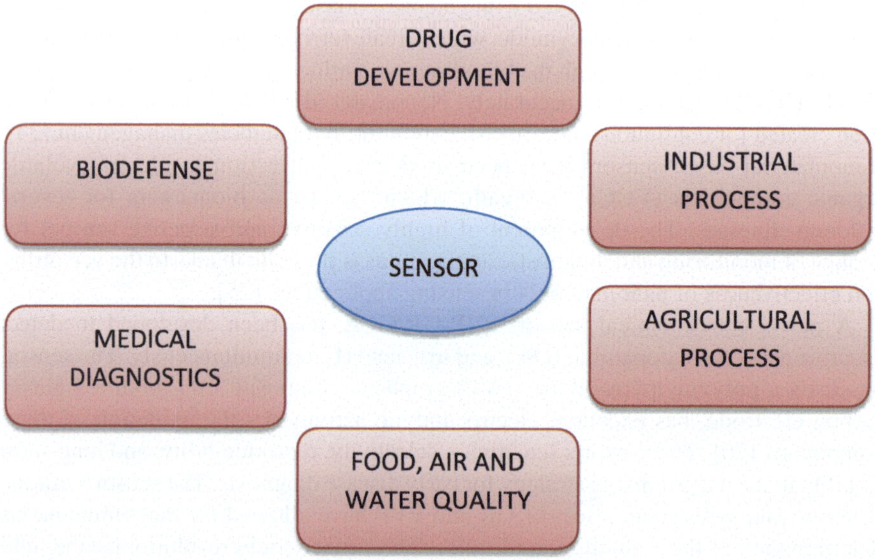

Fig. 1 Applications of nanosensors in different sector

2 Nanosensors

As per the International Union of Pure and Applied Chemistry, a chemical sensor is a device that can convert chemical information into a signal that can be processed. The sensor consists of two primary parts: a transducer, the second key part, and a receptor, which converts chemical information into a form of energy that can be measured. The receptor component, which is responsible for identifying the analyte, can have its foundation in physical and chemical concepts. A sensor is classified as a biosensor if the recognition mechanism is predicated on a biological response [31]. Furthermore, depending on the samples to be analyzed, applying nanomaterials to biosensors offers various detection limits and simplifies the sensitivity level to suit the requirements.

Nanosensors find use in manufacturing, communication, energy, transportation, disease detection, pollution management, and medicine [32]. They are also utilized for process monitoring. By detecting physical attributes, including volume, motion, quantity, speed, gravity, magnetic force, temperature, humidity, and electricity, this technology pinpoints the precise cells at the molecular level to give medications and track the growth of certain body parts. Here, in Fig. 2, various types of nanosensors are explained in schematic presentation.

Since the characteristics of nanomaterials differ from those of bulk materials and hence enable improved device performance, they have received particular attention in the design of sensors and biosensors. The primary feature of NMs is their huge surface area to volume ratio, which improves bioreceptor immobilization. The fact that these macromolecules are similar in size to NMs helps them in terms of lifetime and activity as well [33]. Furthermore, depending on their size and shape, nanostructures have a number of special qualities that will impact the functionality and design of sensing platforms. These attributes include optical, mechanical, electrical, and conductive qualities. Sensing devices have employed a variety of nanomaterials, including metal nanoparticles, nanotubes, nanofibers, and quantum dots (QDs) [34]. These nanosensors are designed to detect specific molecules or ions at the nanoscale level, allowing for highly sensitive and selective measurements in various applications such as environmental monitoring and medical diagnostics. Nanosensors can also be used in food safety testing, drug delivery systems, and even in security applications [35]. Their small size and high sensitivity make them ideal for detecting trace amounts of substances in complex environments. Additionally, nanosensors have the potential to revolutionize industries by providing real-time data and analysis for improved decision-making.

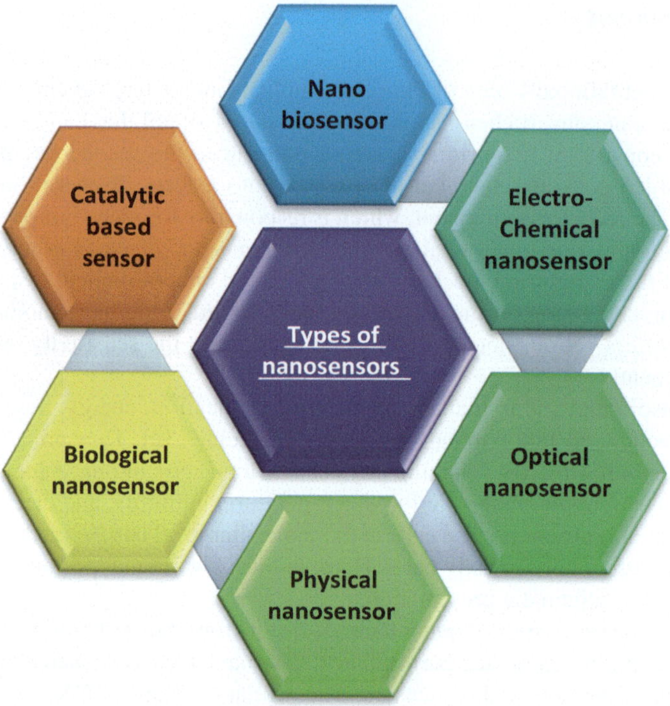

Fig. 2 Various types of nanosensors used in different field

3 Principle and Operation of Nanosensors

Nano biosensors measure biological or biochemical events using any electrical, optical, or magnetic technology in conjunction with a tiny probe. Biological targets such as antibodies, pathogens, nucleic acids, and metabolites, can be identified using a nano-biosensor [36]. A typical biosensor typically consists of following elements:

1. Analyte: A substance of interest that is to be detected in the sample. For instance, glucose is an 'analyte' when we want to detect glucose using a biosensor.
2. A biological receptor (Enzymes, DNA probe, antibodies) that's immobilized on the biosensor and has specificity for the analyte present in the sample.
3. A transducer that is used to convert the (bio)chemical signal into an electronic output. The concentration of the test analyte is directly proportional to the intensity of the generated signal sample.
4. Electronics: The part of the biosensor used to process the transduced signal and prepare it for display. The complex electronic circuit amplifies the signal and converts the analog signals to the digital form.

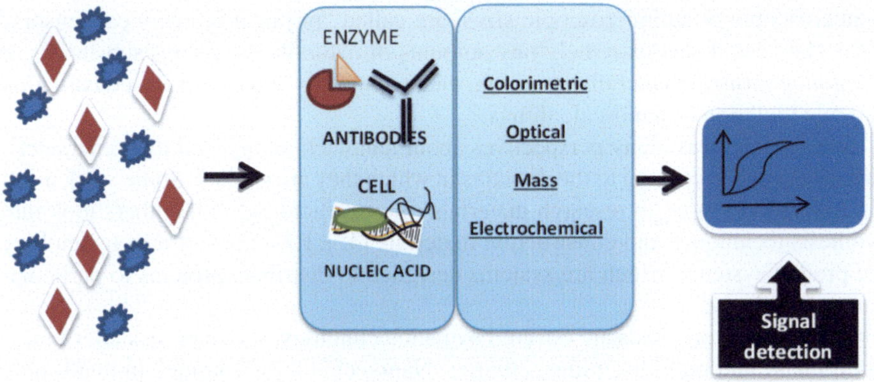

Fig. 3 Schematic representation of operating principle of biosensors

5. Display: The processed signals are quantified and displayed on the display unit of the biosensor. The display unit consists of a user interpretation system, such as the screens or the printer on which a number of curves are generated. A combination of hardware and software is used in the display unit so that the results can be generated in a user-friendly format. The final output can be in the form of numeric graphs, tables, or images. A schematic presentation of the operating principle of biosensor is shown in Fig. 3.

It is monitored as the signal changes in electric potential, current, conductance, impedance, mass, temperature, viscosity, and the intensity and phase of electromagnetic radiation. The differences in some or all of these metrics are used to quantify the presence or absence of biological agents. In nano-biosensors, the nanostructures act as a layer of intermediate between biological agents and either the physicochemical detector elements or the biological agents themselves. The transducer converts this connection into a measurable effect. Precise contact with the target of interest occurs in the biological layer [37–39].

4 Present and Future of Nanosensors in Medical Sector

The field of molecular nanotechnology works on hundreds or thousands of nanometers. The fundamental building block of molecular nanotechnology is the atom, and groupings of atoms can enhance their properties, including tension, conductance, surface area, and power. Nanostructures have the potential to enhance biomedical applications, and the development and completion of nanotechnology require extensive efforts [40]. Over the last ten years, research in a number of sectors, including aircraft, medical, health, and electronics, has been conducted using nanosensors. Nanosensors aim to obtain nuclear-scale information and convert it to readily examined information. Because of their extreme sensitivity, nanocomponents, often found at microscopic or submicroscopic sizes, are called "physical or chemical sensors" [41]. They can detect extremely tiny amounts of naturally dangerous substances or subatomic viruses. Generally speaking, nanotechnology allows for the construction and manipulation of nanoscale things.

Scholars possess many perspectives about nanotechnology, and their characterizations change according to the domains in which they are studied. Some work using electron microscopes to research materials' microstructures, while others look into synthesis techniques and construction materials [42]. Examples of nanotechnology are protein systems, which are systems designed to distribute proteins to particular parts of the body.

Numerous study fields are covered by nanotechnology, which necessitates multidisciplinary and multidiscipline activities. Nanotechnology is unique in that it only shares its small size. Generally speaking, nanotechnology is defined as the design, construction, and use of nanomaterials and nanostructures. The basic knowledge of physical attributes and the phenomena of nanomaterials and nanostructures is covered. This includes the connection between physical attributes and material sizes at the nanoscale.

New capabilities for characterizing, quantifying, and working with nanomaterials and nanostructures at the atom level have been created thanks to advancements in technologies like scanning probe microscopy (SPM), transmission electron microscopy (TEM), and scanning tunneling microscopes (STM) [43]. Because of their potential to create iron nanostructures, iron-reducing bacteria (IRB) can be introduced as advantageous microorganisms in nanobiology technology. These organisms can produce specific nanomaterials with unique properties that are not easily achievable through traditional chemical methods. By harnessing the power of IRB, researchers can explore new avenues for creating innovative nanoscale structures for various applications, such as drug delivery systems, sensors, and environmental remediation [44]. The intersection of biology and nanotechnology holds great promise for developing novel materials and technologies that could revolutionize multiple industries. Figure 4 describes various types of nanosensors that are used in various lethal and other diseases that are not the cause of death but affect the lifestyle of human beings.

4.1 Nano Biosensor-Based Diagnostic Tools in Viral Infections: Special Emphasis on Covid-19

Currently, various diagnostic instruments, including biosensors, lab-on-a-chip devices, and lateral flow tests, use sensors based on nanomaterials. These sensors allow for the quick and accurate identification of infections, illnesses, biomarkers, and analytes such as glucose levels, infectious agents, and cancer indicators. Nanosensors are useful for detecting biomarkers because they can monitor physical quantities and convert them into evident signals [45]. Nanosensors are being employed

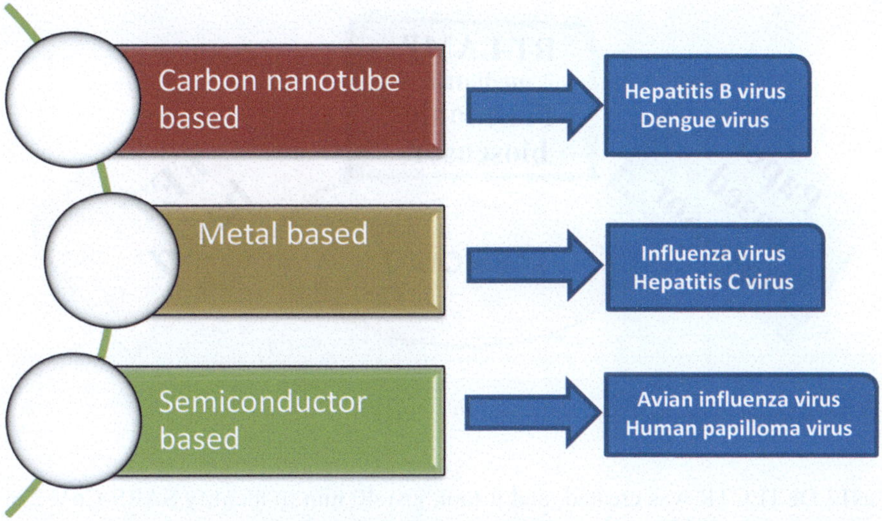

Fig. 4 Different types of nanosensors used in various diseases

in diagnostics to detect tumor masses, sense contaminants in organ implants, and administer drugs. They may offer quick and affordable diagnostic methods for a number of illnesses, including neurological conditions. To facilitate rapid identification and the start of suitable treatments, nanosensors are thought to be highly promising point-of-care diagnostic instruments [46].

Methods based on Nano biosensors offer a more exacting and focused way to identify emerging viral infections. The diagnosis of Covid-19 infection is currently being made using EC-based biosensors, surface-enhanced Raman spectroscopy (SERS)-based biosensors; field-effect transistor (FET)-based biosensors, and surface Plasmon resonance (SPR)-based biosensors [47], though the outcomes are not very good. Currently, early disease diagnosis by the use of nano-biosensors for Covid-19 infections is being explored. In this regard, nanomaterials like carbon and gold have attracted much interest in identifying the virus and its components [48]. By recognizing spike protein S1 in clinical samples, these nanomaterials, linked with analyte-like complementary single-stranded nucleic acid aptamer, may provide a unique technique for sensing SARS-CoV-2 [49]. While RT-PCR and CT scans are widely used as standard procedures for SARS-CoV-2 detection, each technology has certain limitations. Researchers are developing more sophisticated RT-PCR methods, which may be useful for pathogen identification. A reverse transcription loop-mediated isothermal amplification (RT-LAMP) test was recently developed by Park et al. and Yu et al. to detect the genomic RNA of SARS-CoV-2. The authors successfully developed a cross-reactivity-immune RT-PCR method that was highly sensitive [49, 50]. Similarly, Lin et al. improved the previously described method by combining RT-LAMP with a pH indicator. This enables a simple reading of the amplification reaction indicated by a color shift [49]. Additionally, a fluorescent assay called RT-LAMP/

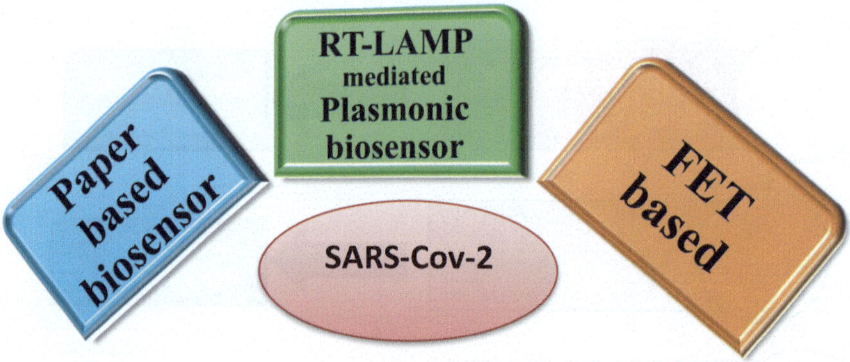

Fig. 5 Different nanosensors used for Covid-19 infection

Cas12 DETECTR was created, and it took just 40 min to identify SARS-CoV-2 in respiratory swab RNA extracts. To detect SARS-CoV-2, this technology was just as sensitive and effective as the US FDA EUA-approved CDC assay [51]. In following Fig. 5, different nanosensors are explained which were used as tool in SARS-Cov-2.

4.2 *Cancer Diagnosis*

Owing to its high mortality rate, cancer has been the focus of extensive and ongoing research worldwide. Nanostructure-based biosensors are very appealing as cancer detection candidates because of their improved sensitivity, selectivity, ease of use, and better analytical performance for quick sensing [52]. A brief review of the literature on nanostructure-based biosensors and bio-sensing methods that use nanostructures for cancer diagnosis is given in the following sections in Table 1.

4.3 *Drug Delivery System*

Nanosensors have been investigated as effective devices for drug delivery systems, with faster response and more sensitivity than planar sensors. Because nanotechnology allows for tailored drug delivery with reduced side effects, controlled dose release, and faster drug absorption, it has the potential to transform the human healthcare industry completely [68]. The goal of nanomedicine, or the medical application of nanotechnology, is to provide novel therapeutics and enhanced drug delivery methods [69]. Drug delivery has been studied using both nanotechnology and microtechnology. The size and form of drug carriers influence a drug's distribution, time in circulation, and cell uptake. Because they provide better specificity,

Table 1 CNT based biosensors

S. No.	Biosensors	Disease detection	Refs.
1	Paper-based bipolar electrode biosensors	Electrochemiluminescent detection of prostate specific antigen (PSA)	[53]
	CNT based biosensor	Drugs aimed at treatment of breast cancer	[54]
2	Screen-printed carbon electrode, modified by CNTs and AuNPs	Berberine, significant antimicrobial and anti-cancer activity	[55]
3	Tricosane-functionalized single-walled CNTs biosensor	VOCs (Volatile Organic Compounds) present in the breath, lung cancer	[56]
4	CNT based biosensor	Cancer marker galactin-3	[57]
5	Folic acid functionalized polydopamine coated CNTs biosensor	HeLa and HL60 cancer cells	[58]
6	Multilayer CNT based biosensor	Liver cancer cells	[59]
7	AuNP immobilized on CNT liquid electrodes	HER2 in breast cancer	[60]
8	Quantum-dot based biosensor	miR-141 in prostate cancer	[61]
9	Quantum-dot based fluorescent probes	Fluorescence imaging of metalloproteinase in cancer cells	[62]
10	Amplified electrochemiluminescent biosensors	Cancer cell detection	[63]
11	Surface-plasmon resonance (SPR) based biosensors	Tumor necrosis factor alpha (TNF-α)	[64]
12	Optical fiber SPR	Prostate specific antigen (PSA), oral cancer biomarkers	[65]
13	Quartz crystal microbalance (QCM) based biosensors	Single-nucleotide polymorphism in the p53 tumor suppressor gene	[66]
14	Magnetic nanoparticles-based biosensors	Cancer cells	[67]

solubility, therapeutic index, and immunogenicity, nanosensors are, therefore, essential to the development of drug delivery systems [70]. The use of nanotechnology in drug delivery has changed and developed quickly during the last few decades. DNA nanotechnology offers a straightforward and effective design method for self-assembling nanostructures, one of the many developing technology types. This method has special benefits and the potential to improve therapeutic targeting and decrease drug toxicity. Numerous sequence planning and optimization techniques have been developed to create DNA nanostructures that increase cells' absorption efficiency. These advancements suggest that DNA-based nanotechnology-based nanomedicines and drug delivery methods have a promising future (Fig. 6) [70].

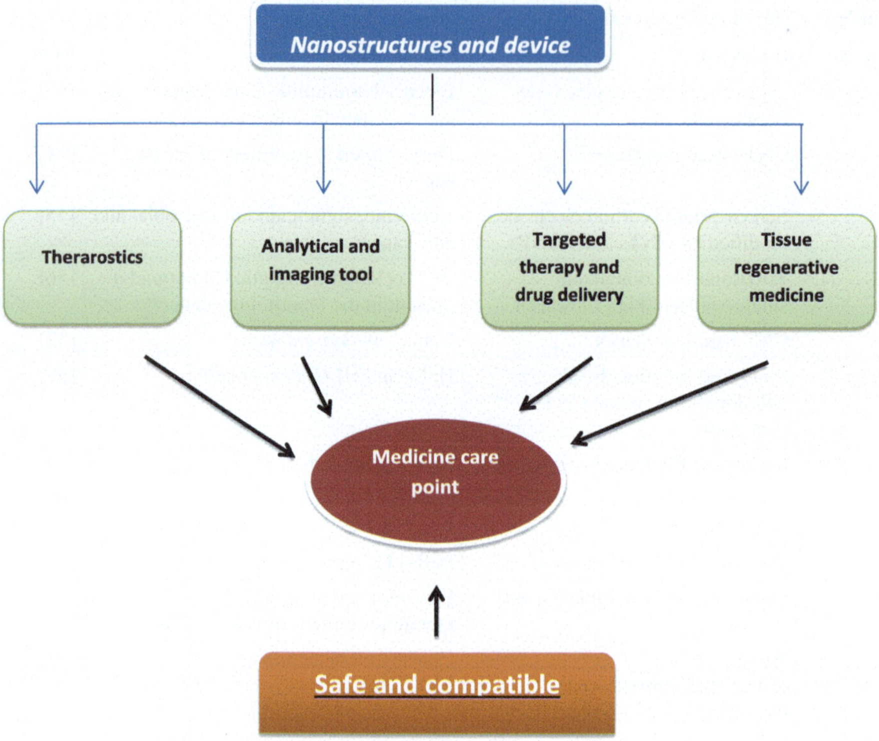

Fig. 6 Nanostructure and its application in medicine

4.4 Magnetic Resonance Imaging (MRI)

MRI contrast agents are iron oxide nanoparticles. They improve imaging quality and aid in the diagnosis of several illnesses, such as tumors and heart ailments [71]. Atomic-scale diamond impurities have been employed as spin sensors to achieve amazing outcomes in nano-MRI by taking advantage of magnetic dipole interactions rather than induction-based detection. The most significant use of nano-MRI is single-biomolecule imaging in vivo under room temperature and on the nanometre scale. This has been achieved by using diamond NV centers as sensors [72]. MRI nanosensors have been created to improve picture sensitivity and allow for non-invasive illness monitoring and diagnosis. These nanosensors offer enhanced diagnostic capabilities and better longitudinal analysis of physiological states since they may react to biological stimuli like pH, enzymes, and biomolecules as well as proteins [73]. Acetylcholine detection in the brain is one particular use of nanosensors in MRI, which can aid in the understanding of neuro signaling processes and brain function [74]. A novel nanosensor has been developed for detecting and discriminating thrombosis, a potential tool for early detection and accurate characterization

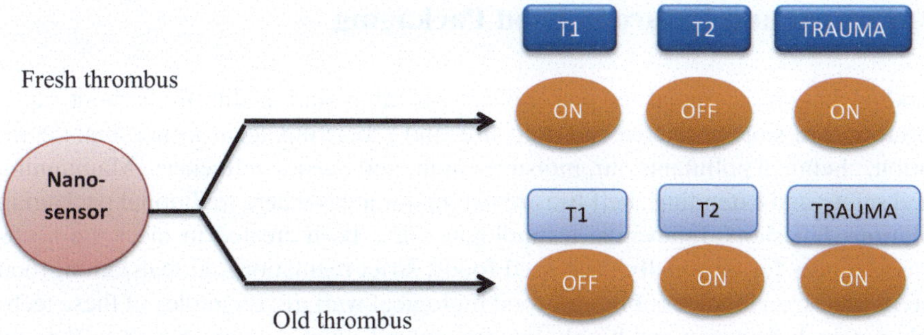

Fig. 7 A magnetic resonance nanosensor that can be activated is being explored as a potential imaging tool for detecting and classifying thrombosis

of cardiovascular diseases and cancer. The nanosensors, functionalized with a fibrin-binding peptide, can switch between T1 and T2 signals based on thrombus age or thrombin presence. They show thrombin-induced activation via NMR and MRI, with higher density resulting in lower activation. The nanosensors can distinguish between fresh and aged thrombi in vitro and are non-toxic to cells. This technology has potential for clinical diagnostics, reducing morbidity and mortality in vulnerable patients, and can be applied to other diseases like cancer (Fig. 7) [75].

4.5 Nanobots

Nanobots are the most important development in nanomedicine. It is possible to use nanobots to replace whole intracellular components and repair damaged cells. Moreover, they may be duplicated to fix a genetic flaw or swapped out for a DNA molecule to eradicate an illness. Medical nanobots have the potential to revolutionize healthcare by performing procedures like organ replacement or artery unblocking [76]. Despite their ability to deliver precise dosage concentrations to targeted areas, nanocarriers' tiny size forces a paradoxical trade-off between manufacturing and volume [77]. Artificial intelligence-enabled coordinated behavior and communication will enable nanobots to do tasks methodically and effectively. It is possible to independently program, search for, and move nanobots in different directions [78]. A single nanobot may communicate with friends and colleagues, each following their own set of rules, and request that they band their coatings, loads, or sizes together in order to more effectively monitor the presence, diagnosis, or targeting of diseased tissues. This collective effort can greatly enhance the capabilities of nanobots in medical applications, such as targeted drug delivery or tissue imaging. By working together, nanobots can overcome the limitations of their individual sizes and function more efficiently in complex environments.

5 Present and Future in Food Packaging

Food packaging is crucial for preserving food safety and quality throughout transportation and storage, increasing shelf life, and protecting against microbial degradation, chemical pollutants, air, moisture, light, and outside influences. Maintaining food safety and extending shelf life are two major areas where traditional packaging solutions fall short. Innovative technologies have been created to meet the needs of consumers for minimally processed foods, strict regulations, globalization, food safety issues, and the possibility of food biological warfare. Examples of these technologies include active and highly intelligent packaging systems. Smart packaging monitors the environment and food quality to forecast a safe preservation period and warns users when infections, dangerous chemicals, or harmful products are present. Additionally, it may identify allergy proteins to stop customers from experiencing negative side effects after consuming particular foods like flour, peanuts, and tree nuts [79]. Food packaging can have biological or chemical sensors for quick, precise examination. But these devices must also be affordable, simple, repeatable, easy to use, safe for food contact, dependable, and ecologically friendly. The selectivity, sensitivity, and simultaneous detection of several analytes or diseases have made nanoscale sensors a high priority. These sensors can evaluate food quality non-destructively and improve sensitivity and reaction times [80]. It is designed to bring attention to the positive and negative aspects compared to conventional techniques, motivating experts to create novel approaches for enhanced food safety and quality.

Nano sensors are being developed to track the effects of the atmosphere on food items from the manufacturing plant to the end user. Salespeople may verify freshness and storage conditions with the use of these sensors, which can measure temperature, moisture content, and other exterior factors [81]. Using inexpensive RFID tags, these sensors—which are still in the prototype stage—transmit data. On the nanoscale, nano sensors for internal conditions can be identified. One example is the production of nano biosensors, which alter the nano sensor output by attaching chemical or bacterial detectors to a probe. More than 1000 nm is detected by these flexible, tiny, nearly undetectable sensors [82].

Since the packaging's nanostructures have the potential to seep into food, the safety of nanotechnology packaging is a complicated subject. There are worries that nanostructures may be hazardous to the environment and people. Policies regarding packaging utilizing nanotechnology differ across nations. Food-contact nanotechnology packaging is regulated in the US by the Food and Drug Administration (FDA). Nanotechnology packaging is likewise governed by EU regulations; however, EU regulations are more extensive than US regulations. As a result, the subject of nanotechnology packaging safety is always changing. To evaluate the advantages and disadvantages of nanostructures, further studies are required [83].

5.1 Nanomaterial-Based Food Packaging

Food packaging is essential for maintaining food's nutritional and organoleptic qualities, halting deterioration, and shielding delicate bioactive substances from adverse environments. By employing functional nanoparticles to increase barrier qualities, thermal stability, strength, and durability and so lengthen food items' shelf lives, nanotechnology is being investigated as a potential improvement for food packaging [84]. It is possible to create nano-based packaging materials by creating nanocoatings and nanocomposite multi-layer materials or by adding nanoparticles to conventional packaging materials. Additionally, better, active, and smart packaging may be made with nanotechnology. A $1.33 trillion worldwide market for smart packaging is anticipated in 2024, with intelligent packaging developing at the quickest rate because of its distinctive, interactive, and user-friendly characteristics.

Food packaging uses nanoparticles to improve mechanical and physical characteristics such as water vapor and oxygen permeability, strength, flexibility, biodegradability, thermal resistivity, and UV absorptivity [85]. These characteristics are enhanced when metal oxides are added to polymers. Ag–Cu nanoparticles and fish skin gelatin have been used to create bio-composite films that exhibit enhanced mechanical strength, thermal stability, UV barrier qualities, and antibacterial activity. Polymers can benefit from cellulose nanoparticles enhancing their oxygen barrier capabilities [86]. It has also been discovered that transglutaminase and nano clay enhance the mechanical and physical characteristics of fish protein films. Nowadays, food packaging uses nano-coating to create thin films or layers of edible and non-edible elements (Fig. 8). The potential of edible coatings to increase of food product shelf life is encouraging and in Table 2, their uses in food packaging are shown.

a. **<u>Active packaging</u>**

An innovative method of extending the shelf life of food products is active packaging, which inhibits oxidation, moisture gain, microbial invasion, and overripening. Reactive oxygen species, which prevent DNA replication and ATP synthesis and cause cell damage or death, are partly produced by nanoparticles. Since metal and nano clays have potent antibacterial qualities, it has been discovered that gelatin film containing nanoparticles works well in active packaging formulations [97]. Organoclay nanoparticles, in combination with polyethylene nanocomposite films, have demonstrated strong antibacterial activity against *S. aureus* and *E. coli*. Films made of polyvinyl alcohol and containing cellulose nanoparticles and fennel seed oil have demonstrated enhanced antioxidant and antibacterial properties. Technologies known as ethylene scavengers have great promise in extending the shelf life of fresh fruits and vegetables. They work by ensnaring and absorbing ethylene. Research has demonstrated that at 1% TiO_2 concentration, ethylene photocatalytic degradation activity is demonstrated by chitosan-TiO_2 nanocomposite films [98]. Zhang et al. created films based on chitosan, TiO_2 nanoparticles, and black plum peel extract with strong antibacterial,

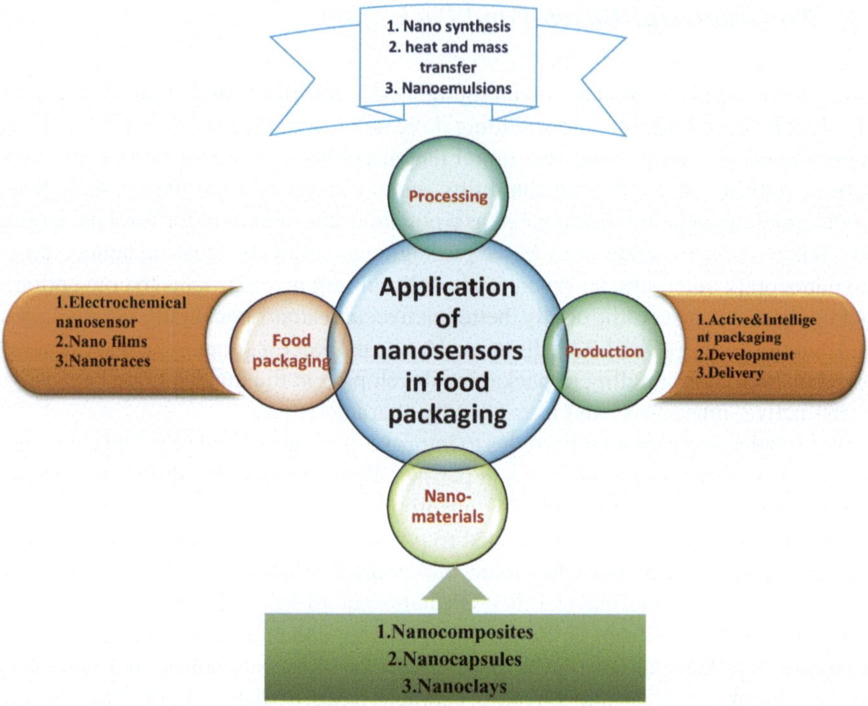

Fig. 8 Application of nanosensors in food packaging

antioxidant, and ethylene-scavenging capabilities [99]. Active packaging technologies, including ethylene-absorbing systems, moisture regulators, oxygen scavengers, and antimicrobial packaging, show promise for food packaging. Packaging with a low oxygen barrier helps prevent mold formation in baked goods. Because of their antibacterial properties, metal nanoparticles, metal oxide nanomaterials, and carbon nanotubes are also used [100]. To prevent migration problems, it is best to avoid direct contact with food. By combining less expensive passive packaging with more expensive active packaging, these technologies can lower the cost of packaging. Because active packaging methods have better quality, safety, and fewer additives, consumers prefer them.

b. **Intelligent packaging**

Smart packaging, or intelligent packaging materials, keeps an eye on the state of packed food or its surroundings. These materials are made up of smart devices that communicate, detect, record, track, and signal information about food safety, quality, and history across the supply chain. These devices include barcodes, RFID tags, sensors, and indicators. An effective use of nanotechnology is the creation of nano sensors, which are used to track the internal and exterior parameters of processed and fresh food items. Food quality may be tracked with the help of indicators like

Table 2 Different nanosensors used in food packaging and detection

S. No.	Biosensor	Action	Refs.
1	Gold nano-textured electrode	Detecting arsenic in food and water	[87]
2	Multi-walled carbon nanotube	Identification of Folic acid in fruit juices, wheat flour, and milk samples	[88]
3	Single-walled carbon nanotube	Sensing amino acids, tryptophan, and vitamin ascorbic acid	[89]
4	Magnetic nanoparticle-modified MWNT	Instantaneous grit on ascorbic acid	[90]
5	Polyaniline based bio-electrode	Determination of H_2O_2 and Azide	[91]
6	Hybrid molecularly imprinted polymer with a surface-enhanced Raman spectroscopy (SERS) sensor	He selective detection of chloramphenicol in milk	[89]
7	Aptamer and silver nanoparticles-based Nano sensor	To detect Acetamiprid	[92]
8	Rhodamine B/UiO-66-N3	H_2S detection via reaction-based ratiometric fluorescent nanosensor	[92]
9	SnO_2 nanowires (gas sensor)	To distinguish methanol from ethanol in alcoholic beverages	[93]
10	Carboxylated (c-MWCNT)-modified screen-printed electrode-based bionanosensor	Detecting the time of ripening of tomato with respect to its malic acid concentration	[94]
11	Glassy carbon electrode (GCE) modified with calixarene and gold nanoparticles	Detecting two toxic food dyes (metanil yellow and fast green)	[95]
12	Molecularly imprinted silica layers appended to quantum dots (fluorescence sensor)	To detect saxitoxin toxin in shellfish	[96]

pH, time–temperature, and leak indicators [101]. For example, spoiling in chicken breast and silver carp may be detected using colorimetric hydrogen sulphide sensors composed of gellan gum-capped silver nanoparticles [102]. Zinc oxide nanoparticles and phytochemicals combined with an alcohol-based starch-poly vinyl matrix have demonstrated potential use in intelligent food packaging. Due to pH variations during spoiling, soluble soybean polysaccharide films containing curcumin and SiO_2 have demonstrated strong antibacterial action as well as unique color changes [103]. Although the phrases "smart" and "intelligent" packaging have different meanings, they are sometimes used synonymously. Smart sensors, for instance, are active substances that include antioxidant and/or antibacterial qualities and can track the freshness and quality of food. Such smart sensors include polyphenols with halochromic characteristics, such as the naturally occurring pigments betalains and anthocyanins [104]. These bioactive substances are natural markers of food spoiling because they have antibacterial and antioxidant qualities and can change color in response to pH variations (Fig. 9).

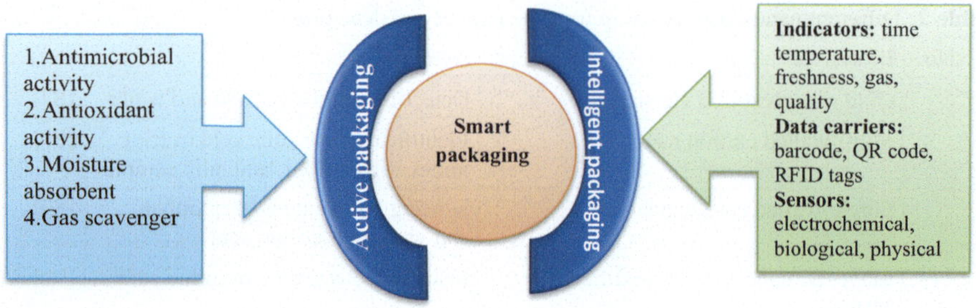

Fig. 9 Active and intelligent packaging and its application

5.2 *Toxicity of Nanomaterials Used for Food Packaging*

Food packaging contains nanoparticles that can seep into food items and enter the body through the skin, food, or ingestion, perhaps leading to illness (reference). A nanoparticle's toxicity rises with an increase in its surface area. Characteristics of nanoparticle size, such as polarity, lipophilicity, hydrophilicity, and catalytic activity, determine their distribution inside the human body [105]. In the experiment by Blasco et al., coating procedures were carried out utilizing polymer functionalization techniques such as adding isocyanates to alcohol, organic chemistry, and ester activation [106]. Its excellent stability and compatibility make it promising, but its isocyanide toxicity and instability present drawbacks. Ingestion of food and beverages is the primary method by which humans are exposed to nanoparticles [106].

Food packaging has discovered silver (Ag) nanoparticles to be beneficial because of their low toxicity and affordability. Although less harmful and more affordable than other nanoparticles, zinc oxide (ZnO) is carcinogenic to humans. Although chemical vapor deposition (CVD) is a natural, robust, stiff, and suitable NP material, using it in packaging poses safety risks. Carbon nanotubes have been transformed into polymers for use in smart and antimicrobial sensors, yet these materials may be highly poisonous. TiO_2 may induce lung cancers, but it also has the ability to alter the properties of cheap, nontoxic, and light-resistant coatings. Food packaging may be accomplished with starch, a non-toxic, renewable, natural, and ecological polysaccharide. On the other hand, particles such as CNTs, silver, zinc, and titanium may infiltrate the bloodstream and gather in organs such as the stomach, heart, and lungs [107].

Nanotechnology is a powerful tool for creating new materials, packaging, and coatings, but concerns remain about the safety of nanostructures [108]. The toxicology of polymeric packaging and coatings is directly related to the polymer matrix and additives. Some countries have established standards and legislation for polymers and additives in packaging, and each country has agencies to regulate and inspect the production, use, and trade of food products and materials. The European Food Safety Authority (EFSA) released a "Guidance on the risk assessment of nanomaterials to be applied in the food and feed chain: human and animal health" in 2021 [109]. The incorporation of nanostructures into polymeric matrices and biopolymers will result in changes in the properties of the packaging produced, which can modify the material's physicochemical properties, such as viscosity, tensile strength, elastic modulus, water solubility, thermal conductivity, electrical conductivity, thermal stability, and opacity. Or, they may add new properties, such as antimicrobial activity, to enlarge food protection.

5.3 Heating Effects on Nanomaterials in Food Packaging

When compared to bulk materials, nanomaterials exhibit special physical and chemical qualities. One polymeric nanoparticle utilized in food packaging is polylactic acid, which is non-toxic, biodegradable, and cost-effective [110]. Its qualities include high thermal processability, antibacterial activity, flame resistance, good heat seal ability, and resistance to water and oil. It is made using sustainable resources and low-carbon emission procedures. An inorganic metal oxide called zinc oxide nanoparticles is utilized in medical and food preservation [111]. It improves the nanomaterials' mechanical, thermal, and barrier qualities when combined with the polymer matrix. Solvent evaporation techniques are used to create the composite films, and flame atomic absorption spectrometry is used to measure the starting ingredient of zinc oxide in PLA nanocomposite films. Isooctane and 10% ethanol are used as food replacements in the general and particular migration tests. A differential scanning calorimeter (DSC) measures the heat of PLA/ZnO composite film [111].

5.4 Edible Food Packaging Using Nanomaterials

In the food sector, edible packaging is becoming more and more popular. It uses biodegradable materials as a covering that consumers may consume. The research looks toward improving its functionality using active substances like nanoparticles. This biodegradable, sustainable substitute increases food quality, extends shelf life, and minimizes waste. The promise of edible coatings to replace synthetic polymers and their positive effects on the environment has made them appealing [112]. Nonetheless, worries regarding the movement of nanoparticles and how they affect the environment and public health still exist.

5.5 Materials Used to Avoid the Environmental Concerns

Thermal resistance, heat resistance, and low flexibility of biodegradable films derived from renewable resources set them apart from packaging materials based on plastic. The goal of the research is to combine agar with other biological polymers, particles, and antioxidants to improve the quality of agar films. Further research is necessary to enhance the characteristics of agar in light of its intended use. Films and coatings have the potential to replace synthetic polymer films and increase the reusability of packaging materials [113]. Sustainable packaging is made from biopolymers, such as starch-based films, sugar cane pulp, and polylactide acid plastics. However, because these materials are neither renewable or biodegradable, they are disposed of in the land or ocean and are not thought to be ecologically benign [114].

5.6 DNA Nanosensors for Detection of Pathogens

To stop the spread of diseases, food samples must be tested for microorganisms, antibiotics, pesticides, metals, and heavy metals. Water and food product contamination has become a global issue, and outbreaks of infectious diseases, epidemics, or conditions akin to pandemics must be avoided by promptly identifying pathogenic microorganisms. *Bacillus subtilis*, *Vibrio cholera*, *Aspergillus* sp., *Escherichia coli*, *methicillin-resistant Staphylococcus aureus (MRSA)*, and *Candida* sp. are just a few of the dangerous microbes that DNA nanosensors can identify [112].

The manufacture of antibiotics and the animal husbandry sector depend on the detection of antibiotics. DNA-based nanosensors are multiplexed analytical techniques that are extremely sensitive, fast, selective, and affordable for detecting different antibiotics. Finding pesticides is also essential to guaranteeing the quality and safety of food. Pesticides such as carbamates, triazines, organophosphorus, neonicotinoids, and organophosphates are detected using DNA-based nanosensors [115].

Identifying heavy metals in food samples is crucial for food safety since elevated concentrations can seriously endanger consumers' health. Toxic pollutants such as arsenic(III), silver ions (Ag^{+}), mercury (Hg^{2+}), lead, and copper can be detected using DNA aptamer-based nanosensors.

DNA nanosensors may also identify toxic contaminants like cadmium and arsenic. DNA aptamer-based nanosensors are used to detect bisphenol A and fluorogenic rhodamine B dye [116]. DNA-CuNPs nanosensors can identify dopamine, and carcinogenic nitrosamines can be detected using chitosan carbon dot electrodes bound with DNA. G-rich DNA-AgNCs nanosensors can detect cyanide, whereas carbon nanotubes and graphene electrochemical sensors are employed to detect mycotoxin [117, 118].

6 Major Challenges Faced in Nano-biosensors

Although launched decades ago, glucose meters and lateral flow pregnancy tests remain the most commercialized biosensors. Recently, several nanomaterial-based biosensors have been developed, but they are limited to the publication of literature and are technically complicated, making commercialization unfeasible. Safety and toxicity issues related to nanomaterials are also a concern, necessitating further studies on these materials and their special effects. Despite these challenges, the development and application of nano-biosensors in real life remain a challenge [119].

In a previous research article, three major issues were pointed out which may be beneficial in the coming years [120]:

- Nanoparticle selection for biological research and tests. Physiological parameters, accumulation and sediment capacities, and other features must be assessed to identify the nanoparticles before, during, and after the studies.
- It is necessary to investigate how man-made nanoparticles are captured by living entities in various situations.
- The organisms and measuring sites that could be employed in experiments

Nano sensor research is often conducted in controlled laboratory environments, which may differ from real-world circumstances. In addition to being able to function in the presence of various chemicals and microbes, sensors must be able to withstand high pressures, temperatures, and pH changes. Infrastructure expenditures are anticipated to be high until the technology gains traction. The packaging industry's whole fabrication process might be altered by technical improvements in nano industries, particularly when biotechnological initiatives are taken into account [121].

Another factor to consider is the stability of nano sensors in the presence of other species. Analytes interacting with several unknown species may change their fingerprint signal, making detection and identification more difficult. To ensure that nano sensors are resilient and operate efficiently, new performance testing must be done.

It is important to remember that biological recognition components like enzymes, antigens, and antibodies are stable. When proteases get access to the recognition component, biosensors based on the enzyme acetylcholinesterase (AChE) have proven successful in identifying pesticide residues [122]. Use of lipid membrane-encased porins can choose target molecules.

Optimizing functional characteristics and achieving efficient, consistent immobilization of biological materials provide the primary technological difficulty in developing new nano sensors, particularly immunological ones. Protein molecules can be absorbed on silicon or metal surfaces, which is frequently unsuccessful.

7 Future Aspects of Nano Sensors

Future prospects for nanosensors in medicine are full of opportunities. For medical applications, they provide the potential for diagnostic instruments with higher sensitivity, specificity, and dependability. Localized Plasmon resonance (LSPR)-based nanoplasmonic biosensors offer several benefits, including multiplexing, high sensitivity, real-time detection, and low cost, speed, and portability [123]. They have been used in precision medicine to evaluate vaccines, deliver drugs, and modify DNA. The high surface area, electrical characteristics, and electro-catalytic activity of nanosensors have led to an increasing interest in this multidisciplinary research field. Nano diagnostics will be accessible within the next ten years, enabling the simultaneous study of thousands of biomarkers. Biofluid analysis is the most widely used clinical diagnostic application. Ultrasensitive cancer biomarker detection can be provided by nanobiosensors, which may help in early diagnosis and treatment [124]. The viability of Nanodevices for this use is being investigated. In the developing field of nano Theranostics, prophylactic implanting of nanotherapeutic devices allows for early diagnosis and therapy [119]. Biosensors are crucial in biomedical research, healthcare, and pharmaceuticals for detecting biomarkers for disease diagnosis. With advancements in biotechnology and nanotechnology, nanobiosensors are becoming more powerful in medicine. Different nano biosensors include nanosensors, immunosensors, enzymatic, and microbial biosensors based on different transduction approaches like electrochemistry, optic, and piezoelectric measurements. Advances have been made in nanomaterials like carbon (graphene, CNT), gold, silver, and semiconductors (QDs, TiO_2, ZnO). Comparing nanomaterials in biosensors is beneficial for understanding their working principles, constructions, advantages, and applications [125].

Food matrix convolution, high prices, and a lack of thorough testing have hindered the widespread application of nanoparticles in food quality and freshness research, which is still in its early phases. It is necessary to address concerns like cost, biocompatibility, and expansibility to guarantee effective deployment. Nanotechnology advances food production, preservation, and analysis; nevertheless, concerns about health and environmental effects may arise because of its toxicity. Bioactive substances can be encapsulated to improve product stability, reduce pesticide use, and add taste and nutrition. Liposomes have demonstrated the greatest outcomes in enhancing food shelf life and adding nutrients and tastes. Nanomaterials or nanobeads can be used to design biodegradable polymers. Natural biopolymers that are utilized to enhance food texture, taste, smell, and color include titanium dioxide and silicon dioxide [126].

8 Conclusion

Over the past ten years, biological research has made great strides in medicine, especially in the area of intracellular signaling transduction, which controls intracellular activity and secretes growth factors and cytokines to maintain tissue organization. Diseases can result from deviations from the regular metabolic pathways. Research constantly reveals the hidden functions of genes and proteins that produce these mistakes. Early illness detection is made possible by sensors' ability to identify particular proteins or signaling chemicals. Consequently, sensors are playing a more and more important role in both general and personalized medicine.

The outstanding mechanical, physic-chemical, thermal, and barrier qualities of nanotechnology make it essential for food packaging. Food packaging can benefit from the usage of nano sensors, which employ tiny enzyme immobilization systems to detect gasses and microbes. Plant microbes may be utilized to produce nanoparticles, and multi-layered packing is employed as a special barrier. Metal oxide nanoparticles are restricted because of their toxicity, whereas metal nanoparticles have antibacterial effects. The food business strives for easily biodegradable, renewable, and recyclable materials, and edible materials improve food safety and quality. Nanocarriers are the most common organic nanoparticles in food and agriculture; nevertheless, in vitro substitutes are needed for toxicity evaluations. Research on the biosynthesis of nanomaterials is being conducted worldwide; edible nanomaterials may be able to function better in food packaging.

Nanotechnology is revolutionizing global technologies, particularly in the food sector, packaging, and coatings. It enables safer food production, longer shelf life, less industrial waste, and higher nutritional value. Packaging with nanostructures promotes intelligent packaging, active packaging with antimicrobial properties, improved physicochemical properties, and drug carrier applications. However, studies are needed to ensure safety and environmental well-being. New standards, legislation, and tests are needed for nanostructures in food. Currently, there are few specific regulations for nanotechnology applied to food. The REACH legislation from the European Union ensures the safety and risk assessment of chemical substances used in the industry, including nanostructures. The FDA regulates nanotechnology in the food packaging sector, while ISO standards establish guidelines for the characterization, measurement, risk assessment, and safety of nanomaterials in products. Brazil lacks specific regulations for using nanotechnology in food, but ISO standards are a reference for evaluating nanomaterials.

The growing demand for food has led to increased research on food packaging, with nanotechnology promising improved physical, mechanical, and functional properties. By incorporating nanomaterials into the polymer matrix, packaging materials can enhance mechanical, water, oxygen, and antimicrobial properties, enhancing shelf life. Nanoparticles are also used in active packaging to prolong shelf life and communicate with retailers and consumers. However, using nanoparticles in food packaging can cause adverse health issues if migrated into food and exposed for extended periods. Therefore, studying the migration, toxicity, and permissible limits of nanoparticles in food packaging is crucial.

References

1. Duncan TV (2011) Applications of nanotechnology in food packaging and food safety: barrier materials, antimicrobials and sensors. J Colloid Interface Sci 363(1):1–24
2. Chaudhry Q, Watkins R, Castle L (2017) Nanotechnologies in food: what, why and how? In: Chaudhry Q, Castle L, Watkins R (eds) Nanotechnologies in food. The Royal Society of Chemistry
3. Roco MC (2003) Nanotechnology: convergence with modern biology and medicine. Curr Opin Biotechnol 14(3):337–346
4. Gonzalez L et al (2013) Nanotechnology in corneal neovascularization therapy—a review. J Ocul Pharmacol Ther 29(2):124–134
5. Wang J et al (2022) Metallic nanomaterials with biomedical applications. Metals 12:2133
6. Curulli A (2023) Functional nanomaterials enhancing electrochemical biosensors as smart tools for detecting infectious viral diseases. 28(9):3777
7. Chen J et al (2023) Emerging nanomaterials to enhance electrochemical impedance spectroscopy for biomedical applications. Front Mater 10
8. Bais R (2022) An overview of the synergy of electrochemistry and nanotechnology for advancements in sensing applications
9. Zhang W et al (2023) Plasmonic nanomaterials in dark field sensing systems. 13(13):2027
10. Yang G (2022) Synthesis, properties, and applications of carbyne nanocrystals. Mater Sci Eng R Rep 151:100692
11. Dimitrievska I, Paunovic P, Grozdanov A (2023) Recent advancements in nano sensors for air and water pollution control. Mater Sci Eng 7:113–128
12. Avaz S et al (2017) Graphene based nanosensor for aqueous phase detection of nitroaromatics. RSC Adv 7(41):25519–25527
13. Hashim H et al (2016) Manipulation and immobilization of a single fluorescence nanosensor for selective injection into cells. 16(12):2041
14. Rashidiani J et al (2018) Nano-aptasensor: strategies and categorizing. Int J Med Rev 5:68–76
15. Ding B et al (2010) Electrospun nanomaterials for ultrasensitive sensors. 13(11):16–27
16. Lachowicz JI et al (2021) Metals and metal-nanoparticles in human pathologies: from exposure to therapy. 26(21):6639
17. Katz E, Willner I (2004) Biomolecule-functionalized carbon nanotubes: applications in nanobioelectronics. ChemPhysChem 5(8):1084–1104
18. Marczin N, Yacoub M (2002) Disease markers in exhaled breath: basic mechanisms and clinical applications, vol 346. IOS Press
19. Zhang H (2022) Applications of advanced nanomaterials in sensor devices. 15(24):8995
20. Holzinger M, Le Goff A, Cosnier S (2014) Nanomaterials for biosensing applications: a review. Front Chem 2:63
21. Huang X, Zhu Y, Kianfar E (2021) Nano biosensors: properties, applications and electrochemical techniques. J Market Res 12:1649–1672
22. Tonelli D et al (2023) Focus review on nanomaterial-based electrochemical sensing of glucose for health applications. 13(12):1883
23. Fiedorova K et al (2022) Review of present method of glucose from human blood and body fluids assessment. 211:114348
24. Zhang S et al (2023) Wearable non-invasive glucose sensors based on metallic nanomaterials. 20:100638
25. Velusamy P et al (2023) Volatile organic compounds as potential biomarkers for noninvasive disease detection by nanosensors: a comprehensive review. 53(8):1828–1839

26. Wu Y et al (2020) Simultaneous and sensitive determination of ascorbic acid, dopamine and uric acid via an electrochemical sensor based on PVP-graphene composite. 18:1–13
27. He Q et al (2020) Sensitive voltammetric sensor for tryptophan detection by using polyvinylpyrrolidone functionalized graphene/GCE. 10(1):125
28. Kanapskyte A et al (2021) Space biology research and biosensor technologies: past, present, and future. 11(2):38
29. Agrawal S, Prajapati R (2012) Nanosensors and their pharmaceutical applications: a review. Int J Pharm Sci Technol 4:1528–1535
30. Sinha A, Behera A (2022) Nanotechnology in the space industry. Nanotechnology-based smart remote sensing networks for disaster prevention. Elsevier, pp 139–157
31. (1991) Chemical sensors: definitions and classification. J Pure Appl Chem 63:1247–1250
32. Javaid M et al (2021) Exploring the potential of nanosensors: a brief overview. 2:100130
33. Malik P et al (2021) Emerging nanomaterials for improved biosensing. 16:100050
34. Agasti SS et al (2010) Nanoparticles for detection and diagnosis. Adv Drug Deliv Rev 62(3):316–328
35. Patel G et al (2020) Application of nanosensors in the food industry. Nanosensors for smart cities. Elsevier, pp 355–368
36. Prasad S (2014) Nanobiosensors: the future for diagnosis of disease? Nanobiosens Dis Diagnos 1–10
37. Xu P et al (2020) Efficient hydrolysis of ammonia borane for hydrogen evolution catalyzed by plasmonic Ag@Pd core-shell nanocubes. ACS Sustain Chem Eng 8(33):12366–12377
38. Wang P et al (2020) A superhydrophobic/electrothermal synergistically anti-icing strategy based on graphene composite. Compos Sci Technol 198:108307
39. Cui D et al (2021) Pyrolysis temperature effect on compositions of basic nitrogen species in Huadian shale oil using positive-ion ESI FT-ICR MS and GC-NCD. J Anal Appl Pyrol 153:104980
40. Ramos AP et al (2017) Biomedical applications of nanotechnology. 9(2):79–89
41. Mousavi SM et al (2018) Nanosensors for chemical and biological and medical applications. 8(8):2161-0444.1000515
42. Suga M et al (2014) Recent progress in scanning electron microscopy for the characterization of fine structural details of nano materials. 42(1–2):1–21
43. Su D (2017) Advanced electron microscopy characterization of nanomaterials for catalysis. Green Energy Environ 2(2):70–83
44. Nadeem M et al (2021) A review of microbial mediated iron nanoparticles (IONPs) and its biomedical applications. 12(1):130
45. Adam T, Gopinath SC (2022) Nanosensors: recent perspectives on attainments and future promise of downstream applications. Process Biochem 117:153–173
46. Thwala LN et al (2023) Nanotechnology-based diagnostics for diseases prevalent in developing countries: current advances in point-of-care tests. 13(7):1247
47. Samson R, Navale GR, Dharne MS (2020) Biosensors: frontiers in rapid detection of COVID-19. 3 Biotech 10(9):385
48. Holzinger M, Le Goff A, Cosnier S (2014) Nanomaterials for biosensing applications: a review. 2
49. Yu L et al (2020) Rapid detection of COVID-19 coronavirus using a reverse transcriptional loop-mediated isothermal amplification (RT-LAMP) diagnostic platform. Clin Chem 66(7):975–977
50. Park GS et al (2020) Development of reverse transcription loop-mediated isothermal amplification assays targeting severe acute respiratory syndrome coronavirus 2 (SARS-CoV-2). J Mol Diagn 22(6):729–735
51. Broughton JP et al (2020) CRISPR-Cas12-based detection of SARS-CoV-2. Nat Biotechnol 38(7):870–874
52. Das S et al (2023) Diagnosis of cancer using carbon nanomaterial-based biosensors. 2(2):268–289

53. Feng Q, Chen H, Xu J (2015) Disposable paper-based bipolar electrode array for multiplexed electrochemiluminescence detection of pathogenic DNAs. 58(5):810–818
54. Baj-Rossi C, Micheli GD, Carrara S (2012) Electrochemical detection of anti-breast-cancer agents in human serum by cytochrome P450-coated carbon nanotubes. 12(5):6520–6537
55. Labuda J (2006) Nanostructured electrochemical DNA biosensors for detection of the effect of berberine on DNA from cancer cells. Anal Bioanal Chem 2006
56. Liu FL et al (2011) Single-walled carbon nanotube-based biosensors for the detection of volatile organic compounds of lung cancer. Physica E 44(2):367–372
57. Park YK et al (2011) D-(+)-galactose-conjugated single-walled carbon nanotubes as new chemical probes for electrochemical biosensors for the cancer marker galectin-3. Int J Mol Sci 12(5):2946–2957
58. Zheng T-T et al (2012) A label-free cytosensor for the enhanced electrochemical detection of cancer cells using polydopamine-coated carbon nanotubes. Analyst 137(6):1316–1318
59. Liu Y et al (2014) Construction of carbon nanotube based nanoarchitectures for selective impedimetric detection of cancer cells in whole blood. Analyst 139(20):5086–5092
60. Arkan E et al (2015) A novel antibody–antigen based impedimetric immunosensor for low level detection of HER2 in serum samples of breast cancer patients via modification of a gold nanoparticles decorated multiwall carbon nanotube-ionic liquid electrode. Anal Chim Acta 874:66–74
61. Jou AF-j et al (2015) Diagnosing the miR-141 prostate cancer biomarker using nucleic acid-functionalized CdSe/ZnS QDs and telomerase. Chem Sci 6(1):659–665
62. Jeong S et al (2017) Cancer-microenvironment-sensitive activatable quantum dot probe in the second near-infrared window. Nano Lett 17(3):1378–1386
63. Jie G, Zhao Y, Niu S (2013) Amplified electrochemiluminescence detection of cancer cells using a new bifunctional quantum dot as signal probe. Biosens Bioelectron 50:368–372
64. Law W-C et al (2011) Sensitivity improved surface plasmon resonance biosensor for cancer biomarker detection based on plasmonic enhancement. ACS Nano 5(6):4858–4864
65. Jang HS et al (2009) Optical fiber SPR biosensor with sandwich assay for the detection of prostate specific antigen. Opt Commun 282(14):2827–2830
66. Wang D et al (2012) Highly selective detection of single-nucleotide polymorphisms using a quartz crystal microbalance biosensor based on the toehold-mediated strand displacement reaction. Anal Chem 84(16):7008–7014
67. Kumar M et al (2010) Image-guided breast tumor therapy using a small interfering RNA nanodrug. Can Res 70:7553–7561
68. Marianecci C, Carafa M (2019) Smart nanovesicles for drug targeting and delivery. Pharmaceutics 11(4)
69. Kapil A, Aggarwal G, Sl H (2014) Nanotechnology in novel drug delivery system. J Drug Deliv Therap 4
70. Sun J, Yang Z, Teng L (2020) Nanotechnology and microtechnology in drug delivery systems. Dose Response 18(2):1559325820907810
71. Rahman MJN (2023) Magnetic resonance imaging and iron-oxide nanoparticles in the era of personalized medicine. Nanotheranostics 7(4):424
72. Boretti A, Rosa L, Castelletto S (2015) Towards single biomolecule imaging via optical nanoscale magnetic resonance imaging. Small 11(34):4229–4236
73. Wu K et al (2020) Magnetic particle spectroscopy: a short review of applications using magnetic nanoparticles. ACS Appl Nano Mater 3(6):4972–4989
74. Ellis CM, Pellico J, Davis JJ (2019) Magnetic nanoparticles supporting bio-responsive T1/T2 magnetic resonance imaging. 12(24):4096
75. Ta HT et al (2018) Activatable magnetic resonance nanosensor as a potential imaging agent for detecting and discriminating thrombosis. Nanoscale 10(31):15103–15115
76. Kong X et al (2023) Advances of medical nanorobots for future cancer treatments. 16(1):74
77. Haleem A et al (2023) Applications of nanotechnology in medical field: a brief review. Glob Health J 7(2):70–77

78. Mohammad H et al (2023) Strategies of artificial intelligence tools in the domain of nanomedicine. 105157
79. Siró I (2014) Intelligent packaging and food safety. Pract Food Saf Contempor Issues Futur Dir 375–394
80. Vanderroost M et al (2014) Intelligent food packaging: the next generation. 39(1):47–62
81. Wang Y, Duncan TV (2017) Nanoscale sensors for assuring the safety of food products. Curr Opin Biotechnol 44:74–86
82. Castillo-Henríquez L et al (2020) Biosensors for the detection of bacterial and viral clinical pathogens. 20(23):6926
83. Mitrano DM, Wohlleben W (2020) Microplastic regulation should be more precise to incentivize both innovation and environmental safety. Nat Commun 11(1):5324
84. Kuswandi B, Moradi M (2019) Improvement of food packaging based on functional nanomaterial. Nanotechnol Appl Energy Drug Food 309–344
85. Jafarzadeh S, Jafari SM (2021) Impact of metal nanoparticles on the mechanical, barrier, optical and thermal properties of biodegradable food packaging materials. Crit Rev Food Sci Nutr 61(16):2640–2658
86. Wang J et al (2018) Moisture and oxygen barrier properties of cellulose nanomaterial-based films. 6(1):49–70
87. Babar N-U-A et al (2019) Highly sensitive and selective detection of arsenic using electrogenerated nanotextured gold assemblage. 4(9):13645–13657
88. Wei S et al (2006) Voltammetric determination of folic acid with a multi-walled carbon nanotube-modified gold electrode. 152:285–290
89. Xiao F et al (2008) Single-walled carbon nanotube-ionic liquid paste electrode for the sensitive voltammetric determination of folic acid. 134(2):895–901
90. Taei M et al (2016) Simultaneous determination of ascorbic acid, acetaminophen and codeine based on multi-walled carbon nanotubes modified with magnetic nanoparticles paste electrode. 69:1–11
91. Singh N, Singh S (2017) A modified mean gray wolf optimization approach for benchmark and biomedical problems. Evol Bioinform 13:1176934317729413
92. Jokar M et al (2016) Design and evaluation of an apta-nano-sensor to detect Acetamiprid in vitro and in silico. 34(11):2505–2517
93. Tonezzer M et al (2022) Nanosensor based on thermal gradient and machine learning for the detection of methanol adulteration in alcoholic beverages and methanol poisoning. 22(15):5554
94. Dalal A, Rana J, Kumar A (2017) Ultrasensitive nanosensor for detection of malic acid in tomato as fruit ripening indicator. Food Anal Methods 10:3680–3686
95. Shah A (2020) A novel electrochemical nanosensor for the simultaneous sensing of two toxic food dyes. ACS Omega 5(11):6187–6193
96. Sun A et al (2018) Development of a selective fluorescence nanosensor based on molecularly imprinted-quantum dot optosensing materials for saxitoxin detection in shellfish samples. 258:408–414
97. Shankar S, Jaiswal L, Rhim J-W (2016) Gelatin-based nanocomposite films: potential use in antimicrobial active packaging. Antimicrobial food packaging. Elsevier, pp 339–348
98. Siripatrawan U, Kaewklin P (2018) Fabrication and characterization of chitosan-titanium dioxide nanocomposite film as ethylene scavenging and antimicrobial active food packaging. Food Hydrocoll 84:125–134
99. Zhang X et al (2019) Development of multifunctional food packaging films based on chitosan, TiO_2 nanoparticles and anthocyanin-rich black plum peel extract. 94:80–92
100. Lee SJ, Rahman AM (2014) Intelligent packaging for food products. Innovations in food packaging. Elsevier, pp 171–209
101. Kalpana S et al (2019) Intelligent packaging: trends and applications in food systems. 93:145–157
102. Zhai X et al (2019) A colorimetric hydrogen sulfide sensor based on gellan gum-silver nanoparticles bionanocomposite for monitoring of meat spoilage in intelligent packaging. 290:135–143

103. Jayakumar A et al (2019) Starch-PVA composite films with zinc-oxide nanoparticles and phytochemicals as intelligent pH sensing wraps for food packaging application. 136:395–403
104. Rodrigues C et al (2023) *Opuntia* spp.: an overview of the bioactive profile and food applications of this versatile crop adapted to arid lands. 12(7):1465
105. Sahu SC, Hayes AW (2017) Toxicity of nanomaterials found in human environment: a literature review. Toxicol Res Appl 1:2397847317726352
106. Blasco E et al (2017) 50th anniversary perspective: polymer functionalization. 50(14):5215–5252
107. Ziarati P et al (2018) A comprehensive review: toxicity of nanotechnology in the food industry. 3(2):1–12
108. Adeyeye SAO, Ashaolu TJ (2021) Applications of nano-materials in food packaging: a review. J Food Process Eng 44(7):e13708
109. Committee ES et al (2021) Guidance on risk assessment of nanomaterials to be applied in the food and feed chain: human and animal health. 19(8):e06768
110. Li G et al (2020) Synthesis and biological application of polylactic acid. 25(21):5023
111. Lu W et al (2021) Changes of thermal properties and microstructure of nano-ZnO/polylactic acid composite films during Zn migration. 34(1):3–10
112. Nunes C et al (2023) Edible coatings and future trends in active food packaging—fruits' and traditional sausages' shelf life increasing. 12(17):3308
113. Mostafavi FS, Zaeim D (2020) Agar-based edible films for food packaging applications—a review. Int J Biol Macromol 159:1165–1176
114. Aguirre-Joya JA et al (2018) Basic and applied concepts of edible packaging for foods. Food packaging and preservation. Elsevier, pp 1–61
115. Joshi S et al (2023) Nanosensors in pesticide residue detection in grapes. 124–136
116. Caglayan MO, Şahin S, Üstündağ Z (2022) An overview of aptamer-based sensor platforms for the detection of bisphenol-A. Crit Rev Anal Chem 1–22
117. Kumar V, Guleria P (2020) Application of DNA-nanosensor for environmental monitoring: recent advances and perspectives. Curr Pollut Rep 1–21
118. Koedrith P et al (2015) Recent trends in rapid environmental monitoring of pathogens and toxicants: potential of nanoparticle-based biosensor and applications. 2015
119. Karki K et al (2019) Nano-biosensors for cancer diagnostics: current status and future prospects. 797–801
120. Kahru A, Dubourguier H-C (2010) From ecotoxicology to nanoecotoxicology. Toxicology 269(2):105–119
121. Malik S, Muhammad K, Waheed Y (2023) Nanotechnology: a revolution in modern industry. Molecules 28(2):661
122. Prasad SN, Bansal V, Ramanathan R (2021) Detection of pesticides using nanozymes: trends, challenges and outlook. Trends Anal Chem 144:116429
123. Csáki A, Stranik O, Fritzsche W (2018) Localized surface plasmon resonance based biosensing. Exp Rev Mol Diagnos 18(3):279–296
124. Khazaei M et al (2023) Nanosensors and their applications in early diagnosis of cancer. 100569
125. Hamideh R, Sajad J (2015) Medical nanobiosensors: a tutorial review. Nanomed J
126. Biswas R et al (2022) Application of nanotechnology in food: processing, preservation, packaging and safety assessment. Heliyon 8(11):e11795

Fabrication and Different Applications for Polymer-Based Nanosensors

Aya Abd El Kareem Abd El Aziz, Mariam Ahmed Sayed, Abd El Rahman Mamdouh Mohamed, Mostafa Gamal Ahmed, Somia Samir Saied, Esraa Said Mohammed, and Riham R. Mohamed

Abstract Nanosensors are nanoscale component devices designed to detect a specific molecule, biological factor, or environmental condition. Compared to their macroscale analogues, these sensors are highly precise, portable, economical, and detect at a far lower level. The need for nanostructured sensors to identify and measure physical and chemical features as well as properties of sophisticated biological and industrial systems has led to a growth in importance in their commercialization. In this chapter, we will discuss different routes to fabricate polymer-based nanosensors and also various applications for their usage in industry like; chemical, optical, medicinal, food, textiles, and electronics. Particularly exciting uses for nanosensors include; chemical detection, food and water quality monitoring, and medical diagnostic applications. 2030 sustainable development goals have the potential to be successfully implemented with the help of nanosensors technology. However, the commercial impact of nanosensors has been diminished by a lack of knowledge about the health impacts of these materials and the expensive cost of some of the basic ingredients needed.

Keywords Nanosensors · Food · Textiles · Medicine · Optical · Chemical detection · Biosensors

1 Introduction

Numerous nanotechnology developments have the potential to improve our planet and make it a more secure and ultimately better place to live. This is also true of the sensor industry. Over more than ten years, numerous institutes have been researching nanosensors. A sensor developed at the atomic level and utilizing readings made at the nanoscale is called a nanosensor [1]. The investigation and progress of nanosensors

A. A. E. K. A. E. Aziz · M. A. Sayed · A. E. R. M. Mohamed · M. G. Ahmed · S. S. Saied · E. S. Mohammed · R. R. Mohamed (✉)
Faculty of Science, Chemistry Department, Cairo University, Giza, Egypt
e-mail: rihamrashad@hotmail.com

I. Uddin (ed.), *Smart Nanosensors*, Smart Nanomaterials Technology,
https://doi.org/10.1007/978-981-96-3878-9_4

for a wide range of uses has advanced significantly. A few notable uses include the aviation, medical, integrated circuits, national security, and many more industries [2].

There are numerous varieties of nanosensors as well as numerous techniques to develop them, in addition to the numerous applications for which they are used. The manufacturing of these nanosensors has a lot of difficulties at the moment, but if they are established for routine usage, they will be superior to the sensors used in technology today in a variety of ways. A nanosensor's primary job is to transform atomic-scale data into simply evaluated data. These devices could be well-defined as physical or chemical sensors built from nanoscale constituents, which are typically tiny or submicroscopic in size. Due to their high sensitivity, these sensors are able to detect even minute quantities of potentially harmful chemicals, such individual virus particles [3].

Components of biological detection covalently linked to the transducer or potentially including biological recognition elements make up the biosensitive layer of a nanosensor. Target sample and bioreceptor interaction is to induce physicochemical changes that can be transformed into an outcome that can be measured, like an electrical signal [4]. In order to fully comprehend this technology, one must recognize that if one or more of a sensor's dimensions fall between 10 and 100 nm, the sensor is given the prefix nano. Sensors and nanosensors represent the state of the art in analytical electronics technology [5].

Therefore, a nanosensor can be any sensitive material that transmits information about molecules and nanoparticles from detection processes in the forms of chemical, physical, or biological data. They may also be referred to as molecular analysis tools because they feature a transduction system linked to a recognition component that allows the signal generated by the interaction of the sensing component and the analyte to be processed [6].

The nature of the identifying element (e.g., biological or inorganic), the transduction system, the mechanism utilized to detect the link, and the type of contact formed between the sensing element and the analyte are the four key variables that can be used to categorize nanosensors. As a result, distinct components, information-transmission signals, and detection targets can be used to identify and classify various types of nanosensors. Biochemical assays, nucleic acids, drug carriers, enzyme assays, antigen-antibodies, medical diagnosis using disease markers, proteins, antibodies, hormones, deoxyribonucleic acid sequencing, DNA, environmental observing (air pollution) and antibiotics are just a few of the applications for which these structures are being studied to find new tools [7–9]. This chapter discusses fabrication methods of nanosensors and some of the applications of nanosensors in different aspects of industry.

2 Synthesis and Fabrication of Nanosensors

I. Fabrication of nanosensor

The fabrication of nanomaterials or distinct nanoscale structures for diverse uses is known as nano-manufacturing. With many orders of magnitude lower production dimensions, this may be viewed as an improved kind of micromanufacturing/ microfabrication. While the terms are occasionally used interchangeably, nanofabrication more closely refers to a nanoscale fabrication technique utilized in financed research projects, while nano-manufacturing more closely relates to the process of producing goods to generate revenue [10]. Nonetheless, we do not find it particularly troubling as the word "nanomanufacturing" has no precise meaning; instead, a broad overview of the prevalent nanomanufacturing procedures used today to produce nanosensors was offered here in this section.

1. Top-down approach

The well-known statue David by Michelangelo is among the most famous sculptures ever created. If one is curious about the process used to create David or any other stone or wooden sculpture, the explanation is straightforward: a sizable block of stone or wood is progressively sculpted into the desired shape. We refer to this approach as top-down. Using this technique, a large block of material is removed and gradually cut away until the required form is obtained in nano-manufacturing. The two phases in the top-down method are pattern transfer and nanolithography.

In nanolithography, a special form of sacrificial layer known as a resist is used to generate the desired design [11]. Nanolithography can be achieved by several techniques, including; photolithography, electron beam lithography, X-ray lithography, soft lithography, and others. Each instance has a similar core principle. Initially, the substrate is covered with a layer of resist. The photoresist is then subjected to an energy source with the aid of a pattern; for instance, ultraviolet light is used in photolithography, electron beams are used in electron beam lithography, and X-rays are used in X-ray lithography. The resist has a chemical reaction as a result of this patterned exposure, and the coating's overall chemical and mechanical characteristics change. Subsequently, a pattern is made by removing either the exposed or unexposed portion of the resist, depending on whether the resist is positive or negative. The metal layer SiO in Fig. 1—is now prepared for etching. Following etching, the resist's pattern is eliminated mechanically or chemically. Figure 1 shows a visual representation of the streamlined procedure [11]. In the scenario manufacturing sector, the top-down technique is now the most well-liked and often applied strategy. However, in nano-manufacturing, the other two methods are also starting to take center stage.

2. Bottom-up approach

The bottom-up strategy is likened to constructing a home, brick by brick. With this method, tiny parts—even molecules—are assembled or joined to form the final structure. Generally speaking, there are a number of bottom-up techniques, such as; coating, imprinting, contact printing, assembling and joining, chemical or physical

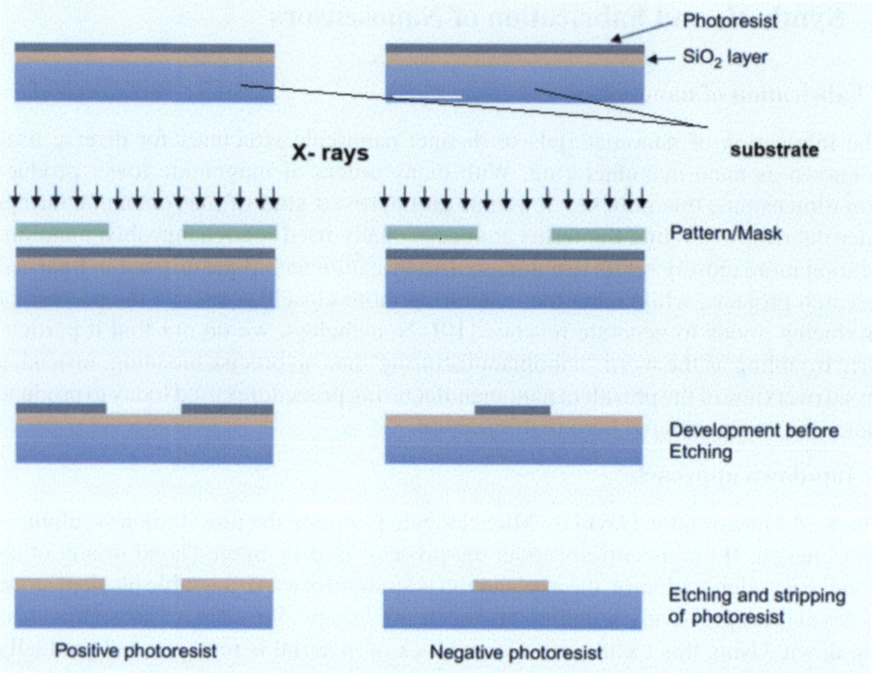

Fig. 1 X-ray lithography image of the positive and negative resist [11]. Copyright 2020. Reproduced with permission from Elsevier

vapor deposition. The medical field and bottom-up approaches have a lot of potential. Using a bottom-up strategy, carbon nanomaterials and carbon nanotubes may be utilized to nanofabricate a device that can function on a single cell [12].

Nanomanufacturing processes for nanosensors

The capacity to measure, forecast, and produce at the atomic and molecular sizes while taking use of the special qualities displayed by nanomaterials at that scale is known as nanomanufacturing. Researcher contributions to the diverse subject of nanomanufacturing come from a variety of backgrounds. The only procedures related to nanomanufacturing that are utilized in the production of nanosensors are covered in this section. In the section before, the two general approaches—the top-down technique and the bottom-up approach—were illustrated.

i. Electron beam lithography

Lithography is the technique of transferring patterns from one medium to another with the use of a material called resist. Before, lithography employed various particle beams; however, the utilization of the electron beam made it feasible to create features that are nanoscale in size. In the production of sensors, electron beam lithography (EBL) is widely employed because of its valuable pattern-making capabilities.

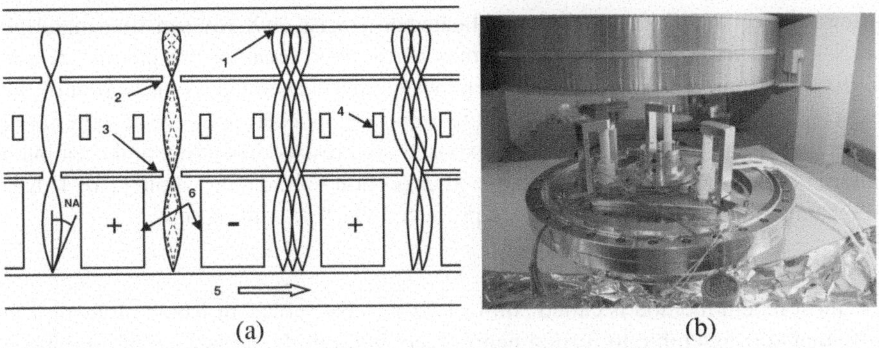

Fig. 2 **a** DiVa's conceptual diagram: 1. wafer, 2. deflection plates, 3. shaping apparatus, 4. deflector, 5. wafer, 6. shaping apparatus second set; **b** The Stanford University DiVa device in experimentation [14]. Copyright 2002. Reproduced with permission from Elsevier

Here, electron beam lithography technique is discussed among other electron beam lithography technologies because of its ease of use and regular usage.

A single wafer pixel is exposed at a time in direct writing EBL, which uses a moving, precisely focused Gaussian round be [13]. Figure 2 depicts the basic setup for direct writing EBL. With the aid of the etching and deposition processes, the beam forms the required pattern on the wafer, enabling the production of an extremely intricate nanostructure. The major disadvantage of this approach, despite its popularity and low cost, is the amount of time it takes. Nonetheless, in an effort to increase the applicability of this method, academics are working to advance technology.

ii. Lithography with focused ion beam

Focused Ion Beam Lithography is another technique for nanomanufacturing that is similar to electron beam lithography, however in this case, ions are utilized to carry out the lithography rather than an electron beam. Focused ion beam lithography has the potential to be more effective than electron beam lithography since ions are substantially heavier than electrons. There are other categories for focused ion beam lithography, but the most straightforward and affordable is direct writing, which is the one covered here. This approach does not need the employment of a resist; instead, the ion dosage is adjusted by adjusting the wafer's distance, which produces a trench on the wafer that varies in depth. To get a more powerful effect, heavy-ion species like Ga^+ and Au^+ can also be employed in this lithography. An inverse Gaussian-shaped trench is created as a beam passes across the wafer. The trench grows increasingly V-shaped, narrow, and acute as strength increases [15]. Complex forms can also be created by making many passes.

iii. X-ray lithography

X-ray lithography (XRL) is an improved type of optical lithography that uses shorter wavelengths. This method uses a particular type of mask with discrete local X-ray absorption zones to define the pattern. After initially being applied on a substrate (often a silicon wafer), this design is replicated on a resist, an X-ray-sensitive material. An X-ray passing through the pattern may cause bond breaking for positive resists or cross-linking for negative resists, depending on the chemical composition of the resist. After exposure, the entire object is immersed in a certain solvent; the kind of solvent used will dictate whether the exposed area resists dissolving and creating a pattern or does the reverse. The resist's other portion will remain whole [16]. This is how nanopatterns on the substrate are produced via X-ray lithography.

3. Molecular self-assembly

The most recent method is called "molecular self-assembly," in which molecules in particular self-assemble to form a nano-object without the assistance of an outside force. Many variables, including form, surface, properties, charge, polarizability, and magnetic dipole of the molecule to drive them to assemble together to form a particular structure. This is still a growing field and various developments are required before this approach is used in industry [11].

Two types of materials are employed to produce sensors: classical materials [17] **and polymers**

I. Classical materials

Solid-state sensors are based on the electrical reaction to the chemical background. The presence of a liquid or gas phase, for example, modifies the electrical properties, which can be utilized to recognize different chemical species. The usage of silicon chemical sensors, like Field Effect Transistor (FET), has been restricted despite their development due to issues with selectivity, stability, reproducibility, and cost as well as technology for production. The catalytic activation effect is also present in metal oxide semiconductor sensors, such as SnO_2 thin films and compact powders [17].

II. Polymers

Over the past ten years, polymeric materials have found widespread application in sensor technologies. These materials offer numerous benefits when used to make sensors, such as the following:

- The correct physical and chemical characteristics for the sensor are created by a variety of polymers with distinct structural features, such as side chains, charged or neutral particles, etc.
- The chemical detector can dissolve or even disperse uniformly in the polymeric tissue.
- Numerous substrates are available for the release of polymers. In addition, chemical deposition of specific chemical species onto polymer substrates can be employed to minimize reactant leakage into the sample solution.
- Polymers are reasonably priced materials with straightforward, uncomplicated production methods. They don't need any unique circumstances. Under certain instances, even under standard laboratory settings, the polymerization necessary to create a sensor layer is carried out [18–20].

Materials and polymers appropriate for the production of sensors need to possess a variety of unique qualities, including:

- Detector solubility in the polymeric tissue.
- Proper lifetime.
- Absence of species migration, reorientation, or crystallization.
- No aging or depreciation impact.
- No intrinsic color or luminosity (when used as a background).
- Biocompatibility (non-toxic substances).
- Transparency versus light.
- Resistance to chemicals such as; acids, bases, and antioxidants [18–20].

Different applications for nanosensors

1. Applications of nanosensors in chemical industry

A sensor is an apparatus that quantifies a changeable amount, often electrically then turns the result into precise signals. Sensors must be diverse, sensitive, accurate, selective, and stable [21]. The prefix nano is assigned to a sensor if one or more spatial characteristics fall between 10 and 100 nm. Nanosensors are sensitive materials that transfer biological, physical, and chemical information about nanoparticles and molecules through events of recognition. They might alternatively be described as molecular analyzers that have a transduction mechanism linked to a recognition element. It makes it possible to process the signal produced when the analyte and sensing element interact [6]. Nanosensors can monitor physical and chemical activities in distant regions, detect cell biochemicals, and measure nanoparticles in industries and the environment [22, 23]. In their antennae, mosquitoes have several sensors, one of which is a chemical sensor. They have a 36-m detection range for lactic acid and carbon dioxide. When they breathe, mammals and birds release these gases. It also seems that certain compounds in perspiration draw mosquitoes. This is the reason why some people perspire more readily and attract much more mosquitoes. Nanosensors are classified into four classes depending on their applications:

(A) Chemical nanosensors, (B) Deployable nanosensor, (C) Electrometer, (D) Biosensor.

A. Chemical nanosensor

One molecule or chemical can be studied using this kind of method. Different concentrations of ions were used to monitor pH using a variety of optical chemical nanosensors.

B. Deployable nanosensor

Applications such as Sniffer STAR and other military and national security uses this type. It is a portable, lightweight chemical detection device, which uses a tiny electromechanical sensor and nanomaterials for sample concentration and collection [24].

C. Electrometer

The device comprises an apparatus for resonating, a gate electrode, and a detector electrode that links charge to the mechanical component.

D. Biosensor

This sensor is widely utilized for early detection of cancer and other disorders [25].

In this section, we will discuss chemical nanosensors.

2.1 Chemical Nanosensor

Chemical sensors are analytical instruments that transform the chemical potential energy of the chosen analyte into a corresponding quantifiable signal [26].

A. Gas nanosensor

A device called a gas nanosensor was created utilizing nanotechnology and made to determine chemical components in a gaseous medium quantitatively [27].

The two fundamental parts of each gas nanosensor are the measurement transducer and the sensing element. The test gaseous medium modifies the optical, mechanical, electrical, and other characteristics of the sensor part, which were documented by the sensor's Transducer. While creation of gas nanosensors are a new invention breakthrough in analytical equipment, a closer look at the structural characteristics of the devices shows a great deal of diversity and variation. For instance, the transducers for nanosensors can be created utilizing the conductometric, the refractometric, the luminescent, or other measurement methods, whilst sensing components able to composed of a range of materials, including polymers, carbon nanotubes, Metals, and Metal oxides [7].

There are three classes of gas nanosensors based on polymer:

(a) Gas nanosensor utilizing sensing substance comprising a covering of polymer only.
(b) Gas nanosensor utilizing sensing substance comprising a polymer covering with another element.
(c) Gas nanosensor with utilizing sensing substance comprising a polymer based composite covering.

A. Gas nanosensor with sensing material containing a polymer coating only

Certain nano sensors that measure ozone have sensing material coated using a thin covering of polypyrrole [28]. A nano sensor involving a sensing element coated with polypyrrole nanoparticles was developed for ammonia detection [29]. A nanosensor utilizing sensing element covered with polypyrrole nanofibers was created to identify the volatile organic substances vapour [30]. A nanosensor, whose sensing substance is covered by Teflon AF 2400 nanoparticles, was recommended for evaluation of carbon dioxide in 0.045. If carbon dioxide is in the air the sensing material is coated with a thin coating of Teflon AF 2400 nanoparticle [31, 32].

If a sensing material is coated with polyaniline, this nanosensor detects either aliphatic amines or volatile ethyl alcohol, this depends on covering thickness of the detecting (sensing) material [33, 34]. A nanosensor with a covering of poly(2-methoxyaniline) nanoparticles was created to measure vaporized water [35]. Furthermore, a nanosensor which has a detecting substance carrying a covering of poly-acrylic acid nanofibers was assumed to determine water vapor too [36]. To identify the acetone vapors, methanol, and isopropanol, a nanosensor was constructed. The detecting element contained a thin coating covering of poly methylmethacrylate [37].

A nanosensor was created to detect water vapor, its sensing element was coated with poly (2-Methyl-2-(prop-2-enamido)propane-1-sulfonate) nanoparticle [38]. Nanosensors coated with poly-3-hexyl thiophene nanoparticle was used to identify volatile organic compound in vapor form chemicals seen in meat deterioration or *Salmonella* contamination [39]. A nanosensor using polydimethylsiloxane nanoparticles as a covering of detecting elements was created to detect ethanol vapors [40]. A nanosensor involving a sensing element coated with poly ethyleneimine nanoparticle was proposed as a means of detecting volatile organic chemicals [41]. Nanosensors with a thin covering of poly[1,1′-bis(ethynyl)-4,4′-biphenyl bis (tributylphosphine) Pd(II)] nanoparticle were designed to detect hydrogen and relative humidity in gaseous mediums [42].

B. Gas nanosensor utilizing sensing substance comprising a polymer covering with another element

Nanosensors feature sensing units coated with a polymer followed by:

(1) Another polymer layer.
(2) Polymer covering with a metal layer.
(3) Polymer that has a carbon nanotube covering.
(4) Polymer comprising organic substance.
(5) Polymer that comprising inorganic substance.

(1) Gas nanosensor with sensing material contains a polymer coating with another polymer

To detect formaldehyde, a nanosensor was created using a sensor component covered in polystyrene nanofibers and a layer of poly ethyleneimine nanoparticle. Poly ethyleneimine nanoparticles ranged in size from 50 to 1200 nm, whereas polystyrene nanofibers were 110–870 nm in diameter [43]. A nanosensor was evolved to measure the humidity level of a gaseous phase. The detecting element was covered with poly (diallyl dimethyl ammonium chloride) nanoparticles and polystyrene sulfonate nanoparticles [44].

(2) Gas nano sensor containing a detecting element of polymer covered with a layer of metal

Nanosensors coated with 3,4-ethylene dioxy thiophene and thiophene-3-acetic acid copolymer nanoparticles, as well as palladium (Pd) or nickel (Ni) nanoparticles immobilized using 4-aminothiophenol, have been suggested for detecting acetone and toluene fumes. Adding copper, silver, or gold nanoparticles to a layer of 3,4-ethylene dioxy thiophene and thiophene-3-acetic acid copolymer nanoparticle allowed specific detection of various volatile organic chemicals [45].

(3) Gas nanosensor with detecting material contains polymer covered with a carbon nanotube layer

A nanosensor, for example, has a sensing component coated with polycarbonate nanoparticles and multi-walled carbon nanotubes. It was created for evaluating the volatile organic chemical vapors. The measuring sensitivity of volatile organic compound vapors corresponded with the relationship S0 (measurement sensitivity) toluene vapor > S0 ethanol vapor > S0 water vapor [46].

(4) Gas nanosensor with detecting material contains a polymer layered with an organic chemical layer

Nanosensor coated by poly(diallyl dimethyl ammonium chloride) nanoparticles and porphyrin pyrrole or porphyrin pyrrole-comprising manganese (Mn) were developed to measure relative moisture content in atmosphere [23].

(5) Gas nanosensors with detecting material contain a polymer covered within an inorganic compound layer

A nanosensor with a covering of polyaniline and cadmium sulfide nanoparticles serves as an example of such a device. The device was designed to determine the amount of petroleum gas that has been liquefied. At a normal temperature, the sensing component produced a greater output signal when exposed to the aforementioned gas than when exposed to nitrogen or carbon dioxide. To illustrate, the influence of gas from liquefied petroleum at a concentration of 0.104% led to 80% shift within the signal output [47].

C. Gas nanosensor with detecting material contains a covered of polymer-based composite material

Nanosensor use a composite substance covering its sensing components made of:

(1) Two polymers.
(2) A metal-containing polymer.
(3) A metal oxide-containing polymer.
(4) A carbon nanotube-containing polymer.
(5) A polymer containing an organic substance.

A composite is a heteromorphy material created by combining two or more separate materials, resulting in increased bulk qualities that differ from their components.

Composites typically consist of two phases: matrix material and dispersed material [48, 49].

(1) Gas nanosensor with sensing material contains a coating of two polymer based composite material

For the detection of ammonia in a gaseous medium with fluctuating humidity, for instance, a two-sensing element nanosensor was presented. A composite material comprising polystyrene sulfonate and poly(3,4-ethylene dioxythiophene) nanoparticles was applied to the nanosensor [50]. A nanosensor was created using a sensing element coated with a nanowire composite substance made up of polystyrene sulfonate and poly(3,4-ethylenedioxythiophene) nanoparticles in order to identify chemical substances that could be considered possible air pollutants. The nanowires exhibit a diameter of 220 nm and roughly 6 μm in length [51]. A nanosensor with a covering of a composite material composed of polyvinyl alcohol and polyethyleneimine nanofibers was created in order to measure formaldehyde. The nanofibers' diameters varied from 40 to 1800 nm [52]. A nanosensor was created to evaluate a gaseous medium's relative humidity. An amalgamated material is applied to the nanosensor's sensing component made of polyvinyl alcohol and acrylic acid nanofibers [36].

(2) Gas nanosensor with sensing element contains a polymer coating and metal based composite material

Occasionally, these devices' sensing components are coated with a polymer that has been doped with gold (Au) For the purpose of measuring ammonia, for instance, a nanosensor was created, with the detecting element covered with a composite material made of gold nanoparticles and polyaniline. The instrument showed incredibly quick reaction and signal recovery times [53]. A nanosensor was created to measure the gaseous medium's relative humidity. The sensing element of the nanosensor was covered in a composite substance made of gold and polyvinyl alcohol nanoparticles [54]. Palladium is also utilized in polymer doping and has applications in humidity and ammonia detection [29, 42].

(3) Gas nanosensor with sensing material contains a coating polymer and metal oxide based composite material

A nanosensor based on zinc oxide (ZnO) and poly(vinyl pyrrolidone) (PVP) nanoparticles was created, and its sensing element was covered by the composite substance to measure acetone vapour. Zinc oxide nanoparticles had a maximum size of 10 nm [55]. To measure carbon dioxide, nitrogen dioxide, and hydrogen, a composite material coating consisting of indium (III) oxide (In_2O_3) nanoparticles and polyaniline nanofibers was applied to a sensing material of a nanosensor.

The nanosensor demonstrated quick indication recovery and reaction times at room temperature, as well as strong measurement repeatability for these compounds [56]. A nanosensor utilizing a thin layer coating of a composite material consisting of palladium oxide (PdO) and polyaniline nanofibers was suggested for the measurement of methane. The methane measurement sensitivity was higher at the sensing element's typical temperature than it was at 50 °C [57].

(4) Gas nanosensor with sensing element contains a coating of polymer and carbon nanotubes based composite material

For the purpose of measuring ammonia, for instance, a nanosensor was created with a sensing element covered in a composite material made of polyaniline nanoparticles and multi-walled carbon nanotubes that had been exposed to oxygen plasma [58]. To monitor the relative humidity of a gaseous medium, for instance, a nanosensor with a sensing element covered in a composite material based on polyimide nanoparticles and multi-walled carbon nanotubes treated with plasma was developed [59]. To measure the vapors of methyl alcohol, dichloromethane, chloroform, toluene, acetone, hexane, and ethyl acetate, sensors with thin layer coatings of composite materials made from multilayer carbon nanotubes and poly(methylmethacrylate) nanoparticles with customized surfaces were created [60].

(5) Gas nanosensor with sensing material contain a coating of polymer and an organic compound based composite material

To determine a gaseous medium's relative humidity, a nanosensor was created. The sensing element of the nanosensor was coated with a composite material made of polyaniline nanofibers and a layer of camphor sulfonic acid nanoparticles [61]. The purpose of the nanosensors was to measure oxygen levels. The sensitive parts of the sensors were covered with a composite material made of polyurethane or polystyrene–polyacrylonitrile copolymer nanoparticles and indigo or triindigo dye nanoparticles. The colorless covering of the sensing element containing indigo dye nanoparticles became permanently red upon contact to this chemical compound, whereas the colorless covering containing triindigo dye nanoparticles turned blue [62]. An element for sensing was proposed in a nanosensor coated with a composite material based on poly(o-anisidine) polystyrene nanofibres and camphor sulfonic acid nanoparticles for the purpose of determining the vapors of volatile highly polar chemical substances [63]. To measure the amount of ethanol vapor in each sample, a nanosensor was created. Its element of sensing featured a thin layer covering a composite material made of poly [(2-bromo-5-hexyloxy1,4-phenylene vinylene) co (1,4-phenylene vinylene)] and dodecyl benzenesulfonic acid nanoparticles biofuel [64]. Nanomaterials have unique properties so they are used in every commercial field such as nano-fluid, nano-structured coatings, nano-remediation, microbial fuel cells, nano-catalysts, nano-phosphors development, removal of environment contaminants, nanoparticles in sun-cream [65]. Nano-fluids are heat transfer fluids of tiny particles spread evenly and stably in a base fluid. Nano-fluid qualities vary based on elements such as; pH, surfactant, ultrasonication, stability, homogeneity, base fluid properties, particle shape, and fluid-particle interfacial interaction.

Nano-fluids outperform ordinary fluids because of their large surface-to-volume ratio, which enhances heat conductivity. Heat exchangers are necessary to transmit the highest quantity of heat. Heat transfer rate is influenced by surface area, working fluid, and heat exchanger material parameters (e.g. thermophysical) [66]. Nanomaterials (NMs) are used as coating materials protecting metals from erosion and corrosion, while also improving their mechanical qualities. NMs coating is a distinct technique to increase the surface qualities of any product and enhance operating effectiveness [67]. Among other uses, nano-remediation is gaining greater attention because of its contaminants' decontamination characteristics [68–70]. Among other uses, nano-remediation is gaining greater attention because of its contaminants' decontamination characteristics [71, 72]. Remediation occurs by the adsorption of xenobiotics for water treatment. In certain circumstances, after nano-remediation, nanoparticles (NPs) can be restored and used in later treatments. The main goal of decontamination, also known as remediation for the environment, is to get rid of harmful pollutants. Without endangering the environment or upsetting it, molecules and elements are collected from in situ environments [3].

2. Applications of nanosensors in optical industry

Contradictory to its name, nanotechnology has completely transformed a wide range of industries worldwide [73]. All forms of matter, such as polymers, metals, semiconductors, dielectrics, ... etc., can be made smaller, and this allows for the creation of intriguing new features, particularly optical ones [74]. The most significant characteristics of materials are their optical absorption and emission, which are highly influenced by their band gap and band structure. These factors are significant from both a basic and an application perspective for the device, especially when determining their ability for prospective optoelectronic applications, such as LED, LASERS, photodetectors, and photovoltaic devices [75–78]. Because of their customizable optical characteristics, nanomaterials—particularly quantum structures such as quantum dots (QDs), quantum wells (QWI), quantum wires (QW), etc.—allow greater opportunities for optical devices. They could be made to have certain optical properties by adjusting the size of grain [79], surface functionalization [80], structure, and core–shell type compositions.

Their tiny scale also aids in the shrinking of devices, thus expanding their field of use [81]. The development of transparent polymer formulations with nanoscale fillers serves as one of the top scientific goals that have been going strong lately. This provides exciting opportunities for optoelectronic and optical devices [82]. The dispersion of nanoparticles inside matrices of polymers can have a significant influence on another physical properties such as magnetic, optical, thermal, electrical, flame retardancy, wear resistance, water resistance, or barrier to diffusion [83].

The following are the primary justifications for the use of polymer optics in various gadgets and applications.

I. Transmittance of light is on balance with premium crown glasses.
II. Low-cost polymer material and affordable manufacturing process providing large quantities.
III. Polymers are highly suited for creating lighter and smaller items in the cutting-edge technique due to their small weight per volume. Equivalent glass products are 2.5–fivefold heavier compared to polymers.

Fig. 3 Optical grade polymers

IV. High impact resistance: Since polymers do not break as glass, they are very cost effective and durable for use in products including eyewear, medical and biomedical recyclable optics, goggles, and more [84] (Fig. 3).

In this section, some applications of nanotechnology in optics will be highlighted.

3 Optical Lenses

Optical lenses are a group of optical elements that are intended to concentrate or deviate light. They can consist of one or more items, depending on their function which can range from laser scanning to microscopy [85]. It has been reported that a changeable smart composition lens with an outstanding optical transmission of 93%, somewhat water-soluble and viscous at 0.001% weight percentage of the filler was created using a poly (diethylene glycol adipate) and cellulose nanocrystal (PDEGA/CNC) combination [86]. A potential bio-nanocomposite microstructure is depicted in Fig. 4. The graphical illustration of PDEGA also depicts the chemical framework, with several hydroxyl groups and lone pairs resulting from the oxygen double-bonding. This is what allows for the bio-nanocomposite's surface reactions. The focal length of this lens may be changed in a matter of seconds despite being exposed to a small electric voltage of 50 V. Because of PDEGA's great thermal stability, it is powerful enough to endure temperatures up to 200 °C. As well as displaying fully reversible and highly repeatable functionality, the biodegradable recyclable tunable microlens may facilitate the future developments in the area of smart optics. The PDEGA/CNC lenses with entirely closed construction and electric field of action enable adjustment across a broad focal length range. Moreover, it can be easily incorporated into panels for potential usage in sensing, lab-on-a-chip, and medical diagnostics.

In another study made by Ahlen and others [87, 88] a prospective application of hydrogel-based contact lenses loaded with nanoparticles used in the Ophthalmic Drug delivery systems (ODDS) was developed. Utilizing in-situ solidified nanoparticles with cellulose nanocrystals (CNC) in PVA lenses and chitosan-poly (acrylic acid) nanoparticles in polyvinyl alcohol (PVA) hydrogels, two prospective contact lens platforms for ODDS have been developed. Due to lysozyme's hydrolysis of the chitosan chains, it was demonstrated that the nanoparticles would dissolve at a physiological enzyme concentration of 0.2 mM. After insertion of nanoparticles

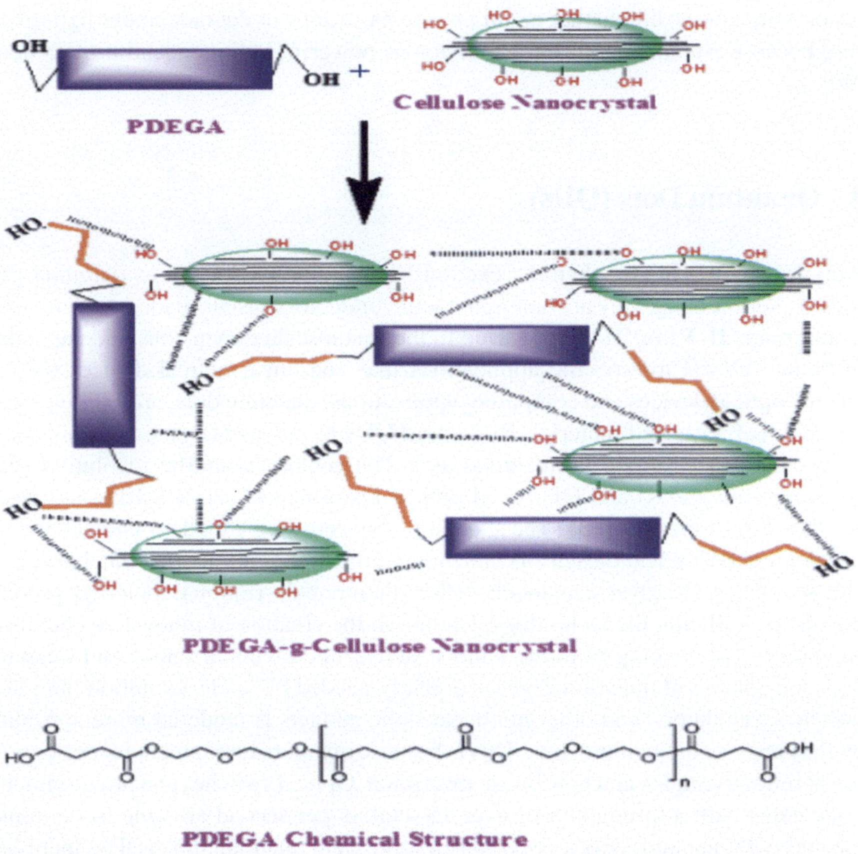

Fig. 4 Microstructure of PDEGA/CNC bionanocomposite [86]. Copyright 2016. Reproduced with permission from ACS Publications

into the composite lenses, the medication release extended over 28 h. Their investigation demonstrated that the nanoparticle-CNC-PVA lenses had even more opportunity to achieve prolonged duration of medication release, which they ascribed to the particles leaking out of the expanding of PVA network throughout the incubation time. By binding the particles to the cellulose nanocrystals and immobilizing the PVA network's gel, the in-situ solidification between the CNC and nanoparticles was also observed to prevent leaking. They concluded that the information presented a plethora of opportunities for the development of innovative controlled medication-release contact lenses and an enzyme-triggered eye system for drug delivery. Recently, a dual soft contact lens provides a laser protecting impact has been invented. The optical density of the hydrogel was significantly increased by the inclusion of gold-based plasmonic silica-shelled nanocapsules within a methacrylate polymer matrix to induce the shielding response [89], while maintaining light transmittance. Applications for these composites in the ophthalmic field appear to be rather promising. Since, this method is even more cost-effective than other less penetrating compositions that might also be hazardous or degrade under light exposure because Au nanocomposites function as powerful light-concentrating systems [90].

4 Quantum Dots (QDs)

Particles smaller in size than the exciton Bohr radius are known as *quantum dots* (QDs), Additionally, they are nanoscale semiconductor crystals made up of elements from groups II–VI or III–V [91]. Due to their nanoscale dimensions and capability of being tailored for specific applications like non-linear optical devices (NLO), electro-optical devices and computing applications, quantum dots, are exciting zero-dimensional advanced material. By joining QDs to polymers, nanocomposites can be created, which could be regarded as a 21st-century scientific revolution [92]. Quantum dots (QDs) and QD-based optical instruments such as LEDs and lasers are the rage in this era [93]. These quantum systems are excellent choices for the development of optical biosensors and bioimaging probes due to their tunable optical characteristics. However, a major obstacle to their popularization is their high production costs. Utilizing the technology, leading to the creation of innovative chemistry techniques for creating ultrafine nano-systems, like quantum cubes and quantum dots, and the use of quantum cages is urgently needed [93]. The evolution of optical switches, regulators, and other nonlinear optic gadgets is rendered more appealing by the presence of quantum dots. This is because quantum entrapment increases non-linear refractive index and non-linear absorption. Optical switches and programmable logic gates with a throughput of over 15 terabits per second are able to be implemented with quantum dots as well [94]. Fluorescent quantum dots (QDs) including core, core–shell, and core multi-shell structures can be introduced into the polymeric matrix to create fluorescent polymeric nanocomposites, which can be utilized to fabricate LEDs. Quantum dots for short, are fluorescent semiconductor nanocrystals with a maximum size of ten nm [95–97]. It was claimed that an extremely transparent and luminous nanocomposite epoxy, likely to be used in white LEDs, was created utilizing fluorescent amido group functionalized CdSe quantum dots. Within the polymeric matrix, the amido-modified quantum dots displayed enhanced optical capabilities and greater dispersion [98]. Nanocomposites of poly(3-hexylthiophene)-cadmium selenide (P3HT-CdSe) were created by grafting vinyl-terminated poly(3-hexylthiophene)-methyl dioctylphosphine oxide (DOPOBr)-functionalized CdSe quantum dot (QD) surfaces throughout a mild palladium-catalyzed Heck coupling [99], eliminating the requirement for ligand exchange chemistry.

It was simple to apply the "grafting onto" method. ^{1}H-NMR and ^{31}P-NMR spectroscopies were used to confirm the coupling's effectiveness. The P3HT-CdSe nanocomposites that were produced have a distinct interface, which greatly enhanced the electrical interplay between these two elements and promoted the distribution of CdSe inside the P3HT matrix. It was discovered that the photophysical characteristics of nanocomposites were different from those of standard composites containing physical mixing of P3HT and CdSe QDs. The solid-state emission spectra of the nanocomposites demonstrated that there is a charge transfer from P3HT to CdSe QDs, whereas the P3HT/CdSe integrates showed that P3HT charges CdSe QDs at a wavelength of 3.5 nm.

A quicker lifetime decay provided additional evidence that charge transfer occurs in P3HTCdSe nanocomposites. It had been supposed to investigate potential opto-electronic device applications by building a single layer solar cell using P3HT- CdSe nanocomposites [99].

5 Optical Nanosensors

Instruments with entire dimensions less than 1000 nm that use an optical signal to transmit a biological or chemical action are referred to as optical nanosensors. Sasaki et al. [100], presented the first optical nanosensor, which used fluorescein encapsulated in a polyacrylamide nanoparticle for pH monitoring. For the specific detection of methamphetamine (METH), a nanosensor based upon graphene quantum dots (GQDs) enclosed inside a molecularly imprinted polymer (MIP) was used. The great fluorescence intensity of GQDs and the high selectivity of the MIP were combined in this method. The GQDs and the acrylate polymer were linked using triethoxyvinylsilane (TEVS). Sensitivity and selectivity towards the METH molecules were supplied by the porous framework of MIP following the removal of the reference molecules. In METH analysis, this optical nanosensor showed a small limit of detection and a great accuracy, which leads to the evolution of selective, inexpensive, and sensitive nanosensors [101]. MIPs are synthetic polymers that are typically created by polymerizing and cross-linking monomers when a template molecule is present. It should be possible for the template and the monomers to produce complexes together. The template is taken out after the reaction is finished and the polymer has been created

[102, 103]. However, the MIPs have gaps that precisely match the structure and chemical makeup of the template, and they are able to identify the template using a method similar to that of an enzyme substrate. In a research done by Mina Mehrzad-Samarin and others [104], another optical nanosensor was developed for the measurement of ornidazole drug in biological specimens by employing graphene-quantum-dot-inserted in silica molecular imprinted polymer (GQD-SMIP) as a selective sensor for this substrate. After being excited at 365 nm, the GQD-SMIP was shown to exhibit an intense fluorescence emission at 450 nm. In the presence of Ornidazole, this emission was observed to quench linearly in the concentration interval of 0.75–30 μM. The probe was utilized to determine the analyte in plasma samples, and it was successful in reaching 0.24 μM as a limit of detection. SO, this luminescence nanosensor exhibits tremendous potential for usage in advanced analytical applications due to its low detection limit and outstanding accuracy when dealing with materials that have complex matrices.

3. Applications of nanosensors in medical diagnosis

Nanomedicine, commonly known as biotechnology applications of nanoscience, is the result of advances in nanotechnology. One could think of nanomedicine as the intersection of the biological, physics, chemistry, and medical disciplines. Additionally, the discipline of nanomedicine's advancement in the analysis and treatment of biological systems at the cellular and subcellular levels has produced groundbreaking techniques for the identification and avoidance of several deadly illnesses. The National Cancer Institute of America, for instance, anticipates using nanotechnology and nanoscience to minimize mortality by controlling cancer. Because nanoscience and fertility are closely related [105]. Such nanosensors' sensitivity and selectivity can be used in biomedical applications to provide many benefits, such as early disease, toxin, or biological threat detection, and to significantly enhance clinical, environmental, and industrial results [106]. In contrast to nanomaterials smaller than 20 nm, which may pass through blood vessels and circulate throughout the body, nanomaterials smaller than 50 nm do not have trouble penetrating numerous cells. Consequently, following specific processes, nanomaterials are frequently employed as drug delivery systems, the function of which is to transport therapeutic genes or chemotherapeutic chemicals to cancerous cells, all the while shielding and preserving healthy cells. Importantly, it should be noted that although nanomaterials are frequently referred to as non-biological nanomaterials in many practical portions of the literature, organisms and biological techniques have been remarkably employed in the manufacture and fabrication of nonbiological nanomaterials [107]. Early diagnosis and treatment are essential for containing the sickness pandemic and limiting the spread of the virus. Rapid, accurate, portable, wide-scale, and virus-detection technologies are critical for the diagnosis of many viral illnesses [108].

6 Polymers in Oncology

The word "nanomedicine" refers to the use of nanotechnology in medicine. It seeks to address issues about illnesses at the nanoscale, where the majority of biological molecules are found and function [109]. The diagnosis, monitoring, treatment, and prevention of cancer have all shown great promise when applied to nanotechnology [110, 111]. A well-designed nanomaterial can target a tumor, detect pathophysiological flaws in tumors, release the agent in reaction to outside stimuli, convey therapeutic medications, genes, or imaging agents, and track how effectively a treatment is working. Moreover, because these nanoparticles are 100–10,000 times smaller than cancer cells, they can just go across cell boundaries. Because of the enhanced effect of retention and permeability, nanoparticulate devices can enable targeted medication delivery to solid tumors (Fig. 5). Tumor vasculature anomalies, such as hypervascularization, aberrant vascular architecture, significant vascular permeability factor production, and lack of lymphatic drainage, are the cause of this [112], which increase the medications' therapeutic efficacy and lessen their negative, non-specific side effects by promoting the accumulation of drugs and the extravasation of nanoparticles in the tumor interstitium [113, 114].

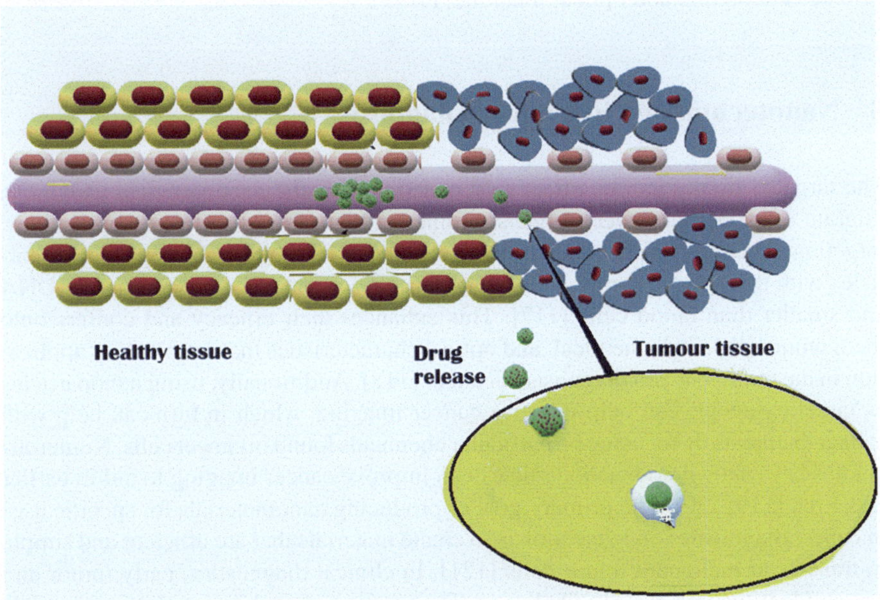

Fig. 5 Enhanced permeability and retention effect

7 Polymeric Nanoparticles in Diagnosis

Among the drawbacks of the traditional imaging agents now utilized in clinics for diagnosis include their toxicity, in vivo instability, and lack of specificity. For instance, there are several important restrictions on imaging molecules, When considering compounds of fluorophores, such as blue, green, or red fluorescent dyes that have a visual range of 400–600 nm, tissue auto-fluorescence and light-absorbing elements are two examples [115]. Image quality is therefore impacted when modest concentrations of contrast agents are applied to areas of interest. They also have a small number of chemical alteration sites that don't substantially alter their biological activity.

The great adaptability of nanoparticulate systems allows them to serve multiple purposes, involving the precise administration of high-concentration imaging probes to bodily disease areas. Imaging probes based on polymers are less toxic, and have better plasma half-lives and stability, better targeting, and broad surface areas that are simple to modify with a variety of imaging moieties. By concentrating and shielding the marker from deterioration, they aid in the tumor cell detection process and provide a more sensitive analysis. Currently available imaging methods for nanoparticles include computed tomography, ultrasonography, nuclear imaging, magnetic resonance imaging, and optical imaging [116].

8 Nanotechnology in Cancer Diagnosis

The targeted tissues can be effectively penetrated by the nanoparticles, which can migrate effortlessly between organs within the body. For diagnostic applications, targeting unhealthy tissues, such cancer cells, is possible by combining the nanoparticles with therapeutic compounds. Nanoparticles are nearly the same size as DNA and smaller than blood cells [117]. This enhances their efficacy and confers onto them unique physical, chemical, and optical characteristics that enable their application in the treatment and diagnosis of cancer [118]. Additionally, using nanoparticles as targeting agents can help improve cancer imaging, which in turn can help with cancer diagnosis by focusing on particular chemicals found in cancer cells. Numerous studies have shown how nanoparticles can improve cancer imaging to aid in earlier detection [119, 120]. The primary goal of producing nanomaterials for specific uses in cancer diagnostics and detection is to create materials that are efficient and simple to transfer to malignant tumor cells [121]. In clinical diagnostics, early tumor and cancer identification is still a challenge. The creation of sophisticated nanomaterials offers hope for the early diagnosis, imaging, and detection of cancer. Additionally, metal nanoclusters can be used to diagnose and cure cancer [122, 123].

a. Nanoparticle Function in Cancer Image

Nanoparticle-specific antibody conjugation for cancer has allowed for the detection of breast and colon cancer in blood or urine samples using metallic nanoparticles. Biomarkers are proteins that are present on inside the cell, the cell membrane, or outside the cell. These biomarkers, which target these proteins, can be used in conjunction with different imaging techniques like MRI and CT to diagnose patients. A novel imaging modality called photoacoustic (PA) imaging has emerged as an effective tool for identifying high-resolution cancer tissues with excellent optical contrast. Furthermore, one of the best methods for identifying cancer cells is now the Raman imaging approach. Better guiding of tumor excision is possible with this imaging technique, and histological investigation can verify the imaging's accuracy [124].

9 Frequently Used Nanoparticles in the Diagnosis of Cancer

Nano diagnostics are the use of nanoparticles in clinical diagnosis [125]. Thanks to nanotechnology, clinical diagnostics have recently improved because of its increased sensitivity and capability for early cancer diagnosis. Numerous types of nanomaterials, including carbon nanotubes, dendrimers, polymeric nanoparticles, and quantum dots, are utilized in the diagnosis of cancer. Nanoparticles can be coupled with aptamers, carbohydrates, peptides, antibodies, and other small molecules that selectively target certain molecules and allow them to reach the appropriate site in order to improve their cancer detection capabilities [126]. Studies on the application of gold nanoparticles in cancer diagnostics have been conducted. When Au-doped nanoparticles were used, the luminescence intensity was remarkably high [127].

I. Gold Nanoparticles

Gold nanoparticles (AuNPs) are important for a variety of biological applications due to their many unique properties [128]. AuNPs are tiny molecules with a variety of special qualities that make them useful for imaging methods. Gold nanoparticles provide superior tumor targeting and longer blood circulation durations for higher-quality cancer imaging approaches. Additionally, AuNPs have a variety of applications, including photothermal ablation, medication delivery, radiation, and the transport of nucleic acids [128].

II. Quantum Dots

Nanoscale crystals known as quantum dots (QDs) can move electrons. When exposed to UV light, QDs can release highly energetic light in a variety of hues. The creation of solar cells, fluorescent biological markers, and nanocomposites are just a few of the numerous uses for these QDs. With the use of high-resolution cellular imaging, cancer diagnosis has improved. It has been discovered that exposure to certain substances alters the fluorescence characteristics of quantum dots. Furthermore, the surface of QDs contains a passive site that facilitates the easy conjugation of particular antibodies. QDs have been used in imaging for several bio applications. Additionally, QDs have been applied to cancer treatment and medication delivery [118].

III. Graphene

A flat mesh of regular hexagonal rings is called graphene, or isolated atomic layers of graphite. Its remarkable thinness, transparency, light weight, and exceptional thermal and electrical conductivity make it a desirable material for cancer imaging and detection. Among graphene's numerous beneficial characteristics are its enormous quantum oscillations and bipolar transistor effect [129], making it a strong contender for efficient imaging of cancer. Because graphene exhibits hydrophobic interactions and π-π stacking, it can also be loaded with anticancer medicines due to its enormous specific surface area [129].

IV. Fullerene

As an allotrope of the carbon family, fullerene is a member of the Buckminsterfullerene family. Its molecule is made up of carbon atoms joined by single or double bonds to create a structure resembling a mesh. In chemical applications, fullerene (C60) has been employed to trap drug molecules in its mesh for effective drug administration, as well as in the detection and diagnostics of cancer. For instance, fullerene has been useful in the creation of biosensors that measure blood serum glucose levels [118].

10 Polymer Based Biosensors

Turner just released a thorough overview of the development of biosensors [130]. Thirty years ago, Lyons and Clark began developing biosensors [131]. They gave an account of the first enzyme-electrode [132]. Biosensors quickly gained significant attention from researchers. The number of articles released each year detailing the advancement of a biosensor is increasing. Though the industry started more slowly, it has already grown to be worth several billions of US dollars [132]. Biosensors are used in food safety, environmental monitoring, and defense, but their primary use is still in medical diagnostics, where glucose biosensors dominate the market. The majority of biosensors are created to be utilized outside labs—for example, as field-based or point-of-care (POC) sensors—at home, by general practitioners, or in the field. As a result, they must be affordable, simple to use, and produce incredibly accurate results [114].

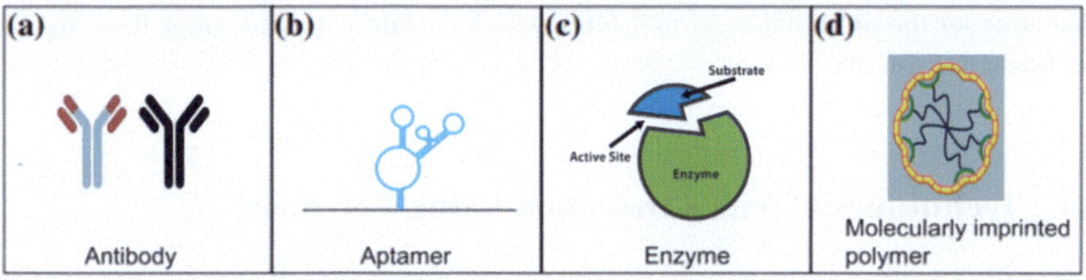

Fig. 6 A biosensor with four distinct bioelements: **a** antibodies are frequently utilized in biosensors and other immunoassay applications, **b** aptamers can bind and detect different targets with specificity, **c** enzyme is very desirable for biosensor applications due to its great target selectivity and **d** molecular imprinted polymers (MIPs) created by polymerizing monomers around a target, then monomers can be separated, leaving an imprint that has a very high specificity for capturing the target

11 What Are Biosensors?

A device that primarily uses electrical, thermal, or optical signals to detect chemical molecules by means of certain biochemical processes mediated by immune systems, organelles, tissues, or entire cells [133]. The analyte, the element that recognizes the analyte, and the transducer, which detects the recognition event and initiates signal processing, are the three main parts of a biosensor. An event of binding or recognition takes place when the left sensor is exposed to an analyte for which it is specific, resulting in a signal that the transducer detects. An additional analyte is exposed to the sensor on the right. There will be no processing of the detection signal since the bioelement is unable to identify it [114].

12 Recognition Elements

A biosensor can employ a variety of bioelement kinds. A few of the various kinds of bioelements are depicted in (Fig. 6). A bioelement must have a high affinity and be highly specific for the analyte to be considered good. The biosensor's sensitivity is limited by the bioelement's affinity for the analyte [114].

13 Sensitivity and Selectivity of Biosensors

The assessment of biosensors involves evaluating their sensitivity and selectivity. Selectivity refers to the sensor's ability to accurately identify the analyte within the complex sample matrix, such as blood, urine, saliva, etc. On the other hand, sensitivity refers to the extent of signal variation in response to changes in analyte concentration. It establishes the lowest analyte concentration that the sensor is capable of detecting, or the limit of detection (LOD), which is widely used to gauge the sensor's performance quality [134].

14 Techniques of Virus Detection Using Polymers

Controlling viral infections and illnesses requires diagnosis at the point of care methods that are portable, precise, and accurate in detecting viruses. Until now, polymers have been used as simple and practical tools for scientific research because of the fundamental physicochemical changes that lead to rapid advances in the electronic structures, orientation, and direction of charge distribution in polymers. The speedy conversion of innovations into polymers made it possible to use them as a rapid identification test that contrasted biorecognition components including chemicals, proteins, and antibodies and responded to the objective's intermolecular communication [108]. The polymer-based biosensor in particular is one of the most promising developments for the early detection of virus infections due to its shown potential and advantages. Thus, it is expected that in the near future, research pertaining to prospective studies associated with various forms of viral detection will demonstrate a noteworthy interest in the creation of novel polymer-based biosensor technology [108]. The hepatitis E virus (HEV) is a major contributor to cases of acute viral hepatitis globally. To develop a highly sensitive electrochemical sensor, this study employed quantum dots, graphene, and gold-embedded polyaniline nanowires. These nanomaterials were created by self-assembly after interfacial polymerization, which produced a pulse-triggered sensor with remarkable sensitivity [135].

4. Applications of nanosensors in food and water quality sensing

Water is becoming more and more important for us, especially when we think about drinking water. We still have some toxic components, such as pesticides or heavy metal ions, so sensors have been developed that can detect such contaminants in the drinking water.

Nanosensors are favored due to their fantastic qualities, which include high selectivity, sensitivity, and low detection limit values, which help us detect contaminants in water [136]. Furthermore, current methods like chemical analysis, biological assays, and physical measurements, for example, are used to monitor water quality. Water samples are collected and analyzed in a laboratory using these techniques. However, these techniques have disadvantages such as limited detection limits, time-consuming analysis, being expensive, and the inability to monitor in real time [137].

On the other hand, according to some statistics, more than 130 million people in Europe have some health problems because of food, so the existence of sensors is necessary in our lives to help us face this thorny problem, where the sensors can be used in the processing of food and detection of contaminants. So the world is now moving towards nanosensors in food for the numerous features, for example:

1. Nanosensors can detect pathogenic microorganisms, food contaminants, vitamins, fertilizers, colors, adulterants, pesticides, and taste/odor changes.
2. Sensors can cause analyses, enzymes, and reporter dyes to interact cooperatively.
3. Nanosensors are not selectively incorporated into food.
4. Nanosensors enhance the ability to perform radiometric measurements.
5. The nanosensor matrix shielded food from any contaminants and hazardous impacts of the sensing dye and enzymes [138].

Food items devoid of harmful toxins and adulteration are becoming increasingly popular. Adulteration occurs when inferior components are mixed into food or drink, resulting in a product that does not meet standards. Adulterants are unwanted substances added to food products to increase profits. Food adulteration has become a pervasive issue for the food processing industry. To ensure product quality and safety, regulatory authorities in many countries have established compositional standards for a wide range of food products. Despite these measures, some producers prioritize profits over consumer well-being and resort to adulterating food with cheaper substitutes. This is particularly concerning when toxic chemicals like formalin and calcium carbide are used to adulterate fish [139], meat [140], fruits [141], and dairy products [142] as it threatens human health.

Additionally, the use of coloring dyes, urea, and certain preservatives can have negative effects on regions of the human body, causing persistent health difficulties such as liver diseases, diarrhea, vomiting, cancer, and heart disease. To protect consumers and maintain food quality across the entire supply chain, it is crucial to monitor food products closely [143]. New, simple, and rapid methods are essential to safeguarding human life by determining the adulterant concentrations in food products. Previously, researchers have employed techniques like spectroscopy, chromatography, and ELISA to determine adulterants. However, these procedures are costly, time-consuming, and demand specialized workers. Fortunately, recently developed nanotechnology-based techniques have proven successful in identifying adulterates and contaminants. These methods are straightforward, sensitive, and do not rely on costly equipment. This study will discuss current nanotechnology-based approaches to detect adulterants, focusing on their advantages and limitations [144], as shown in Table 1.

- **Molecularly imprinted polymers (MIPs)**

Molecularly imprinted polymers (MIPs) are synthetic receptors, or antibody mimics, that may attach to certain molecules. Molecular imprinting has arisen in the last 30 years as an exceptionally adaptable approach for synthesizing networks with high affinity and selectivity for a chemical species, which is utilized as a molecular template during their synthesis [145]. Many drawbacks of traditional detection methods are solved by MIP-based sensors, which enable efficient, sensitive, and low-cost detection employing smart miniaturized equipment. MIPs may also be employed as target molecule sensors, especially for optical and electrochemical processes, among other uses. MIPs can give information on the presence of target molecules (nutrients or pollutants) in food, both qualitatively and quantitatively. MIPs can be

Table 1 Names of common adulterants used for the adulteration of food products [144]

Products	Name of Adulterant
Fruit Juice, Pickles, Jam	Sugar, Dilution with water
Milk and Milk Products	Urea, Vegetable oil, banana, Starch, Detergent, annatto, Addition of water
Coffee	tamarind seeds, Chicory
Ice cream	Starch, rice powder or wheat flour

employed for the detection of foodborne pathogens, toxins, and adulterants. They can assist assure the quality and safety of food.

Products by offering quick and dependable detection techniques. It may be applied in order to reduce the negative effects of some compounds that harm food quality while keeping organoleptic qualities [146].

Here are some important examples;

- MIP-based Nanosensors are capable of detecting and monitor contaminants and adulterants in food products. For example, they can be designed to selectively bind to specific toxins, such as aflatoxins or mycotoxins, commonly found in grains and nuts [147].
- MIP nanosensors can detect pesticide residues in fruits, vegetables, and other agricultural goods. The sensors may be programmed to recognize and attach to specific pesticide compounds, enabling the measurement and monitoring of pesticide levels in food samples. This allows for the adoption of effective quality control procedures as well as regulatory compliance [148].

15 Nanosensors for Foodborne Pathogen Detection

A pathogen is a creature, such as a virus, bacterium, or protozoan, that may infect its host. Virulence refers to the degree of illness symptoms generated by a pathogen [149]. Polymer nanosensors can detect and identify foodborne pathogens such as bacteria, viruses, and parasites. These sensors can provide quick and precise detection, allowing for early action to avoid foodborne diseases and outbreaks. Nanotechnology research has advanced to the point that it has Encouraged the development of nanosensors for detecting foodborne diseases that surpass traditional methods. Only polyaniline and polypyrole are extensively used to identify pathogens in food. Conducting polymers are good immobilizing substrates for biologic molecules at electrodes, that lead to improved signal transmission, biocompatibility, selectivity, high sensitivity, durability, and flexibility [150]. There are other types and examples, such as:

A. Electrochemical nanosensors

Electrochemical detection techniques obtained widespread use because of their simplicity, high sensitivity, and compatibility with reduction for portable device applications [151]. An electrochemical immunosensor with screen-printed gold electrodes was developed for monitoring *S. typhimurium* at concentrations as minimal as 20 cells/mL using chronoamperometry. Label-free electrochemical detection presents a highly appealing alternative, which has been validated through the application of electrochemical impedance spectroscopy such as: graphene oxide/gold nanoparticles were used to create an electrochemical impedance aptasensor for *S. typhimurium*, allowing for the detection. Moreover, *E. coli* O157:H7 and *S. typhimurium* were rapidly detected in food samples using a screen-printed interdigitated microelectrode paired with magnetic-bead sample separation [152]. The previously mentioned electrochemical techniques were limited because they required a redox probe, such as ferro/ferricyanide, in the measuring solution. Researchers devised a label-free aptasensor for detecting *S. typhimurium* by using a poly[pyrrole-co-3-carboxyl-pyrrole] copolymer-aptamer bioconjugate. This innovative sensor is based on the intrinsic changes in the electrical properties of the polymeric surface, eliminating the need for redox probes. Calibration, selectivity, and recovery investigations supported the aptasensor's specificity, sensitivity, and usefulness for detecting *S. typhimurium* in food samples, such as apple juice [152].

Furthermore, nanomaterial-based biosensors have recently gained a lot of interest as they can distinguish different pesticide compounds that are present in minute concentrations by dramatically magnifying the observed signals. Metal–organic framework (MOF) sensors are one form of biosensor that is now employed for pesticide detection [153, 154]. MOFs are sophisticated porous coordination polymers that self-assemble by combining metal ions or metal ion clusters with organic ligands via controlled structural characteristics of coordinating bonds. MOFs are easy to employ due to the many links between inorganic fragments and the characteristics of organic binders. They have a high absorption and catalytic activity due to their large specific surface area and pore volume. MOFs are able to utilized as surface modifications for sensing [155]. Moreover, Because of their enormous surface areas, extreme porosities, and adjustable sizes, which demonstrate potential for chemical sensing, drug delivery, gas storage, and catalysis [156]. MOFs are emerging class of adsorbents that have the ability to adsorb pollutants from the environment and food. MOFs were found to display binary functions by Shirin et al. They work well as efficient mercury ion adsorbents as well as quick colorimetric and fluorescence probes. The sorbent's hue changes from pale cream to brilliant yellow as a result of the adsorption of Hg^{2+} [157]. Thanks to advancements mesoporous substance and metal–organic frameworks (MOFs), notable advancements have been made recently in this category of porous solids. MOFs considered to be crystalline molecular materials with exceptional chemical and physical characteristics, as well as an interior porosity that is both incredibly large and permanent. Functional MOFs have attracted significant scientific interest due to their various uses include gas storage, medication delivery, catalysis, optical imaging, and sensing. Flexible MOFs like MIL-53(Al), Zn-MOF, or Zr-MOF, with adjustable host–guest characteristics, were especially beneficial for detecting levels of antibiotics in wastewater exhibit structural alterations in response to environmental stimuli. However, the fluorescence sensors that are based on MOFs have inadequate sensing abilities when it comes to identifying antibiotics and pathogen at concentrations ranging from 2.4 to 2.4×10^3 CFU mL^{-1} [158].

B. Metal–organic framework (MOF)

Furthermore, nanomaterial-based biosensors have recently gained a lot of attention since they can detect various pesticide chemicals at low concentrations by greatly amplifying the recorded signals. Metal–organic framework (MOF) sensors are a type of biosensor that is now used for pesticide detection [153, 154]. MOFs are sophisticated porous coordination polymers that self-assemble by combining metal ions or metal ion clusters with organic ligands via controlled structural characteristics of coordinating bonds. MOFs are easy to employ due to the many links between inorganic fragments and the characteristics of organic binders. They have a high absorption and catalytic activity due to their large specific surface area and pore volume. MOFs can serve as surface modifiers in detection [155]. Moreover, because of enormous surface areas, high porosities, and adjustable sizes, which demonstrate potential for chemical sensing, drug delivery, gas storage, and catalysis [156]. MOFs are emerging class of adsorbents that have the ability to adsorb pollutants from the environment and food. MOFs were found to display binary functions by Shirin et al. They work well as efficient mercury ion adsorbents as well as quick colorimetric and fluorescence probes. The sorbent's hue changes from pale cream to brilliant yellow as a result of the adsorption of Hg^{2+} [157]. Recent advances in the study of porous solids have been made possible by breakthroughs in mesoporous materials and metal–organic frameworks (MOFs). MOFs are widely recognized as crystalline molecular materials with Outstanding chemical and physical qualities, in addition to a significant and persistent interior porosity. Functional MOFs have sparked widespread scientific attention because of their multiple uses in gas storage, drug delivery, catalysis, optical imaging, and sensing. Flexible MOFs having tunable host–guest properties, such as MIL-53(Al), Zn-MOF, or Zr-MOF, were very beneficial for detecting antibiotic residues in wastewater due to structural changes in response to external stimuli. However, MOF-based fluorescence sensors are insufficient for detecting antibiotics [158].

C. Conducting polymer (CP) sensors

Conducting polymers, or CPs, are a type of polymeric material distinguished by conjugated main chains. CPs' electrical and optical properties may be adjusted by modifying their doping states. Due of their long-term water stability, these materials are commonly employed in watery settings. They have been shown to be electroactive biointerfaces and electrode materials for bioelectronic and device integration. CPs can also operate as organic electrodes and multifunctional interfaces, giving them a suitable alternative for a variety of bioelectronic applications [159]. For example, milk analysis: one of the world's most popular drinks, milk is high in calcium, minerals, proteins, and other nutrients. In addition, milk is a component of many healthy meals. Premium milk is in high demand since it is so necessary in daily life. However, milk adulterants can be hazardous to human health, thus it's necessary to have detection tools for them [160]. As a metabolic byproduct of protein metabolism, urea's concentration in milk is thought to provide important information for understanding cow care and supervision [161].

Conductive polymer composite components Electrochemical sensing devices capable of detecting urea and melamine levels have been developed. Das and Sarkar created a urea sensor by encapsulating PANI-based hydrogels on a functioning electrode and immobilizing urease enzyme. Because of the hydrogel's wide surface area and the electroactivity of PANI, the sensor proved sensitive and useful in detecting urea in materials such as milk, puffed rice, dirt, and human blood [162]. A Fe_3O_4/MWCNT/PANI-Nafion composite film was presented by Singh and co-workers as a urea sensor that captured the urease enzyme. In contrast to other groups, Fe_3O_4 and CNT functioned as two efficient mediators of electron transfer, effectively amplifying their current signals [163]. Regasa and colleagues developed a thin PAA composite sheet imprinted with melamine. To detect melamine accurately, hydrogen bonding the PANI-PAA composite structure resulted in a donor–acceptor sequence of 4.8 Å [164].

D. Water quality sensing

Water is a necessary ingredient for all species on Earth, and a valuable resource for human civilization. Access to safe and inexpensive water is seen as a basic humanitarian aim and a critical problem for the twenty-first century. Pollution and other human activities exacerbate water scarcity in both developed and poor nations. Stringent water quality requirements, as well as the advent of new pollutants, have put strain on industrialized countries' current water treatment and distribution infrastructure. Recent advancements in nanotechnology provide prospects for developing the water supply systems of the future. Our current water treatment, distribution, and disposal systems, which are mostly based on centralized infrastructure, are no longer practical. Nanotechnology's incredibly efficient, versatile, and multifunctional processes are expected to give low-cost water and wastewater treatment alternatives that rely less on massive infrastructures. Nanotechnology-enabled water and wastewater treatment has the potential to overcome fundamental obstacles faced by present treatment technologies, as well as give novel treatment capabilities that might allow for the economic use of unusual water sources to extend water supply.

E. Polymeric nano-adsorbents

Adsorption is a common polishing technique used in the treatment of water and wastewater to eliminate both organic and inorganic pollutants. Conventional adsorbents are often limited in their efficiency by their adsorption kinetics, lack of selectivity, and surface area or active sites. The remarkable specific surface area and sorption sites that they possess, together with their short intra-particle diffusion distance, variable pore size and surface chemistry, make nanoadsorbents a major advance [165]. Dendrimers are adsorbents with branching structures that have a specified shape, size, and molecular weight to extract organic molecules and heavy metals. Polymers known as dendrimers, which are composed of monomers extending from a central core, are becoming more and more common. Due to their many functional groups in the pocket, periphery, core, and surface, dendrimers are a very effective nanoscale multidentate chelating agent/ligand for ion separation technology. Conversely, the dendrimer scaffold's microenvironment could serve as a suitable host for a variety of pollutants. The potential of dendrimers to clean up the environment has also been investigated. The effectiveness of dendrimers in eliminating both organic and inorganic pollutants has been well-established. Another appealing chemical for the creation of electrochemical sensors is dendrimers. Because of their adjustable topologies and properties, dendritic polymer-based nanocomposites have piqued the curiosity of researchers. It is possible to modify dendritic polymers to specifically target pollutants by adding the right functional moieties. Dendritic materials are suitable for capturing chemical contaminants due to their high specific surface area and pore volume. They may be used with a variety of materials and supports. Multistep synthesis presents a difficulty for field and large-scale applications, requiring deep knowledge of polymer evolution. Researchers in separation science and organic chemists can work together to find a solution to this problem. Dendrimers have demonstrated good potential in aqueous matrix cleanup of both organic and inorganic contaminants [166]. Heavy metal decontamination of contaminated water using dendritic polymeric materials is affected by pH because it affects the moieties on the surface of these polymers. Additionally, by changing the composition and morphology of metal ions, solution pH affects how well dendritic polymeric materials adsorb substances. Different research employed the density functional theory (DFT) method to examine the adsorption of mercury (Hg) onto salicylaldehyde-modified PAMAM dendrimers that were supported by silica gel. Because oxygen and nitrogen species exist, mercury ions have the ability to chelate with dendritic polymers. While the results of the Dubinin-Radushkevich (D-R) isotherm model indicate that mercury was removed via chemisorption onto sorbent pores, thermodynamic tests showed the process of sorption was endothermic and spontaneous [167–169].

Polycyclic aromatic hydrocarbons (PAHs) are basically hydrocarbons that have two or more fused aromatic rings. There are hundreds of PAHs. Hazardous PAHs are produced when carbon-based compounds burn partially or completely. They are carcinogenic and have the potential to alter hormones. To extract pyrene and phenanthrene from an aqueous solution. Arkas et al. combined hyperbranched polyethylene imine with silicic acid. The dendrimer-silica nanoparticle combination was created using sol–gel techniques. It was shown that the adsorption of PAHs was enhanced by the addition of dendrimer to the silica nanospheres. The capacity of PEI-silica nanoparticles to cure water was boosted by the creation of complexes with transferable charges between the dendrimer's tertiary amino groups and PAHs [170–172].

F. Carbonized polymer dot (CPD)

While water is required for daily activities, it may also be seen as an impurity in some circumstances. Carbonized polymer dot (CPD), which was created using a one-pot hydrothermal process with ethylenediamine and catechol, allowed us to create a fluorescent nanoprobe for measuring water content. Because of its several surface functional groups, this carbon dot reacts differently in various organic solvents, which are governed by hydrogen bonding. This nanoprobe detects water content extremely well in a range of chemical solvents, particularly 1,4-dioxane. The analytical technique gave a for 0.1 vol% water content (n = 3), the limit of detection (LOD) was 0.0460 vol% and the relative standard deviation (RSD) was 0.198%. More importantly, our CPD-based fluorescence analytical approach has the widest detection range (0.0–100.0 vol%) when tracing water in heavy water when compared to other probes. Its low LOD (5.51 vol%), reasonable repeatability, and strong reproducibility (RSD = 2.25% at 5.0 vol% water content, n = 3) are other noteworthy features. With this work, the uses of carbon compounds in analytical domains and gives a fresh perspective on real-time water content monitoring [173]. Deuterium oxide, often known as heavy water (D_2O), is an isotopically labeled form of ordinary water (H_2O) that is essential for many uses in chemical analysis and medicine, among other fields. However, distinguishing between these two forms of water is difficult since they are virtually identical in size and shape to regular water. Currently, there are few reliable techniques to identify light water from heavy water, and fluorescent probes aren't commonly accessible [174]. A novel luminous substance called carbon dots (or CDs) was unintentionally found in 2004. The benefits of CDs over standard quantum dots are low cost, low synthesis complexity, strong stability, and good biocompatibility. Owing to these benefits, CDs are now often employed in fluorescence sensors, photovoltaic cells, lighting displays, and catalysis, among other applications. Carbonized polymer dots (CPDs) are a visually pleasing kind of CD. It has a core–shell structure with numerous polymer chains or functional groups on the exterior and a high level of carbonization inside. CPDs are chemically stable and easy to functionalize because of their complex surface features. They also work extremely well for monitoring responses to metal ions, pH, and biomolecules since their fluorescence is quenched (an "on–off" approach). The CDs' PL quenching may be explained by three separate mechanisms: the inner filter effect (IFE), photo-induced electron transfer (PET), and fluorescence resonance energy transfer. An additional sensing method relies on the "on–off–on" rule. With this approach, the addition of the investigated analytes can progressively restore the fluorescence of CDs that were previously quenched by quenchers. In addition to intensity fluctuation, the CDs' ability to sense band shifting can also be achieved by varying the energy level of the CDs. Peak-shifting can also be visually detected by fluorescence hues, making it easier to observe with the unaided eye. The water content of certain solvents, such as D_2O, can trigger a response from the novel fluorescent CPD. Catechol and ethylenediamine were used as precursors in the green hydrothermal synthesis of the CPD, which involved dehydration, polymerization, and carbonization. The water content response curves in many organic solvents were investigated, and ultra-sensitive water in 1,4-dioxane was detected at trace levels as low as 0.0460%. This nanoprobe, This, in comparison to older water content sensors, offers a larger detection range, was used to determine the quantity of water in heavy water [173, 175].

G. Harmful Algal Blooms (HABs)

Harmful algal blooms (HABs) are a phenomenon that may happen in fresh, marine, and brackish water sources all over the world. They are caused by a variety of environmental conditions, including eutrophication, which is defined as an excess of nutrients in the water, especially fixed nitrogen (nitrates, ammonia, and urea) and phosphate. Other elements that impact HAB development and persistence are sunshine, water temperature, nutrition, salinity, wind, and water currents [176–178]. Harmful algal blooms (HABs) have become more intense and frequent worldwide in the last few years. HABs are present in a variety of freshwater environments in the United States, including rivers, small inland lakes, and the Great Lakes. As a result, new laws have been created to safeguard the public against HABs [179].

Unmanned equipment in the air (RIGHT) and water (CENTER panel) may be capable of monitoring HABs in real time. The unmanned devices were calibrated to respond to a fluorescein dye that was fired and deployed as a substitute for HABs [180]. Polymer nanosensors were used to detect hazardous algal blooms (HABs). For example, a dual-emitting ratiometric fluorescence nano-sensor was built utilizing molecularly imprinted polymers (MIPs) for detecting HABs [181]. Toxins generated by harmful algal blooms (HABs), such as microcystins, were found using molecularly imprinted polymers (MIPs). MIPs are manufactured receptors with specific holes designed to fit a template molecule, in this example, the target toxin. Because of the tridimensional cavities created during the polymerization process, they are able to adsorb target molecules, giving the target analytes exceptional selectivity and specificity.

MIPs have been used to selectively identify the cyanotoxins generated by freshwater cyanobacteria, known as microcystins. Numerous health consequences, such as liver disease, primary liver cancer, and gastrointestinal disorders, have been linked to these poisons. Because MIPs are inexpensive to make, durable to harsh environmental conditions, and have recognition skills similar to natural receptors, making them an effective tool for detecting HAB.

MIPs were developed for the particular detection of microcystins in a study by Mbisana et al., with good sensitivity and selectivity for the targeted toxins. In a separate work, Tian et al. developed a MIP for microcystin-LR extraction utilizing magnetic graphene oxide (GO) and Fe_3O_4. This MIP displayed better selectivity and less aggregation than standard GO [182–184].

To sum up, MIPs have proven to be an efficient and sensitive approach for analyzing these hazardous substances in a variety of water sources by effectively detecting HABs and associated toxicities.

16 Heavy Metals in Water

Scientists developed and tested a surface plasmon resonance (SPR) sensor made of optical fiber to find heavy formation of hemoglobin, resulting in anemia. To overcome this issue, Ghosh et al. created a fiber biosensor that detects heavy metal ions utilizing long-period grating (LPG) and fiber Bragg grating (FBG) technology. To improve selectivity for Pb^{2+} detection, the biosensor's surface was coated with graphene oxide, a cross-linked chitosan composite, and polypropionic acid. The biosensor captures heavy metals by selective adsorption, allowing it to detect Pb^{2+} ions by evaluating the way the surrounding refraction and the high-order cladding mode fading field interact.

A range of sensing probes coated with pyrrole/chitosan/ITO/Ag, pyrrole/ITO/Ag, pyrrole/chitosan/Ag, and pyrrole were created by researchers. The probes were assessed for the ions of Cd^{2+}, Pb^{2+}, and Hg^{2+} heavy metals [185].

Deshmukh et al. developed another biosensor for Pb^{2+} detection by altering polypyrrole/single-walled carbon nanotubes produced electrochemically using ethylene diamine tetraacetic acid. With a 0.07 μM detection limit, the biosensor demonstrated high sensitivity [186–188] metal ions in drinking water.

Lead (Pb), a common heavy metal, can cause significant harm to the human body by impacting the central nervous system and leading to issues such as behavioral changes and hearing impairment, potentially even resulting in brain damage. Even at low concentrations, mercury is a very hazardous contaminant that harms ecosystems and humans over the long term by building up in organisms all throughout the food chain. This is so that Hg^{2+}-specific biosensors may be built because Hg^{2+} stabilizes the thymine-thymine base pair. Pi et al. developed a Hg^{2+} sensor that works by selectively binding Hg^{2+} to thymine-rich DNA. The interaction of DNA with SYBR Green I generates the fluorescence signal. The findings show that DNA-based Hg^{2+} sensor technology may be successfully applied in freshwater settings. Using magnetic silica nanocomposites, Wang et al. enclosed CdTe QDs and rhodamine 6G to generate multifunctional inorganic–organic nanocomposites. The Hg^{2+} test had a linear range of 7–900 nM, and these nanocomposites were able to regenerate ratiometric fluorescence sensing of Hg^{2+}. This approach offers significant reusability, good sensitivity, and selectivity since it is based on magnetic mesoporous silica nanocomposites [189, 190]. The human body may suffer from an accumulation of heavy metals such as zinc (Zn), chromium (Cr), and copper (Cu).

High levels of copper can cause neurological conditions like Alzheimer's disease. Lung cancer and mitochondrial damage can result from an excess of Cr in the human body. While the body requires trace amounts of zinc to help with DNA and protein synthesis, too much zinc can cause headaches and diarrhoea. Wong et al. developed a biosensor for the detection of heavy metals, such as Al, Cu, Pb, and Zn, using polyurethane-wrapped β-carotene. The biosensor detects changes in optical density when β-carotene and heavy metals react. Heavy metals may be identified with a basic spectrophotometric device by measuring the optical density at $\lambda = 450$ nm [191–193].

5. Applications of Nanosensors in Food Packaging

Nanotechnology is one of the most fascinating fields of study nowadays, with the potential to completely change the global food supply system. Developing and manufacturing novel packaging solutions that guarantee product safety and quality is necessary to meet the significant difficulties posed by the requirement for safe food products. Several notable applications of nanotechnology in the food packaging sector include nanocomposites, nanosensors, and assays utilizing nanomaterials to identify food-related analytes (e.g., gasses, microorganisms, and viruses). Food safety and risk assessment are major topics in the realm of food research. While nanotechnology is still a relatively young field, safety concerns are highlighting the need for international regulations to guarantee scalability, protect the industry's and consumers' acceptance of this instrument. We will address these issues and possible solutions in this section [194]. Nanosensors are platforms that resemble small nanometers in size. They function similarly to conventional sensors, sensing minute quantities and translating them into signals that need analysis. Since food is essential to life, one of nanosensors' main objectives is to stop or postpone unintended changes in texture, color, and odor. The nanodevices give direct information on the interior microbial growth and chemical changes occurring in food. Intelligent packaging is the use of various gadgets, such as nanosensors, to sense the safety and quality of food both inside and outside the meal. The goal is to quickly detect food that is contaminated and pinpoint the procedures that impact product quality. One of the best options in cases when oxygen exposure causes food to deteriorate quickly by oxidizing lipids and vitamins and encouraging the growth of bacteria is vacuum packing. The basic principles of nanosensors, their importance, and the innovative materials used in them for the food packaging industry are all covered in this chapter [195].

17 Nanosensors for Detection of External Conditions

Reducing size, a significant advancement in computer technology, opens up the possibility of using sensors in food, allowing for the monitoring of all atmospheric effects on the product from the manufacturing plant to the end user. If temperature, moisture, or other measuring devices were affixed to newly harvested fruit using micro- or nanosensors, salespeople might use the data to confirm the product's freshness and the conditions of storage or transit [196]. They might choose to accept or reject the items in question based on this information.

18 Nanosensors for Internal Condition Detection

Depending on the kind of detection, two more types of nanosensors found in food items can be identified at the nanoscale: nanosensors and nanobiosensors. Similar procedures are used to create chemical and biosensors: reagents, chemical or bacterial detectors, are placed on the probe, and when the desired chemical element or bacterium reacts with them, the nanosensor's output changes [197].

One of the most promising smart packaging technologies is antimicrobial nanocomposite packaging, which helps to extend food shelf life by reducing or preventing food rotting and the growth of harmful microbes that contaminate food [198]. When compared to their bulk counterparts, microorganisms can be killed or rendered inactive by the enhanced surface reactivity of the nanosized antimicrobial agents and the high surface-to-volume ratio of the nanofillers, which contribute to the antimicrobial properties of polymer nanocomposite food packaging materials [199]. During the polymer manufacturing process, antimicrobial ingredients are blended with the polymeric matrix to create antimicrobial packaging films. The antibacterial properties of polymer nanocomposite materials have been studied as growth inhibitors [200], antimicrobial agents, antimicrobial carriers or antimicrobial food packaging films [198].

Metal and metal oxide nanoparticles are among the most widely used nanomaterials for antibacterial food packaging applications because to their high specificity and large surface area. They have strong antimicrobial action as well. Food packaging, biotechnological, medical, and other industries have shown a great deal of interest in the antibacterial properties of metals and their oxide nanoparticles, such as silver, copper, titanium, and zinc nanoparticles [201, 202].

Shankar et al. [198] developed an agar polymer nanocomposite containing six different types of nanoparticles, and they discovered that it exhibited potent antibacterial activity against foodborne pathogens that were both Gram-positive and Gram-negative. Antimicrobial packaging films have also been made using metal oxides, such as TiO_2, ZnO, and MgO, due to their high antibacterial activity and excellent stability [203–205].

Excitingly, polymer nanocomposite films made with organically modified nanoclay have been known to have profound antimicrobial activity [206, 207]. Rhim et al. [206] the antibacterial action against pathogenic microorganisms was demonstrated by produced chitosan/clay nanocomposite films combined with two kinds of nanoclay. The researchers discovered that, in contrast to the natural MMT, which showed no antimicrobial action, the nanocomposite film made with the organically modified MMT (CloisiteR 30B) had antimicrobial activity against Gram-positive bacteria like Listeria monocytogenes and Staphylococcus aureus. Even though it has long been known that chitosan alone has antibacterial properties against a variety of microorganisms.

Hong and Rhim [208] showed the potent antibacterial activity of organically modified clays and came to the conclusion that the quaternary ammonium salt of the organically modified nanoclay (Cloisite 30B) is primarily responsible for the antimicrobial activity of the chitosan/clay nanocomposite.

19 Nanotechnology in Food Packing and Security

The packing process is one of several steps required to guarantee food safety. There is not a packing material that is completely impermeable to airborne gases, water vapor, and natural chemicals [209, 210]. When it comes to packaging fresh fruits and vegetables, it is not ideal to totally stop gas movement and permeability since these items go through a process known as cellular respiration [211]. On the other hand, carbonated beverages must never be exposed to air or carbon dioxide (CO_2) while still in their container. This prevents oxidation and decarbonization [211]. Different methods of exchange of oxygen (O_2), water vapor (H_2O), and carbon dioxide (CO_2) occur depending on the food matrices and the types of packaging materials employed.

As a result, various nanocomposite materials—like polymers—may be used to identify and address these issues with food packaging [212]. A hundred times thinner than a human hair's approximate thickness of 10,000 nm, and a thousand times thinner than a book page's approximate thickness of 100,000 nm, are nanoparticles having a diameter of less than 100 nm. If the proper regulatory processing clearance is secured, different nanoparticle shapes might be extremely useful in a range of fields related to the food and pharmaceutical sectors, especially in research projects involving food science. Food packaging's ability to act as a barrier against the buildup of gasses has improved as a consequence of recent developments in the use of nano-biocomposites [213]. Current advancements in the food packaging sector are encouraging the use of biodegradable polymers enhanced with environmentally friendly nano-fillers [214]. Nevertheless, there is a good reason to be concerned about ingesting these nano-compounds during meal time. Therefore, it is imperative to carry out research on these nanoparticles' movement throughout the human body in addition to their immunogenic and hazardous consequences [215]. The biological decomposability of these nano-filled decomposable polymers has also been questioned [216]. Scientists that are searching for environmentally and human-safe nanomaterials take these worries extremely seriously [217, 218].

Nanosensors are able to identify the emitted gases during degradation in addition to a change in color. The gases that the sensors are particularly sensitive to are sulfur dioxide (SO_2), hydrogen (H_2), nitrogen oxides, hydrogen sulfide (H_2S), and ammonia (NH_3). They combine the functions of a sensor and processor in one, producing electrical impulses in response to variations in heat, light, gas, humidity, and chemicals. Due to their high sensitivity and selectivity, nanosensors outperform conventional sensors in terms of effectiveness.

Gold, palladium, and platinum are among the metals utilized to make these gas sensors [219]. There have been cases when mycotoxin B1 in milk has been identified using gold nanoparticles. Sometimes the use of deoxyribonucleic acid (DNA) and single-walled carbon nanotubes (CNTs) in the construction of the sensors increases their sensitivity. It has been possible to identify peanut-allergenic proteins in infections and chocolate by using liposome nano-vesicles.

Salmonella species, *Listeria monocytogenes*, and *E. coli* were identified using immunomagnetic bead sandwich testing and universal protein G-liposomal nanovesicles [220]. Nanosensors are being utilized to monitor soil conditions in order to improve agricultural production. They are also helpful in locating pesticide residue on food items. It has proven possible to use nanosensors to detect carcinogens in food as well as pesticides. In the food packaging industry, two common types of sensors used are gas detectors and time–temperature integrators. Numerous applications, including array biosensors, nanoparticles in solution, nanoparticle-based sensors, nano-test strips, electronic noses, and nano cantilevers, make use of nanosensors.

"Electronic noses" are one type of sensor that combines an information processing system and several chemical sensors. Because the sensor replicates the functions of the human nose, it is known as an electronic nose. Similar to the electronic nose, there have also been discussions of electronic tongue sensors. It alerts the customer that the food is no longer safe to eat by changing color when it finds a rotten ingredient [218].

20 Nano-polymer with Risks of Migration into Food

When it comes to the hazards associated with using nanocomposites, it's critical to carefully evaluate, control, and govern risks [221]. It is extremely important to properly evaluate the impact of the engineered nanoparticles' size, shape, surface charge, concentration, and other characteristics when using them to create nanocomposites. Risk management should also be controlled and monitored [222, 223].

Even though nanotechnology is used in many different sectors, more in-depth research is still needed to fully comprehend toxicity and determine its true nature. In this because polymer-based nanoparticles have the potential to contaminate food and induce allergies and toxicity, using them in nanocomposites poses a concern. This is made feasible by the migration of packing materials into food; under controlled settings, the migration test of nanomaterials is crucial [224]. Food and nanomaterials' respective physicochemical characteristics affect how they migrate from polymer nanocomposites to food. These include food type, pH, size, shape, concentration, solubility, diffusion, and length of contact with the nanoparticles used in packaging, among other things [225]. Researchers also raised a number of topics that require attention with regard to the toxicity caused by The physicochemical characterization of nanoparticles and their migration; techniques to evaluate the migration of nanomaterials; advanced techniques to identify and characterize nanomaterials; the relationship between toxicity and the physical characteristics of nanoparticles (size, shape, surface charge, etc.).

Toxicokinetics following ingestion [222] warned that there might be disastrous consequences from the toxicity of nanomaterials. Thus, to truly comprehend the issue of nanotoxicity, it is imperative to conduct a thorough investigation on the properties of nanomaterials, including their size, shape, solubility, and agglomeration. In a similar vein [223], have examined in detail the potential genotoxicity of nanoparticles on biological processes such as cell division and DNA damage. Nanoparticles have been used in many different industries due to their appealing and distinctive qualities, but they also pose a risk to human health and the environment [226]. Some researchers suggest that appropriately encapsulating nanomaterials in polymers prevents their migration into meals [227]. US Food and Drug Administration (FDA) and the European Commission (EU) have laws in place to address the migration of nanomaterials into food products that regulates [228]. As shown by EU regulation 85/572/EEC. According to this guideline, the migration of nanomaterials must exceed the limit of 10 mg/dm^2 [229].

6. Application of nanosensors in textiles

The development of smart fabrics has benefited greatly from the increased use of nanotechnology in recent years. In all domains of industrial design, "designing" refers to a process that begins with an analysis of the design requirements and concludes with the transformation of the design into some novel forms. "Needs for design" refer to the requirements that both manufacturers and consumers have for the product being designed in terms of its appearance, usability, design, safety, and cost. Some studies have examined design demands (or requirements) in the context of smart clothing using their various definitions. The demands in terms of heat management, moisture control, movement, adaptability, sizing and fit, robustness, and garment care have been investigated by Dunne et al. [230]. Nanomaterials have been used to give fabrics and clothing durable antimicrobial, resistant to UV, conductive to electricity, optical, hydrophobic and flame-resistant qualities.

Nowadays, a variety of tasks are being carried out by smart textile-based nanomaterial-based systems, such as drug release sensing, power production, optics, and power storage. These advancements are being developed for broader usage in healthcare, defense, and on-body energy-producing applications, in addition to their widespread applications in the fashion industry [231]. So-called "smart fabrics," made by fusing smart nanoparticles with traditional materials, have been developed in recent years. A smart fabric is one that can adjust one or more of its properties to perform a function in response to environmental changes [232].

The incorporation of nanotechnology enables the production of intelligent, multifunctional fabrics with several cutting-edge applications in the fields of health, medicine, sports, fashion, the military, improved protection, and transportation. Now, fabrics and apparel can have a high added-value feature while retaining other desirable characteristics like comfort, flexibility, lightness, and aesthetic appeal by using the special capabilities of nanoparticles [233, 234]. Up until now, textiles and microelectronic devices have been able to work together seamlessly. Nanomaterials have been incorporated into textiles using a variety of methods. The fibres needed to create fabrics are first produced using a "bottom-up" method. On the other hand, the "top-down" methodology is utilized throughout the final stages, such as printing and spray coating or penetration processes. It has been demonstrated that the comparatively recent electrospinning technique, which turns processed raw materials into fibres and fabrics, is ideal for producing nanofibers [231, 233].

Different inorganic and organic substances could be synthesized as nanoscale units and used directly in coating processes. Potential uses for these nanomaterial-modified textiles include wound healing [235, 236], ventilation [237], medicine delivery [238], cosmetics, the generation of renewable energy and electronic uses like the development of on-body circuits, diodes, and transistors [238].

Among the nanomaterials that have been employed are polyaniline, polyacetylene, polypyrrole [239], Au [234], Ag [238], Pd [240], Cu [241], Si [242], CuO [243], ZnO [244], carbon nanotube (CNT) [245, 246], TiO_2 [247, 248], chitosan [249] and graphene oxide (GO) [250] nanoparticles.

Some classes of smart fabrics will be stated in details.

I. Antimicrobial fabrics

The cloth's antibacterial barrier is both intriguing and healthy for people. Numerous antimicrobial agents, such as TiO_2 [251], chitosan [252], *N*-Halamine [253], Ag [254], Cu_2O [255], and metal/hemp fibres [256], have been added to textiles [257–259]. The active nanoparticles can be chemically or physically incorporated into the fibres to create an antibacterial textile [260]. Muoz-Bonilla and Fernández Garca [261] used a variety of techniques, including electrospinning, nano-precipitation, and self-assembly, to create antibacterial nanomaterials. They investigated the bactericidal properties and surface properties of nanostructured polymeric coatings. Silver (Ag) was one of the first antibacterial nanoparticles put to textile surfaces. Without altering its mechanical characteristics, it serves as a doping antibacterial agent and demonstrates exceptional antimicrobial efficacy [243].

As shown in Fig. 7 Ag nanoparticles interact with bacterial proteins through their huge surface area and small size, which inhibits the growth of bacteria's cells.

Riaz and colleagues [262] looked into the possible uses of TiO_2 in the textile sector in addition to 3-glycidoxypropyl trimethoxysilane and 3-(trimethoxysilyl) propyl *N,N,N*-dimethyl octadecyl ammonium chloride. The researchers came to the conclusion that the treated cotton exhibited antibacterial activity, self-cleaning properties, and long-lasting super-hydrophobicity. ZnO nanoparticles can clean and disinfect clothes contaminated with aerobic gram-positive *Staphylococcus aureus* and gram-negative *Escherichia coli*, and they function similarly to TiO_2. Nanomaterial face masks have garnered a lot of interest since the COVID-19 pandemic. Antiviral face masks and PPE kits were developed by researchers across multiple disciplines, with the ability to filter out multiple infections, including SARS-CoV-2.

In 2020, Talebian et al. [263] proposed two methods of controlling COVID-19: biosensors and disinfectants based on nanomaterials attached to masks or personal protective equipment (PPE) textiles, respectively. They suggested that metallic nanoparticles as Cu, Ag, TiO_2, and others, with their potent antiviral qualities, could displace more traditional disinfectants including alcohols, peroxides, quaternary amines, and chlorides. Furthermore, they proposed that extremely effective biosensors may be combined into face masks or personal protective equipment tools

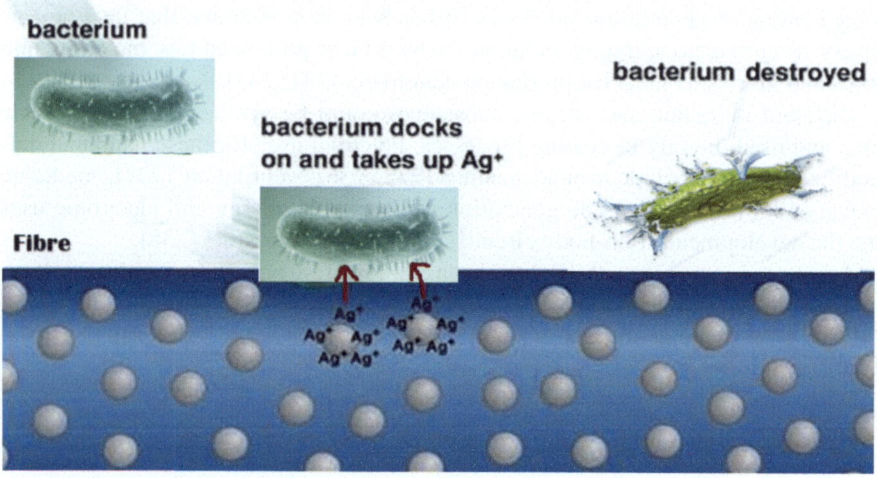

Fig. 7 Silver nanoparticles intermingle with bacterium to inhibit their cells growth

to facilitate the prompt identification of SARS-CoV-2 or other viruses. A range of multi-layer face masks with alternating hydrophilic and hydrophobic layers were created by Lustig et al. in 2020. They discovered that the hydrophilic layer cannot flow through it because the hydrophobic layer repels the aqueous aerosol from it. Stopping the infection is the aim of these face masks. These face masks are intended to stop the virus from spreading through coughs and sneezes [264].

II. Ultraviolet-resistant textiles

UV-blocking nanoparticles, which block UVB and UVA rays, are applied to fabrics to improve UV shielding and produce UV protection materials. An instrument for measuring the effectiveness of UV protection is the ultraviolet protection factor (UPF), which varies based on the composition of the fabric. Nanomaterials like ZnO and TiO_2 that are sensitive to UV light can either absorb or scatter UV radiation [265]. Both of these chemicals are safe and could maintain their stability even at greater temperatures. The size of the particles and the radiation's wavelength determine how much UV light is scattered by nanoparticles. Cotton has been treated with TiO_2 nanoparticles as UV shields. The robustness of TiO_2 finishing was proven to be stable even after 50 washings [266].

According to Dhineshbabu and Bose [267], MnO_2-$FeTiO_3$ nanoparticles and thermoplastic polyurethane cotton fabrics are claimed to block UV rays. The results demonstrate that nano-coated materials on textiles have a high UV-shielding ability compared to uncoated materials, making them more durable and intelligent fabrics. Because the UV-absorbing fact can help protect humans from harmful UV exposure, textiles can be used for a variety of purposes.

III. Electrically conductive textiles

The conductive qualities of textile materials are the main factor driving the advancement of sensors and detectors in the fabrics industry. There are many uses for conducting polymers in the textile industry. These materials' textile surfaces react electrically to external stimuli because their resistivity has been tuned. By mixing in various nanoparticles, these polymers' matrix can be changed to give them a particular characteristic. For example, polymers conducting electricity such as nanostructured polypyrrole (PPy), polyaniline (PANI), and polythiophene (PT) have superior conducting qualities, optical features and mechanical strength. These polymers are useful because of their low production costs, flexibility, and lightweight nature, which make them ideal for usage in the textile sector. The conductivity of fibres is increased by an order of magnitude when conductive polymers are added to their surfaces [268]. For example, SiO_2 nanoparticles and polyimidoamide fibres have been combined via spinning.

Electrically conductive channels were created in polyacrylonitrile (PAN) fibres as a consequence of the insertion of nanoparticles. More mechanical and antistatic qualities were the outcome [269].

Electrical conductivity has been increased in cotton by coating it with various conductive matrices. Shim et al. [270] established a polyelectrolyte-based layer in conjunction with multi-walled carbon nanotubes for application in fabrics. Mattana and colleagues utilized a combination of several conformally distributed metal nanoparticles across the uneven surface of cotton fibres in their investigation [271]. Because in-situ polymerization forms flexible bridges between the nanoparticles, it can be employed to lessen mechanical deformations in transistors based on cotton. Additionally, graphene has been added to textile fibers to incorporate electrical properties.

Another study looked into a simple method for using leftover natural fiber to create knitted fabric that is electrically conductive and may be used in a variety of applications. Wet-spun blank PAN and wool/PAN composite fibers were used to make the garments. A straightforward and practical procedure was then used to cover the created knitted textiles with graphene oxide (GO). To add electrical conductivity, a chemical reduction technique was applied to the GO-coated fibers.

In comparison to the blank PAN woven fabrics and GO, the wool/PAN composite fabrics may have had a higher affinity for GO as a result of amino and hydroxyl groups existence. This may help to explain why, in comparison to the control PAN/reduced GO knitted textiles, the wool/PAN/low GO woven textiles had stronger color power and minor lightness (L*). The wool/PAN/low GO woven textiles proven minor resistance (50 Ω) and greater conductivity (1.67 S/cm) in comparison to the blank PAN/low GO knitted fabrics (285 Ω and 0.35 S/cm, respectively), according to the electrical conductivity research. The enhanced GO adhesion and decrease in the wool/PAN woven textiles compared to the blank PAN woven fabrics could be the reason for this difference in conductivity. Thus, the electrically conductive wool/PAN/low GO woven textile developed in this research with waste wool threads and a modest yet practical method can be said to have potential applications in actuators, sensors, super-capacitors, wearable textiles, and cars as a conductive textile to decrease the metals application, particularly in the vehicle seat.

IV. Sensors on textile

Numerous types of sensors, such as pressure, chemical, optical, thermal, touch, and fragrance sensors, can be incorporated into textiles for a range of applications. In the fields of smart clothing, health monitoring, and human motion detection, carbon-based nanomaterials including carbon nanotubes (CNT), graphene, and carbon nanofibers have all been extensively investigated for application as lightweight, flexible, and high-strain sensors [272–274]. To act as strain sensors, carbon-based nanoparticles have been synthesized in a number of ways and evenly distributed within polymers. Direct film casting and electrospinning were used to create strain sensors. Human hair coated with graphene has also been used to construct strain sensors [275].

Cotton and silk fabrics were also utilized as strain sensors after spray coating, carbonization, and stabilization [276, 277]. Applications for smart textiles are increasingly using plasmon-based sensors. It has been discovered that plasmonic sensors have extremely high sensitivity for biological detection. Flexible capacitors integrated into the fibre have also been used to generate touch sensor textiles [278]. Using a Dobby loom and capacitor fibres, a 1D sensor array was produced and incorporated into a wool matrix. Fifteen capacitor fibres were used in the design of the touch sensor fabric. When a finger touches the capacitor fibre, the local current and voltage distribution change; this change is recorded to detect touch. By joining these fibre capacitors with other conductive or battery-powered fibres, an effective electrical circuit may also be constructed on the clothing. This capability may find use in programmable fabrics, safety gear, and fashion. Furthermore, pressure-responsive fabrics have been created [279].

Using a dye-coating technique, the fibres intended for use as pressure sensors were coated with an organic conductive polymer, such as poly (styrene sulfonate), and a dielectric perfluoropolymer layer. The warps and wefts of the modified threads were utilized, while the remaining spaces in the matrix were filled with pure nylon fibres. Capacitors were created where the fibres converged.

The cloth's capacitance ranged from 0.22 to 0.63 pF with a sensitivity range of 0.98–9.80 N cm^2 when 4.9 N cm^2 of pressure was applied [279]. Similar humidity and temperature sensors are found in the textiles [280].

7. Applications of nanosensors in electronics

Numerous capabilities are provided by nanosensors, such as remarkable performance, high sensitivity and specificity, fast response, compact size, portability, and real-time analysis [281]. Consequently, a variety of applications have made use of nanosensors for target identification [282]. These gadgets are significant because they have the ability to facilitate low-complexity, low-power technologies, especially within the domain of sensors [283].

21 Wireless Nano Sensor Networks (WNSN)

The multidisciplinary field of nanotechnology will improve our quality of life, make it safer, and lessen our environmental impact. We have the chance and capacity to study and find previously undiscovered places and phenomena thanks to the nanoscale dimension. Numerous scientific discoveries were made possible by this technology, open up new product opportunities in a wide range of industries and domain [284]. It is anticipated that nanotechnology will permeate every part of our lives and spread throughout the globe [285].

Nanotechnology, which includes tiny basic units called nanodevices (nanosensors and nano-controllers) and ranges in size from one to a few hundred nanometers [286], facilitates the amazing development of nanoscale devices. These nanodevices, which are integrated with other devices and have storage, processing, sensing, and communication units nanodevices, create Wireless Nano Sensor Networks (WNSN), which have a wider coverage area [287].

Numerous light weight, Wireless sensor networks (WSNs) made up of low-power sensor nodes in the field [25]. They transmit the gathered data, each node is outfitted with a variety comprising sensors, a wireless communication module, and interface electronics. A wide range of fields have proposed applications for wireless sensor networks (WSNs), including health care [288], industrial equipment monitoring [289], precision horticulture [290], volcanology [291], habitat observation [292], and engine monitoring [293].

Since human health is directly impacted by air quality, issues related to air pollution have received more attention, monitoring urban air pollution [294]. Wireless sensor networks (WSN) offer a great, low-cost, and simple way to use automatic real-time monitoring and control for air quality. As a result, numerous WSN-based systems for monitoring air quality have been put forth. Among them were sensor networks installed on automobiles.

For example, the Mobile Air Quality Monitoring Network (MAQUMON) [295] is made up of sensor nodes that are installed on automobiles and contain a number of gas sensors that measure the concentrations of ozone, CO, and NO_2. Samples are taken frequently while the vehicle is in motion, but only a when it's stationary, a few times per hour. The samples' location and time are marked on the labels. then stored. When a car is in a Wi-Fi hotspot's coverage area that is open, every piece of data is sent to the servers, where it is processed and made available on the Sensor Map website.

However, as a target gas corresponds to each gas sensor, necessitating installing numerous sensors to keep an eye on various gasses, it has proven to be impractical for general applications. The detection, analysis, and recognition of odors have demonstrated significant promise for electronic nose (E-Nose) systems, and they were made to resemble the olfactory systems of mammals [296]. A WSN and an E-Nose combined to create an array of sensors would significantly increase the variety of odor categories and real-world applications.

One of the best options for achieving polymer-coated surface acoustic wave (SAW) array offers great sensitivity in situations where the detection of organic gases is necessary [297]. To create a gas detector fitted with the benefits little in size, inexpensive, highly sensitive, and quick response, a SAW array was used. Multiplexing was used as a switching element to control the 2 × 2 non-continuous working oscillators. The sensor's film functioned as a kind of smart skin, producing chemical cues through interactions between the film and the molecules [298]. The polymer sheet that senses served as the nasal cavity's mucous membrane and was crucial in the identification of gases. Seven polymers were chosen as the components of the spin-coated sensitive film, the array's resonator surface to improve sensitivity and selectivity; Poly-*N*-vinyl pyrrolidone (PNVP), Poly-4-vinyl phenol (P4VP), polystyrene (PS), polystyrene-co-maleic anhydride (PSMA), polyvinyl acetate (PVAc), polyethylene-glycol (PEG), and polysulfone (Psu).

22 An Internet of Things (IOT)

What does the Internet of Things entail? Our lives have changed significantly throughout the previous few years due to the widespread use of internet-connected computers, cellphones, and smart wearable electronics that interact with our surroundings and carry enormous amounts of data. Thus, the Internet of Things is born. The Internet of Things (IoT) is a fascinating field that is defined as a network of physical objects (things) that are embedded with sensors and information transmitters to allow data exchange between these physical objects [299, 300].

So that many fields have seen the application of the Internet of Things (IoT), including wearables [301], smart cities [302], smart farming [303], smart homes [304], connected cars [305], and connected drones [306].

23 IOT in Agriculture

Traditional or antiquated farming methods are finding it increasingly difficult Keeping up with the swift increase in population that comes with the expansion of humanity. Therefore, to address the expanding population's need for food, advanced farming techniques are imperative. intelligent agricultural systems that leverage The Internet of Things and embedded systems (IoT) have gained popularity in recent years as a means of improving food production for human consumption. These systems are designed for use in livestock farms and agriculture farm fields, and they rely on wireless sensor networks, embedded systems, and IoT Smart-phones, PCs, and so forth. Adequate irrigation, fertilization, and use of pesticides are essential for better farming and food production. If we create a mechanized method to carry out These tasks are all completed automatically, it will be much easier. The recommended systems include quadcopter fertilization and pH monitoring, automated pesticide spraying and insect detection, an alarm system for animals or intruders for the agriculture field, and automated moisture monitoring and watering. The agribusiness sector saw significant progress [307].

These upgrades have brought in an automated system that allows devices to be controlled via WSN (Wireless Sensor Network) and allows for the observation of crop growth. First and main function of wireless sensor networks (WSNs) is to gather data from a remote location and send it over wireless networks that the recipient can monitor. The agricultural sector can benefit from the application of WSN technology [308], particularly when it comes to managing and, more crucially, providing farmers with up-to-date information about their field. IoT devices allow farmers to monitor continuously and vigilantly, enabling them to identify delicate modifications to crucial factors that could signal the beginning of potential challenges. Precision monitoring functions as a sentinel that is always on guard, catching any detail or anomaly that may arise, whether it be changes soil temperature, moisture content, and humidity swings, or early warning indications of agricultural diseases. Equipped with the capacity to promptly identify and evaluate emerging problems, farmers are able to act quickly by taking focused and defensive action [309]. Agriculturalists are better able to reduce risks, avoid needless losses, and maximize yields when possible issues are identified early on.

Hence, IoT-enabled precision monitoring acts as a trustworthy ally by giving farmers access to predictive and practical data insights required to successfully guide their agricultural endeavors [310]. In order to enable farmers to utilize vital resources like water, fertilizer, and pesticides as efficiently as possible, resource optimization is a critical component of the application of IoT in agriculture. Growers may learn a great deal about the moisture content of their crops' overall health and the quality of their soil by utilizing real-time data. Equipped with this understanding, they able to strategically implement nutrients and water just where and when they're most needed, guaranteeing maximum productivity and growth. This focused strategy reduces needless waste, saving farmers a substantial amount of money while also protecting limited resources. By limiting the use of dangerous chemicals and averting too much water use, the capacity to carefully control the distribution of resources helps to preserve the environment. The application of IoT technology in agriculture gives farmers a previously unheard-of degree of control and precision in resource management.

Internet of Things, or IoT, is becoming popular in many fields, including security and agriculture. This is because devices can transmit data over the internet without needing direct communication between people or between people and computers. IoT is therefore developing quickly to satisfy the needs of various applications. Due to a shortage of water resources, water use must be managed and controlled to ensure that crops are consistently irrigated without running the risk of tank overflow. IoT-based water level monitoring systems have just been used. To operate without maintenance for an extended period of time, a low-power, wireless system is necessary. Because of this, the system's microcontroller and sensor must have low power consumption. Regarding water several types of sensors, including pressure, hydrostatic, ultrasonic, and capacitive sensors, are used for level monitoring [311–316].

The comparison of various sensors reveals that pressure sensors, particularly pressure sensors Using polymer/carbon nanotube as a basis, are the most appropriate because of their low power, low sensitivity, and lower cost consumption. Actually, the change in electrical resistance brought on by the external load is the basis for this sensor's operation [317].

This work's objective is to create an autonomous mobile system that measures and controls. the tank's water level This system will allow the farmer or user to use the pump with ease using his phone as well as complete the process of packing up without wasting any water. When there is less than 25% of water in the overhead tank, the system warns the user to start the engine and turns it off when the level is reached.

24 Hardware Components

In order to control the water level in the field, the system for controlling water level that is movable and made up of an I2C LCD (16 $\times$ 2) display, a battery shields for the microcontroller, a pump, and a pressure sensor based on polymer carbon nanocomposite (PCN). There will be an input and an outlet for water in the tank. The water inlet will be used to fill the tank, and the water outlet will be used to irrigate the field.

25 Conclusion

Numerous key characteristics set the nanosensor devices apart from other sensor technologies currently available on the market, including ultrahigh sensitivity, exquisite selectivity, direct, label-free, and real-time electrical signal transduction, as well as the possibility of massive addressable array integration. This chapter includes examples of anything from protein, virus, and DNA detection to the study of tiny chemical compounds attaching to proteins. These examples show distinct characteristics that could significantly influence genetic screening, drug discovery, and disease diagnosis. They can also be powerful new tools for research across a wide range of industry areas.

References

1. Datta SPA (2016) Future healthcare: bioinformatics, nano-sensors, and emerging innovations. In: Nanosensors: theory and applications in industry, healthcare and defense, vol 247
2. Urban GA (2009) Micro-and nanosensors for medical applications. In: World congress on medical physics and biomedical engineering, September 7–12, 2009, Munich, Germany: vol 25/8 Micro-and nanosystems in medicine, active implants, biosensors. Springer
3. Riu J, Maroto A, Rius FX (2006) Nanosensors in environmental analysis. Talanta 69(2):288–301
4. Colombo M et al (2009) Femtomolar detection of autoantibodies by magnetic relaxation nanosensors. Anal Biochem 392(1):96–102
5. Cullum BM, Griffin GD, Vo-Dinh T (2002) Nanosensors: design and application to site-specific cellular analyses. In: Biomedical diagnostic, guidance, and surgical-assist systems IV. SPIE
6. Velasco-Garcia MN, Mottram T (2003) Biosensor technology addressing agricultural problems. Biosys Eng 84(1):1–12
7. Arroyo-Hernández M (2010) Micro-and nanomechanical biosensors. Handb Nanophys Nanomed Nanorobot
8. Zhao G, Wang H, Liu G (2015) Advances in biosensor-based instruments for pesticide residues rapid detection. Int J Electrochem Sci 10(12):9790–9807
9. Steffens C et al (2012) Atomic force microscopy as a tool applied to nano/biosensors. Sensors 12(6):8278–8300
10. Mohamed K (2019) 2.16 Nanoimprint lithography for nanomanufacturing. Compr Nanosci Nanotechnol 2:357–386
11. Pal K, Gomes F (2020) Nanofabrication for smart nanosensor applications. Elsevier Science
12. Elbert DL (2011) Bottom-up tissue engineering. Curr Opin Biotechnol 22(5):674–680
13. Huang C-F, Tseng Y-C (2003) The coverage problem in a wireless sensor network. In: Proceedings of the 2nd ACM international conference on wireless sensor networks and applications
14. Groves T et al (2002) Maskless electron beam lithography: prospects, progress, and challenges. Microelectron Eng 61:285–293
15. Arshak K et al (2004) Focused ion beam lithography-overview and new approaches. In: 2004 24th international conference on microelectronics (IEEE Cat. No. 04TH8716). IEEE
16. Maldonado JR, Peckerar M (2016) X-ray lithography: some history, current status and future prospects. Microelectron Eng 161:87–93
17. Mahmoud Z et al (2023) Organic chemical nano sensors: synthesis, properties, and applications. Braz J Biol 84:e268893
18. Freeman MH, Hall JR, Leopold MC (2013) Monolayer-protected nanoparticle doped xerogels as functional components of amperometric glucose biosensors. Anal Chem 85(8):4057–4065
19. Chen X et al (2014) Advances in enzyme-free electrochemical sensors for hydrogen peroxide, glucose, and uric acid. Microchim Acta 181:689–705
20. Lin J et al (2020) Anti-liquid-interfering and bacterially antiadhesive strategy for highly stretchable and ultrasensitive strain sensors based on Cassie-Baxter wetting state. Adv Func Mater 30(23):2000398
21. Saleh Ahammad A, Lee J-J, Rahman MA (2009) Electrochemical sensors based on carbon nanotubes. Sensors 9(4):2289–2319
22. Mousavi SM et al (2018) Nanosensors for chemical and biological and medical applications. Med Chem (Los Angeles) 8(8):2161-0444.1000515
23. Swierczewska M et al (2012) High-sensitivity nanosensors for biomarker detection. Chem Soc Rev 41(7):2641–2655
24. Agrawal S, Prajapati R (2012) Nanosensors and their pharmaceutical applications: a review. Int J Pharm Sci Technol 4:1528–1535
25. Abdel-Karim R, Reda Y, Abdel-Fattah A (2020) Nanostructured materials-based nanosensors. J Electrochem Soc 167(3):037554

26. Thangadurai D et al (2021) Nanosensors for detection and evaluation of organic compounds in soil. Nanomaterials for soil remediation. Elsevier, pp 205–219
27. Buzanovskii V (2014) Gas nanosensors based on polymers. Rev J Chem 4(1):33–51
28. Myslík V et al (2007) Organic active layers for chemical sensors prepared by Matrix Assisted Pulse Laser Deposition (MAPLD). J Phys: Conf Ser (IOP Publishing)
29. Hong L, Li Y, Yang M (2010) Fabrication and ammonia gas sensing of palladium/polypyrrole nanocomposite. Sens Actuators B Chem 145(1):25–31
30. Pirsa S, Alizadeh N (2010) Design and fabrication of gas sensor based on nanostructure conductive polypyrrole for determination of volatile organic solvents. Sens Actuators B Chem 147(2):461–466
31. Wang W et al (2007) A novel wireless, passive CO_2 sensor incorporating a surface acoustic wave reflective delay line. Smart Mater Struct 16(4):1382
32. Kebabian PL, Freedman A (2006) Fluoropolymer-based capacitive carbon dioxide sensor. Meas Sci Technol 17(4):703
33. Ayad MM, Torad NL (2010) Quartz crystal microbalance sensor for detection of aliphatic amines vapours. Sens Actuators B Chem 147(2):481–487
34. Choudhury A (2009) Polyaniline/silver nanocomposites: dielectric properties and ethanol vapour sensitivity. Sens Actuators B Chem 138(1):318–325
35. Patil D et al (2010) Poly (o-anisidine)–tin oxide nanocomposite: synthesis, characterization and application to humidity sensing. Sens Actuators B Chem 148(1):41–48
36. Wang X et al (2009) A highly sensitive humidity sensor based on a nanofibrous membrane coated quartz crystal microbalance. Nanotechnology 21(5):055502
37. Frank M, Moon KS, Kassegne S (2010) A PMMA coated PMN–PT single crystal resonator for sensing chemical agents. Smart Mater Struct 19(3):035015
38. Su P-G et al (2011) Fully transparent and flexible humidity sensors fabricated by layer-by-layer self-assembly of thin film of poly (2-acrylamido-2-methylpropane sulfonate) and its salt complex. Sens Actuators B Chem 153(1):29–36
39. Khot LR, Panigrahi S, Lin D (2011) Development and evaluation of piezoelectric-polymer thin film sensors for low concentration detection of volatile organic compounds related to food safety applications. Sens Actuators B Chem 153(1):1–10
40. Martínez-Hipatl C et al (2010) Detection of volatile organic compounds by an interferometric sensor. Sens Actuators B Chem 147(1):37–42
41. Viespe C, Grigoriu C (2010) Surface acoustic wave sensors with carbon nanotubes and SiO_2/Si nanoparticles based nanocomposites for VOC detection. Sens Actuators B Chem 147(1):43–47
42. Caliendo C et al (2007) Nanostructured organometallic polymer and palladium/polymer hybrid: surface investigation and sensitivity to relative humidity and hydrogen in surface acoustic wave sensors. Nanotechnology 18(12):125504
43. Zhang C et al (2011) Nanoporous polystyrene fibers functionalized by polyethyleneimine for enhanced formaldehyde sensing. Sens Actuators B Chem 152(2):316–323
44. Korposh S, Selyanchyn R, Lee S-W (2010) Nano-assembled thin film gas sensors. IV. Mass-sensitive monitoring of humidity using quartz crystal microbalance (QCM) electrodes. Sens Actuators B Chem 147(2):599–606
45. Vaddiraju S, Gleason KK (2010) Selective sensing of volatile organic compounds using novel conducting polymer–metal nanoparticle hybrids. Nanotechnology 21(12):125503
46. Lu J et al (2009) Vapour sensing with conductive polymer nanocomposites (CPC): polycarbonate-carbon nanotubes transducers with hierarchical structure processed by spray layer by layer. Sens Actuators B Chem 140(2):451–460
47. Dhawale D et al (2010) Room temperature LPG sensor based on n-CdS/p-polyaniline heterojunction. Sens Actuators B Chem 145(1):205–210
48. Schultze J, Karabulut H (2005) Application potential of conducting polymers. Electrochim Acta 50(7–8):1739–1745
49. Bakhshi A, Bhalla G (2004) Electrically conducting polymers: materials of the twentyfirst century

50. Pacher P et al (2010) A novel concept for humidity compensated sub-ppm ammonia detection. Sens Actuators B Chem 145(1):181–184
51. Dan Y et al (2009) Gas sensing properties of single conducting polymer nanowires and the effect of temperature. Nanotechnology 20(43):434014
52. Wang X et al (2010) Nanofibrous polyethyleneimine membranes as sensitive coatings for quartz crystal microbalance-based formaldehyde sensors. Sens Actuators B Chem 144(1):11–17
53. Jiang S et al (2009) Au nanoparticles-functionalized two-dimensional patterned conducting PANI nanobowl monolayer for gas sensor. Sens Actuators B Chem 140(2):520–524
54. Yao W, Chen X, Zhang J (2010) A capacitive humidity sensor based on gold–PVA core–shell nanocomposites. Sens Actuators B Chem 145(1):327–333
55. Nath S et al (2010) Acetone sensing property of ZnO quantum dots embedded on PVP. Sens Actuators B Chem 148(2):353–357
56. Sadek A et al (2006) A layered surface acoustic wave gas sensor based on a polyaniline/In_2O_3 nanofibre composite. Nanotechnology 17(17):4488
57. Xie G et al (2010) Fabrication of methane gas sensor by layer-by-layer self-assembly of polyaniline/PdO ultra thin films on quartz crystal microbalance. Sens Actuators B Chem 145(1):373–377
58. Yoo K-P et al (2009) Effects of O_2 plasma treatment on NH_3 sensing characteristics of multiwall carbon nanotube/polyaniline composite films. Sens Actuators B Chem 143(1):333–340
59. Yoo K-P et al (2010) Novel resistive-type humidity sensor based on multiwall carbon nanotube/polyimide composite films. Sens Actuators B Chem 145(1):120–125
60. Philip B et al (2003) Carbon nanotube/PMMA composite thin films for gas-sensing applications. Smart Mater Struct 12(6):935
61. Wu T-T, Chen Y-Y, Chou T-H (2008) A high sensitivity nanomaterial based SAW humidity sensor. J Phys D Appl Phys 41(8):085101
62. Wilhelm S, Wolfbeis OS (2011) Irreversible sensing of oxygen ingress. Sens Actuators B Chem 153(1):199–204
63. Aussawasathien D et al (2011) Poly (o-anisidine)–polystyrene composite fibers via electrospinning process: surface morphology and chemical vapor sensing. Sens Actuators B Chem 151(2):341–350
64. Benvenho AR, Li RW, Gruber J (2009) Polymeric electronic gas sensor for determining alcohol content in automotive fuels. Sens Actuators B Chem 136(1):173–176
65. Dogra V et al (2021) Nanomaterials; applications; implications and management. New Front Nanomater Environ Sci 23–45
66. Peyghambarzadeh S et al (2011) Improving the cooling performance of automobile radiator with Al_2O_3/water nanofluid. Appl Therm Eng 31(10):1833–1838
67. Hidalgo VH et al (2001) High temperature erosion wear of flame and plasma-sprayed nickel–chromium coatings under simulated coal-fired boiler atmospheres. Wear 247(2):214–222
68. Keshavamurthy R et al (2014) Tribological properties of hot forged Al2024-TiB2 in-situ composite. Adv Mater Manuf Charact 4(2):87–92
69. Kusmoko A, Dunne D, Li H (2015) Effect of heat input on Satellite 6 coatings on a medium carbon steel substrate by laser cladding. Mater Today Proc 2(4–5):1747–1754
70. Shukla V, Jayaganthan R, Tewari V (2015) Degradation behavior of HVOF-sprayed Cr_3C_2-25% NiCr cermet coatings exposed to high temperature environment. Mater Today: Proc 2(4–5):1805–1813
71. Bartke S et al (2018) Market potential of nanoremediation in Europe–market drivers and interventions identified in a deliberative scenario approach. Sci Total Environ 619:1040–1048
72. Corsi I et al (2018) Ecofriendly nanotechnologies and nanomaterials for environmental applications: key issue and consensus recommendations for sustainable and ecosafe nanoremediation. Ecotoxicol Environ Saf 154:237–244
73. Malik S, Muhammad K, Waheed Y (2023) Nanotechnology: a revolution in modern industry. Molecules 28(2):661

74. Nayfeh MH (2016) Optics in nanotechnology. Opt Our Time 223–264
75. Wood V, Bulović V (2010) Colloidal quantum dot light-emitting devices. Nano Rev 1(1):5202
76. Cao J, Liu H (2011) Terahertz semiconductor quantum well photodetectors. Semiconductors and semimetals. Elsevier, pp 195–242
77. Sayed I, Bedair S (2019) Quantum well solar cells: principles, recent progress, and potential. IEEE J Photovolt 9(2):402–423
78. Corzine SW, Yan R-H, Coldren LA (1993) Optical gain in III–V bulk and quantum well semiconductors. In: Quantum well lasers, vol 23
79. Lin K-F et al (2005) Band gap variation of size-controlled ZnO quantum dots synthesized by sol–gel method. Chem Phys Lett 409(4–6):208–211
80. Bankole OM, Achadu OJ, Nyokong T (2017) Nonlinear interactions of zinc phthalocyanine-graphene quantum dots nanocomposites: investigation of effects of surface functionalization with heteroatoms. J Fluoresc 27:755–766
81. Xu JX et al (2020) Quantification of the photon absorption, scattering, and on-resonance emission properties of CdSe/CdS core/shell quantum dots: effect of shell geometry and volumes. Anal Chem 92(7):5346–5353
82. Borshchov V et al (2023) Nanopolymer optically transparent structures, systems and devices. Radiotekhnika 213:56–64
83. Šupová M, Martynková GS, Barabaszová K (2011) Effect of nanofillers dispersion in polymer matrices: a review. Sci Adv Mater 3(1):1–25
84. Doushkina V, Fleming E (2009) Optical and mechanical design advantages using polymer optics. In: Advances in optomechanics. SPIE
85. Bormashenko E et al (2004) Polyvinylidene fluoride—piezoelectric polymer for integrated infrared optics applications. Opt Mater 27(3):429–434
86. Sadasivuni KK et al (2016) Electroactive and optically adaptive bionanocomposite for reconfigurable microlens. J Phys Chem B 120(20):4699–4705
87. Orasugh JT et al (2019) Utilization of cellulose nanocrystals (CNC) biopolymer nanocomposites in ophthalmic drug delivery system (ODDS). J Nanotechnol Res 1(3):75–87
88. Åhlén M, Tummala GK, Mihranyan A (2018) Nanoparticle-loaded hydrogels as a pathway for enzyme-triggered drug release in ophthalmic applications. Int J Pharm 536(1):73–81
89. Bhagyaraj S, Oluwafemi OS, Krupa I (2020) Polymers in optics. Polymer science and innovative applications. Elsevier, pp 423–455
90. Marino-Lopez A et al (2019) Laser-protective soft contact lenses: keeping an eye on the eye through plasmonics. Appl Mater Today 15:1–5
91. Jamieson T et al (2007) Biological applications of quantum dots. Biomaterials 28(31):4717–4732
92. Mansur HS (2010) Quantum dots and nanocomposites. Wiley Interdiscip Rev: Nanomed Nanobiotechnol 2(2):113–129
93. Nair SS, Philip R (2021) Nanomaterials for luminescent devices, sensors, and bio-imaging applications. Springer
94. Kang H-Y (2010) A review of the emerging nanotechnology industry: materials, fabrications, and applications, vol 64. Department of Toxic Substances Control Pollution Prevention and Green Technology, California
95. Mohan S et al (2014) Facile synthesis of transparent and fluorescent epoxy–CdSe–CdS–ZnS core–multi shell polymer nanocomposites. New J Chem 38(1):155–162
96. Mohan S et al (2015) Green synthesis of yellow emitting PMMA–CdSe/ZnS quantum dots nanophosphors. Mater Sci Semicond Process 39:587–595
97. Mohan S, Oluwafemi OS (2016) Simple synthesis of orange fluorescent CdSe–polycaprolactone nanofiber via a completely non-phosphine based route. Mater Lett 174:157–161
98. Zou W et al (2011) A transparent and luminescent epoxy nanocomposite containing CdSe QDs with amido group-functionalized ligands. J Mater Chem 21(35):13276–13282
99. Xu J et al (2007) Organic−inorganic nanocomposites via directly grafting conjugated polymers onto quantum dots. J Am Chem Soc 129(42):12828–12833

100. Aylott JW (2003) Optical nanosensors—an enabling technology for intracellular measurements. Analyst 128(4):309–312
101. Masteri-Farahani M, Mashhadi-Ramezani S, Mosleh N (2020) Molecularly imprinted polymer containing fluorescent graphene quantum dots as a new fluorescent nanosensor for detection of methamphetamine. Spectrochim Acta Part A Mol Biomol Spectrosc 229:118021
102. Mehrzad-Samarin M et al (2017) A novel metronidazole fluorescent nanosensor based on graphene quantum dots embedded silica molecularly imprinted polymer. Biosens Bioelectron 92:618–623
103. Wei X et al (2015) Highly-controllable imprinted polymer nanoshell at the surface of silica nanoparticles based room-temperature phosphorescence probe for detection of 2,4-dichlorophenol. Anal Chim Acta 870:83–91
104. Mehrzad-Samarin M, Faridbod F, Ganjali MR (2019) A luminescence nanosensor for Ornidazole detection using graphene quantum dots entrapped in silica molecular imprinted polymer. Spectrochim Acta Part A Mol Biomol Spectrosc 206:430–436
105. Yih W-Y, Kratochvil FJ, Stewart JC (2005) Intraoral minor salivary gland neoplasms: review of 213 cases. J Oral Maxillofac Surg 63(6):805–810
106. Munawar A et al (2019) Nanosensors for diagnosis with optical, electric and mechanical transducers. RSC Adv 9(12):6793–6803
107. Varadan VK, Chen L, Xie J (2008) Nanomedicine: design and applications of magnetic nanomaterials, nanosensors and nanosystems. Wiley
108. Puttananjegowda K, Takshi A, Thomas S (2023) Polymers for biosensing applications in viral detection and diagnosis. Viral drug delivery systems: advances in treatment of infectious diseases. Springer, pp 193–217
109. Wang M, Thanou M (2010) Targeting nanoparticles to cancer. Pharmacol Res 62(2):90–99
110. Ferrari M (2005) Cancer nanotechnology: opportunities and challenges. Nat Rev Cancer 5(3):161–171
111. Srinivas PR, Barker P, Srivastava S (2002) Nanotechnology in early detection of cancer. Lab Invest 82(5):657–662
112. Maeda H et al (2000) Tumor vascular permeability and the EPR effect in macromolecular therapeutics: a review. J Control Release 65(1–2):271–284
113. Greish K (2010) Enhanced permeability and retention (EPR) effect for anticancer nanomedicine drug targeting. Cancer Nanotechnol Methods Protocols 25–37
114. Puoci F (2015) Advanced polymers in medicine. Springer
115. Park K et al (2009) New generation of multifunctional nanoparticles for cancer imaging and therapy. Adv Func Mater 19(10):1553–1566
116. Sato N et al (2001) Pharmacokinetics and enhancement patterns of macromolecular MR contrast agents with various sizes of polyamidoamine dendrimer cores. Magn Reson Med: Off J Int So Magn Reson Med 46(6):1169–1173
117. Dobson P, Jarvie H, King S (2019) Nanoparticle. Encycl Britannica 14
118. Alrushaid N et al (2023) Nanotechnology in cancer diagnosis and treatment. Pharmaceutics 15(3):1025
119. Swain S et al (2016) Nanoparticles for cancer targeting: current and future directions. Curr Drug Deliv 13(8):1290–1302
120. Zhang W et al (2022) Photodynamic therapy and multi-modality imaging of up-conversion nanomaterial doped with AuNPs. Int J Mol Sci 23(3):1227
121. Abed A et al (2022) Platinum nanoparticles in biomedicine: preparation, anti-cancer activity, and drug delivery vehicles. Front Pharmacol 13:797804
122. Ramanathan A (2019) Toxicity of nanoparticles challenges and opportunities. Appl Microsc 49(1):2
123. Patel P, Patel V, Patel PM (2022) Synthetic strategy of dendrimers: a review. J Indian Chem Soc 99(7):100514
124. Yang H et al (2022) A test strip electrochemical disposable by 3D MXA/AuNPs DNA-circuit for the detection of miRNAs. Microchim Acta 189(1):50

125. Wang X et al (2022) Evaluating diabetic ketoacidosis via a MOF sensor for fluorescence imaging of phosphate and pH. Chem Commun 58(18):3023–3026
126. Pavitra E et al (2021) Engineered nanoparticles for imaging and drug delivery in colorectal cancer. In: Seminars in cancer biology. Elsevier
127. Teodor ED et al (2018) Functionalized magnetic nanostructures for anticancer therapy. Curr Drug Targets 19(3):239–247
128. Aygun A et al (2020) Biogenic platinum nanoparticles using black cumin seed and their potential usage as antimicrobial and anticancer agent. J Pharm Biomed Anal 179:112961
129. Abbasi E et al (2016) Graphene: synthesis, bio-applications, and properties. Artif Cells Nanomed Biotechnol 44(1):150–156
130. Turner A (2013) Biosensors: then and now. Trends Biotechnol 31(3):119–120
131. Clark LC, Lyons C (1962) Electrode systems for continuous monitoring in cardiovascular surgery. Ann N Y Acad Sci 102(1):29–45
132. Turner AP (2013) Biosensors: sense and sensibility. Chem Soc Rev 42(8):3184–3196
133. McNaught A, Wilkinson A (1997) Amorphous carbon. In: IUPAC compendium of chemical terminology: the "Gold Book", 2nd edn. Blackwell Scientific Publications, Oxford
134. Daniels JS, Pourmand N (2007) Label-free impedance biosensors: opportunities and challenges. Electroanal Int J Devot Fundam Pract Aspects Electroanal 19(12):1239–1257
135. Chowdhury AD et al (2019) Electrical pulse-induced electrochemical biosensor for hepatitis E virus detection. Nat Commun 10(1):3737
136. Graboski AM et al (2020) Nanosensors for water quality control. Nanotechnology in the beverage industry. Elsevier, pp 115–128
137. Vikesland PJ (2018) Nanosensors for water quality monitoring. Nat Nanotechnol 13(8):651–660
138. Jafarizadeh-Malmiri H et al (2019) Nano-sensors in food nanobiotechnology. In: Nanobiotechnology in food: concepts, applications and perspectives, pp 81–94
139. Gu D-C et al (2019) A novel method for rapid quantitative evaluating formaldehyde in squid based on electronic nose. LWT 101:382–388
140. Li J et al (2018) Validation of UPLC method on the determination of formaldehyde in smoked meat products. Int J Food Prop 21(1):1246–1256
141. Pandey S (2016) Prevailing practices of artificial ripening of mango, banana and papaya through calcium carbide in Jharkhand. Agric Eng Today 40(3):35–39
142. Moosavy M et al (2019) Assessment of the chemical adulteration and hygienic quality of raw cow milk in the northwest of Iran. Qual Assur Saf Crops Foods 11(5):491–498
143. Roy M, Yadav B (2022) Electronic nose for detection of food adulteration: a review. J Food Sci Technol 1–13
144. Singh R et al. Nanotechnology-based approaches for detection of food adulterants
145. Akgönüllü S, Yavuz H, Denizli A (2020) SPR nanosensor based on molecularly imprinted polymer film with gold nanoparticles for sensitive detection of aflatoxin B1. Talanta 219:121219
146. Filipe-Ribeiro L, Cosme F, Nunes FM (2020) New molecularly imprinted polymers for reducing negative volatile phenols in red wine with low impact on wine colour. Food Res Int 129:108855
147. Basak S, Venkatram R, Singhal RS (2022) Recent advances in the application of molecularly imprinted polymers (MIPs) in food analysis. Food Control 139:109074
148. Mahmoudpour M et al (2020) Nanomaterial-based molecularly imprinted polymers for pesticides detection: recent trends and future prospects. TrAC Trends Anal Chem 129:115943
149. Pirofski L-A, Casadevall A (2012) Q&A: what is a pathogen? A question that begs the point. BMC Biol 10(1):6
150. Devraj L et al (2023) Nanotechnology-oriented sensors for the quick recognition of foodborne microbes and pathogens. Nanotechnology applications for food safety and quality monitoring. Elsevier, pp 93–112
151. Zhu Y et al (2015) Building an aptamer/graphene oxide FRET biosensor for one-step detection of bisphenol A. ACS Appl Mater Interfaces 7(14):7492–7496

152. Sheikhzadeh E et al (2016) Label-free impedimetric biosensor for Salmonella typhimurium detection based on poly [pyrrole-co-3-carboxyl-pyrrole] copolymer supported aptamer. Biosens Bioelectron 80:194–200
153. Jalili R et al (2020) Detection of penicillin G residues in milk based on dual-emission carbon dots and molecularly imprinted polymers. Food Chem 314:126172
154. Gholami P et al (2020) Photocatalytic degradation of gemifloxacin antibiotic using Zn-Co-LDH@ biochar nanocomposite. J Hazard Mater 382:121070
155. Sohrabi H et al (2022) MOF-based sensor platforms for rapid detection of pesticides to maintain food quality and safety. Food Chem Toxicol 165:113176
156. Mortada B et al (2017) Postmetalated zirconium metal organic frameworks as a highly potent bactericide. Inorg Chem 56(8):4739–4744
157. Safaei S, Kazemian H, Junk PC (2021) Dual functional MOF as a selective fluorescent naked-eye detector and effective sorbent for mercury ion. J Solid State Chem 300:122267
158. Wang S et al (2019) Bimetallic cerium/copper organic framework-derived cerium and copper oxides embedded by mesoporous carbon: label-free aptasensor for ultrasensitive tobramycin detection. Anal Chim Acta 1047:150–162
159. Lin C-H et al (2021) Conducting polymer-based sensors for food and drug analysis. J Food Drug Anal 29(4):544
160. Erkmen C, Kurbanoglu S, Uslu B (2020) Fabrication of poly (3, 4-ethylenedioxythiophene)-iridium oxide nanocomposite based Tyrosinase biosensor for the dual detection of catechol and azinphos methyl. Sens Actuators B Chem 316:128121
161. Khan M, Ahommed M, Daizy M (2020) Detection of xanthine in food samples with an electrochemical biosensor based on PEDOT: PSS and functionalized gold nanoparticles. RSC Adv 10(59):36147–36154
162. Azad T, Ahmed S (2016) Common milk adulteration and their detection techniques. Int J Food Contam 3(1):1–9
163. Albaaj A, Foucras G, Raboisson D (2017) Changes in milk urea around insemination are negatively associated with conception success in dairy cows. J Dairy Sci 100(4):3257–3265
164. Peng Y-H, Wu T, Lin Y-W (2020) Detection of melamine based on the suppressed anodic response of uric acid by a Au–Ag nanoparticles modified glassy carbon electrode. J Food Drug Anal 28(3):475
165. Khan NA et al (2019) Applications of nanotechnology in water and wastewater treatment: a review. Asian J Water Environ Pollut 16(4):81–86
166. Adeola AO, Nomngongo PN (2022) Advanced polymeric nanocomposites for water treatment applications: a holistic perspective. Polymers 14(12):2462
167. Niu Y et al (2014) Synthesis of silica gel supported salicylaldehyde modified PAMAM dendrimers for the effective removal of Hg (II) from aqueous solution. J Hazard Mater 278:267–278
168. Sajid M et al (2018) Removal of heavy metals and organic pollutants from water using dendritic polymers based adsorbents: a critical review. Sep Purif Technol 191:400–423
169. Wang Q et al (2022) A review: adsorption and removal of heavy metals based on polyamide-amines composites. Front Chem 10:814643
170. Munyeza CF et al (2020) Polycyclic aromatic hydrocarbon gaseous emissions from household cooking devices: a Kenyan case study. Environ Toxicol Chem 39(3):538–547
171. Hardonnière K et al (2016) The environmental carcinogen benzo [a] pyrene induces a Warburg-like metabolic reprogramming dependent on NHE1 and associated with cell survival. Sci Rep 6(1):30776
172. Arkas M et al (2005) Alkylated hyperbranched polymers as molecular nanosponges for the purification of water from polycyclic aromatic hydrocarbons. J Appl Polym Sci 97(6):2299–2305
173. Huo Z et al (2021) A carbonized polymer dot (CPD) nanosensor for trace water detection with a wide detection range. Dyes Pigm 196:109805
174. Dunning SG et al (2017) A sensor for trace H_2O detection in D_2O. Chem 2(4):579–589

175. Tao S et al (2017) The polymeric characteristics and photoluminescence mechanism in polymer carbon dots: a review. Mater Today Chem 6:13–25
176. Kudela RM et al (2015) The monitoring of harmful algal blooms through ocean observing: the development of the California Harmful Algal Bloom Monitoring and Alert Program. Coastal Ocean observing systems. Elsevier, pp 58–75
177. Hallegraeff G (2003) Harmful algal blooms: a global overview. Man Harmful Mar Microalgae 33:1–22
178. Brown AR et al (2020) Assessing risks and mitigating impacts of harmful algal blooms on mariculture and marine fisheries. Rev Aquac 12(3):1663–1688
179. Schmale DG III et al (2019) Perspectives on harmful algal blooms (HABs) and the cyberbiosecurity of freshwater systems. Front Bioeng Biotechnol 7:128
180. Powers C, Hanlon R, Schmale DG III (2018) Tracking of a fluorescent dye in a freshwater lake with an unmanned surface vehicle and an unmanned aircraft system. Remote Sens 10(1):81
181. Lian Z et al (2021) Dual-emission ratiometric fluorescent sensor based molecularly imprinted nanoparticles for visual detection of okadaic acid in seawater and sediment. Sens Actuators B Chem 346:130465
182. Mbisana M et al (2024) Molecularly imprinted polymers for the selective recognition of microcystins: an African perspective. Chem Africa 7(1):13–33
183. Fernando PAI et al (2022) Analytical methods incorporating molecularly imprinted polymers (MIPs) for the quantification of microcystins: a mini-review. Crit Rev Anal Chem 52(6):1244–1258
184. Cengiz N et al (2022) Application of molecularly imprinted polymers for the detection of volatile and off-odor compounds in food matrices. ACS Omega 7(18):15258–15266
185. Verma R, Gupta BD (2015) Detection of heavy metal ions in contaminated water by surface plasmon resonance based optical fibre sensor using conducting polymer and chitosan. Food Chem 166:568–575
186. Deshmukh MA et al (2018) Nanocomposite platform based on EDTA modified Ppy/SWNTs for the sensing of Pb (II) ions by electrochemical method. Front Chem 6:451
187. Ghosh S et al (2022) Lead (Pb^{2+}) ion sensor development using optical fiber gratings and nanocomposite materials. Sens Actuators B Chem 364:131818
188. Sarlak Z et al (2023) Pb exposure from plant foods in Iran: a review. Int J Environ Anal Chem 103(19):7395–7416
189. Wang Y et al (2018) Recyclable multifunctional magnetic mesoporous silica nanocomposite for ratiometric detection, rapid adsorption, and efficient removal of Hg (II). ACS Sustain Chem Eng 6(2):1744–1752
190. Pi K, Liu J, Van Cappellen P (2020) A DNA-based biosensor for aqueous Hg (II): performance under variable pH, temperature and competing ligand composition. J Hazard Mater 385:121572
191. Brewer GJ (2012) Copper toxicity in Alzheimer's disease: cognitive loss from ingestion of inorganic copper. J Trace Elem Med Biol 26(2–3):89–92
192. Ray RR (2016) Adverse hematological effects of hexavalent chromium: an overview. Interdiscip Toxicol 9(2):55
193. Natasha N et al (2022) Zinc in soil-plant-human system: a data-analysis review. Sci Total Environ 808:152024
194. Bhattacharya T, Sengupta B, Chopra H (2023) Polymeric composite sensors for food packaging applications. Polymeric nanocomposite materials for sensor applications. Elsevier, pp 459–478
195. Mohammadpour Z, Naghib SM (2021) Smart nanosensors for intelligent packaging. Nanosensors for smart manufacturing. Elsevier, pp 323–346
196. Vergara A et al (2007) An RFID reader with onboard sensing capability for monitoring fruit quality. Sens Actuators B Chem 127(1):143–149
197. Ileš D, Martinović G, Kozak D (2011) Review of potential use, benefits and risks of nanosensors and nanotechnologies in food. Strojarstvo: časopis za teoriju i praksu u strojarstvu 53(2):127–136

198. Shankar S et al (2014) Effect of reducing agent concentrations and temperature on characteristics and antimicrobial activity of silver nanoparticles. Mater Lett 137:160–163
199. Rhim J, Wang L, Hong S (2013) Preparation and characterization of agar/silver nanoparticles composite films with antimicrobial activity. Food Hydrocoll 33(2):327–335
200. Cioffi N et al (2005) Copper nanoparticle/polymer composites with antifungal and bacteriostatic properties. Chem Mater 17(21):5255–5262
201. Shankar S, Rhim J-W (2014) Effect of copper salts and reducing agents on characteristics and antimicrobial activity of copper nanoparticles. Mater Lett 132:307–311
202. Hostynek JJ, Maibach HI (2004) Copper hypersensitivity: dermatologic aspects. Dermatol Ther 17(4):328–333
203. Kanmani P, Rhim J-W (2014) Physical, mechanical and antimicrobial properties of gelatin based active nanocomposite films containing AgNPs and nanoclay. Food Hydrocoll 35:644–652
204. Emamifar A et al (2011) Effect of nanocomposite packaging containing Ag and ZnO on inactivation of Lactobacillus plantarum in orange juice. Food Control 22(3–4):408–413
205. Zhang L et al (2010) Mechanistic investigation into antibacterial behaviour of suspensions of ZnO nanoparticles against E. coli. J Nanopart Res 12:1625–1636
206. Rhim J-W et al (2006) Preparation and characterization of chitosan-based nanocomposite films with antimicrobial activity. J Agric Food Chem 54(16):5814–5822
207. Wang X et al (2006) Preparation, characterization and antimicrobial activity of chitosan/layered silicate nanocomposites. Polymer 47(19):6738–6744
208. Hong S-I, Rhim J-W (2008) Antimicrobial activity of organically modified nano-clays. J Nanosci Nanotechnol 8(11):5818–5824
209. Singh A et al (2021) Recent advances in electrochemical biosensors: applications, challenges, and future scope. Biosensors 11(9):336
210. Jagtiani E (2022) Advancements in nanotechnology for food science and industry. Food Front 3(1):56–82
211. Robertson GL (2005) Food packaging: principles and practice. CRC Press
212. Abbaspour A et al (2015) Aptamer-conjugated silver nanoparticles for electrochemical dual-aptamer-based sandwich detection of staphylococcus aureus. Biosens Bioelectron 68:149–155
213. Ghanbarzadeh B, Oleyaei SA, Almasi H (2015) Nanostructured materials utilized in biopolymer-based plastics for food packaging applications. Crit Rev Food Sci Nutr 55(12):1699–1723
214. Abdollahi M, Rezaei M, Farzi G (2012) A novel active bionanocomposite film incorporating rosemary essential oil and nanoclay into chitosan. J Food Eng 111(2):343–350
215. Lan T, Pinnavaia TJ (1994) Clay-reinforced epoxy nanocomposites. Chem Mater 6(12):2216–2219
216. Klaine SJ et al (2012) Paradigms to assess the environmental impact of manufactured nanomaterials. Citeseer 3–14
217. Klaine SJ (2009) Considerations for research on the environmental fate and effects of nanoparticles. Environ Toxicol Chem 28(9):1787
218. Singh B et al (2022) Electrochemical sensing and photocatalytic degradation of 2, 4-dinitrophenol via bismuth (III) oxide nanowires. J Mol Struct 1255:132379
219. Cheng H et al (2022) Recent advances in the application of nanotechnology to create antioxidant active food packaging materials. Crit Rev Food Sci Nutr 1–16
220. Rashidi L, Khosravi-Darani K (2011) The applications of nanotechnology in food industry. Crit Rev Food Sci Nutr 51(8):723–730
221. Iavicoli I et al (2017) Nanotechnology in agriculture: opportunities, toxicological implications, and occupational risks. Toxicol Appl Pharmacol 329:96–111
222. Amini SM, Gilaki M, Karchani M (2014) Safety of nanotechnology in food industries. Electron Phys 6(4):962
223. Evans SJ et al (2017) Critical review of the current and future challenges associated with advanced in vitro systems towards the study of nanoparticle (secondary) genotoxicity. Mutagenesis 32(1):233–241

224. Honarvar Z, Hadian Z, Mashayekh M (2016) Nanocomposites in food packaging applications and their risk assessment for health. Electron Phys 8(6):2531
225. Huang J-Y, Li X, Zhou W (2015) Safety assessment of nanocomposite for food packaging application. Trends Food Sci Technol 45(2):187–199
226. Onyeaka H et al (2022) The safety of nanomaterials in food production and packaging. Curr Res Food Sci 5:763–774
227. Bott J, Franz R (2018) Investigation into the potential migration of nanoparticles from laponite-polymer nanocomposites. Nanomaterials 8(9):723
228. Paidari S et al (2021) Migration of various nanoparticles into food samples: a review. Foods 10(9):2114
229. Hannon JC et al (2015) Advances and challenges for the use of engineered nanoparticles in food contact materials. Trends Food Sci Technol 43(1):43–62
230. Dunne L, Ashdown S, McDonald E (2002) Smart systems: wearable integration of intelligent technology. In: International centre of excellence for wearable electronics and smart fashion products (ICEWES) conference
231. Shah MA et al (2022) Applications of nanotechnology in smart textile industry: a critical review. J Adv Res 38:55–75
232. Pereira C et al (2020) Nanoengineered textiles: from advanced functional nanomaterials to groundbreaking high-performance clothing. Handbook of functionalized nanomaterials for industrial applications. Elsevier, pp 611–714
233. Ahmadi Z (2019) Epoxy in nanotechnology: a short review. Prog Org Coat 132:445–448
234. Fateixa S et al (2019) Gold loaded textile fibres as substrates for SERS detection. J Mol Struct 1185:333–340
235. Zhou Y, Tang R-C (2017) Natural flavonoid-functionalized silk fiber presenting antibacterial, antioxidant, and UV protection performance. ACS Sustain Chem Eng 5(11):10518–10526
236. Khan MR et al (2020) Tunable color coating of e-textiles by atomic layer deposition of multilayer TiO_2/Al_2O_3 films. Langmuir 36(11):2794–2801
237. Chen J et al (2021) Nature-inspired hierarchical protrusion structure construction for washable and wear-resistant superhydrophobic textiles with self-cleaning ability. ACS Appl Mater Interfaces 13(15):18142–18151
238. Butola B, Verma D (2019) Facile synthesis of chitosan-silver nanoparticles onto linen for antibacterial activity and free-radical scavenging textiles. Int J Biol Macromol 133:1134–1141
239. Lim T, Kim S, Oh K (2015) Fabrication of organic materials for electronic textiles. In: Handbook of smart textiles, pp 739–773
240. Sriramulu M, Sumathi S (2018) Biosynthesis of palladium nanoparticles using Saccharomyces cerevisiae extract and its photocatalytic degradation behaviour. Adv Nat Sci: Nanosci Nanotechnol 9(2):025018
241. Rezaie AB, Montazer M (2019) In situ incorporation and loading of copper nanoparticles into a palmitic–lauric phase-change material on polyester fibers. J Appl Polym Sci 136(3):46951
242. Sriram S, Kumar A (2019) Separation of oil-water via porous PMMA/SiO_2 nanoparticles superhydrophobic surface. Colloids Surf A 563:271–279
243. Vasantharaj S et al (2019) Synthesis of ecofriendly copper oxide nanoparticles for fabrication over textile fabrics: characterization of antibacterial activity and dye degradation potential. J Photochem Photobiol B 191:143–149
244. Wang M et al (2019) Fabrication of highly durable polysiloxane-zinc oxide (ZnO) coated polyethylene terephthalate (PET) fabric with improved ultraviolet resistance, hydrophobicity, and thermal resistance. J Colloid Interface Sci 537:91–100
245. Kubley A et al (2019) Smart textiles and wearable technology innovation with carbon nanotube technology. Nanotube superfiber materials. Elsevier, pp 263–311
246. Lu Y et al (2014) Characterization and cytotoxicity study of nanofibrous mats incorporating rectorite and carbon nanotubes. RSC Adv 4(63):33355–33361
247. Ferreira V, Goddard A, Monteiro O (2018) In situ synthesis and modification of cotton fibers with bismuthoxychloride and titanium dioxide nanoparticles for photocatalytic applications. J Photochem Photobiol A 357:201–212

248. Pirzada BM et al (2015) Synthesis, characterization and optimization of photocatalytic activity of TiO_2/ZrO_2 nanocomposite heterostructures. Mater Sci Eng B 193:137–145
249. Yan K et al (2013) Coding for hydrogel organization through signal guided self-assembly. Soft Matter 10(3):465–469
250. Hu J et al (2019) Surface-modified graphene oxide-based cotton fabric by ion implantation for enhancing antibacterial activity. ACS Sustain Chem Eng 7(8):7686–7692
251. Li Y et al (2020) Electrospun Ag-doped SnO_2 hollow nanofibers with high antibacterial activity. Electron Mater Lett 16:195–206
252. Butola B (2019) Recent advances in chitosan polysaccharide and its derivatives in antimicrobial modification of textile materials. Int J Biol Macromol 121:905–912
253. Tian H et al (2017) Durable antibacterial cotton fabrics containing stable acyclic N-halamine groups. Ind Eng Chem Res 56(28):7902–7909
254. Xu Q et al (2019) One-pot fabrication of durable antibacterial cotton fabric coated with silver nanoparticles via carboxymethyl chitosan as a binder and stabilizer. Carbohyd Polym 204:42–49
255. Ren J et al (2011) Crystallography facet-dependent antibacterial activity: the case of Cu_2O. Ind Eng Chem Res 50(17):10366–10369
256. Kostic MM et al (2014) Preparation and characterization of silver-loaded hemp fibers with antimicrobial activity. Fibers Polym 15:57–64
257. Yeo SY, Lee HJ, Jeong SH (2003) Preparation of nanocomposite fibers for permanent antibacterial effect. J Mater Sci 38:2143–2147
258. Rehman S et al (2019) Biocompatible tin oxide nanoparticles: synthesis, antibacterial, anticandidal and cytotoxic activities. ChemistrySelect 4(14):4013–4017
259. Naikoo GA et al (2021) Bioinspired and green synthesis of nanoparticles from plant extracts with antiviral and antimicrobial properties: a critical review. J Saudi Chem Soc 25(9):101304
260. Sun D, Siddiqui MOR, Iqbal K (2019) Specialty testing techniques for smart textiles. Smart textile coatings and laminates. Elsevier, pp 99–116
261. Muñoz-Bonilla A, Fernández-García M (2015) The roadmap of antimicrobial polymeric materials in macromolecular nanotechnology. Eur Polymer J 65:46–62
262. Riaz S et al (2019) Fabrication of robust multifaceted textiles by application of functionalized TiO_2 nanoparticles. Colloids Surf A 581:123799
263. Talebian S et al (2020) Nanotechnology-based disinfectants and sensors for SARS-CoV-2. Nat Nanotechnol 15(8):618–621
264. Lustig SR et al (2020) Effectiveness of common fabrics to block aqueous aerosols of virus-like nanoparticles. ACS Nano 14(6):7651–7658
265. Yang H, Zhu S, Pan N (2004) Studying the mechanisms of titanium dioxide as ultraviolet-blocking additive for films and fabrics by an improved scheme. J Appl Polym Sci 92(5):3201–3210
266. Daoud WA, Xin JH (2004) Low temperature sol-gel processed photocatalytic titania coating. J Sol-Gel Sci Technol 29:25–29
267. Dhineshbabu NR, Bose S (2019) UV resistant and fire retardant properties in fabrics coated with polymer based nanocomposites derived from sustainable and natural resources for protective clothing application. Compos B Eng 172:555–563
268. Anbarasan R et al (1999) Chemical grafting of aniline and o-toluidine onto poly (ethylene terephthalate) fiber. J Appl Polym Sci 73(1):121–128
269. Wang D et al (2004) Polyacrylonitrile fibers modified by nano-antimony-doped tin oxide particles. Text Res J 74(12):1060–1065
270. Shim BS et al (2008) Smart electronic yarns and wearable fabrics for human biomonitoring made by carbon nanotube coating with polyelectrolytes. Nano Lett 8(12):4151–4157
271. Mattana G et al (2011) Organic electronics on natural cotton fibres. Org Electron 12(12):2033–2039
272. Kim H, Ahn J-H (2017) Graphene for flexible and wearable device applications. Carbon 120:244–257

273. Suzuki K et al (2016) Rapid-response, widely stretchable sensor of aligned MWCNT/elastomer composites for human motion detection. ACS Sens 1(6):817–825
274. Naikoo GA et al (2021) Nanomaterials-based sensors for respiratory viral detection: a review. IEEE Sens J 21(16):17643–17656
275. Wang C et al (2016) Carbonized silk fabric for ultrastretchable, highly sensitive, and wearable strain sensors. Adv Mater 28(31):6640–6648
276. Wang C et al (2017) Carbonized silk georgette as an ultrasensitive wearable strain sensor for full-range human activity monitoring. J Mater Chem C 5(30):7604–7611
277. Zahid M et al (2017) Strain-responsive mercerized conductive cotton fabrics based on PEDOT: PSS/graphene. Mater Des 135:213–222
278. Gorgutsa S, Gu JF, Skorobogatiy M (2011) A woven 2D touchpad sensor and a 1D slide sensor using soft capacitor fibers. Smart Mater Struct 21(1):015010
279. Takamatsu S et al (2012) Fabric pressure sensor array fabricated with die-coating and weaving techniques. Sens Actuators A 184:57–63
280. Cherenack K et al (2010) Woven electronic fibers with sensing and display functions for smart textiles. Adv Mater 45(22):5178–5182
281. Adhikari B, Majumdar S (2004) Polymers in sensor applications. Prog Polym Sci 29(7):699–766
282. Bakhshpour M et al (2019) Quartz crystal microbalance biosensor for label-free MDA MB 231 cancer cell detection via notch-4 receptor. Talanta 204:840–845
283. Blum AS et al (2004) Cowpea mosaic virus as a scaffold for 3-D patterning of gold nanoparticles. Nano Lett 4(5):867–870
284. Omanović-Mikličanin E, Maksimović M, Vujović V (2015) The future of healthcare: nanomedicine and internet of nano things. Folia Medica Facultatis Medicinae Universitatis Saraeviensis 50(1):23–28
285. Nasrollahzadeh M et al (2019) Applications of nanotechnology in daily life. Interface Sci Technol 28:113–143
286. Dabhi K, Maheta A, Student A (2017) Internet of nano things-the next big thing. Int J Eng Sci Comput 7(4):10602–10604
287. Akyildiz IF, Jornet JM (2010) The internet of nano-things. IEEE Wirel Commun 17(6):58–63
288. Corchado JM et al (2009) Using heterogeneous wireless sensor networks in a telemonitoring system for healthcare. IEEE Trans Inf Technol Biomed 14(2):234–240
289. Gungor VC, Hancke GP (2009) Industrial wireless sensor networks: challenges, design principles, and technical approaches. IEEE Trans Industr Electron 56(10):4258–4265
290. López JA et al (2009) Development of a sensor node for precision horticulture. Sensors 9(5):3240–3255
291. Werner-Allen G et al (2006) Deploying a wireless sensor network on an active volcano. IEEE Internet Comput 10(2):18–25
292. Szewczyk R et al (2004) Habitat monitoring with sensor networks. Commun ACM 47(6):34–40
293. Lee D (2008) Wireless and powerless sensing node system developed for monitoring motors. Sensors 8(8):5005–5022
294. Cai C-J et al (2010) Characteristics of ambient volatile organic compounds (VOCs) measured in Shanghai, China. Sensors 10(8):7843–7862
295. Völgyesi P et al (2008) Air quality monitoring with SensorMap. In: 2008 international conference on information processing in sensor networks (IPSN 2008). IEEE
296. Gardner J (1988) Pattern recognition in the Warwick electronic nose. In: 8th international biennial congress of the European chemoreception research organisation
297. Wen W et al (2007) Enhanced sensitivity of SAW gas sensor coated molecularly imprinted polymer incorporating high frequency stability oscillator. Sens Actuators B Chem 125(2):422–427
298. Dermody DL, Crooks RM, Kim T (1996) Interactions between organized, surface-confined monolayers and vapor-phase probe molecules. 11. Synthesis, characterization, and chemical sensitivity of self-assembled polydiacetylene/calix [n] arene bilayers. J Am Chem Soc 118(47):11912–11917

299. Al-Fuqaha A et al (2015) Internet of things: a survey on enabling technologies, protocols, and applications. IEEE Commun Surv Tutor 17(4):2347–2376
300. Pawlick J, Chen J, Zhu Q (2018) iSTRICT: an interdependent strategic trust mechanism for the cloud-enabled internet of controlled things. IEEE Trans Inf Forensics Secur 14(6):1654–1669
301. Ometov A et al (2016) Facilitating the delegation of use for private devices in the era of the internet of wearable things. IEEE Internet Things J 4(4):843–854
302. Mohammadi M et al (2017) Semisupervised deep reinforcement learning in support of IoT and smart city services. IEEE Internet Things J 5(2):624–635
303. Ahmed N, De D, Hussain I (2018) Internet of Things (IoT) for smart precision agriculture and farming in rural areas. IEEE Internet Things J 5(6):4890–4899
304. Song T et al (2017) A privacy preserving communication protocol for IoT applications in smart homes. IEEE Internet Things J 4(6):1844–1852
305. Lu N et al (2014) Connected vehicles: solutions and challenges. IEEE Internet Things J 1(4):289–299
306. Chen Y-J, Wang L-C (2018) Privacy protection for internet of drones: a network coding approach. IEEE Internet Things J 6(2):1719–1730
307. Baldovino RG, Valenzuela IC, Dadios EP (2018) Implementation of a low-power wireless sensor network for smart farm applications. In: 2018 IEEE 10th international conference on humanoid, nanotechnology, information technology, communication and control, environment and management (HNICEM). IEEE
308. Hu Z et al (2019) Application of non-orthogonal multiple access in wireless sensor networks for smart agriculture. IEEE Access 7:87582–87592
309. Ranganathan V et al (2022) Re-inventing the food supply chain with IoT: a data-driven solution to reduce food loss. IEEE Internet Things Mag 5(1):41–47
310. Sarker MNI et al (2019) Big data driven smart agriculture: pathway for sustainable development. In: 2019 2nd international conference on artificial intelligence and big data (ICAIBD). IEEE
311. Vikram N et al (2017) A low cost home automation system using Wi-Fi based wireless sensor network incorporating Internet of Things (IoT). In: 2017 IEEE 7th international advance computing conference (IACC). IEEE
312. Khriji S et al (2014) Precision irrigation based on wireless sensor network. IET Sci Meas Technol 8(3):98–106
313. Khriji S et al (2019) Energy-efficient routing algorithm based on localization and clustering techniques for agricultural applications. IEEE Aerosp Electron Syst Mag 34(3):56–66
314. Loizou K et al (2015) A low-cost capacitive sensor for water level monitoring in large-scale storage tanks. In: 2015 IEEE international conference on industrial technology (ICIT). IEEE
315. Wang S-W et al (2018) A continuous water-level sensor based on load cell and floating pipe. In: 2018 IEEE international conference on applied system invention (ICASI). IEEE
316. Sengupta D et al (2012) Sensing of hydrostatic pressure using FBG sensor for liquid level measurement. Microw Opt Technol Lett 54(7):1679–1683
317. Ramalingame R et al (2019) Flexible piezoresistive sensor matrix based on a carbon nanotube PDMS composite for dynamic pressure distribution measurement. J Sens Sens Syst 8(1):1–7

Carbon-Based Smart Sensors for Environmental Pollution Detection

Shoaib Mukhtar, Abid Ali, Sameer Aryan, Shumaila Shahid, Md Nadim Akhtar, Juan Santiago Hidalg, Mohammad Umar, A. M. El-Khawaga, and Mohd Shoeb Khan

Abstract This chapter studies the advancement and utilization of carbon-based smart sensors for detecting environmental pollutants, emphasizing its capacity to transform environmental monitoring and repair endeavours. Carbon-based materials are analysed for their distinctive characteristics, such as their substantial surface area, electrical conductivity, and chemical stability, which render them well-suited for sensing applications. The chapter explores several detection approaches that use these materials, demonstrating their capacity to identify a wide range of environmental

---

S. Mukhtar (✉)
Research Group of Environmental and Inorganic Photochemistry, Center for Natural Sciences, Faculty of Engineering, University of Pannonia, P.O. Box 1158, Veszprém H-8210, Hungary
e-mail: mukhtar.shoaib@phd.mk.uni-pannon.hu

A. Ali
Department of Applied Chemistry, ZHCET, Aligarh Muslim University, Aligarh, India

S. Aryan
Department of Chemistry, Khandelwal College of Management Science and Technology, Bareilly, Mahatma Jyotiba Phule Rohilkhand University, Bareilly, U.P., India

S. Shahid
Division of Plant Pathology, ICAR-Indian Agricultural Research Institute, New Delhi, India

Md N. Akhtar
Department of Chemistry, Aligarh Muslim University, Aligarh, U.P., India

J. S. Hidalg
Research and Development Center, University of Pannonia, Nagykanizsa, Hungary

M. Umar
Institute of Genetics and Biotechnology, Hungarian University of Agriculture and Life Sciences, Gödöllő, Hungary

A. M. El-Khawaga
Department of Basic Medical Sciences, Faculty of Medicine, Galala University, Galala City, Suez, Egypt

M. S. Khan
Department of Applied Sciences and Humanities, Faculty of Engineering and Technology, Jamia Millia Islamia, New Delhi, India

I. Uddin (ed.), *Smart Nanosensors*, Smart Nanomaterials Technology,
https://doi.org/10.1007/978-981-96-3878-9_5

contaminants, including heavy metals, organic pollutants and biological contaminants. The chapter discusses the economic and environmentally friendly potential of carbon-based sensors, emphasizing their importance in promoting sustainable environmental restoration. This chapter seeks to serve as a significant resource for academics, engineers, and policymakers interested in furthering the subject of environmental pollution detection by offering a complete overview.

Keywords Carbon-based sensors · Pollutant detection · Heavy metals · Biological contaminant · Environmental remediation

1 Introduction

One of the main worldwide issues influencing human sustainable development is environmental degradation. When it comes to controlling pollution in the environment, detecting pollutants is a crucial first step [1]. An ideal detection technique would be one that is very sensitive and accurate, as well as easy and efficient, adaptable, and practical, according to the need and development trend in detecting technology. There have been several attempts to employ different types of analysis, such as spectrophotometry [2], gas chromatography [3], high performance liquid chromatography [4], mass spectrometry [5] etc., to identify environmental pollutants. The detection accuracy is somewhat lacking in spectrophotometry, despite its simplicity. Although HPLC, GC, and MS provide accurate results, they are not practical for use in real-time because of their operational complexity. The sensor's intrinsic qualities, such as sensitivity, selectivity, simplicity of operation, and in situ detection, make it look like a potential alternative. Numerous sensing systems including chemical, biological, materials, electrical, and other interdisciplinary fields have been developed with the aim of real-time online in-situ detection, attracted by sensors' sensitivity, speed, and specific response. Sensing technology has advanced significantly throughout the last few decades [6]. Electrochemical sensors [7], fluorescent sensors [8], colorimetric methods [9], surface plasmon resonance (SPR) sensors [10], field-effect transistor (FET) sensors [11], etc., evolved one after the other, progressing from chemical sensors to biosensors [12], and from single detectors to chip techniques and high-throughput detection. The environmental pollution detection sensors are becoming better understood by the public due to attributes such as sensitivity, ease of operation, affordability, miniaturization, portability and etc., compared to other types of pollution detection (such as those used in medicine, food safety, etc.), environmental pollutant detection faces more dynamic, complicated, and unexpected situations.

In order to conduct very sensitive and selective quantitative and qualitative analyses, sensor approaches must meet the problem of creating a stable and dependable strategy with a specific level of environmental resistance [13]. Because of the massive size and volume of implementation in environmental monitoring, cost is another important consideration that may significantly impact sensor use. Environmental pollutant detection calls for a sensing technique that is both reasonable and inexpensive, and one that is both practicable and affordable. In order to fulfill a range of low/trace requirements and environmental safety warning demands, sensor research is driven by the goal of increasing sensitivity and accuracy [14]. Environmental pollution sensors are no exception to this rule. High throughput detection is also progressing in line with the trend of future sensing, as is automation, downsizing, etc. [15]. That is to say, it is essential to thoroughly assess the performance, efficiency, simplicity, and practicability of sensor technology while developing sensors for environmental pollutant detections [16].

Many academics have put a lot of work into developing different sensing techniques that are either directly or indirectly linked to environmental analysis, as it is an useful and efficient detection approach [17]. But, as far as we are aware, it has never provided a comprehensive chapter of the sensing techniques used for environmental contaminant detection, thus the future of the field is unclear. Thus, for the purpose of this study, we compiled and categorized all of the new and exciting sensing techniques that have emerged in the last few decades with an eye on their potential usefulness in the field of environmental pollutant detection. This chapter's overarching goal is to provide a cognitive system and framework for environmental monitoring sensing strategies, with the hope that readers will be better able to understand how to put these strategies to use and spot trends in the data.

2 Overview of Carbon-Based Smart Sensors

Carbon-based smart materials have gained significant traction in sensing along with various applications due to their distinct features, including high adsorption, wide surface area, better optical qualities, and cost-effectiveness in manufacture. The structural conformation and hybridization state of carbon govern its chemical, physical, and electrical properties in a variety of carbonaceous substances [18]. The layer of sp^2-connected carbon atoms, which makes up the majority of the newly formed carbon forms, is associated to three additional atoms of carbon in the same plane as well as a weakly delocalized pi electron cloud in the plane perpendicular to it. Because of their structure, these materials have exceptional electrical conductivity, enhanced charge transfer capabilities, and π-plasmon resonances. Carbon-based smart materials are available in a variety of configurations and possess distinctive qualities due to the atomic structure of carbon being unique. Carbon materials are classified into different categories based on their form, size, and dimensionality: zero-dimensional (carbon dots), one-dimensional (carbon nanotubes and carbon nanofibers), two-dimensional (graphene), and three-dimensional (carbon nanomaterials) [19].

In general, the physical and chemical characteristics of carbon nanostructures are determined by their structure and the interactions they have at the interface with surrounding bulk materials. Graphene, with its single-layered graphite structure, shows great promise for use in environmental cleanup processes. Graphene and nanotubes of carbon are commonly used in various sensing applications. These materials are ideal for detection applications due to their exceptional optical properties, rapid charge transfer reactions resulting from their high conductivity, heightened sensitivity thanks to their high surface-to-volume ratio, effortless functionalization with a range of biomolecules and polymer compounds to ensure selectivity in different conditions, chemical stability, and biocompatibility [20]. Excellent mechanical strength and chemical stability shown by carbon-based materials guarantee the long-term functioning of sensors under different environmental circumstances. Particularly in high temperature uses, sensors' great thermal conductivity and resistance to thermal deterioration help to provide their stability. Natural biocompatibility of carbon-based materials reduces the danger of biofouling and degradation by usage in biological sensing systems. Stopping oxidation or chemical degradation can help carbon materials with surface changes or protective coatings become even more stable.

2.1 Types of Carbon-Based Electrode for Sensing

2.1.1 Graphene

Graphene consists of a single layer of carbon atoms arranged in a two-dimensional hexagonal lattice, giving it a flat, planar sheet morphology. It is one of the thinnest materials available, typically 1–3 atomic layers thick, measured in nanometers, yet it can cover large surface areas. This structure grants graphene a very high surface area, ideal for applications requiring extensive interaction surfaces, such as sensors and catalysts [21]. Graphene is renowned for its extremely high tensile strength, surpassing that of steel, making it valuable for structural reinforcement in composite materials. Its exceptional thermal conductivity makes it an excellent choice for heat dissipation in electronic devices and other thermal management applications. Additionally, graphene's excellent electrical conductivity is utilized in flexible electronics, conductive inks, and advanced composites. These unique properties make graphene suitable for various applications, including flexible electronics, conductive composites, energy storage devices like batteries and supercapacitors, sensors, and as a reinforcing agent in composites [22].

2.1.2 Carbon Nanofibers

Carbon nanofibers (CNFs) consist of continuous or discontinuous carbon fibers, which can be straight, helical, or tangled, depending on their synthesis method. This versatile morphology allows CNFs to be tailored to specific applications requiring different mechanical and structural properties. Their diameters range from nanometers to micrometers, while their lengths can vary from micrometers to millimeters, providing a high aspect ratio beneficial for reinforcement purposes. The fibrous morphology of CNFs results in a high surface area, enhancing their effectiveness in applications like catalysis and filtration. CNFs exhibit high tensile strength and modulus, making them suitable for reinforcing materials in advanced composites and structural applications. They also possess high thermal conductivity, which is useful in thermal interface materials and heat sinks for efficient heat dissipation. Additionally, CNFs offer excellent electrical conductivity, making them suitable for use in conductive composites, electronic devices, and energy storage systems. Because of their high strength, conductivity, and surface area, CNFs are thus used in composite materials for thermal interface materials, electronic devices, filtration systems, and catalytic supports. These applications provide unique benefits for sensing applications [23].

2.1.3 Carbon Nanotubes

Carbon nanotubes (CNTs) are hollow cylindrical tubes of carbon atoms, which can be single walled (SWCNTs) or multi-walled (MWCNTs), with each type exhibiting different properties. Their tubular structure influences their mechanical strength and electrical conductivity, with single-walled or multi-walled variants affecting these attributes. CNTs vary in length from micrometers to millimeters, with diameters dependent on their wall structure, allowing for customization in various applications. The high surface area of CNTs, resulting from their tubular form, makes them suitable for applications requiring extensive surface interactions, such as in supercapacitors and batteries. CNTs are exceptionally strong and stiff, ideal for use in high-strength materials and nanocomposites. They also exhibit high thermal conductivity, beneficial for electronics and thermal management applications, including heat sinks and thermal interface materials. Their excellent electrical conductivity makes them valuable for nanoelectronics, sensors, conductive materials, and energy storage systems. Consequently, CNTs are utilized in a wide range of applications, including nanocomposites, conductive materials, sensors, electronic devices, energy storage systems, and thermal management solutions, owing to their superior mechanical, thermal, and electrical properties [24].

2.1.4 Carbon Aerogels

Carbon aerogels (CA) are mesoporous carbon materials characterized by an interconnected three-dimensional network, resulting in extremely low density and high surface area. Their highly open-porous nanostructure contributes to these properties, with most pores in the mesoporous range of 2–50 nm, enhancing their suitability for high-performance applications [25]. CA are renowned for their large specific surface area, making them ideal for catalysis, adsorption, and thermal insulation. They also possess excellent mechanical properties, with tensile strengths reaching up to 58.45 MPa, making them suitable for structural applications requiring both low density and high strength. CA exhibit high thermal conductivity, useful in thermal insulation, heat exchangers, and other heat-related applications. This high thermal conductivity often correlates with good electrical properties, making CA suitable for electrodes and conductive composites [26]. Consequently, CA are used in a variety of applications, including catalysis, adsorption, thermal insulation, energy storage, and as lightweight structural materials due to their high surface area, mechanical strength, and thermal conductivity [27].

Other carbon-derived materials encompass a diverse range of substances, including fullerenes, carbon black, carbon fibers, and carbon quantum dots, each with unique structures such as spherical, irregular, fibrous, or crystalline formations. Their morphology, dimensions, and surface area vary widely, influencing their effectiveness and applications. These materials exhibit a broad spectrum of mechanical properties, thermal conductivity, and electrical conductivity, ranging from high to low, depending on their specific composition and structure. This versatility makes them valuable in numerous fields, including energy storage, environmental remediation, electronics, and composite materials, where their distinct properties can be tailored to meet specific needs. Properties and application of carbon-based materials have been shown in Fig. 1 [28].

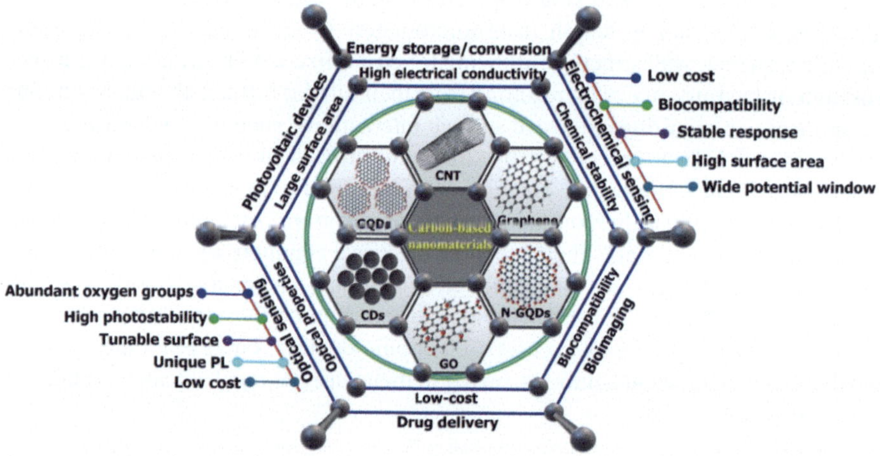

Fig. 1 Properties and application of carbon-based smart materials

3 Mechanisms of Pollutant Detection

3.1 Electrochemical Sensing

Electrochemical sensing is a method that makes use of the concepts of electrochemistry, the study of the connection between electricity and chemical processes, to identify and quantify chemical compounds via their electrochemical interactions. It includes elements like the analyte, an electrolyte, and working, counter, and reference electrodes. Redox processes, in which the analyte receives or loses electrons at the working electrode to produce an electrical current proportionate to the analyte concentration, are the basis of the approach [29]. Different kinds of electrochemical sensors assess variations in voltage, current, or conductivity. These types of sensors include potentiometric, amperometric, conductometric, and voltammetric. Applications include quality control in the food sector, chemical monitoring in industrial processes, environmental monitoring for pollutants, and medical diagnostics [30]. High sensitivity and specificity, quick detection, mobility, and cost-effectiveness are among the advantages of electrochemical sensing.

Dey et al. (2024) presented an effective electrochemical approach for detecting chlorpyrifos in vegetable and fruit samples, as well as a systematic hydrothermal synthesis process for the creation of Fe-PyDA MOFs on the surface of CNF. Important basic characteristics of the MOF anchored electrode include large surface area, plenty of active sites, a wide linear operating spectrum (1.0–150 μM), low detection threshold (15.1 nM), and increased sensitivity to chlorpyrifos. In addition to exhibiting even more remarkable electrochemical properties, this Fe-PyDA/CNF/GCE electrode's low detection threshold and wide linear detection range highlight its suitability for practical uses with respectable recoveries [31].

Cobalt ferrous oxides modified multiwalled carbon nanotube (CNT) fiber-based microelectrode were proposed by Nazir et al. (2024) and launched as a flexible and wearable uric acid and ascorbic acid sensor with outstanding electrocatalytic activity. A simple hydrothermal technique was used to successfully manufacture the $CoFe_2O_4$ microspheres on top of the surface of CNT fibers. Because of the interaction between the CNT fibers and $CoFe_2O_4$ microspheres, the prepared microelectrode demonstrated a strong electrochemical response. A flexible microelectrode sensor based on $CoFe_2O_4$@CNTs demonstrated good selectivity, a broad linear range, and a reduced detection limit. These results demonstrated the potential for using these kinds of microstructure sensors as disposable, effective, and reasonably priced microelectrodes for wearable sensing applications [32].

Gaber et al. in 2024 conducted a study on the electrochemical detection of catechol (CC) using a green carbon material (HZ) derived from waste hazelnut shells through the hydrothermal carbonization technique (HTC). They also incorporated SnO_2 nanoparticles (NPs) into the material. Catechol is a substance that is both carcinogenic to humans and a periodic environmental pollutant commonly found in food and beverages. It is known for its low degradability and high toxicity in the ecological environment. The suggested nanosensor's limit of detection (LOD) for

CC was found to be 8.51 nM, and its linearity range was determined to be 0.80–80 μM under ideal circumstances utilizing adsorptive stripping differential pulse voltammetry (AdsDPV). The newly created green approach has been used to accurately and precisely identify the presence of CC in green tea, fruit juice, and urine samples [33].

Mutić et al. (2024) have developed a quick and direct electrochemical method for measuring the amount of the anticonvulsant medicine carbamazepine (CBZ). The carbon paste electrode (CPE) was enhanced by including a two-dimensional conjugated polymer called graphitic carbon nitride (g-C_3N_4), along with an ionic liquid known as tetrabutylammonium chloride (TBACl). This modification resulted in the development of a highly sophisticated electrochemical sensor for CBZ. The sensor that was created was used to measure the presence of CBZ in medicines and human urine samples that were intentionally contaminated. The results showed that the sensor had a high level of accuracy and consistency, demonstrating its effectiveness in monitoring CBZ in various substances [34].

3.2 *Optical Sensing*

The optical detection and analysis of carbon-based materials, such as graphene and carbon nanotubes, relies on the use of light to examine their structural, chemical, and physical characteristics by means of interactions such as absorption, reflection, scattering, and transmission. Methods like Raman spectroscopy, UV-Vis spectroscopy, infrared spectroscopy, and photoluminescence are used to get comprehensive data on molecular vibrations, electronic transitions, and material integrity. This technology may be used for many purposes like as analyzing materials, monitoring the environment, sensing biological data, storing energy, and monitoring the health of structures. Optical sensing has many benefits, including the ability to analyze without causing damage, a high level of sensitivity, and immediate response. Optical sensing is a very effective and adaptable method for investigating and harnessing carbon-based materials.

Rutin, a naturally occurring flavone derivative, is well recognized as one of the most potent bioactive flavonoids and is sometimes referred to as vitamin P. It is found in 70 herbal medicines. Additionally, it serves as a therapeutic medication that effectively addresses various circulatory issues, such as lowering blood pressure and minimizing capillary permeability. Moreover, it demonstrates a broad range of physiological properties such as anti-inflammatory, antibacterial, and antioxidant activities. In addition, several healthy foods and other drinks often include rutin. Rutin's advantageous properties have led to its use in both medical applications and human dietary practices [35]. Sinduja et al. (2017) explain the process of creating carbon dots (CDs) by utilizing asparagine as a starting material. They then demonstrate the use of these CDs, which emit a bluish green light, for detecting rutin (optical sensing). The authenticity of the produced CDs was verified by UV-vis spectroscopy, FT-IR spectroscopy, XRD, and spectrofluorimetry. Upon the addition of 0.5 μM rutin to CDs, the emission intensity at 441 nm exhibited a drop. In addition, when the concentration of rutin increased, the emission intensity consistently dropped. The reduction in emission intensity at 441 nm was ascribed to the hydrogen bonding occurring between the hydroxyl and amine functional groups of CDs and rutin, respectively, as well as the π–π contact between the aromatic stacks. In addition, CDs exhibited specific rut determination in the presence of a 1000-fold surplus of Mg^{2+}, K^{+}, Ca^{2+}, Na^{+}, NO^{3-}, and oxalate, a 500-fold surplus of Cu^{2+} and Fe^{3+}, and a 600-fold surplus of glycine, glucose, histidine, proline, and cysteine. The detection limit was determined to be 1×10^{-7} M with a signal-to-noise ratio of three [36].

According to Ivan et al. (2017), researching graphene quantum dots (GQDs) is a compelling approach to creating innovative optical and electrochemical sensors for the quick and accurate identification of hazardous heavy metals (HMs), such Cd, Hg, and Pb. They demonstrated how, upon complexation with neutral and charged HM species, the solvent–solute interaction modifies the geometrical configuration, stability, and absorption spectra of zigzag/armchair-edged GQDs. Adatoms of Cd and Hg are seen to exhibit physisorption behavior on the sp^2 surface, exhibiting a solvent-mediated augmentation in the binding energy as solvent polarity increases. Additionally, they noticed that when the solvent's dielectric constant increases, the chemisorption strength in the HM cation–π system weakens as a result of the solvent effect. Therefore, achieving the appropriate binding energy in the HM@GQDs and enabling high sensitivity and selectivity of both optical and electrochemical sensors to harmful HMs may be accomplished by solvent engineering strategies based on modification of the dielectric permittivity [37].

4 Applications in Pollutant Detection

4.1 Heavy Metal Ions

The issue of pollution caused by heavy metal ions is progressively growing more important with each passing year. Arsenic, cadmium, chromium, lead, and mercury, which are heavy metal ions, possess significant importance in terms of public health owing to their considerable toxicity. These metal ions have the ability to spread throughout the body and are poisonous, causing harm to many organs even when exposed to low amounts. Hence, it is essential to devise novel methods that enable rapid, effortless, specific, dependable, and cost-effective detection of heavy metal ions. Table 1 presents a list of typical Heavy metal ions that might potentially cause health hazards. It also includes the symptoms that are linked to excessive exposure to these interfaces [38].

Optical sensors have been used in the detection of heavy metal ions by using different types of materials. The crucial factor in the detection of heavy metal ions is the sensitive material, which plays a significant role in enhancing sensor sensitivity,

Table 1 The hazard of different Heavy metal for the human system

HMI types	Hazard for excess	Refs.
Mercury (Hg^{2+})	Damage to the visual nerves of the brain	[39]
Chromium (Cr^{2+})	Numbness in the extremities, psychosis	[40]
Cadmium (Cd^{2+})	Hypertension, cardiovascular, cerebrovascular diseases, Destroys bone calcium, kidney dysfunction	[41]
Lead (Pb^{2+})	Congenital brain channel shallow, low intelligence, dementia, brain death	[42]
Cobalt (Co^{2+})	Radiation damage to the skin	[43]
Vanadium (V^{2+})	Hurts the heart and lungs, abnormal cholesterol metabolism	[44]
Antimony (Sb^{3+})	Strong damage to the skin	[45]
Thallium (TI^{+})	Polyneuritis	[46]
Manganese (Mn^{2+})	Hyperthyroidism	[47]
Stannum (Sn^{2+})	Freeze into a lump and cause death	[48]
Zinc (Zn^{2+})	Dry cough, headache, dizziness, high fever, chills	[49]

selectivity, limit of detection (LOD), and reaction time. Low-dimensional materials (LDM), nucleic acid (NA), and polymers have been extensively combined with optical techniques to create HMI sensors. The present progress of these sensors is summarized in Table 2.

Table 2 The comparison of materials used for Heavy metal detection

Material type	Target ion	Optical method	LOD	Refs.
LDM/NA-AuNP + DNA	Hg^{2+}/Pb^{2+}	FL	51 M/27 pM	[50]
LDM-Oxalic acid + AuNPs	Pb^{2+}	Absorbance	10.14 nM	[51]
Polymer-ppy/CS	Cd^{2+}	SPR	0.129 nM	[52]
Polymer-hydrogel	Cu^{2+}	MZI + FBG	3:00 pM	[53]
LDM/polymer-CS + GO + PAA	Pb^{2+}	LPG + FBG	0.5 nM	[54]
Polymer-BSA + CS	Hg^{2+}	Absorbance	0.499 nM	[55]
NA-DNA	Hg^{2+}/Pb^{2+}	FL	22 pM/20 nM	[56]
LDM/polymer-Polyaniline + CDs	As^{3+}	FL	1.3 pM	[57]

4.2 Organic Pollutants

Serious environmental contamination has come from worldwide population increase and industrial expansion over the last several decades. The environment has been dumped significant volumes of dangerous organic chemicals like antibiotics, pesticides, and phenolics. As the harmful compounds, organic pollutants will compromise human health and the surroundings worldwide. Studies on the relationship between long-term low-level exposure to organic chemicals and certain illnesses have become more frequent.

To detect 4-NP in wastewater specimens under varied matrices with sensitivity and selectivity, Chang et al. (2019) devised a novel and simple synthesis method. The sulfur-doped graphene quantum dots, or S-GQDs, had a consistent diameter spanning from 1 to 5 nm with a peak fluorescence around 450 nm. The quantum yield (QY) of the fluorescence emission is enhanced by S element doping. The π–π contact between 4-NP and the "large" graphitic carbon planar backbone of GQDs, which allows 4-NP to attenuate the fluorescence, served as the foundation for the response mechanism. In wastewater, the limit of detection (LOD) for GQDs was 3.5 nmol/L, whereas in purified water, it was 0.7 nmol/L. In the meanwhile, 4-NP may be found in different water matrices without influence from the matrix. This technique would provide a perceptive means of creating a GQD-based sensor with varying matrix effects for the detection of nitroarenes in solution [58]. In order to detect 4-NP using a post-synthetic modification technique, Zhang et al. presented a workable method of fixing amine-modified carbon dots (amine-CQDs) into the cavities of UiO-66 ($Zr_6O_4(OH)_4(bdc)_6$). Two factors improved the response performance in terms of sensitivity and selectivity: on the one hand, the UiO-66 composite was able to selectively enrich the targeted compounds by capturing large amounts of target molecules. Conversely, fluorescent amine-CQDs acting as functional monomers were successful in detecting the weak chemical force that formed between UiO-66 and the molecules under test. Consequently, a sensitive test for a trace quantity of 4-NP in the linear concentration range of 0.01–20.0 μmol/L was developed, with a relatively low detection limit of 3.5 nmol/L. This method used the amine-CQDs@UiO-66 fluorescent nano-probe to provide a quick and precise detection method. Its research of MOF-based composite (amine-CQDs@UiO-66 composite) suggests that it has a bright future in the study of dangerous substances [59]. Using a one-step hydrothermal process, Bogireddy et al. (2020) synthesized a kind of N doped CDs (NOCDs) with excitation wavelength independent fluorescence characteristics. Their fluorescent NOCDs exhibited excellent sensing capabilities, detecting hazardous dyes such as the bromophenol blue, methyl orange, rhodamine 6G, and methylene blue with a limit of detection (LOD) of 2.0 μmol/L, 38.3 nmol/L, 1.0 μmol/L, and 1.5 μmol/L, respectively. The fluorescence of nitrogen-doped carbon dots (NOCDs) has a significant correlation with both temperature and pH levels. Following incubation with NOCDs at concentrations ranging from 0.1 to 1000 μg/mL, the CDs demonstrated excellent biocompatibility. The exceptional sensitivity of the generated NOCDs offers exciting opportunities to use them as nanoprobes for fluorescently detecting and imaging hazardous organic dyes, both in laboratory research and practical sample detection applications [60].

Wu et al. (2017) synthesized chlorophyll-based carbon quantum dots (CQDs) that exhibit adjustable fluorescence properties. These CQDs were specifically designed to detect paraoxon. The synthesized CQDs exhibit a high degree of uniformity and have a spherical shape. The fluorescence emitted by CQDs may be significantly attenuated by the mechanism of Fluorescence Resonance Energy Transfer, when interacting with Au NPs. Through the effective and committed action of butyrylcholinesterase, acetylthiocholine undergoes hydrolysis, resulting in the production of thiocholine. This process might cause the aggregation of Au NPs, leading to the restoration of the initially quenched fluorescence in a Fluorescence Resonance Energy Transfer system. The OPs would permanently hinder the catalytic activity of butyrylcholinesterase and diminish the restorative impact. Paraoxon, out of all the other OPs, demonstrated the greatest percentage of inhibition. As a result, it was chosen as the representative of OPs with a limit of detection (LOD) of 0.05 μg/L. This sensing platform can detect concentrations of paraoxon ranging from 0.05 to 50 μg/L. The results of the sample test conducted in both river and tap water indicate that the fluorescence sensor exhibited high levels of repeatability and accuracy. This makes it a valuable tool for assessing food safety and environmental conditions [61]. Srinivasan et al. (2024) demonstrated a notable advancement by using a binder-free Zn-MOF@CNF composite as a modifier on a screen-printed electrode (SPE) to produce a detection method for tartrazine (TRZ) that is both extremely sensitive and selective. This composite, consisting of a distinctive rod-like structure for Zn-MOF and a coiled thread-like morphology for carbon nanofibers (CNF), exhibited a significant threefold augmentation in the sensing current of TRZ at a reduced anodic oxidation potential (0.92 V) in comparison to the unmodified SPE. The Zn-MOF@CNF/SPE demonstrated a decreased charge transfer resistance of 61.9 Ω, an increased electrochemical active surface area of 0.416 cm^2, and a high heterogeneous electron transfer rate constant (k_0) of 9.77×10^{-5} cm s^{-1}. In addition, it exhibited a wide range of measurement from 0.2 to 1000 μM, with a notable sensitivity of 0.136 $\mu A \mu M^{-1}$ cm^{-2} and an excellent detection limit of 54 nM. The Zn-MOF@CNF/SPE sensor exhibited exceptional selectivity in detecting TRZ among comparable compounds, along with remarkable repeatability, reproducibility, and stability. The practical effectiveness of Zn-MOF@CNF/SPE was confirmed by achieving recovery rates ranging from 97 to 100.9% in various food samples, including jelly, ice cream, soft drinks, and chocolates [62].

4.3 Biological Contaminants

A non-enzymatic glucose sensor using sugarcane bagasse-activated carbon-modified CPE was described by Malode et al. in 2024. The SBAC-modified CPE has shown significant electrocatalytic activity in the process of oxidizing glucose. The structural features of the SBAC were investigated using AFM, FTIR, SEM–EDS, XRD, and other characterization techniques. The sensor has effectively shown exceptional selectivity, consistency, replicability, and durability. Additionally, it has shown a significant level of sensitivity, measuring 0.858×10^{-8} M. The glucose measurement spanned a wide range, from 2.5×10^{-8} to 50×10^{-8} M. The SBAC/CPE was used to examine the electrooxidation of glucose in an alkaline solution containing halide ions (Cl^- and I^-). When glucose was subjected to halide-ion electrooxidation, the modified electrode exhibited enhanced electrocatalytic activity and produced more chemicals. Elevated levels of halide ions result in the oxidation of glucose, whereas low levels of halide ions cause contamination of the modified electrode by glucose oxidation. Halide ions increase the electrocatalytic activity of SBAC/CPE in an alkaline environment, causing a shift towards larger negative potential values during glucose electrooxidation compared to samples lacking halide ions [63].

Fatemi et al. (2024) reported that aptasensors based on carbon nanomaterials are used to quickly detect foodborne pathogenic bacteria. These biosensors, which combine aptamers with carbon nanomaterials, provide a dependable method for identifying pathogens in food safety. It has a high level of specificity and sensitivity in accurately detecting infections. The combination of aptamer and carbon nanomaterials technology shows great potential in several applications for detecting bacterial infections in food samples. Addressing the important challenges of stability and sensitivity is essential for advancing biosensor technology [64]. To quickly and on-site monitor *E. coli* O157:H7 in actual sample matrices, such as milk and water samples, Luo et al. (2017) developed a portable NMR biosensor. The pNMR biosensor used antibody-functionalized polymer-coated magnetic nanoparticles (MNP) to identify the bacteria, which facilitated the interaction between the antibody and the pathogen. The MNP accelerates the decay of NMR resonance signals. The biosensor has the capability to detect *E. coli* O157:H7 at concentrations as low as 76 CFU/mL in water samples and as low as 92 CFU/mL in milk samples. By altering the specificity of the antibodies, the biosensor may be expanded to detect various bacteria [65].

Carbon-based electrode materials for sensor application: a review was published by Kangmennaa et al. in 2024. These next-generation carbon-based sensors are very reliable and efficient, which will enhance industrial applications, environmental monitoring, and healthcare. It is anticipated that further advancements in synthesis and functionalization methods would result in a boom of electroanalytical applications in a variety of domains, such as medication quality monitoring, medical analysis, and guaranteeing food and environmental security. The uses of carbon-based materials as sensors are listed in Table 3 [66].

Table 3 Utilizing carbon-based materials for sensor applications

Carbon-based material	Detected material	Detection limit (nM)	References
AUNPs/AC/GCE	Ciprofloxacin	0.20	[67]
MIP/SA-AC/GCE	Morphine	48	[68]
Ni-Pd@AC/GCE NCs	Glucose	0.014	[69]
Gox-PtNPs-PAA-aSPCEs	Glucose	7.6	[70]
CVE	Salivary uric acid	0.05	[71]
TiO_2/ACE	Hydroquinone	3.3	[72]
Aptamer/graphene gold electrode	*S. aureus*	0.0041	[73]
Graphene FET	NH_3 and CO_2	0.3 and 0.4	[74]
Graphene/SnO_2 NP	H_2	0.1	[75]
Graphene/1-Octadecanethiol	Hg^b	0.1	[76]
GO/AnNPs	Pathogen	0.5	[77]
PPy/MWCNT	NH_3	0.3	[78]
PANI (SWCNT)	DCM	481	[79]
PPy (Co_3O_4@Au/MWCNT)	DMMP	10	[80]
(P(L-P) MCPS)	Indigo carmine (IC)	0.0187	[81]
(CeO_2/ZnO/GCE)	Norepinephrine	0.1	[82]
(PNMCPE)	Catechol (CC) and hydroquinone (HQ)	1.917 and 1.902 respectively	[83]

5 Ecofriendly Prospectives of Carbon-Based Smart Sensors for Environmental Pollutant Detection

Carbon-based smart sensors offer significant eco-friendly advantages in the detection of environmental pollutants, aligning with the global push for sustainable and green technologies. These sensors utilize materials like graphene, carbon nanotubes, and carbon quantum dots, which are derived from carbon, a naturally abundant and non-toxic element. This makes them inherently more environmentally friendly compared to traditional sensors that often rely on rare, hazardous, or resource-intensive materials. One of the key eco-friendly aspects is their ability to operate with low power consumption. Carbon-based sensors typically require less energy to function, which reduces their overall environmental footprint, particularly when deployed on a large scale. This is especially important in remote or mobile monitoring systems, where energy efficiency is crucial. Additionally, carbon-based sensors can contribute to waste reduction. Their high durability and long lifespan minimize the need for frequent replacements, leading to less electronic waste. Moreover, these materials can potentially be sourced from renewable or recycled carbon sources, further enhancing their sustainability. Furthermore, the use of carbon-based sensors in pollutant detection supports environmental protection efforts by enabling more precise and timely monitoring of pollutants. This contributes to quicker responses and better-informed decisions, ultimately helping to reduce environmental damage and promote a healthier ecosystem.

6 Economical Aspects of Carbon-Based Smart Sensors for Environmental Pollutant Detection

Carbon-based smart sensors have significant commercial potential for detecting environmental pollutants and have the capacity to revolutionize pollution management and monitoring. These sensors make use of carbon materials, which were formerly expensive but are now becoming more widely available because of improvements in production processes and rising demand. The long-term cost-effectiveness of carbon-based sensors provides a significant economic benefit. Because of their extreme durability and long-term stability, these materials don't need to be replaced as often. Additionally, by reducing the need for costly, large-scale equipment and laboratory studies, their great sensitivity and selectivity may reduce the total costs of environmental monitoring. The economic attraction of carbon-based sensors is further enhanced by their scalability. The cost per unit is anticipated to drop as manufacturing methods grow more efficient, opening up these sensors to a wider variety of sectors, including small firms and underdeveloped nations. Because of its scalability, environmental monitoring may become more widely used, which would lower its total cost.

7 Future Aspects Carbon Based Smart Sensors for Environmental Pollutant Detection

The outlook for carbon-based smart sensors in detecting environmental contaminants is positive, fueled by the pressing need for efficient and accurate monitoring systems. Carbon-based materials are now leading the way in terms of their outstanding characteristics. These materials possess remarkable qualities such as extensive surface areas, extraordinary electrical conductivity, and high chemical resistance, which make them very suitable for sensor applications. One important upcoming trend is the integration of carbon-based sensors with advanced technologies such as the Internet of Things (IoT) and artificial intelligence (AI). This integration has the potential to provide real-time, remote monitoring with enhanced precision, hence enabling faster reactions to environmental fluctuations. These sensors have a broad range of capabilities, since they can detect many pollutants, including gases like carbon dioxide and nitrogen oxides, as well as dangerous elements in water, such as heavy metals and organic compounds. A notable advancement is the fabrication of pliable, wearable sensors using carbon-based substances. These gadgets have the capability to continuously monitor an individual's exposure to contaminants, hence improving personal health and safety. With ongoing progress, these sensors are anticipated to grow to be more cost-effective and adaptable, rendering them indispensable for extensive environmental monitoring and worldwide pollution control initiatives.

8 Conclusion

Carbon-based smart sensors provide a revolutionary method for detecting environmental pollutants. They provide unmatched sensitivity, selectivity, and adaptability. Carbon-based materials have distinctive characteristics that enable the creation of sophisticated sensing systems capable of accurately detecting a diverse range of contaminants. In summary, carbon-based smart sensors are positioned to have a vital impact on the future of environmental monitoring and remediation. They provide a robust set of instruments to tackle the intricate issues of pollution management and environmental protection. This chapter emphasizes the need of ongoing research and development in this subject, since these technologies have the potential to greatly enhance our capacity to protect the environment for future generations.

Acknowledgements For Fig. 1 and Tables 1 and 2, copyright permissions have been provided.

Funding Not applicable.

Ethics Statements

Acceptance of Participation and Ethical Clearance Not applicable.

Availability of Information and Resources The corresponding author may provide you with the data used in this research upon request.

Competing Interests The writers affirm that they have no financial or other conflicts of interest.

References

1. Shiferaw N, Kim J, Seo D (2023) Identification of pollutant sources and evaluation of water quality improvement alternatives of a large river. Environ Sci Pollut Res 30(11). https://doi.org/10.1007/s11356-022-24431-5
2. Saxena R, Saxena M, Lochab A (2020) Recent progress in nanomaterials for adsorptive removal of organic contaminants from wastewater. https://doi.org/10.1002/slct.201903542
3. Di S et al (2020) Recent advances and applications of magnetic nanomaterials in environmental sample analysis. https://doi.org/10.1016/j.trac.2020.115864
4. Ajala OJ, Tijani JO, Salau RB, Abdulkareem AS, Aremu OS (2022) A review of emerging micro-pollutants in hospital wastewater: environmental fate and remediation options. https://doi.org/10.1016/j.rineng.2022.100671

5. Chun S, Muthu M, Gopal J (2022) Mass spectrometry as an analytical tool for detection of microplastics in the environment. https://doi.org/10.3390/chemosensors10120530
6. Sharma A, Singh A, Gupta V, Arya S (2022) Advancements and future prospects of wearable sensing technology for healthcare applications. https://doi.org/10.1039/d2sd00005a
7. Baranwal J, Barse B, Gatto G, Broncova G, Kumar A (2022) Electrochemical sensors and their applications: a review. https://doi.org/10.3390/chemosensors10090363
8. De Acha N, Elosúa C, Corres JM, Arregui FJ (2019) Fluorescent sensors for the detection of heavy metal ions in aqueous media. https://doi.org/10.3390/s19030599
9. Ryu WK, Kim HW, Kim GD, Rhee HI (2017) Rapid determination of capsaicinoids by colorimetric method. J Food Drug Anal 25(4). https://doi.org/10.1016/j.jfda.2016.11.007
10. Das S, Devireddy R, Gartia MR (2023) Surface plasmon resonance (SPR) sensor for cancer biomarker detection. https://doi.org/10.3390/bios13030396
11. Falina S et al (2021) Ten years progress of electrical detection of heavy metal ions (HMIS) using various field-effect transistor (FET) nanosensors: a review. https://doi.org/10.3390/bios11120478
12. Le PG, Choi SH, Cho S (2023) Alzheimer's disease biomarker detection using field effect transistor-based biosensor. https://doi.org/10.3390/bios13110987
13. Bharath G et al (2016) Enhanced electrocatalytic activity of gold nanoparticles on hydroxyapatite nanorods for sensitive hydrazine sensors. J Mater Chem A Mater 4(17). https://doi.org/10.1039/c6ta01528j
14. Adla S, Bruckmaier F, Arias-Rodriguez LF, Tripathi S, Pande S, Disse M (2024) Impact of calibrating a low-cost capacitance-based soil moisture sensor on AquaCrop model performance. J Environ Manage 353. https://doi.org/10.1016/j.jenvman.2024.120248
15. Mahmud MS, Zahid A, Das AK (2023) Sensing and automation technologies for ornamental nursery crop production: current status and future prospects. https://doi.org/10.3390/s23041818
16. Hussein SM, López Ramos JA, Ashir AM (2022) A secure and efficient method to protect communications and energy consumption in IoT wireless sensor networks. Electronics (Switzerland) 11(17). https://doi.org/10.3390/electronics11172721
17. Liu Q, Zhou Q, Jiang G (2014) Nanomaterials for analysis and monitoring of emerging chemical pollutants. https://doi.org/10.1016/j.trac.2014.02.014
18. Teng F et al (2021) Piezoresistive/piezoelectric intrinsic sensing properties of carbon nanotube cement-based smart composite and its electromechanical sensing mechanisms: a review. https://doi.org/10.1515/ntrev-2021-0112
19. Horváth O, Mukhtar S, Szabó-Bárdos E (2024) Sustainable synthesis of carbon-based nanocomposite for dye degradation. In: Uddin I (ed) Sustainable nanomaterials: synthesis and environmental applications. Springer Nature Singapore, Singapore, pp 81–101. https://doi.org/10.1007/978-981-97-2761-2_4
20. Islam S, Shaheen Shah S, Naher S, Ali Ehsan M, Aziz MA, Ahammad AJS (2021) Graphene and carbon nanotube-based electrochemical sensing platforms for dopamine. https://doi.org/10.1002/asia.202100898
21. Uddin I, Mukhtar S, Horváth O, Pósfai M (2024) Room-temperature synthesis of rGO/Ag nanocomposite as a photocatalyst for the degradation of organic pollutants in wastewater. Materialia (Oxf) 36:102179. https://doi.org/10.1016/j.mtla.2024.102179
22. Bressi AC, Dallinger A, Steksova Y, Greco F (2023) Bioderived laser-induced graphene for sensors and supercapacitors. https://doi.org/10.1021/acsami.3c07687
23. Sengupta D et al (2020) Single and bundled carbon nanofibers as ultralightweight and flexible piezoresistive sensors. npj Flex Electron 4(1). https://doi.org/10.1038/s41528-020-0072-2
24. Rais Saman S, Mukhtar (2024) Facile synthesis of carbon nanotubes and its application in food science and wastewater treatment. In: Uddin I (ed) Sustainable nanomaterials: synthesis and environmental applications. Springer Nature Singapore, Singapore, pp 301–316. https://doi.org/10.1007/978-981-97-2761-2_12
25. Weng B et al (2016) Hierarchical Nafion enhanced carbon aerogels for sensing applications. Nanoscale 8(6). https://doi.org/10.1039/c5nr08631k

26. Mahmoudpour M, Dolatabadi JEN, Hasanzadeh M, Soleymani J (2021) Carbon-based aerogels for biomedical sensing: advances toward designing the ideal sensor. https://doi.org/10.1016/j.cis.2021.102550
27. Chen Z et al (2020) Wood-derived lightweight and elastic carbon aerogel for pressure sensing and energy storage. Adv Funct Mater 30(17). https://doi.org/10.1002/adfm.201910292
28. Gowthaman NSK et al (2023) Zero-, one- and two-dimensional carbon nanomaterials as low-cost catalysts in optical and electrochemical sensing of biomolecules and environmental pollutants. https://doi.org/10.1016/j.microc.2023.109291
29. Tian H et al (2024) Electrochemical sensing fibers for wearable health monitoring devices. https://doi.org/10.1016/j.bios.2023.115890
30. Shahid S, El-Khawaga AM, Mukhtar S (2024) Sustainable nanomaterials as promising antibacterial agents. In: Uddin I (ed) Sustainable nanomaterials: synthesis and environmental applications. Springer Nature Singapore, Singapore, pp 203–225. https://doi.org/10.1007/978-981-97-2761-2_8
31. Dey B, Ahmad MW, Syed A, Bahkali AH, Verma M, Choudhury A (2024) Iron metal organic framework decorated carbon nanofiber-based electrode for electrochemical sensing platform of chlorpyrifos in fruits and vegetables. Mater Sci Semicond Process 181:108669. https://doi.org/10.1016/J.MSSP.2024.108669
32. Nazir A et al (2024) A flexible carbon nanotubes-based microelectrode for non-enzymatic electrochemical uric acid and ascorbic acid sensing. Mater Sci Eng B 307:117480. https://doi.org/10.1016/j.mseb.2024.117480
33. Gaber A, Bilge S, Osman Donar Y, Sınağ A (2024) Waste hazelnut shell based effective hydrothermal carbon/SnO_2 nanoparticles: towards electrochemical sensing of catechol in green tea, fruit juice, and urine samples. Microchem J 201:110545. https://doi.org/10.1016/j.microc.2024.110545
34. Mutić S, Đurđić S, Petrović S, Gemeiner P, Stanković D, Anojčić J (2024) Electrochemical sensing platform for anticonvulsant drug carbamazepine detection based on graphitic carbon nitride and tetrabutylammonium chloride ionic liquid. Electrochim Acta 500:144755. https://doi.org/10.1016/J.ELECTACTA.2024.144755
35. Middleton E, Drzewiecki G (1984) Flavonoid inhibition of human basophil histamine release stimulated by various agents. Biochem Pharmacol 33(21). https://doi.org/10.1016/0006-2952(84)90102-3
36. Sinduja B, Abraham John S (2018) Sensitive determination of rutin by spectrofluorimetry using carbon dots synthesized from a non-essential amino acid. Spectrochim Acta A Mol Biomol Spectrosc 193. https://doi.org/10.1016/j.saa.2017.12.067
37. Shtepliuk I, Khranovskyy V, Yakimova R (2017) Insights into the origin of the excited transitions in graphene quantum dots interacting with heavy metals in different media. Phys Chem Chem Phys 19(45). https://doi.org/10.1039/c7cp04711h
38. Fu R et al (2024) Optical fiber sensors for heavy metal ion sensing. https://doi.org/10.1016/j.jmst.2023.12.020
39. Ashrafi AM, Koudelkova Z, Sedlackova E, Richtera L, Adam V (2018) Review—electrochemical sensors and biosensors for determination of mercury ions. J Electrochem Soc 165(16). https://doi.org/10.1149/2.0381816jes
40. Manoj SR et al (2020) Understanding the molecular mechanisms for the enhanced phytoremediation of heavy metals through plant growth promoting rhizobacteria: a review. J Environ Manage 254. https://doi.org/10.1016/j.jenvman.2019.109779
41. Li Z et al (2022) FeOOH nanosheet assisted metal ion coordination with porphyrins for rapid detection and removal of cadmium ions in water. Anal Methods 14(47). https://doi.org/10.1039/d2ay01508k
42. Wu D, Hu Y, Cheng H, Ye X (2023) Detection techniques for lead ions in water: a review. https://doi.org/10.3390/molecules28083601
43. Vivas EL, Cho K (2021) Efficient adsorptive removal of Cobalt(II) ions from water by dicalcium phosphate dehydrate. J Environ Manage 283. https://doi.org/10.1016/j.jenvman.2021.111990

44. Lu Q et al (2024) A bioactive tannic acid/vanadium ions co-deposited coating on various surfaces for antifouling, photothermal anti-bacterial, and antioxidant effects. Prog Org Coat 186. https://doi.org/10.1016/j.porgcoat.2023.108010
45. Li J et al (2018) Antimony contamination, consequences and removal techniques: a review. https://doi.org/10.1016/j.ecoenv.2018.03.024
46. Jin X et al (2023) Evaluation of thallium ion as an effective ion in human health using an electrochemical sensor. Environ Res 238. https://doi.org/10.1016/j.envres.2023.117026
47. Omar NAS et al (2021) A sensing approach for manganese ion detection by carbon dots nanocomposite thin film-based surface plasmon resonance sensor. Optik (Stuttg) 243. https://doi.org/10.1016/j.ijleo.2021.167435
48. Naughton DP, Petróczi A (2009) Erratum: heavy metal ions in wines: meta-analysis of target hazard quotients reveal health risks (Chem Central J (2008) 2 (22)). https://doi.org/10.1186/1752-153X-3-6
49. Chasapis CT, Spiliopoulou CA, Loutsidou AC, Stefanidou ME (2012) Zinc and human health: an update. https://doi.org/10.1007/s00204-011-0775-1
50. Chung CH, Kim JH, Jung J, Chung BH (2013) Nuclease-resistant DNA aptamer on gold nanoparticles for the simultaneous detection of Pb^{2+} and Hg^{2+} in human serum. Biosens Bioelectron 41(1). https://doi.org/10.1016/j.bios.2012.10.026
51. Boruah BS, Biswas R (2018) Localized surface plasmon resonance based U-shaped optical fiber probe for the detection of Pb^{2+} in aqueous medium. Sens Actuators B Chem 276. https://doi.org/10.1016/j.snb.2018.08.086
52. Verma R, Gupta BD (2015) Detection of heavy metal ions in contaminated water by surface plasmon resonance based optical fibre sensor using conducting polymer and chitosan. Food Chem 166. https://doi.org/10.1016/j.foodchem.2014.06.045
53. Liu Z, Li G, Zhang A, Zhou G, Huang X (2021) Ultra-sensitive optical fiber sensor based on intermodal interference and temperature calibration for trace detection of copper (II) ions. Opt Express 29(15). https://doi.org/10.1364/oe.434687
54. Ghosh S, Dissanayake K, Asokan S, Sun T, Rahman BMA, Grattan KTV (2022) Lead (Pb^{2+}) ion sensor development using optical fiber gratings and nanocomposite materials. Sens Actuators B Chem 364. https://doi.org/10.1016/j.snb.2022.131818
55. Sadani K, Nag P, Mukherji S (2019) LSPR based optical fiber sensor with chitosan capped gold nanoparticles on BSA for trace detection of Hg(II) in water, soil and food samples. Biosens Bioelectron 134. https://doi.org/10.1016/j.bios.2019.03.046
56. Han S et al (2016) Practical, highly sensitive, and regenerable evanescent-wave biosensor for detection of Hg^{2+} and Pb^{2+} in water. Biosens Bioelectron 80. https://doi.org/10.1016/j.bios.2016.01.070
57. Saikia A, Karak N (2018) Polyaniline nanofiber/carbon dot nanohybrid as an efficient fluorimetric sensor for As(III) in water and effective antioxidant. Mater Today Commun 14. https://doi.org/10.1016/j.mtcomm.2017.12.020
58. Ngoc Anh NT, Chang PY, Doong RA (2019) Sulfur-doped graphene quantum dot-based paper sensor for highly sensitive and selective detection of 4-nitrophenol in contaminated water and wastewater. RSC Adv 9(46). https://doi.org/10.1039/c9ra04414k
59. Yang JM, Hu XW, Liu YX, Zhang W (2019) Fabrication of a carbon quantum dots-immobilized zirconium-based metal-organic framework composite fluorescence sensor for highly sensitive detection of 4-nitrophenol. Micropor Mesopor Mater 274. https://doi.org/10.1016/j.micromeso.2018.07.042
60. Bogireddy NKR, Lara J, Fragoso LR, Agarwal V (2020) One-step hydrothermal preparation of highly stable N doped oxidized carbon dots for toxic organic pollutants sensing and bioimaging. Chem Eng J 401. https://doi.org/10.1016/j.cej.2020.126097
61. Wu X et al (2017) Carbon quantum dots as fluorescence resonance energy transfer sensors for organophosphate pesticides determination. Biosens Bioelectron 94. https://doi.org/10.1016/j.bios.2017.03.010
62. Srinivasan P et al (2024) Nanomolar detection of organic pollutant-tartrazine in food samples using zinc-metal organic framework/carbon nano fiber-composite modified screen-printed electrode. J Food Compos Anal 133:106462. https://doi.org/10.1016/j.jfca.2024.106462

63. Malode SJ, Bilagi S, Shetti NP (2024) Synthesis of biowaste-derived nanostructured material for the nanomolar detection of glucose: biological sample analysis. J Pharmaceut Biomed Anal Open 3. https://doi.org/10.1016/j.jpbao.2024.100026
64. Fatemi K et al (2024) Carbon nanomaterial-based aptasensors for rapid detection of foodborne pathogenic bacteria. Anal Biochem 115639. https://doi.org/10.1016/j.ab.2024.115639
65. Luo Y, Alocilja EC (2017) Portable nuclear magnetic resonance biosensor and assay for a highly sensitive and rapid detection of foodborne bacteria in complex matrices. J Biol Eng 11(1). https://doi.org/10.1186/s13036-017-0053-8
66. Kangmennaa A, Forkuo RB, Agorku ES (2024) Carbon-based electrode materials for sensor application: a review. Sens Technol 2(1). https://doi.org/10.1080/28361466.2024.2350174
67. Gissawong N et al (2021) An electrochemical sensor for voltammetric detection of ciprofloxacin using a glassy carbon electrode modified with activated carbon, gold nanoparticles and supramolecular solvent. Microchimica Acta 188(6). https://doi.org/10.1007/s00604-021-04869-z
68. Salajegheh M, Kazemipour M, Foroghi MM, Ansari M (2019) Morphine sensing by a green modified molecularly imprinted poly L-lysine/sodium alginate-activated carbon/glassy carbon electrode based on computational design. Electroanalysis 31(3). https://doi.org/10.1002/elan.201800395
69. Koskun Y, Şavk A, Şen B, Şen F (2018) Highly sensitive glucose sensor based on monodisperse palladium nickel/activated carbon nanocomposites. Anal Chim Acta 1010. https://doi.org/10.1016/j.aca.2018.01.035
70. Jiménez-Fiérrez F, González-Sánchez MI, Jiménez-Pérez R, Iniesta J, Valero E (2020) Glucose biosensor based on disposable activated carbon electrodes modified with platinum nanoparticles electrodeposited on poly(Azure a). Sensors (Switzerland) 20(16). https://doi.org/10.3390/s20164489
71. Bukharinova MA, Stozhko NY, Novakovskaya EA, Khamzina EI, Tarasov AV, Sokolkov SV (2021) Developing activated carbon veil electrode for sensing salivary uric acid. Biosensors (Basel) 11(8). https://doi.org/10.3390/bios11080287
72. Hayat A et al (2016) An electrochemical sensor based on TiO_2/activated carbon nanocomposite modified screen printed electrode and its performance for phenolic compounds detection in water samples. Int J Environ Anal Chem 96(3). https://doi.org/10.1080/03067319.2015.1137910
73. Lian Y, He F, Wang H, Tong F (2015) A new aptamer/graphene interdigitated gold electrode piezoelectric sensor for rapid and specific detection of *Staphylococcus aureus*. Biosens Bioelectron 65. https://doi.org/10.1016/j.bios.2014.10.017
74. Inaba A, Yoo G, Takei Y, Matsumoto K, Shimoyama I (2013) A graphene FET gas sensor gated by ionic liquid. In: Proceedings of the IEEE international conference on micro electro mechanical systems (MEMS). https://doi.org/10.1109/MEMSYS.2013.6474408
75. Zhang C et al (2014) Synthesis of RuO_2 decorated quasi graphene nanosheets and their application in supercapacitors. RSC Adv 4(22). https://doi.org/10.1039/c3ra47641c
76. Chen F, Xu M, Wang L, Li J (2011) Preparation and characterization of organic aerogels from a lignin-resorcinol-formaldehyde copolymer. Bioresources 6(2). https://doi.org/10.15376/biores.6.2.1262-1272
77. Jung JH, Cheon DS, Liu F, Lee KB, Seo TS (2010) A graphene oxide based immuno-biosensor for pathogen detection. Angew Chem Int Ed 49(33). https://doi.org/10.1002/anie.201001428
78. Bachhav SG, Patil DR (2015) Study of polypyrrole-coated MWCNT nanocomposites for ammonia sensing at room temperature. J Mater Sci Chem Eng 03(10). https://doi.org/10.4236/msce.2015.310005
79. Chang CP, Yuan CL (2009) The fabrication of a MWNTs-polymer composite chemoresistive sensor array to discriminate between chemical toxic agents. J Mater Sci 44(20). https://doi.org/10.1007/s10853-009-3766-3
80. Ramesh S, Lee YJ, Msolli S, Kim JG, Kim HS, Kim JH (2017) Synthesis of a Co_3O_4@gold/MWCNT/polypyrrole hybrid composite for DMMP detection in chemical sensors. RSC Adv 7(80). https://doi.org/10.1039/c7ra09161c

81. Prinith NS, Manjunatha JG (2023) Voltammetric analysis of hazardous azo-dye indigo carmine at simple and cost-effective carbon paste sensor modified with L-phenylalanine. Sens Technol 1(1). https://doi.org/10.1080/28361466.2023.2218877
82. Sarbandian Z, Beitollahi H (2023) CeO_2/ZnO nanocomposite-modified glassy carbon electrode as an enhanced sensing platform for sensitive voltammetric determination of norepinephrine. Sens Technol 1(1). https://doi.org/10.1080/28361466.2023.2255132
83. Arpitha SB, Swamy BEK, Banu R (2023) Electrochemical studies of catechol and hydroquinone at poly(nigrosine) modified carbon paste electrode: a cyclic voltammetric study. Sens Technol 1(1). https://doi.org/10.1080/28361466.2023.2258789

Substrates for Colorimetric Sensors Array

Virendra Patil, Tejansh Chandole, and Tushar Borase

Abstract This chapter outlines colorimetric sensor arrays and their significance in identifying and measuring analytes by observing changes in colorimetric reactions. Further, it explores the characteristics of different substrates commonly employed in these arrays, such as paper, glass, polymer, nanoporous material substrates, and nanoparticle-packed surfaces. Each substrate is thoroughly examined including manufacturing techniques, surface alterations, and vital features affecting the sensors effectiveness. The chapter investigates the basic principles behind substrate selection refinement to improve the performance and versatility of colorimetric sensor arrays across various sensing tasks. Additionally, it surveys the diverse applications of these arrays in environmental monitoring, biomedical diagnostics, food safety, industrial monitoring, and other explored domains. Furthermore, it delves into ongoing trends, and advancements in substrate design and creation, addressing obstacles and prospects in the field.

Keywords Colorimetric sensor · Substrate · Colorimetric sensor array · Paper · Glass · Polypropylene · PVDF · Polyethylene terephthalate · Nanoporous silica · Nanoparticle packed surface

V. Patil (✉) · T. Borase (✉)
Swentek Nordic Private Limited, Bld-G1, Atal Incubation Centre, IISER Campus, Dr Homi Bhabha Road, Pashan, Pune 411008, Maharashtra, India
e-mail: virendra.patil@swenteknordic.com

T. Borase
e-mail: tushar.borase@swenteknordic.com

T. Chandole
Sanguine Transformation Innovatives LLP, Sinhgad Road, Pune 411051, Maharastra, India

I. Uddin (ed.), *Smart Nanosensors*, Smart Nanomaterials Technology,
https://doi.org/10.1007/978-981-96-3878-9_6

1 Introduction of Colorimetric Sensor and Colorimetric Sensor Array

Colorimetric sensors play a crucial role in chemical sensing by identifying and measuring target substances through visible color alterations. This straightforward and easily understandable detection method makes colorimetric sensors highly valuable in diverse fields like environmental monitoring [1], healthcare diagnostics [2], food safety [3], and industrial control [4]. Molecular recognition forms the core principle of colorimetric sensing, where specific interactions between the target substance and a receptor molecule prompt a physical or chemical change, leading to a detectable color shift. This alteration can occur through various mechanisms such as chemical reactions, pH level fluctuations, or molecular structure changes [5]. The simplicity and adaptability of colorimetric sensors render them suitable for quick, on-the-spot analysis, often eliminating the necessity for complex equipment. Furthermore, their potential for downsizing and integration into portable devices enhances their usefulness for point-of-care diagnostics and applications in the field [6]. Colorimetric sensor arrays offer a sophisticated method for chemical sensing by harnessing the collective reaction of numerous selective receptors to identify and differentiate a broad spectrum of substances [7] (Fig. 1).

Unlike single-component colorimetric sensors, arrays exploit varied interactions between substances and an assortment of chemically distinct receptors to create unique, fingerprint-like patterns, facilitating swift and precise identification of target compounds. At the heart of colorimetric sensor arrays lies multidimensional chemical sensing, where the amalgamation of multiple sensing elements enhances sensitivity, selectivity, and specificity [8]. Each receptor within the array selectively reacts to different substances, resulting in distinct alterations in color or spectral shifts. Through comprehensive analysis of the entire spectrum of color changes, either through visual observation or computational methods, the presence and concentration of various substances can be concurrently determined [9]. The development and fine-tuning of colorimetric sensor arrays necessitate careful selection of

Fig. 1 Schematic showing a colorimetric sensor array containing 25 chemically responsive dyes printed on a paper substrate

receptor molecules, optimization of sensor composition, and calibration against known analyte concentrations [10]. Furthermore, the utilization of data analysis techniques, such as chemometric methods and machine learning algorithms, is essential for deciphering intricate sensor responses and deriving significant insights [11]. Colorimetric sensor arrays have diverse applications spanning environmental monitoring, quality control in food and beverage industries, biomedical diagnostics, and homeland security. Their capability to swiftly detect and distinguish multiple analytes in intricate samples renders them indispensable for real-time monitoring, early disease diagnosis, and ensuring the safety and authenticity of products [12].

2 Chemo-responsive Dyes of Colorimetric Sensor Array

Commercially available chemo-responsive dyes offer diverse response mechanisms tailored to interact with specific analytes. These dyes change color based on the polarity of their environment [13]. They exhibit significant color changes depending on the solvent polarity. Examples include merocyanines, azobenzenes, and pyridinium N-phenolate betaine dyes. Among the common types are pH indicators, such as litmus and phenolphthalein, which change color in response to pH level alterations. Redox indicators, like methylene blue and potassium permanganate, undergo color changes reflecting oxidation–reduction reactions. Solvatochromic dyes, exemplified by merocyanines and azobenzenes, exhibit significant color shifts based on solvent polarity [14]. Metalloporphyrins, including compounds like hemoglobin, detect metal-ligating vapors due to their open coordination sites. Optical brighteners fluoresce under UV light, aiding in detecting trace substances, commonly employed in laundry detergents and paper industries [10]. Optical brighteners fluoresce under UV light, aiding in detecting trace substances, commonly employed in laundry detergents and paper industries. Fluorescent probes emit fluorescence upon interacting with specific analytes, prevalent in biological and environmental sensing applications [15]. Each dye type presents unique advantages and mechanisms, enabling the detection of a broad array of analytes across different fields.

3 Data Extraction from Colorimetric Sensor Array

Colorimetric sensor arrays are valuable tools for detecting and quantifying various analytes based on changes in colorimetric responses. Data extraction from these sensor arrays involves capturing, processing, and interpreting the color changes to derive meaningful information about the presence and concentration of target analytes [16]. This section explores the methods, challenges, and recent innovations in data extraction from colorimetric sensor arrays. Image acquisition is the initial stage in data extraction, involving the capture of high-resolution images of the colorimetric sensor array using various devices such as flatbed scanners, digital cameras, and cell phones. Consistent illumination is crucial to ensure accurate color reproduction, necessitating proper lighting conditions during image capture. Once acquired, the images undergo image processing techniques aimed at enhancing quality, including color correction, noise reduction, and image registration to align spots accurately. Subsequently, image segmentation algorithms isolate individual colorimetric spots for further analysis [17].

Color analysis follows image processing, where colorimetric data is extracted from the segmented spots by measuring attributes like intensity, hue, and saturation. This process may involve analyzing the red, green, and blue values of the color spots and could extend to spectral analysis to capture the full spectrum of light reflected from each spot [18]. A difference map can be created by digitally subtracting the sensor array image values before and after exposure to the analyte on a pixel-by-pixel basis. This entails subtracting the red, green, and blue values after exposure from those before exposure. Averaging the pixel values within the central areas of the spots, typically around 200 pixels, helps alleviate artifacts arising from non-uniform dye spot distributions, particularly at their edges.

Utilizing the RGB color differences has the added benefit of mitigating printing variations, as these differences are less affected by fluctuations in dye concentration or spot intensity across different arrays. Typically, the ImageJ application is employed for this task [19]. The extracted colorimetric data is then interpreted using calibration curves or machine learning algorithms to establish correlations between color changes and analyte concentrations. Statistical methods like principal component analysis (PCA) or clustering algorithms aid in pattern recognition and analyte classification [20] (Fig. 2).

Despite advancements, challenges persist in data extraction from colorimetric sensor arrays. Variability in color perception due to differences in lighting conditions and camera settings can lead to inaccuracies in data. Spot-to-spot variability,

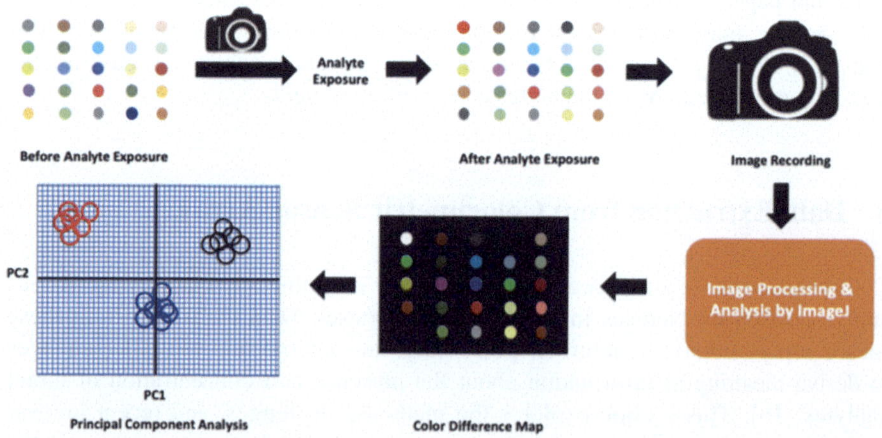

Fig. 2 Schematic showing the overall process involved in colorimetric sensor array data extraction to data processing

cross-sensitivity between analytes and chemically responsive dyes, and background interference from environmental factors further complicate accurate data extraction [21]. Innovations such as machine learning algorithms, multimodal sensing integration, smartphone-based analysis, and microfluidic integration are addressing these challenges, automation in analysis, improving detection efficiency, and enabling real-time, field-deployable sensing applications [22]. These innovations are paving the way for the widespread adoption of colorimetric sensor arrays in diverse fields, from environmental monitoring to biomedical diagnostics, although continued research and development efforts are needed to enhance performance and reliability further [23].

4 Substrate Consideration for Colorimetric Sensor Array

Colorimetric sensor arrays have gained widespread attention for their effectiveness in detecting various analytes. The choice of solid support is crucial in determining the performance and applicability of these sensor arrays. Various solid supports are available for colorimetric sensor arrays, differing in number, nature, and structure [24]. The essential properties of solid support include optical transparency or high reflectivity, facilitating efficient interaction with the analyte and enabling the observation of color changes in the sensor array, along with a homogeneous structure ensuring consistent performance and reliable detection. General chemical compatibility is also vital to prevent interference or reaction with the analyte [25]. For vapor sensing applications, additional properties are desirable, such as accessible substrate microstructure, providing enhanced analyte diffusion to the chromophore, a high surface area for increased loading of the chromophore, and hydrophobicity to reduce the effects of ambient humidity, thereby improving vapor-sensing measurement accuracy [26]. Ensuring these properties in solid supports can effectively enhance the performance of colorimetric sensor arrays in vapor-sensing applications [27]. By optimizing the solid support to meet these criteria, researchers can enhance the sensitivity, selectivity, and reliability of colorimetric sensor arrays, expanding their potential applications in various fields, including environmental monitoring, medical diagnostics, and food safety [24].

5 Types of Substrates for Colorimetric Sensor Array

Paper substrates offer a myriad of advantageous properties that render them ideal for colorimetric sensor arrays. Firstly, their inherent porosity facilitates efficient analyte diffusion to the immobilized dyes, thereby enhancing sensitivity and response time [2]. Additionally, the high absorbency of paper ensures uniform distribution of analytes and reagents, ensuring consistent and reproducible sensor responses. Moreover, paper is cost-effective and readily available, facilitating large-scale production of sensor arrays and enabling the development of disposable and portable sensor devices [28]. In various applications, paper substrates play a pivotal role in colorimetric sensor arrays. They are extensively utilized in environmental monitoring for detecting pollutants, enabling rapid on-site analysis of air and water quality [29].

In biomedical diagnostics, paper-based sensor arrays facilitate the detection of biomarkers and pathogens, offering potential applications in point-of-care testing and disease screening [29]. Furthermore, in food safety and industrial monitoring, paper substrates enable rapid analysis of food products for quality control and detection of chemical leaks and process parameters in industrial settings [30].

Recent innovations in paper-based sensor arrays have focused on enhancing their functionality and applicability [7]. Surface modification techniques, including chemical functionalization and nanoparticle deposition, improve substrate properties such as analyte adsorption and sensor sensitivity [7, 30, 31]. Integration of paper microfluidic devices with sensor arrays enables automated and high-throughput analysis of complex samples [32]. Moreover, coupling paper-based sensor arrays with smartphone readout systems facilitates remote monitoring and real-time analysis, enhancing their utility in point-of-care and field applications [33]. Additionally, advancements in sustainable paper production, improved life cycle, and recycling technologies promote environmental sustainability in sensor development, reducing environmental impact and fostering eco-friendly practices [34].

Glass substrates offer a multitude of properties ideal for colorimetric sensor arrays [27]. Firstly, their optical transparency allows for easy observation of color changes in the sensor array, facilitating real-time monitoring of analyte detection reactions. Secondly, their chemical inertness ensures stability over time by minimizing interference with immobilized dyes, making them suitable for long-term sensing applications [35]. Thirdly, glass substrates provide excellent mechanical stability, crucial for maintaining sensor performance under various environmental conditions. Additionally, their smooth and uniform surfaces enable the precise deposition of chemically responsive dyes, ensuring the fabrication of high-quality sensor arrays with reproducible performance. Finally, their high thermal stability permits integration into thermal cycling and high-temperature applications without degradation [36].

Applications of glass substrates in colorimetric sensor arrays span diverse fields. They are used in environmental monitoring for the rapid on-site analysis of air and water quality, detecting pollutants like heavy metals and volatile organic compounds [37]. In biomedical diagnostics, glass-based sensor arrays detect biomarkers, pathogens, and disease-related molecules, offering potential applications in point-of-care testing and disease screening [38]. For food safety, these sensor arrays enable sensitive analysis of food products for quality control and safety assurance by detecting foodborne pathogens, contaminants, and spoilage indicators. Additionally, glass substrates find applications in industrial monitoring, detecting chemical leaks, emissions, and process parameters in manufacturing and industrial settings [39].

The integration of glass substrates into colorimetric sensor arrays presents exciting avenues for future research and development. Surface modification techniques can enhance analyte adsorption, improve dye immobilization, and increase sensor sensitivity. Integration of nanomaterials like nanoparticles and nanocomposites can enhance analyte detection capabilities and enable real-time monitoring [40]. Combining glass microfluidic devices with sensor arrays creates lab-on-a-chip systems for automated, miniaturized, and high-throughput analysis. Furthermore, coupling glass-based sensor arrays with wireless communication technologies enables remote monitoring, data transmission, and real-time alerts for environmental and health monitoring applications [41].

Impermeable Polyethylene Terephthalate (PET) substrates offer a range of beneficial properties crucial for the effectiveness of colorimetric sensor arrays. Firstly, their chemical inertness ensures minimal interference with immobilized dyes, maintaining array stability over time. Additionally, PET's impermeability to gases and liquids prevents analyte diffusion and environmental contamination, ensuring the reliability of sensor arrays during operation. Moreover, their mechanical strength and dimensional stability enable the fabrication of robust arrays capable of withstanding handling and environmental stresses [42].

PET substrates serve diverse applications in colorimetric sensor arrays, particularly in environmental monitoring, biomedical diagnostics, food safety, and industrial monitoring. These arrays enable the rapid and sensitive detection of various analytes, including environmental pollutants, biomarkers, foodborne pathogens, and industrial process parameters [43]. Furthermore, PET's optical transparency allows for easy visualization of color changes in the array, facilitating real-time monitoring and analysis of analyte detection reactions. Its smooth and uniform surfaces ensure precise deposition of chemically responsive dyes, ensuring the reproducibility and reliability of array performance [44].

Recent innovations in PET-based sensor arrays focus on enhancing their performance and functionality. Surface modification techniques, such as plasma treatment and chemical functionalization, can improve PET substrate properties like analyte adsorption and sensor sensitivity [45]. Integrating nanomaterials onto PET substrates enhances sensor performance by increasing surface area and enabling multiplexed sensing. Microfluidic integration enables the creation of lab-on-a-chip systems for automated and high-throughput analysis. Moreover, combining PET-based sensor arrays with wireless communication technologies enables remote monitoring and real-time alerts for environmental and health applications [46].

Polypropylene substrates (PP) boast a range of advantageous properties crucial for colorimetric sensor arrays. First and foremost is their robust chemical resistance, which ensures minimal interference with immobilized dyes and maintains array stability over time. Additionally, PP's inherent hydrophobicity prevents moisture absorption, making it ideal for outdoor and high-humidity environments [47]. Its exceptional mechanical strength and dimensional stability enable the fabrication of durable sensor arrays capable of withstanding handling and environmental stresses. Moreover, PP's high thermal stability permits integration into thermal cycling and high-temperature applications without deformation or degradation. Furthermore, their low cost and widespread availability make PP substrates an economical choice for large-scale production of colorimetric sensor arrays [48].

The applications of polypropylene substrates in colorimetric sensor arrays span various fields. In environmental monitoring, PP-based sensor arrays detect pollutants like volatile organic compounds (VOCs), gases, and particulate matter, facilitating rapid and sensitive analysis of air and water quality [47]. Biomedical diagnostics utilize PP substrates for detecting biomarkers, pathogens, and disease-related molecules in clinical samples, offering potential applications in point-of-care testing and disease screening [1]. Furthermore, PP-based sensor arrays find utility in ensuring food safety by detecting foodborne pathogens, contaminants, and spoilage indicators, enabling on-site analysis for quality control and safety assurance. Additionally, PP substrates are employed in industrial monitoring, detecting chemical leaks, emissions, and process parameters in manufacturing and industrial settings [49].

Recent innovations in PP-based sensor arrays have opened new avenues. Surface modification techniques, including plasma treatment and chemical functionalization, enhance PP substrate properties such as analyte adsorption, dye immobilization, and sensor sensitivity. Integration of nanomaterials like nanoparticles and nanocomposites onto PP substrates enhances sensor performance by increasing surface area, improving analyte diffusion, and enabling multiplexed sensing [50]. Moreover, PP microfluidic devices integrated with sensor arrays create lab-on-a-chip systems for automated and high-throughput analysis of complex samples. Combining PP-based sensor arrays with wireless communication technologies enables remote monitoring, data transmission, and real-time alerts for environmental and health monitoring applications [51].

Polyvinylidene Fluoride substrate (PVDF) membranes exhibit a host of properties ideal for colorimetric sensor arrays. Firstly, their porosity, ranging from microporous to ultrafiltration, allows tailored analyte diffusion and the immobilization of chemically responsive dyes. Secondly, PVDF boasts chemical resistance against a wide array of solvents and chemicals, ensuring minimal interference with immobilized dyes and maintaining sensor array stability over time. Thirdly, these membranes demonstrate remarkable mechanical strength and dimensional stability, facilitating the creation of robust sensor arrays capable of withstanding handling and environmental stresses [17]. Additionally, PVDF's inherent hydrophobicity prevents moisture absorption, making it advantageous for outdoor and high-humidity environments. Lastly, its high thermal stability enables integration into thermal cycling and high-temperature applications without deformation or degradation [52].

PVDF membranes find extensive use in various applications within colorimetric sensor arrays. They are pivotal in environmental monitoring, detecting pollutants like heavy metals and organic compounds in air and water swiftly and sensitively. In biomedical diagnostics, PVDF-based sensor arrays contribute to detecting biomarkers and pathogens in clinical samples, holding promise for point-of-care testing and disease screening [52]. Furthermore, they play a crucial role in ensuring food safety by detecting foodborne pathogens and contaminants and facilitating rapid on-site analysis for quality control [17, 53]. Additionally, in industrial settings, PVDF membranes aid in monitoring chemical leaks, emissions, and process parameters, contributing to overall safety and quality assurance [54].

Recent innovations in PVDF-based sensor arrays have opened up exciting avenues for future development. Surface modification techniques, including plasma treatment and chemical functionalization, enhance membrane properties like analyte adsorption and sensor sensitivity [55]. Integration of nanomaterials onto PVDF membranes improves sensor performance by increasing surface area and enabling multiplexed sensing. Moreover, the integration of PVDF microfluidic devices with sensor arrays facilitates automated and high-throughput analysis of complex samples [56]. Finally, combining PVDF-based sensor arrays with wireless communication technologies enables remote monitoring, data transmission, and real-time alerts, advancing environmental and health monitoring applications to new levels [57].

Nanoporous Silica substrates present unique attributes that position them as promising contenders for colorimetric sensor arrays. This section delves into the design principles, applications, and recent strides in utilizing nanoporous silica substrates for the advancement of colorimetric sensor arrays. Design principles are fundamental, encompassing meticulous control over porosity, surface functionalization, modification, as well as size and shape. These principles enable tailored analyte diffusion, chemical dye immobilization, and enhanced recognition through specific ligands or biomolecules [58].

Applications of nanoporous silica substrates in colorimetric sensor arrays are diverse and impactful. They span from environmental monitoring, where they detect pollutants, to biomedical diagnostics for identifying biomarkers and pathogens. Furthermore, they contribute to ensuring food safety by detecting pathogens and contaminants, as well as industrial monitoring, particularly in identifying chemical leaks and process parameters [59]. These applications underscore the versatility and relevance of nanoporous silica substrates in various domains.

Recent innovations in nanoporous silica-based sensor arrays are shaping the future landscape of sensor technology. Advances focus on surface engineering techniques, integrating multifunctional nanoparticles, microfluidic integration, and the development of smart sensing platforms [59]. Surface engineering enhances analyte recognition, while the integration of nanoparticles amplifies sensitivity and multiplexing capabilities. Additionally, combining nanoporous silica substrates with microfluidic devices facilitates automated analysis, while smart sensing platforms enable real-time monitoring and remote data transmission. These innovations promise to further enhance the efficiency and applicability of nanoporous silica-based sensor arrays in diverse settings [60].

Nanoparticle-packed substrate design as a solid support is pivotal in determining the efficacy and versatility of sensor arrays. Among the myriad options, nanoparticle-packed surfaces stand out for their unique attributes, rendering them promising candidates for colorimetric sensor arrays [61]. This segment delves into the design principles, applications, and burgeoning trends surrounding nanoparticle-packed surfaces within the realm of colorimetric sensor array development. In understanding the design principles, nanoparticle-packed surfaces are meticulously crafted based on several foundational elements aimed at optimizing their role as solid supports for colorimetric sensor arrays [30]. Key considerations encompass nanoparticle selection, surface functionalization, packing density, surface morphology, and the imperative attributes of stability and durability. These facets are intricately interwoven to facilitate enhanced analyte recognition, binding affinity, and accessibility, laying the groundwork for proficient sensor array performance [62].

Exploring the applications of nanoparticle-packed surfaces within colorimetric sensor arrays reveals a broad spectrum of utility. From environmental monitoring to biomedical diagnostics, food safety assurance, and industrial surveillance, these surfaces find extensive employment [63]. Moreover, recent strides have propelled innovations and future perspectives, focusing on multifunctional nanoparticles, nanostructured surfaces, biomimetic sensors, and the integration of smart sensing platforms. This dynamic landscape not only augments sensitivity, selectivity, and multiplexing capabilities but also paves the way for real-time monitoring and remote data transmission, ushering in a new era of sensor array functionality and applicability [64].

6 Substrates Challenges for Fabrication of Colorimetric Sensor Array

Challenges for Paper Substrate: Utilizing paper as a substrate for colorimetric sensor arrays offers numerous benefits, including affordability, disposability, and ease of transport [2]. However, it presents distinct challenges primarily concerning wettability and absorption due to paper's porous nature, which can cause uneven distribution of analytes and reagents, thereby compromising the accuracy and consistency of sensor readings [24]. Additionally, diffusion and migration of substances within the paper substrate may lead to spatial dispersion and cross-contamination among sensor elements (spider-web effect), undermining the specificity of sensor responses [65]. Furthermore, ensuring the stability and durability of paper substrates, particularly in varying environmental conditions, poses a significant hurdle, impacting the reliability of sensor arrays over time [66]. Integration with electronic systems further complicates the process, requiring careful consideration of compatibility and usability without sacrificing portability or functionality [28].

Challenges for Glass Substrates: Using glass substrates for colorimetric sensor arrays offers several advantages, such as optical clarity, mechanical stability, and ease of surface functionalization [67]. However, there are also some challenges associated with glass substrates in this context [24]. Surface functionalization poses a significant hurdle as glass surfaces are generally inert, necessitating functionalization to immobilize sensing elements or receptor molecules [68]. Achieving uniform and stable attachment of these molecules without altering the substrate's optical properties can be challenging. Moreover, ensuring uniform deposition of sensing elements or reagents on the glass substrate is crucial for consistent sensor performance [26]. Variations in coating thickness or density can lead to non-uniform sensor responses and reduced reproducibility. Additionally, glass substrates may exhibit background interference, such as autofluorescence or light scattering, which can affect the accuracy and sensitivity of colorimetric measurements [38]. Minimizing background interference while maintaining substrate transparency is vital for reliable sensor operation. Furthermore, glass substrates can undergo thermal expansion and contraction, especially when subjected to rapid temperature changes, necessitating maintaining dimensional stability and optical clarity over a wide temperature range for sensor performance in various operating conditions [69]. Lastly, some sensing applications may require exposure to harsh chemical environments or organic solvents, which can degrade or etch the glass substrate [70]. Ensuring chemical compatibility between the substrate and sensor components is essential for long-term stability and reliability.

Challenges for Polyethylene Terephthalate Substrate (PET): Fabricating colorimetric sensor arrays present distinct challenges despite PET's extensive utility across various applications. One primary hurdle lies in the surface properties of PET, characterized by low surface energy and limited chemical reactivity [71]. These properties hinder the effective immobilization and stability of sensing molecules or receptor materials, necessitating meticulous attention to achieving uniform and stable surface functionalization for reliable sensor performance [24]. Additionally, PET's intrinsic hydrophobic nature poses another obstacle, impacting the spreading and distribution of aqueous solutions and reagents on the substrate surface. Ensuring proper wetting and uniform coverage of sensing elements becomes imperative for accurate and reproducible colorimetric measurements [71]. Furthermore, PET substrates may exhibit limited optical clarity compared to alternatives like glass, potentially impeding the transmission and detection of colorimetric signals. To mitigate this, efforts must focus on minimizing substrate opacity and light scattering effects to uphold accuracy and sensitivity in measurements [44]. Additionally, ensuring robust adhesion and stability of sensing layers or coatings to the PET substrate is crucial in preventing delamination or detachment during sensor operation. This necessitates the optimization of surface treatments and coating techniques to enhance the adhesion and durability of sensor materials on PET substrates, thereby bolstering overall sensor reliability [72]. Lastly, considering PET's sensitivity to environmental factors such as temperature, humidity, and UV radiation, ensuring the resilience of PET-based sensor arrays under varying conditions emerges as a key consideration for practical applications, highlighting the importance of robust design and engineering in sensor development [73].

Challenges for Polypropylene Substrate (PP): Polypropylene substrates offer significant advantages for colorimetric sensor arrays, including cost-effectiveness, flexibility, and resistance to various chemicals. These attributes make them appealing for a range of sensing applications [33]. However, several challenges must be addressed when utilizing polypropylene in this context. One primary challenge is the substrate's surface inertness, which makes it difficult to immobilize sensing molecules effectively [74]. Overcoming this hurdle requires innovative approaches to enhance molecular binding and interaction on the substrate surface. Another obstacle associated with polypropylene substrates is their limited adhesion properties, which can compromise the durability and stability of sensor coatings [74]. Developing robust adhesion techniques is crucial to ensure the longevity and reliability of sensor arrays. Additionally, variations in surface roughness pose a challenge to the uniformity and reproducibility of sensor coatings, necessitating careful control and optimization of surface preparation processes [24]. Addressing these challenges is essential to enhance the performance and reliability of polypropylene-based colorimetric sensor arrays for various applications. Furthermore, the hydrophobic nature of polypropylene substrates can impede the distribution of aqueous solutions and reagents, affecting the accuracy and reproducibility of colorimetric measurements [74]. Achieving proper wetting and coverage of sensing elements is critical for optimal sensor performance. Moreover, ensuring thermal stability and chemical compatibility of polypropylene substrates with sensor components is essential to maintain sensor integrity and functionality over prolonged use [75]. By overcoming these challenges through innovative materials and fabrication techniques, polypropylene-based sensor arrays can realize their full potential in diverse sensing applications.

Challenges for PVDF Substrate: PVDF membranes offer unique advantages for colorimetric sensor arrays due to their porous structure, flexibility, and chemical resistance. However, there are also several challenges associated with using PVDF membranes in this context. Surface functionalization poses a significant hurdle as PVDF membranes have relatively inert surfaces, making it difficult to immobilize sensing molecules or receptor materials uniformly [76]. Achieving stable functionalization is crucial for consistent sensor performance. Additionally, the pore size and distribution of PVDF membranes vary, requiring careful attention to ensure uniformity for reliable sensor responses and measurements [77]. Moreover, the membrane thickness, which can affect analyte diffusion rates and sensor response times, must be optimized while maintaining mechanical integrity. Furthermore, background interference, such as autofluorescence or light scattering, is a concern with PVDF membranes, potentially impacting the accuracy and sensitivity of colorimetric measurements [78]. Addressing background interference while preserving membrane transparency is crucial for dependable sensor operation. Hydrophobicity presents another challenge as PVDF membranes' inherent hydrophobic nature can influence the spreading and distribution of aqueous solutions and reagents on the surface. Proper wetting and coverage of sensing elements are vital for precise and reproducible colorimetric measurements [79]. Additionally, chemical compatibility is paramount as PVDF membranes may react with certain analytes or reagents, compromising membrane properties or sensor response over time. Especially it shows reactivity with strong base solutions [30]. Ensuring compatibility between the membrane and sensor components is essential for long-term stability and reliability. Lastly, mechanical stability is critical as PVDF membranes are flexible and susceptible to deformation or damage under mechanical stress or handling, requiring measures to maintain dimensional integrity for consistent sensor performance [80].

Challenges for Nanoporous Silica Substrates: Nanoporous silica substrates present numerous advantages for colorimetric sensor arrays, including their high surface area, tunable pore size, and chemical stability. However, harnessing these benefits comes with its own set of challenges. Surface functionalization poses a significant hurdle due to the inherent chemical inertness of silica surfaces, necessitating the achievement of uniform and stable immobilization of sensing molecules or receptor materials [81]. Moreover, the variability in pore size and distribution among nanoporous silica substrates, influenced by synthesis methods and conditions, requires meticulous attention to ensure consistent sensor responses and reliable measurements. Additionally, the hydrophilic nature of silica surfaces complicates the spreading and distribution of hydrophobic analytes or reagents, highlighting the importance of proper wetting and coverage for accurate and reproducible colorimetric readings [82]. Furthermore, while silica substrates typically exhibit chemical stability, surface modifications or degradation under certain conditions could compromise the long-term reliability of silica-based sensor arrays. Mechanical fragility is another concern, necessitating measures to ensure dimensional integrity and stability under operational conditions. Lastly, interference from solvents or solutes in sample solutions could potentially impact sensor responses, emphasizing the need to minimize such interference for precise and dependable colorimetric measurements [83].

Challenges of Nanoparticle-Packed Surface: Using nanoparticle-packed surfaces for colorimetric sensor arrays offers unique advantages such as high surface area, tunable properties, and enhanced sensitivity [30]. However, several challenges are associated with this approach. Surface functionalization is a primary concern, as nanoparticle surfaces often require modification to immobilize sensing molecules or receptor materials uniformly. Achieving consistent sensor responses relies on a uniform and stable functionalization across the packed nanoparticle surface [30]. Additionally, pore size and distribution play a crucial role, in impacting analyte diffusion rates and sensor response times. Ensuring uniformity in nanoparticle packing and pore size distribution is vital for reliable and reproducible measurements [30]. Moreover, surface accessibility is a challenge as pore blockage or limited surface accessibility can hinder analyte penetration into the packed nanoparticle layers, affecting sensor sensitivity and response kinetics [84]. Hydrophilicity/hydrophobicity is another consideration, as the surface properties of nanoparticles influence the wetting behavior of the sensor surface and the distribution of aqueous solutions and reagents [85]. Balancing hydrophilicity and hydrophobicity is essential for proper analyte spreading and coverage on the sensor surface. Chemical Stability is critical, as nanoparticles may undergo aggregation, leaching, or surface modifications under certain chemical environments, impacting sensor performance and stability. Ensuring the chemical stability of nanoparticle-packed surfaces in the presence of analytes or reagents is vital for long-term reliability [86]. Additionally, mechanical stability is essential, as nanoparticle-packed surfaces may be susceptible to deformation or damage under mechanical stress or handling. Maintaining mechanical stability and dimensional integrity of nanoparticle-packed sensor arrays under operating conditions is crucial for reliable performance [30].

Addressing these challenges necessitates interdisciplinary efforts in materials science, surface chemistry, sensor technology, and engineering. Collaboration across these fields is crucial for developing strategies to overcome obstacles related to surface functionalization, pore size and distribution, surface accessibility, hydrophilicity/hydrophobicity balance, chemical stability, and mechanical stability of nanoparticle-packed sensor arrays.

References

1. Alberti G, Zanoni C, Magnaghi LR, Biesuz R (2020) Disposable and low-cost colorimetric sensors for environmental analysis. Int J Environ Res Public Health 17(22):8331
2. Ko A, Liao C (2023) Paper-based colorimetric sensors for point-of-care testing. Anal Methods 15(35):4377–4404
3. Calabretta MM, Gregucci D, Desiderio R, Michelini E (2023) Colorimetric paper sensor for food spoilage based on biogenic amine monitoring. Biosensors 13(1):126
4. Wu R, Selvaganapathy P (2023) Porous biocompatible colorimetric nanofiber-based sensor for selective ammonia detection on personal wearable protective equipment. Sens Actuators B Chem 393:134270
5. Ahmad NA, Yook Heng L, Salam F, Mat Zaid MH, Abu Hanifah S (2019) A colorimetric pH sensor based on *clitoria* sp and *brassica* sp for monitoring of food spoilage using chromametry. Sensors 19(21):4813
6. Yang J, Wang K, Xu H, Yan W, Jin Q, Cui D (2019) Detection platforms for point-of-care testing based on colorimetric, luminescent and magnetic assays: a review. Talanta 202:96–110
7. Kangas MJ, Burks RM, Atwater J, Lukowicz RM, Williams P, Holmes AE (2017) Colorimetric sensor arrays for the detection and identification of chemical weapons and explosives. Crit Rev Anal Chem 47(2):138–153
8. Tomita S, Nomoto H, Yoshitomi T, Iijima K, Hashizume M, Yoshimoto K (2018) Noninvasive fingerprinting-based tracking of replicative cellular senescence using a colorimetric polyion complex array. Anal Chem 90(11):6348–6352
9. Suslick KS, Bailey DP, Ingison CK, Janzen M, Kosal ME, McNamara WB, Rakow NA, Sen A, Weaver JJ, Wilson JB, Zhang C, Nakagaki S (2007) Seeing smells: development of an optoelectronic nose. Quim Nova 30(3):677–681
10. Rakow NA, Suslick KS (2000) A colorimetric sensor array for odour visualization. Nature 406(6797):710–713
11. Liang Y, Lin H, Kang W, Shao X, Cai J, Li H, Chen Q (2023) Application of colorimetric sensor array coupled with machine-learning approaches for the discrimination of grains based on freshness. J Sci Food Agric 103:6790–6799
12. Ma Y, Li Y, Ma K, Wang Z (2018) Optical colorimetric sensor arrays for chemical and biological analysis. Sci China Chem 61:643–655
13. Xiao-Wei H, Xiao-Bo Z, Ji-Yong S, Zhi-Hua L, Jie-Wen Z (2018) Colorimetric sensor arrays based on chemo-responsive dyes for food odor visualization. Trends Food Sci Technol 81:90–107
14. Xie X, Gutiérrez A, Trofimov V, Szilagyi I, Soldati T, Bakker E (2015) Charged solvatochromic dyes as signal transducers in pH independent fluorescent and colorimetric ion selective nanosensors. Anal Chem 87(19):9954–9959
15. Zhang X, Yin J, Yoon J (2014) Recent advances in development of chiral fluorescent and colorimetric sensors. Chem Rev 114(9):4918–4959
16. Adampourezare M, Nikzad B, Sajedi-Amin S, Rahimpour E (2024) Colorimetric sensor array for versatile detection and discrimination of model analytes with environmental relevance. BMC Chem 18(1):80
17. Cho B, Charoensri K, Doh H, Park HJ (2023) Preparation of colorimetric sensor array system to evaluate the effects of alginate edible coating on boiled-dried anchovy. Foods 12(3):638
18. Hemmateenejad B, Bordbar MM, Shojaeifard Z (2023) Data acquisition and data analysis in colorimetric sensor arrays. Chemom Intell Lab Syst 241:104975

19. Schneider CA, Rasband WS, Eliceiri KW (2012) NIH Image to ImageJ: 25 years of image analysis. Nat Methods 9(7):671–675
20. Zwicker VE, Sergeant GE, New EJ, Jolliffe KA (2021) A colorimetric sensor array for the classification of biologically relevant tri-, di- and mono-phosphates. Org Biomol Chem 19(5):1017–1021
21. Phuangsaijai N, Jakmunee J, Kittiwachana S (2021) Investigation into the predictive performance of colorimetric sensor strips using RGB, CMYK, HSV, and CIELAB coupled with various data preprocessing methods: a case study on an analysis of water quality parameters. J Anal Sci Technol 12:19
22. Chang B (2012) Smartphone-based chemistry instrumentation: digitization of colorimetric measurements. Bull Korean Chem Soc 33(2):549–552
23. Gu Z, Wu M, Yan B, Wang H, Kong C (2020) Integrated digital microfluidic platform for colorimetric sensing of nitrite. ACS Omega 5(19):11196–11201
24. LaGasse MK, Rankin JM, Askim JR, Suslick KS (2014) Colorimetric sensor arrays: interplay of geometry, substrate and immobilization. Sens Actuators B Chem 197:116–122
25. Morbioli GG, Mazzu-Nascimento T, Stockton AM, Carrilho E (2017) Technical aspects and challenges of colorimetric detection with microfluidic paper-based analytical devices (μPADs)—a review. Anal Chim Acta 970:1–22
26. Palinski TJ, Guan B, Bradshaw-Hajek BH, Lienhard MA, Priest C, Miranda FA (2022) Fast vapor detection by a micropillar array-integrated colorimetric sensor. IEEE Sens 1–4
27. Ebralidze II, Laschuk NO, Poisson J, Zenkina OV (2019) Colorimetric sensors and sensor arrays. Nanomater Des Sens Appl 1–39
28. Li Y-Q, Feng L (2020) Progress in paper-based colorimetric sensor array. Chin J Anal Chem 48(11):1448–1457
29. Kannan B, Jahanshahi-Anbuhi S, Pelton RH, Li Y, Filipe CDM, Brennan JD (2015) Printed paper sensors for serum lactate dehydrogenase using pullulan-based inks to immobilize reagents. Anal Chem 87(18):9288–9293
30. Patil VS, Lee MG, Yun J, Lee J, Lim SH, Yi GR (2018) Chemically resistant perfluoroalkoxy nanoparticle-packed porous substrates and their use in colorimetric sensor arrays. Langmuir 34(43):13014–13024
31. Patel S, Jamunkar R, Sinha D, Monisha N, Patle TK, Kant T, Dewangan K, Shrivas K (2021) Recent development in nanomaterials fabricated paper-based colorimetric and fluorescent sensors: a review. Trends Environ Anal Chem 31:e00136
32. Soga T, Jimbo Y, Suzuki K, Citterio D (2013) Inkjet-printed paper-based colorimetric sensor array for the discrimination of volatile primary amines. Anal Chem 85(19):8973–8978
33. Li Z, Askim JR, Suslick KS (2018) The optoelectronic nose: colorimetric and fluorometric sensor arrays. Chem Rev 119(1):231–292
34. Immanuel PN, Huang S-J, Adityawardhana Y, Yen Y-K (2023) A review of paper-based sensors for gas, ion, and biological detection. Coatings 13(8):1326
35. Gugliandolo G, Pilato G, Donato N (2021) Compact system for colorimetric sensor arrays characterization based on computer screen photo-assisted technology. Electronics 10(21):2587
36. Piriya VSA, Joseph P, Daniel SCGK, Lakshmanan S, Kinoshita T, Muthusamy S (2017) Colorimetric sensors for rapid detection of various analytes. Mater Sci Eng C 78:1231–1245
37. Pastore A, Badocco D, Cappellin L, Tubiana M, Pastore P (2024) Real-time monitoring of the pH of white wine and beer with colorimetric sensor arrays (CSAs). Food Chem 139513
38. Li F, Stewart C, Yang S, Shi F, Cui W, Zhang S, Wang H, Huang H, Chen M, Han J (2022) Optical sensor array for the early diagnosis of alzheimer's disease. Front Chem 10:874864
39. Peveler WJ (2024) Food for thought: optical sensor arrays and machine learning for the food and beverage industry. ACS Sens 9(4):1656–1665
40. Scroccarello A, Della Pelle F, Del Carlo M, Compagnone D (2023) Optical plasmonic sensing based on nanomaterials integrated in solid supports. A critical review. Anal Chim Acta 1237:340594
41. Kim C, Kang MS, Raja IS, Oh J, Joung YK, Han D (2024) Current issues and perspectives in nanosensors-based artificial olfactory systems for breath diagnostics and environmental exposure monitoring. Trends Anal Chem 174:117656

42. Feng L, Musto CJ, Kemling JW, Lim SH, Zhong W, Suslick KS (2010) Colorimetric sensor array for determination and identification of toxic industrial chemicals. Anal Chem 82(22):9433–9440
43. Yoon S, Park H, Lee W (2021) Fabrication of inverse opal photonic gel sensors on flexible substrates by transfer process. Lab Chip 21(15):2997–3003
44. Szczurek A, Tran TNL, Kubacki J, Gąsiorek A, Startek K, Mazur-Nowacka A, Dell'Anna R, Armellini C, Varas S, Carlotto A, Chiasera A, Łukowiak A, Krzak J, Ferrari M (2023) Polyethylene terephthalate (PET) optical properties deterioration induced by temperature and protective effect of organically modified SiO_2–TiO_2 coating Mater. Chem Phys 306:128016
45. Zhang H, Wei X, Chan-Park MB, Wang M (2022) Colorimetric sensors based on multifunctional polymers for highly sensitive detection of food spoilage. ACS Food Sci Technol 2(4):703–711
46. Yin S, Dong L (2024) Plant tattoo sensor array for leaf relative water content, surface temperature, and bioelectric potential monitoring. Adv Mater Technol 2024:2302073
47. Wujcik EK, Duirk SE, Chase GG, Monty CN (2016) A visible colorimetric sensor based on nanoporous polypropylene fiber membranes for the determination of trihalomethanes in treated drinking water. Sens Actuators B Chem 223:1–8
48. Lanzalaco S, Gil P, Mingot J, Àgueda A, Alemán C, Armelin E (2022) Dual-responsive polypropylene meshes actuating as thermal and SERS sensors. ACS Biomater Sci Eng 8(8):3329–3340
49. Cho SH, Suh JM, Eom TH, Kim T, Jang HW (2021) Colorimetric sensors for toxic and hazardous gas detection: a review. Electron Mater Lett 17:1–17
50. Raja DA, Munir F, Shah MR, Bhanger MI, Malik MI (2021) Colorimetric sensing of cephradine through polypropylene glycol functionalized gold nanoparticles. R Soc Open Sci 8(5):210185
51. Ma X, Guo G, Wu X, Wu Q, Liu F, Zhang H, Shi N, Guan Y (2023) Advances in integration, wearable applications, and artificial intelligence of biomedical microfluidics systems. Micromachines 14(5):972
52. Hou J, Liu X, Hou C, Hou D, Li J (2023) A PVDF-based colorimetric sensor array for noninvasive detection of multiple disease-related volatile organic compounds. Anal Bioanal Chem 415:6647–6661
53. Pastore A, Badocco D, Pastore P (2022) Determination of the relevant equilibrium constants working in pH colorimetric sensor arrays (CSAs). Microchem J 177.107288
54. Kaewnu K, Boonna S, Kongkaew S, Phonchai A, Chaisiwamongkhol K, Thongprajukaew K, Limbut W (2023) A portable colorimetric device based on PVDF indicator gel for formaldehyde detection in food and wood products. Microchem J 184:108162
55. Correia DM, Ribeiro C, Sencadas V, Botelho G, Carabineiro SAC, Ribelles JLG, Lanceros-Méndez S (2015) Influence of oxygen plasma treatment parameters on poly(vinylidene fluoride) electrospun fiber mats wettability. Prog Org Coat 85:151–158
56. Bang JH, Lim SH, Park E, Suslick KS (2008) Chemically responsive nanoporous pigments: colorimetric sensor arrays and the identification of aliphatic amines. Langmuir 24(22):13168–13172
57. Bobinger M, Keddis S, Hinterleuthner S, Becherer M, Kluge F, Schwesinger N, Salmerón JF, Lugli P, Rivadeneyra A (2018) Light and pressure sensors based on PVDF with sprayed and transparent electrodes for self-powered wireless sensor nodes. IEEE Sens J 19(3):1114–1126
58. Li Z (2018) Nanoporous silica-dye microspheres for enhanced colorimetric detection of cyclohexanone. Chemosensors 6(3):34
59. Kemling JW, Qavi AJ, Bailey RC, Suslick KS (2011) Nanostructured substrates for optical sensing. J Phys Chem Lett 2(22):2934–2944
60. Lin C, Du Z, Tao N, Wang D (2020) Gradient-based colorimetric array sensor for continuous monitoring of multiple gas analyte. ACS Sens 6(2):439–442
61. Cui Y, Zhao J, Li H (2023) Chromogenic mechanisms of colorimetric sensors based on gold nanoparticles. Biosensors 13(8):801
62. Li D, Dong Y, Li B, Wu Y, Wang K, Zhang S (2015) Colorimetric sensor array with unmodified noble metal nanoparticles for naked-eye detection of proteins and bacteria. Analyst 140(22):7672–7677

63. Yu J, Huang M, Tian H, Xu X (2023) Silver nanoparticle sensor array-based meat freshness inspection system. Foods 12(20):3814
64. Xu N, Jin S, Wang L (2020) Metal nanoparticles-based nanoplatforms for colorimetric sensing: a review. Rev Anal Chem 40(1):1–11
65. Bordbar MM, Nguyen T, Tran AQ, Bagheri H (2020) Optoelectronic nose based on an origami paper sensor for selective detection of pesticide aerosols. Sci Rep 10(1):17302
66. Soda Y, Bakker E (2019) Quantification of colorimetric data for paper-based analytical devices. ACS Sens 4(12):3093–3101
67. Janzen MC, Ponder JB, Bailey DP, Ingison CK, Suslick KS (2006) Colorimetric sensor arrays for volatile organic compounds. Anal Chem 78(11):3591–3600
68. Suslick KS, Rakow NA, Sen A (2004) Colorimetric sensor arrays for molecular recognition. Tetrahedron 60(49):11133–11138
69. Djemli A, Ghebouli M, Bouferrache K, Slimani Y, Habila MAMF, Chihi T, Ghebouli B, Sillanpää M (2023) Effect of temperature and glass content on crystalline phases in porcelain sintered with recovered automotive glass. Heliyon 9(12):e22554
70. Amaro AM, Reis PNB, Neto MA, Louro C (2013) Effects of alkaline and acid solutions on glass/epoxy composites. Polym Degrad Stab 98(4):853–862
71. Lomae A, Chaiyo S, Chailapakul O, Siangproh W, Panchompoo J (2022) Fully integrated colorimetric sensor based on transparency substrate for salbutamol determination. MethodsX 9:101913
72. Twu R-C, Li K-H, Lin B-L (2021) Low-cost polyethylene terephthalate fluidic sensor for ultrahigh accuracy measurement of liquid concentration variation. Sensors 21(21):7410
73. Joseph TM, Azat S, Ahmadi Z, Jazani OM, Esmaeili A, Kianfar E, Haponiuk J, Thomas S (2024) Polyethylene terephthalate (PET) recycling: a review. Case Stud Chem Environ Eng 9:100673
74. Li X, Wang J, Chan SP, Teong SP, Zhang Y (2024) Scalable superhydrophobic coatings with recycled polypropylene. Green Chem 26:4493–4497
75. Hernández-Fernández J, Rayón E, López J, Arrieta MP (2019) Enhancing the thermal stability of polypropylene by blending with low amounts of natural antioxidants. Macromol Mater Eng 304:1900379
76. Pâslaru E, Baican MC, Hitruc EG, Nistor MT, Poncin-Epaillard F, Vasile C (2014) Immunoglobulin G immobilization on PVDF surface. Colloids Surf B 115:139–149
77. Baroud TN (2023) Tuning PVDF membrane porosity and wettability resistance via varying substrate morphology for the desalination of highly saline water. Membranes 13(4):395
78. Li M, Katsouras I, Piliego C, Glasser G, Lieberwirth I, Blom PWM, de Leeuw DM (2013) Controlling the microstructure of poly(vinylidene-fluoride) (PVDF) thin films for microelectronics. J Mater Chem C 1(46):7695
79. Hou C, Du L, Li Y, Guo M, Zhou J, Qiao S (2023) Superhydrophobic PVDF membrane formed by crystallization process for direct contact membrane distillation. iScience 26(5):106464
80. Braga FJC, Rogero SO, Couto AA, Marques RFC, Ribeiro AA, Campos JS de C (2007) Characterization of PVDF/HAP composites for medical applications. Mater Res 10(3):247–251
81. Nayl AA, Abd-Elhamid AI, Aly AA, Bräse S (2022) Recent progress in the applications of silica-based nanoparticles. RSC Adv 12(22):13706–13726
82. Pastore A, Badocco D, Pastore P (2020) High accuracy OrMoSil (Polyvinylidene Fluoride)-supported colorimetric sensor: novel approach for the calculation of the pH prediction error. Talanta 213:120840
83. Rimsza JM, Du J (2013) Structural and mechanical properties of nanoporous silica. J Am Ceram Soc 97(3):772–781
84. Liu S, He X, Hu X, Pu Y, Mao X (2024) Porous nanomaterials for biosensing and related biological application in in vitro/vivo usability. Mater Adv 5(2):453–474
85. Ku C-A, Chung C-K (2023) Advances in humidity nanosensors and their application: review. Rev Sens 23(4):2328
86. Jiang C, Guo AJ, Li Y, Wang Y, Sun J, Chen Z, Chen X, Ma Q (2024) Inspired by human olfactory system: deep-learning-assisted portable chemo-responsive dye-based odor sensor array for the rapid sensing of shrimp and fish freshness. Chem Eng J 484:149283

Sensor Development for Heavy Metal Ion Sensing: New Insights on Translating Prevalent Analysis to Portable Smart Nanodevices

Atirah Tauseef and Imran Uddin

Abstract Heavy metal exposure poses a threat to both the natural habitat and human health due to their persistence, ability to accumulate in organisms, and resistance to degradation. Hence, the imperative task of sensing and eliminating heavy metals from the soil ecosystem is vital in order to establish a stable, sustainable, and waste-free ecosystem. This chapter highlights progress in sensing technologies for field mobility. Further, we discussed the revolutionary method for detecting heavy metal ions in biological and environmental samples using low-cost electrochemical and colorimetric nanosensors. It explores advancements in smartphone-operated screen-printed electrodes, plastic chip sensors, carbon fiber paper-based sensors, and microfluidic-based nanosensors. The chapter also discusses future possibilities and the advantages and drawbacks of current nanosensor techniques.

Keywords Heavy metal ion · Mobile smart nanodevices · Traditional techniques · Enzyme-based biosensors

1 Introduction

Over the last decade, the Industrial breakthrough and the rapid growth of technology have been closely linked to issues such as gentrification, food production and distribution, and access to the health care system. Although technological expansion on one hand has substantially improved standard and quality of human life but on other hand imposing unplanned negative influences on environment and natural resources. The inadvertent costs of industrial expansion and anthropogenic actions have incited

A. Tauseef
Section of Plant Pathology and Nematology, Botany Department, Aligarh Muslim University, Aligarh, India

I. Uddin (✉)
Department of Conservative Dentistry and Endodontics, Saveetha Dental College and Hospital, Saveetha Institute of Medical and Technical Sciences, Saveetha University, Chennai 600077, Tamil Nadu, India
e-mail: usmani.imran@gmail.com

I. Uddin (ed.), *Smart Nanosensors*, Smart Nanomaterials Technology,
https://doi.org/10.1007/978-981-96-3878-9_7

the production and liberation of heavy metal ions along with other environmental waste into the ecosystem. Moreover, discharge of municipal wastes, agriculture bases surface runoff, untreated industrial effluent, landfilled electronic and metal waste materials from automobiles have added grim to the scenario by posing serious threat to the ecosystem in respect of ecological damage and heavy metal pollution.

The primary factors contributing to environmental contamination with heavy metal ions are aging infrastructure and corroding municipal water and sewage distribution pipes, electroplating, smelting, batteries, petrochemicals, paints, inorganic dyes, agrochemical, and chemical fertilizer industries [1–3]. Heavy metal ions can enter human bodies through both direct and indirect means such as consumption of heavy metal-contaminated edibles and water, inhalation of the same from air particles, and occupational exposure. As a result of long-term exposure of heavy metals, ecological and public health menace aspects are alarming because of persistent nature of heavy metals. Different studies have unveiled and defined the acute heavy metal concentration accretion in underground and surface water, soil sediment, plants, food crops, aquatic and terrestrial animals [4–6]. The long-term exposure has posed serious threats to the human and animal health like high blood pressure, waning children's intelligent quotient and development challenges, rising carcinoma cases, weakened immune system, cardiovascular ailments, myocardial infraction, miscarriages and still birth, neurological disorders, oxidative cellular stress, and cytotoxicity. Heavy metals are bio-accumulated and bio-magnified in flora, fauna and ultimately human beings by passing on different trophic levels of food chains.

The escalating Flint, Michigan, tap water crisis and the accompanying health endangerment, including social anxiety and combat stress reaction serve as a warning to be ready for potentially common heavy metal contaminations. Creating functional environmental heavy metal ion detectors is essential for protecting both global native land security and public health (Table 1) [7–9].

2 Traditional Techniques for Trace Element Detection and Heavy Metal Ion Analysis

The development of rigorous, efficient techniques for the detection, assessment, and screening of heavy metal ions in environmental samples has made great strides. The detection of heavy metals and trace metals has been made possible by a variety of techniques, such as inductively coupled plasma-mass spectroscopy (ICP-MS), flame atomic absorption spectroscopy (FAAS), inductively coupled plasma-optical emission spectroscopy (ICP-OES), traditional atomic emission spectroscopy (AES), X-ray fluorescence, and diffractometry. Nevertheless, these methods have certain shortcomings and difficulties that limit their use.

For instance, FAAS involves sample size in large amount, ignitable fuels, and a moderately exclusive cathode lamp for separate element. however, a cathode lamp based on multiple elements is obtainable, but element interference occurs. Beside this

Table 1 Permissible limits of heavy metal ions as framed by different regulatory authorities

S. No.	Toxic heavy metal ion	Common sources of contamination	Health risks	WHO	EU	FDA	References
1	Arsenic	Volcanic eruptions, soil erosion, insecticides	Encephalopathy Water diarrhea Heart disease Abdominal pain	0.01	0.01	0.01	[10]
2	Cadmium	Smoking, pigments, metal industries, paints	Salivation, muscle cramps, liver injury, convulsion	0.003	0.003	0.005	[11]
3	Chromium	Synthetic rubber, paints, pigments, Electroplating	Depression, anxiety, kidney and liver disease	0.05	0.05	0.100	[12]
4	Copper	Industrialization, electroplating	Hypotension, jaundice, hemolytic anemia, melena	2.00	2.00	1.000	[13]
5	Lead	Fossil fuel burning, jewelry, paints	Memory issues, coma, infertility, depression	0.01	0.01	0.005	[14]
6	Zinc	Electroplating, industries	Nausea, vomiting, cramps	–	5.00	5.00	[15]
7	Mercury	Industries, paints	Tremors, insomnia, memory loss, retarded CNS, PNS	0.001	–	–	[16]
8	Manganese	Industries, paints, agriculture	Glucose intolerance, skin ailments, blood clotting, fatness	–	0.3	0.05	[17]
9	Nickel	Smoking, industries, pigments	Bronchitis, lung, carcinoma, allergy	–	0.02	0.1	[18]
10	Iron	Chemical industries	Hypovolemic shock Stomach ache, vomiting, metabolic acidosis	–	0.2	0.300	[19]

(continued)

Table 1 (continued)

S. No.	Toxic heavy metal ion	Common sources of contamination	Health risks	WHO	EU	FDA	References
11	Cobalt	Agriculture industries	Thyroid damage Impaired vision Pneumonia	0.005	–	–	[20]
12	Antimony	Industries	Lung diseases Vomiting Skin irritation	0.005	0.005	0.006	[21]

chemical ionization interferences, background, along with spectra are challenging in FAAS. In GF-AAS, the oxidation of carbon graphite founds a task to manage.

In case of ICP-OES and ICP-MS instruments necessitates the application of undiluted argon gas. X-ray diffractometry is costly and not as much of sensitive for metal ion analysis in minute concentration. Beside this running and maintenance cost of these instrument is high with requisite of specialized personnel, lack of portability, limits their practical implication for in-situ and field studies in heavy metal detection. Electrochemical and colorimetric nano-sensors are now emerged as substitute strategies that provide solutions to some of the tasks associated with available HMI detection approaches.

In this facet, different electrochemical nano-sensors for heavy metal sensing have enticed significant heed in current years. The underlying reasons of profound success are greater sensitivity and selection, reasonably priced, fast and easy detection, ease of portability, sometimes disposable option is also present and thus facilitating in-situ and on demand field detection [22–24].

Hu et al. (2023) have extensively investigated the developments and feasibilities associated with moveable heavy metal detection sensors [25]. Furthermore, sensors based on electrochemical and colorimetric principle have been in practical use because of fast screening, on spot visual detection, thereby facilitating field detection of heavy metals.

In order to detect heavy metal ions in biological and environmental samples quickly, precisely, sensibly, and specifically, this chapter presents a revolutionary method to low-cost electrochemical and colorimetric nano-sensors. This chapter examines the recent advancements in smartphone-operated screen-printed electrodes (SPEs), plastic chip SPEs, carbon fiber paper-based sensors, and microfluidic-based nano-sensors for the detection of specific environmental heavy metals. Furthermore, it examines the cost-effectiveness of portable electrochemical nano-sensors. Arsenic(III), Arsenic(IV), Chromium(III), Chromium(IV), Iron(II), Iron(III), Mercury(I), Mercury(II), Lead(II), Lead(IV), and Manganese(II). This chapter comprehensively addresses ions in environmental and biological samples, along with consumable items. The chapter also looks at possible future possibilities in this subject as well as the advantages and drawbacks of current electrochemical and colorimetric nano-sensor techniques for detecting heavy metal ions. Electrochemical nano-sensors that are portable and can be used to identify heavy metal ions in environmental samples.

Because heavy metal pollution can have a negative impact on aquatic life, human health, and ecosystems, it is a serious environmental concern. When heavy metals are introduced into aquatic systems in large quantities—often as a result of industrial processes, urban runoff, and human activity—they are categorized as pollutants. Often found in wastewater are chromium, lead, mercury, cadmium, arsenic, and nickel heavy metals. Developing effective mitigation techniques requires ongoing monitoring of the quality of the water and research into the causes and effects of heavy metal contamination. A widely used analytical technique for assessing and monitoring heavy metals in various environmental samples, such as soil, water, and air, is electrochemical detection [26, 27]. Because they are affordable, easy to use, and suitable for on-site monitoring, disposable and portable electrochemical sensors have garnered a lot of attention [28]. Because of its user-friendly design, the sensors provide accurate and timely measurements without the need for sophisticated equipment or a lengthy sample preparation process. Effective method development requires that these techniques be sensitive to and selective against heavy metal pollution. Electrochemical sensor production requires the use of nanomaterials. Nanomaterials present promising opportunities for enhancing the selectivity and sensitivity of electrodes in heavy metal detection, as they provide robust support structures and feature highly active sites for functionalization [29].

According to recent studies, there is a growing focus on using nanomaterials to enhance electrode surfaces' ability to detect heavy metals. Nanomaterial-modified electrodes have proven to be highly successful in electroanalytical methods for different heavy metal identification. Examples of these nanomaterials include metal nanoparticles, metal oxides, materials based on graphene, carbon nanotubes, and metal–organic frameworks (MOFs). A variety of techniques, such as drop casting, dip coating, spin coating, electrochemical deposition, direct growth, and screen printing, are commonly employed to fabricate nanomaterials on disposable electrodes [30]. With limits of detection of 0.036 nM, 0.003 nM, and 0.011 nM, respectively, Huang et al. (2023) have shown that phosphorus-doped biochar–attapulgite/bismuth film electrodes coated with magnetic Fe_3O_4 nanoparticles (MBABiFE) may be used to detect Cd(II), Pb(II), and Hg(II). The study illustrated the application of data from a portable wireless smart sensor integrated with a modified screen-printed electrode for multi-metal analysis, utilizing a machine learning model grounded in an artificial neural network (ANN) [31]. A composite consisting of thymine acetic acid (Fe_3O_4@Au/CA/T-COOH) has been developed for the detection of Hg(II) in wastewater samples, with a detection range of 1–200 g/L. This composite utilizes core–shell Fe_3O_4@Au nanoparticles that are anchored with cysteamine [32]. Additionally, disposable solid-phase extraction devices have been successfully employed to detect cadmium (Cd II), lead (Pb II), copper (Cu II), and mercury (Hg II) at parts per billion concentrations using silver nanowires and butterfly-shaped silver nanoparticles [33, 34]. Graphene and carbon nanotubes are two common carbonaceous nanomaterials used to alter screen-printed electrodes. Based on epitaxial graphene-modified SiC paper, Hajzus et al. (2022) studied a sensitive platform for the selective voltammetric detection of $CdCl_2$, $CuSO_4$, $HgCl_2$, and $PbCl_2$ in seawater. Using cyclic square wave voltammograms as a basis, they applied machine learning models to precisely detect different kinds of heavy metals [35].

For the purpose of quickly detecting Cu(II) and Hg(II) ions, Bao et al. (2021) built a SPE modified with chitosan/PANi–Bina noparticle@graphene oxide multi-walled carbon nanotubes (CS/PANi–Bi NP@GO–MWCNT). Their portable detection platform consists of an electrode sleeve interface, a photocured 3D-printed shell, a Bluetooth chip, and an in-situ signal analysis circuit. The Hg(II) and Cu(II) detection limits of this portable electrochemical sensor were tested to be 10 ppb and 0.998 ppm, respectively [36].

3 Modish Techniques of Heavy Metal Detection

In the decade of upsurging technology, speedy, high-efficiency, and field-transmittable sensing equipment are commonly recognized owing to their rapid diagnostics and economical. On-the-spot portable sensors have been developed taking into account their high demand, despite of the fact the technology deprives of sensitivity and selectivity in comparison to traditional methods.

This snippet offers an overview of advancements in methodologies for detecting heavy metals via a detector that converts one signal into another, functioning in a physicochemical manner—potentially electrochemical, optical, piezoelectric, or electrochemical luminescence. The signals resulting from the interaction of the analyte with the biological or chemical component facilitate the measurement and calculation of the target analyte. Designed chemical or bio-sensors address the limitations of conventional heavy metal sensing methods by enabling rapid and on-site detection, requiring minimal operational skills, and being economical and field-portable [37].

In fact, on-site hasty analyte detection is the foremost standard that render advantage to these biosensors and chemical sensors over the time and cost intensive, unadventurous detection techniques. Currently, the different kinds of bio and chemical based sensors are in practice for heavy metal detection like ion-imprinted sensors, DNA aptamers, RNA-mediated biosensors, protein-mediated biosensors etc.

The application of aptamer biosensors and enzymatic-based sensors utilizing DNAzymes is particularly notable [38]. An extensive collection of imprinted electrochemical sensors is available and can be easily enhanced for analyte detection at a reasonable cost in a short timeframe with minimal effort. Single-stranded DNA molecules known as Aptamers needs to be handled with caution, they exhibit selectivity towards analytes like heavy metals with colorimetry or fluorimetry [39]. Besides, enzymatic biosensors have also been explicitly applied in heavy metals detection which specially involves DNAzymes for specificity.

3.1 Enzyme-Based Biosensors

Biosensors involve the application of a transducer, which functions as a conduit between the analyte and the transducer. These sensors illustrate two benefits: enhanced selectivity and the ability to detect heavy metals. The presence of heavy metals predominantly inhibits enzyme activity. This contact is very precise for specific ions and henceforth depicts increased signal production. The underlying criterion for choosing enzyme is their capacity of transferring electrons. Following the investigation conducted by Ilangovan et al. (2006), a biosensor utilizing a thick film interdigitated electrode was created [40]. The engineered bio-sensor has been utilized for the detection of heavy metals and demonstrated a positive response to changes in urea concentration between 1 and 15 mM. The standardization of the urea sensor is achieved through the explicit detection of heavy metal ions at different concentrations. Concentrations of heavy metals ranging from 0.1 to 10 mM were identifiable. The highest degree of inhibition occurs with Cd, followed by Cu and Pb among the three metals examined. The sensitivity measured in the spectrometric technique was 1 Mm, while the electrical method exhibited a sensitivity of 5 Mm.

Moyo utilized a horseradish peroxidase (HRP) biosensor to detect Zn ions in aqueous solutions [41]. This study discusses the use of a Maize tassel-multiwalled carbon nanotube (MT-MWCNT) composite as a matrix for the adsorption of horseradish peroxidase (HRP) onto the surface of a glassy carbon electrode through electrostatic interactions. The HRP/MT-MWCNT biosensor was employed for the detection of Zn^{2+} in aqueous solutions, demonstrating a range of 0.35–12 mg l^{-1} and a detection limit of 7.5 μg/L.

The inhibition was characterized as uncompetitive and reversible, consistent with the data modeling conducted using Dixon and Cornish-Bowden plots. The developed sensor demonstrates excellent copiability and remarkable selectivity. This method may also be applied to identify other trace metal ions that hinder HRP activity.

One of mentioned recent advancement is the application of DNAzymes based technology. The sensing based on DNAzymes could be colorimetric, fluorometric and conductometric sensitivity. Some studies involve coating of DNAzymes with nanoparticles which exhibit particular binding site of heavy metals, followed by tagging with quencher and incubation with specific fluorophore tagged with DNA strands [42]. Since the quencher lies in proximation with fluorophore, fluorescence is being quenched, but then again upon incubation with heavy metals, substrate DNA is cleaved efficiently by DNAzymes thereby quencher is freed from fluorophore, imparting fluorescence.

Another DNAzymes based sensor depicting features like simple, hasty, delicate, and specific for lead ion detection was described by Li and Lu [43]. The results indicated that the detection range spanned from 10 nM to 4 μM, with selectivity for Pb ions being 80 times greater than for other elements. While enzyme-immobilized biosensors exhibit subtlety and selectivity, they are more time-consuming than ion-imprinted polymer-based or aptamer-based sensing methods (Table 2).

Table 2 Enzyme based sensor for heavy metal detection

S. No.	Sensor designing	Transduction principal	Metals detected	LOD	References
1	Pb^{2+}-specific DNAzyme immobilized on magnetic beads, coupled to rolling circle amplification (RCA)	pH metry	Pb^{2+}	0.91 nM	[44]
2	Peroxidase-mimicking DNAzyme	Chemiluminescence	Hg(II)	12 nM	[45]
3	Cu^{2+}-dependent DNA ligation DNAzyme	Detection through personalized glucose meter	Cu^{2+}	1 nM	[46]
4	Urease and glucose oxidase based biosensor	Photoluminescence	Cu^{2+}, Pb^{2+}, Cd^{2+}	10 nM	[47]
5	Ultrathin polypyrrole–glucose oxidase (PPy–GOx) potentiometric biosensor	Potentiometry	Cu^{2+}, Pb^{2+}, Cd^{2+}	10 nM	[48]
6	8–17 DNAzyme (the recognition element) immobilized on carboxyl-functionalized graphene modified electrode	Electrochemical	Pb^{2+}	1.74×10^{-14} mol L^{-1}	[49]
7	Fluorescent Cu^{2+} sensor (PL-Cu 1.0) based on the relationship between glutathione/Cu^{2+} and catalytic activity of Pistol-like DNAzyme (PLDz)	Fluorimetry	Cu^{2+}	21.1 nM	[50]
8	Cu^{2+}-specific RNA-cleaving DNAzymes (PSCu10)	Fluorimetry	Cu^{2+}	1.6 nM	[51]

3.2 Biosensors Utilizing Aptamers

Aptamers are classified as oligos or short-chain nucleotides, created based on their ability to bind heavy metals. The production of these oligos primarily involves a systematic advancement of ligands through a serial application of the exponential enrichment (SELEX) procedure. The core principles that underlie aptamer-based sensing can be categorized into two main areas: the first focuses on the immobilization of aptamers onto electrochemical electrode surfaces to improve sensitivity, while the second relates to the interaction of aptamer-based substrates, like gold nanoparticles, which produces a colorimetric response [52].

In fact, specific metal ion detection in sample can be done by either of the method. Surprisingly their LOD, and sensitivity are much higher than conventional techniques hence are much efficient with greater output in detection within contaminated samples. They depict quick detection and portable nature which imparts them benefit of field-moveable technology.

A recent investigation conducted by Wang and Si presented an innovative aptamer biosensor that employs multiwalled carbon nanotubes (MWCNT). This design enables the transfer of fluorescent energy upon binding with heavy metals such as Pb^{+2}, thus allowing for rapid detection [53]. The reported limit of detection for Pb^{+2} was 20 nM. Likewise, this aptamer-based technique was also deployed in cadmium detection.

Wu et al. reported that Cd-specific aptamer [Cd-4] can also be explicitly used in detection of cadmium ions on the basis of colorimetric reaction [54]. The gold nanoparticles were mixed poly (diallyl dimethylammonium chloride) (PDDA) leading to the duplex formation. The wine-red color of these nanoparticles shifted to blue in the presence of cadmium ions, indicating the absence of duplex formation and, consequently, the lack of aggregation of gold nanoparticles. The reported limit of detection was 4.6 nmol L^{-1}. A research study developed an aptamer-sensor utilizing an electrochemical mode, which is based on carboxylic acid functionalized multi-walled carbon nanotubes (MWNTs–COOH) for the detection of lead in contaminated water samples [55].

Methylene blue (MB) served a dual purpose as both a redox probe and a G4-binding indicator. The sensor has the capability to detect Pb concentrations ranging from 0.01 to 50 pmol L^{-1}, with a limit of detection at 4.3 fmol L^{-1}. Moreover, Zhou et al. devised a technique for detecting copper ions utilizing azide and terminal alkyne gold nanoparticles in an aqueous environment through the application of click chemistry. The conversion of Cu^{2+} to Cu^{1+} takes place in the presence of sodium ascorbate. This method enables unaided eye visualization of copper ions through the agglomeration of gold nanoparticles, resulting from copper-mediated conjugation between distinct functional groups. This detection method relies on Cu(I)-mediated catalysis of the 1,3-dipolar cycloaddition of azides and alkynes on the surface of gold nanoparticles, resulting in their aggregation. The catalyst, Cu(I), was obtained through the reduction of Cu(II) using sodium ascorbate. This method enables visual detection of Cu^{2+} ions by the aggregation of AuNPs, observable by the naked eye without advanced instruments, due to the Cu(I)-catalyzed conjugation of the two functional groups. This method for detecting Cu^{2+} ions is based on the Cu(I)-catalyzed 1,3-dipolar cycloaddition of alkynes and azides on the surface of functionalized AuNPs, leading to the aggregation of AuNPs [56].

A straightforward, rapid, and portable colorimetric method for detecting Cr(III) in aqueous solutions has been reported. This technique relies on a color change from wine red to blue due to aggregation, which can be observed with the naked eye and easily documented through the extinction spectrum of the particles. The reaction components consisted of dithiocarbamate-modified N-benzyl-4-(pyridine-4-ylmethyl) aniline ligand (BP-DTC) and functionalized gold nanoparticles, facilitated by Cr(III). The current sensor is capable of detecting Cr(III) with a limit of detection of 31 ppb and can also facilitate the indirect detection of Cr(VI) in wastewater [57].

3.3 Ion Imprinted Sensors

Ion imprinted polymers comprise of particular cavities for specific ions. Generally, in the presence of ions, monomers carried out polymerization to form polymers. The extraction of ions in successive manner produce generates cavities within the polymer matrix. The binding of specific ions in the cavities produced on the polymer matrix paves the way for further sensing studies. These ion-imprinted polymers could be simply polymerized on surface of electrodes for the detection of analyte molecules based on electrochemical reactions. Oftenly ion-imprinted polymers have found various applications such as catalysis, chemical separations, or molecular sensors etc. [58]. Ion-imprinted polymer-based sensors depicts the range from 0.01 to 50 pmol L^{-1} with the limit of detection 4.3 fmol L^{-1}. The transduction principle is based on techniques such as impedance spectroscopy, differential pulse voltammetry (DPV), and square wave voltammetry (SWV). The ion-imprinted technique exhibits distinctive properties, including sensitivity, selectivity, and accuracy, which provide advantages over conventional methods [59]. A polyaniline derivative, Poly(diphenylamine-co-2-amino benzonitrile) (p(DPA-co-2ABN)), was deposited on an electrode through electrodeposition and utilized for the detection of Cd^{2+} ions employing differential pulse voltammetry (DPV) [60]. The presence of cyano groups in p(DPA-co-2ABN) facilitates the pre-deposition of Cd ions, resulting in the formation of a cyano chelation complex. This electrochemical method for Cd detection demonstrates a limit of detection (LOD) of 0.255 mg L^{-1}.

They illustrated the utilization of data from a portable wireless smart sensor integrated with the modified screen-printed electrode for conducting multi-metal analysis via a machine learning model grounded in an artificial neural network (ANN) [31]. A composite including thymine acetic acid (Fe_3O_4@Au/CA/T-COOH) has been developed for the detection of Hg(II) in wastewater samples, within a concentration range of 1–200 g/L, utilizing core–shell Fe_3O_4@Au nanoparticles anchored with cysteamine [32]. Moreover, the application of disposable solid-phase extraction devices has enabled the successful detection of Cd(II), Pb(II), Cu(II), and Hg(II) at parts per billion concentrations using silver nanowires and butterfly-shaped silver nanoparticles [61]. This exceptionally low detection limit is achieved through two methods. The initial approach utilizes SAM-modified electrodes to generate a minimal background current, thereby enhancing electron transport to the underlying polycrystalline electrode through gold nanoparticles attached to the terminus of the SAM. Gold nanoparticles are functionalized with cysteine to enhance copper accumulation. The low limit of detection (LOD) is achieved by incorporating cysteine-modified nanoparticles into the sample solution, in conjunction with the self-assembled monolayer (SAM) modified electrode. The complexation of copper with nanoparticles in solution induces their aggregation, leading to the accumulation of these aggregates on the electrode. As a result, a higher concentration of Cu in the solution accumulates at the electrode, leading to enhanced detection limits. This technique has not yet achieved limit of detection (LOD) values comparable to traditional methods or established toxicity thresholds. However, it offers advantages such as high throughput, cost efficiency, and rapid sample screening, which could be enhanced for portability.

3.4 Other Types of Sensors

The occurrence of heavy metals in water could be detected through numerous electrical and chemical sensors that comprise, impedimetric, electrochemical, electronic tongue (e-tongue), electronic nose (e-nose) facilitated detection. This segment outlines few important detection evidences of some heavy metals in water by techniques discussed above in portable fashion.

Wei et al. (2015) deployed noble metal-based electrodes for detection of Cr(IV) presence in water by application of the voltametric technique. In fact, they deployed spectroscopy based on electrochemical impedance for detection of chromium even in traces in water samples. The underlying tactic was crown ether binds specifically with $HCrO_4$ resulting in the formation of sandwich complexes which in turn retards the accessibility of redox probe for gold electrode as a result of which resistance in transfer of electron is increased. The reported range of detection was 1–100 ppb with LOD as 0.0014 ppb at pH 5 [62]. Likewise, in a different study conducted by Canhoto and Magan (2003) presence of heavy metals were sensed deploying bacterial cultures such as *Pseudomonas aeruginosa*, *Escherichia coli*, and *Enterobacter aerogenes*. In fact, these bacteria detect very minute concentration of heavy metals such as Zn, Cd, Pb and As using e-nose. The activity of these microbes could be altered even with the presence of minute heavy metals as low as 0.5 ppm [63].

Teodoro et al. (2019) have described a improved e-tongue-based heavy metals sensing in the water samples where nanocomposites based on electro spun nanofibers, cellulose nanowhiskers (CNW), and silver nanoparticles have been exploited as sensing material. The described materials are joined in six different forms as sensing materials which are deployed in fabrication of impedimetric e-tongue. They have successfully demonstrated that array of sensors can detect traces of Pb concentration as low as 10 nmol L^{-1} in drinking water [64].

In the similar fashion, another electronic tongue was designed for detection of array of heavy metals such as Cd^{2+}, Cu^{2+} and Pb^{2+} in water samples using sequential injection technique for best results. The designed sensor array comprised of four Ion-Selective Electrode (ISE) sensors, two for detection of Cd, and Cu and remaining for Pb and Zn. Artificial Neural Networking (ANN) and Fast Fourier Transform (FFT) were employed for procuring best results from sensor array [65].

Men et al. (2005) deployed e-tongue along with stripping voltammetry for the detection of environmental contaminants. Multiple light-addressable potentiometric sensors (MLAPS) in twinning with electrochemical sensors were used for the detection of Fe(III) and Cr(VI) ions. Anode and cathode-based stripping voltammetry was deployed for the detection of trace heavy metal concentration in water in addition to field-portable electronic tongue [66]. Also, Wilson et al. (2012) has described an e-tongue coupled with automatic flow injection potentiometric system for rapid detection of heavy metals in water. The e-tongue was a display of five sensors which comprise of selective electrodes like Cu^{2+} and Ca^{2+}. Real time monitoring of heavy metals is possible with these grouped sensor [67].

In addition to contemporary sensor-based detection methods, research has investigated the application of nanomaterials for the identification of heavy metal traces in food. Wu et al. (2019) emphasized the significance of Fe_3O_4 nanoparticles in conjunction with multiwall carbon nanotubes and their fluorinated variants for detecting Cu^{2+}, Pb^{2+}, and Hg^{2+} in food samples. The Fe_3O_4/F-MWCNTs electrochemical sensor exhibited detection limits of 0.08 nM for Pb^{2+}, 0.02 nM for Cu^{2+}, 0.05 nM for Hg^{2+}, and 0.05 nM for Cd^{2+} [68].

Pungjunun et al. (2020) utilized a bismuth nanoparticle modified graphene-coated electrode for the electrochemical detection of Sn(II) and Pb(II) in canned food samples. The square-wave anodic stripping voltammetry technique, in conjunction with CTAB and oxalic acid, was employed to mitigate interferences. A robust linear range was noted for both metals, with detection limits of 0.26 ng mL^{-1} for Sn(II) and 0.44 ng mL^{-1} for Pb(II) under optimal conditions [69].

A different ion-selective sensor utilizing FET technology was developed by Tatavarthi et al. (2020) for the detection of Arsenite As(III) in water and food samples. The developed sensor demonstrated significant sensitivity for arsenite, exhibiting a detection range from 10^{-10} M to 10^{-4} M. The sensor's reported limit of detection is 10^{-10} M, which corresponds to the standard detection capabilities of ICP-MS [70].

Tan et al. (2020) created a fluorinated graphene/gold nanocage nanocomposite aimed at detecting heavy metals in food derived from agriculture. The fluorinated graphene/gold nanocage (FGP/AuNC) electrode was employed for the rapid detection of heavy metals such as Hg^{2+}, Cd^{2+}, Cu^{2+}, Zn^{2+}, and Pb^{2+} using anodic stripping voltammetry in real samples, including ripe bolt, tea, and peanuts, among others. The findings aligned with established detection methods like AAS and ICP-MS [71].

Similarly, Zhang et al. (2020) have created an electrode-based smart device utilizing gold nanoparticles to detect the presence of heavy metals in orange juice, apple juice, and cow milk, among other substances. For the sample digestion, a solid stage microwave digestion device has been formulated. The device demonstrates commendable sensitivity and stability for detection lasting up to 72 h [72].

Palisoc et al. (2019) have devised a sensor for the identification of trace amounts of Cd and Pb in herbal food supplements. The sensor that was developed employed an electrode based on graphite, which was sourced from a zinc battery.

The electrodes underwent a coating process involving bismuth nanoparticles, and the detection method utilized was anodic stripping voltammetry. The linear range observed was from 5 to 1000 ppb, with a limit of detection for Cd^{2+} at 1.06 ppb and for Pb^{2+} at 0.72 ppb [73].

Sensors and biosensors present a range of benefits and drawbacks. Biosensors demonstrate significant affinity and specificity. The high specificity and affinity of biosensors can be attributed to the use of bioreceptors. Conversely, they come with high costs and are fundamentally unstable due to their biological characteristics. Bio-receptors enhance the production efficiency of biosensor devices. Maintaining activity poses a considerable challenge due to the susceptibility of functions to complex immobilization processes and harsh environmental conditions, resulting in diminished shelf lives.

Biomimetic ion-imprinted polymers offer a promising alternative to aptamers and enzymes, characterized by their stability and cost-effectiveness. Speculation exists regarding the high vigor and stability of ion-imprinted polymers across various conditions. Their ease of designing recognition sites for analytes, combined with relatively low production costs, may position them as a more promising sensor for detecting heavy metals.

A further inherent limitation of biosensors is their limited ability for multiplexing. Biosensors are limited to particular analytes, which restricts their transferability across different applications. For the analysis of multiple analytes, the necessity for collateral biosensors increases the technical costs involved.

On the other hand, for the multiplexed detection of analytes, physical and chemical sensors can be readily adjusted, although there is a challenge of interference stemming from the lack of precise recognition elements in complex samples. The choice of an appropriate sensing method is primarily influenced by the characteristics of the water samples, the surrounding environment, accessibility, as well as the requirements for selectivity, sensitivity, and the desired output.

4 Commercialized Instruments: Field Transportability and Accessibility

This snippet presents picturesque of the portable instruments available in the market which are widely used and distributed. Assembly of variable units comprising analyte recognition, signal production and amplification, signal reception and processing elements leads to formation of heavy metal detection system which is an effective substitute for the current signal detectors. Different devices and materials have been designed for miniaturizing conventional units [74].

In fact, for explaining the quantitative heavy metal analysis digital conversation has been applied parallelly. The intent of miniaturization is to apply high computational application like cloud computing, big data, and neural network with artificial intelligence [75]. The advent of charge-coupled devices in smartphones, light-emitting diodes (LEDs), and complementary metal-oxide semiconductor (CMOS) starters has made way for portable sensors that rely on optical and electrical signals.

This produces an innovative approach to displace conventional massive and affluent components by integrating the smartphone or advanced camera or any imaging system to develop a transportable, swift, and profitable quantitative detection. Still, analytical parameters procured by these portable devices are compromised in comparison to laboratory-based instruments. Thus, there is inevitable need of developing modish sensors with augmented characteristics to bring under control these limitations.

One hopeful plan of action is the combination of nanomaterials with biosensors and sensors for upgrading their performance. Fluorescence assays and colorimetric detection are among the most common signal transduction procedures for optical and bio-sensors. Quantum dots have been explored owing fine emission, band width thereby greater sensitivity. The efficacy of any electrochemical sensing device is relied on apt transducer selection.

Application of nanomaterials such as metallic nanoparticles, carbon-based nanomaterials on electrodes elicit catalytic activity and conductivity which increases signal transduction leading to amplified signals. We put confidence in further outstanding platforms that join the collaborative effects of nanoparticles and sensors/biosensors will be seemed to available in market for heavy metal detection.

5 Opportunities and Obstacles of Current Devices

Recent advancements in system integration and automation have facilitated the design of various analytical instruments for the detection of heavy metal ions. Many reported methods have shown to depict designing simplicity besides high accuracy yet, there are some lacks concerning the ideal designing of sensing device for heavy metal detection. Signal strength, stability and reproducibility are among the challenges in sensor development functioning in low concentration of analytes.

Amalgamation of nanotechnology with sensors/biosensors have led the designing modish sensitive platforms, so far, the problems of controlling size, distribution, and homogenization on macro-scale still remains a challenge. Another issue is multiplex detection and selection of current devices.

The marketed systems are primarily deprived of the bio-recognition elements thereby lacks selectivity and restrict manifold quantification capacity. The on-site monitoring of water contaminants presents a significant challenge due to the considerable variations in water quality and pH across different sample locations, rendering the use of a uniform device challenging; for instance, data obtained in Africa may not align with those in Asia. There remains an urgent necessity to create a cost-effective and completely automated, portable device for the real-time detection of heavy metals on-site.

6 Conclusion

The present chapter highlighted developments in sensing potentiality to field portability. The conventional approaches are confined to sensing only. Conventional approaches must go hand in hand with field portable applicability, facilitating quick screening. Earlier column sensitivity was used as a part of conventional system but rapid screen and field portability were remained as challenges to be addressed. Therefore, researchers have developed sensors or biosensors as a system which are equipped with receptors, transducers, decision support systems (DSS) which ensures speedy detection and portability.

The receptors used in modern sensors are aptamers and ion-imprinted receptors unlike conventional receptors which are antibodies and enzymes etc. and has shown successful results. Handful studies are available now a days which explains stable receptors designing, their immobilization led to the advancement in sensing in field conditions. Mostly used receptors for heavy metal detection work on the presence of label markers irrespective of the fact whether the signal is optical or electrochemical. Initially it was very difficult to extract heavy metals present in plants, soil or any other object. However, the advancements in process of heavy metals extraction have enabled ease on working with soil, fruits and vegetables etc. Some factors which need special attention in present day sensors are orientation of receptors, presence of other ions which needs to be managed for addressing noise and interference.

Taking into account all the developments of portable field sensors for heavy metal detection, electrochemical and optics-based sensors are making progress reason being cost-effective and ease of operation in comparison to other methods. Future research shall comprise of enhancing reproducing capability and anti-interference ability of these developed sensors.

References

1. Abeywickrama CJ, Wansapala J (2019) Review of organic and conventional agricultural products: heavy metal availability, accumulation and safety. Int J Food Sci Nutr 4(1):77–88
2. Briffa J, Sinagra E, Blundell R (2020) Heavy metal pollution in the environment and their toxicological effects on humans. Heliyon 6(9)
3. Joseph L, Jun BM, Flora JR, Park CM, Yoon Y (2019) Removal of heavy metals from water sources in the developing world using low-cost materials: a review. Chemosphere 229:142–159
4. Li C, Zhou K, Qin W, Tian C, Qi M, Yan X, Han W (2019) A review on heavy metals contamination in soil: effects, sources, and remediation techniques. Soil Sedim Contam Int J 28(4):380–394
5. Samanta S, Kumar V, Nag SK, Saha K, Sajina AM, Bhowmick S et al (2021) Assessment of heavy metal contaminations in water and sediment of River Godavari, India. Aquat Ecosyst Health Manag 24(4):23–33
6. Köse E, Emiroğlu Ö, Çiçek A, Aksu S, Başkurt S, Tokatli C et al (2020) Assessment of ecologic quality in terms of heavy metal concentrations in sediment and fish on Sakarya River and Dam Lakes, Turkey. Soil Sedim Contam Int J 29(3):292–303
7. Pieper KJ, Martin R, Tang M, Walters L, Parks J, Roy S et al (2018) Evaluating water lead levels during the Flint water crisis. Environ Sci Technol 52(15):8124–8132

8. Kruger DJ, Cupal S, Kodjebacheva GD, Fockler TV (2017) Perceived water quality and reported health among adults during the Flint, MI water crisis. Calif J Health Promot 15(1):56–61
9. Ramdani S, Amar A, Belhsaien K, El Hajjaji S, Ghalem S, Zouahri A, Douaik A (2018) Assessment of heavy metal pollution and ecological risk of roadside soils in Tlemcen (Algeria) using flame-atomic absorption spectrometry. Anal Lett 51(15):2468–2487
10. Ferguson MA, Fernandez DP, Hering JG (2007) Lowering the detection limit for arsenic: implications for a future practical quantitation limit. J Am Water Works Ass 99(8):92–98
11. Nawrot TS, Staessen JA, Roels HA, Munters E, Cuypers A, Richart T et al (2010) Cadmium exposure in the population: from health risks to strategies of prevention. Biometals 23:769–782
12. Codex Alimentarius Commission (2016) General standard for contaminants and toxins in food and feed. CODEX STAN 193–1995. Codex Alimentarius Commission
13. Olivares M, Uauy R (1996) Limits of metabolic tolerance to copper and biological basis for present recommendations and regulations. Am J Clin Nutr 63(5):846S-852S
14. Verberk MM, Willems TE, Verplanke AJ, Wolff FAD (1996) Environmental lead and renal effects in children. Arch Environ Health Int J 51(1):83–87
15. Hecq P, Hulsmann A, Hauchman FS, McLain JL, Schmitz F (2006) Drinking water regulations. In: Analytical methods for drinking water: advances in sampling and analysis
16. Water S, World Health Organization (2006) Guidelines for drinking-water quality [electronic resource]: incorporating first addendum, vol 1, Recommendations
17. Edition F (2011) Guidelines for drinking-water quality. WHO Chronicle 38(4):104–108
18. Levallois P, Villanueva CM (2019) Drinking water quality and human health: an editorial. Int J Environ Res Public Health 16(4):631
19. Kumar M, Puri A (2012) A review of permissible limits of drinking water. Indian J Occup Environ Med 16(1):40–44
20. Anderson AJ, Meyer DR, Mayer FK (1973) Heavy metal toxicities: levels of nickel, cobalt and chromium in the soil and plants associated with visual symptoms and variation in growth of an oat crop. Aust J Agric Res 24(4):557–571
21. Wu F, Fu Z, Liu B, Mo C, Chen B, Corns W, Liao H (2011) Health risk associated with dietary co-exposure to high levels of antimony and arsenic in the world's largest antimony mine area. Sci Total Environ 409(18):3344–3351
22. Jaswal BBS, Rai PK, Singh T, Zorba V, Singh VK (2019) Detection and quantification of heavy metal elements in gallstones using X-ray fluorescence spectrometry. X-Ray Spectrom 48(3):178–187
23. Mohamed R, Zainudin BH, Yaakob AS (2020) Method validation and determination of heavy metals in cocoa beans and cocoa products by microwave assisted digestion technique with inductively coupled plasma mass spectrometry. Food Chem 303:125392
24. Chunqiang LU, Sun D, Sun K, Liu J, Luo CJFI (2019) Determination of the three harmful heavy mental ions in plant fiber molded products by ICP-MS. Food Ind 40:299–301
25. Hu T, Lai Q, Fan W, Zhang Y, Liu Z (2023) Advances in portable heavy metal ion sensors. Sensors 23(8):4125
26. GadelHak Y, Hafez SH, Mohamed HF, Abdel-Hady EE, Mahmoud R (2023) Nanomaterials-modified disposable electrodes and portable electrochemical systems for heavy metals detection in wastewater streams: a review. Microchem J 109043
27. Malik LA, Bashir A, Qureashi A, Pandith AH (2019) Detection and removal of heavy metal ions: a review. Environ Chem Lett 17:1495–1521
28. Ferrari AGM, Carrington P, Rowley-Neale SJ, Banks CE (2020) Recent advances in portable heavy metal electrochemical sensing platforms. Environ Sci Water Res Technol 6(10):2676–2690
29. Thakur A, Kumar A (2022) Recent advances on rapid detection and remediation of environmental pollutants utilizing nanomaterials-based (bio) sensors. Sci Total Environ 834:155219
30. Mohamad Nor N, Ramli NH, Poobalan H, Qi Tan K, Abdul Razak K (2023) Recent advancement in disposable electrode modified with nanomaterials for electrochemical heavy metal sensors. Crit Rev Anal Chem 53(2):253–288

31. Huang P, Xiong Y, Ge Y, Wen Y, Zeng X, Zhang J et al (2023) Magnetic Fe_3O_4 nanoparticles decorated phosphorus-doped biochar-attapulgite/bismuth film electrode for smartphone-operated wireless portable sensing of ultra-trace multiple heavy metal ions. Microchim Acta 190(3):94
32. Butmee P, Mala J, Damphathik C, Kunpatee K, Tumcharern G, Kerr M et al (2021) A portable selective electrochemical sensor amplified with Fe_3O_4@ Au-cysteamine-thymine acetic acid as conductive mediator for determination of mercuric ion. Talanta 221:121669
33. Naseri M, Mohammadniaei M, Ghosh K, Sarkar S, Sankar R, Mukherjee S et al (2023) A robust electrochemical sensor based on butterfly-shaped silver nanostructure for concurrent quantification of heavy metals in water samples. Electroanalysis 35(2):e202200114
34. Saenchoopa A, Klangphukhiew S, Somsub R, Talodthaisong C, Patramanon R, Daduang J et al (2021) A disposable electrochemical biosensor based on screen-printed carbon electrodes modified with silver nanowires/hpmc/chitosan/urease for the detection of mercury (II) in water. Biosensors 11(10):351
35. Hajzus JR, Shriver-Lake LC, Dean SN, Erickson JS, Zabetakis D, Golden J et al (2022) Modifications of epitaxial graphene on SiC for the electrochemical detection and identification of heavy metal salts in seawater. Sensors 22(14):5367
36. Bao Q, Li G, Yang Z, Pan P, Liu J, Li R et al (2021) In situ detection of heavy metal ions in sewage with screen-printed electrode-based portable electrochemical sensors. Analyst 146(18):5610–5618
37. Buledi JA, Amin S, Haider SI, Bhanger MI, Solangi AR (2021) A review on detection of heavy metals from aqueous media using nanomaterial-based sensors. Environ Sci Pollut Res 28:58994–59002
38. Liu Y, Deng Y, Dong H, Liu K, He N (2017) Progress on sensors based on nanomaterials for rapid detection of heavy metal ions. Sci China Chem 60:329–337
39. Yola ML, Eren T, Atar N (2015) A sensitive molecular imprinted electrochemical sensor based on gold nanoparticles decorated graphene oxide: application to selective determination of tyrosine in milk. Sens Actuators B Chem 210:149–157
40. Ilangovan R, Daniel D, Krastanov A, Zachariah C, Elizabeth R (2006) Enzyme based biosensor for heavy metal ions determination. Biotechnol Biotechnol Equip 20(1):184–189
41. Moyo M (2014) Horseradish peroxidase biosensor to detect zinc ions in aqueous solutions. Open J Appl Biosens 2014
42. Zhou Y, Tang L, Zeng G, Zhang C, Zhang Y, Xie X (2016) Current progress in biosensors for heavy metal ions based on DNAzymes/DNA molecules functionalized nanostructures: a review. Sens Actuators B Chem 223:280–294
43. Li J, Lu Y (2000) A highly sensitive and selective catalytic DNA biosensor for lead ions. J Am Chem Soc 122(42):10466–10467
44. Tang D, Xia B, Tang Y, Zhang J, Zhou Q (2019) Metal-ion-induced DNAzyme on magnetic beads for detection of lead (II) by using rolling circle amplification, glucose oxidase, and readout of pH changes. Microchim Acta 186:1–9
45. Gribas AV, Sakharov IY (2018) Homogeneous chemiluminescent determination of mercury (II) using a peroxidase-mimicking DNAzyme assay. Anal Lett 51(9):1280–1290
46. Ming J, Fan W, Jiang TF, Wang YH, Lv ZH (2017) Portable and sensitive detection of copper (II) ion based on personal glucose meters and a ligation DNAzyme releasing strategy. Sens Actuators B Chem 240:1091–1098
47. Syshchyk O, Skryshevsky VA, Soldatkin OO, Soldatkin AP (2015) Enzyme biosensor systems based on porous silicon photoluminescence for detection of glucose, urea and heavy metals. Biosens Bioelectron 66:89–94
48. Ayenimo JG, Adeloju SB (2015) Inhibitive potentiometric detection of trace metals with ultrathin polypyrrole glucose oxidase biosensor. Talanta 137:62–70
49. Yu Z, Li N, Hu X, Dong Y, Lin Y, Cai H et al (2019) Highly efficient electrochemical detection of lead ion using metal-organic framework and graphene as platform based on DNAzyme. Synth Met 254:164–171

50. Wang S, Liu C, Li G, Sheng Y, Sun Y, Rui H et al (2017) The triple roles of glutathione for a DNA-cleaving DNAzyme and development of a fluorescent glutathione/Cu^{2+}-dependent DNAzyme sensor for detection of Cu^{2+} in drinking water. ACS Sens 2(3):364–370
51. Huang PJJ, Liu J (2016) An ultrasensitive light-up Cu^{2+} biosensor using a new DNAzyme cleaving a phosphorothioate-modified substrate. Anal Chem 88(6):3341–3347
52. Wu Y, Liu L, Zhan S, Wang F, Zhou P (2012) Ultrasensitive aptamer biosensor for arsenic (III) detection in aqueous solution based on surfactant-induced aggregation of gold nanoparticles. Analyst 137(18):4171–4178
53. Wang SE, Si S (2013) Aptamer biosensing platform based on carbon nanotube long-range energy transfer for sensitive, selective and multicolor fluorescent heavy metal ion analysis. Anal Methods 5(12):2947–2953
54. Wu Y, Zhan S, Wang L, Zhou P (2014) Selection of a DNA aptamer for cadmium detection based on cationic polymer mediated aggregation of gold nanoparticles. Analyst 139(6):1550–1561
55. Zhu Z (2017) An overview of carbon nanotubes and graphene for biosensing applications. Nano-micro Lett 9(3):25
56. Zhou Y, Wang S, Zhang K, Jiang X (2008) Visual detection of copper (II) by azide-and alkyne-functionalized gold nanoparticles using click chemistry. Angew Chem 120(39):7564–7566
57. Zhao L, Jin Y, Yan Z, Liu Y, Zhu H (2012) Novel, highly selective detection of Cr (III) in aqueous solution based on a gold nanoparticles colorimetric assay and its application for determining Cr (VI). Anal Chim Acta 731:75–81
58. Ravikumar R, Chen LH, Xin Hui MM, Chan CC (2019) Ion-imprinted chitosan-based interferometric sensor for selective detection of heavy metal ions. J Lightwave Technol 37(11):2778–2783
59. Sebastian M, Mathew B (2018) Carbon nanotube-based ion imprinted polymer as electrochemical sensor and sorbent for Zn (II) ion from paint industry wastewater. Int J Polym Anal Charact 23:18–28
60. Philips MF, Gopalan AI, Lee KP (2012) Development of a novel cyano group containing electrochemically deposited polymer film for ultrasensitive simultaneous detection of trace level cadmium and lead. J Hazard Mater 237:46–54
61. Gooding JJ, Shein J, Lai LM (2009) Using nanoparticle aggregation to give an ultrasensitive amperometric metal ion sensor. Electrochem Commun 11(10):2015–2018
62. Wei J, Guo Z, Chen X, Han DD, Wang XK, Huang XJ (2015) Ultrasensitive and ultraselective impedimetric detection of Cr (VI) using crown ethers as high-affinity targeting receptors. Anal Chem 87(3):1991–1998
63. Canhoto OF, Magan N (2003) Potential for detection of microorganisms and heavy metals in potable water using electronic nose technology. Biosens Bioelectron 18(5–6):751–754
64. Teodoro KB, Shimizu FM, Scagion VP, Correa DS (2019) Ternary nanocomposites based on cellulose nanowhiskers, silver nanoparticles and electrospun nanofibers: use in an electronic tongue for heavy metal detection. Sens Actuators B Chem 290:387–395
65. Mimendia A, Legin A, Merkoçi A, del Valle M (2010) Use of sequential injection analysis to construct a potentiometric electronic tongue: application to the multidetermination of heavy metals. Sens Actuators B Chem 146(2):420–426
66. Men H, Zou S, Li Y, Wang Y, Ye X, Wang P (2005) A novel electronic tongue combined MLAPS with stripping voltammetry for environmental detection. Sens Actuators B Chem 110(2):350–357
67. Wilson D, Del Valle M, Alegret S, Valderrama C, Florido A (2012) Potentiometric electronic tongue-flow injection analysis system for the monitoring of heavy metal biosorption processes. Talanta 93:285–292
68. Wu W, Jia M, Zhang Z, Chen X, Zhang Q, Zhang W et al (2019) Sensitive, selective and simultaneous electrochemical detection of multiple heavy metals in environment and food using a lowcost Fe_3O_4 nanoparticles/fluorinated multi-walled carbon nanotubes sensor. Ecotoxicol Environ Saf 175:243–250
69. Pungjunun K, Nantaphol S, Praphairaksit N, Siangproh W, Chaiyo S, Chailapakul O (2020) Enhanced sensitivity and separation for simultaneous determination of tin and lead using paper-based sensors combined with a portable potentiostat. Sens Actuators B Chem 318:128241

70. Tatavarthi SS, Wang SL, Wang YL, Chen JC (2020) Rapid and highly sensitive extended gate FET-based sensors for arsenite detection using a handheld device. ECS J Solid State Sci Technol 9(11):115014
71. Tan Z, Wu W, Feng C, Wu H, Zhang Z (2020) Simultaneous determination of heavy metals by an electrochemical method based on a nanocomposite consisting of fluorinated graphene and gold nanocage. Microchim Acta 187:1–9
72. Zhang W, Liu C, Liu F, Zou X, Xu Y, Xu X (2020) A smart-phone-based electrochemical platform with programmable solid-state-microwave flow digestion for determination of heavy metals in liquid food. Food Chem 303:125378
73. Palisoc S, Vitto RIM, Natividad M (2019) Determination of heavy metals in herbal food supplements using bismuth/multi-walled carbon nanotubes/nafion modified graphite electrodes sourced from waste batteries. Sci Rep 9(1):18491
74. Mehrotra P (2016) Biosensors and their applications—a review. J Oral Biol Craniof Res 6(2):153–159
75. Dash S, Shakyawar SK, Sharma M, Kaushik S (2019) Big data in healthcare: management, analysis and future prospects. J Big Data 6(1):1–25

Nanotechnology for Precision Sensing: Innovations in Nanosensors Synthesis, Mechanisms and Strategic Applications

Sathyabama Kowsalya, Rajkumar Divya, Radhakrishnan Sreena, Maheswaran Saravanakumar, Theivasigamani Parthasarathi, and Saiyyeda Firdous

Abstract The development of sensitive and selective nanosensors used in various industries, including manufacturing, transportation systems, disease detection, pollution control, and medicine, is greatly facilitated by nanomaterials (NMs). These instruments not only measure physical properties such as volume, concentration, mobility, and temperature but are also able to identify individual cells at the molecular level. Because nanoparticles (NPs) have so many applications in science and technology, their use in biosensors is becoming increasingly common. Numerous nanomaterials have been thoroughly investigated for their high sensitivity, selectivity, and ease of use in the detection and measurement of hazardous industrial chemicals, pesticides, toxic metal ions, and gold nanoparticles. These materials include silicon nanowires, carbon nanotubes, and quantum dots. Concerns about biocompatibility and environmental impact are among the difficulties faced by nanotechnology in sensing, as are problems with reproducibility, stability, and scalability of manufacture. Despite the drawbacks, their potential applications in industry, environmental monitoring, and healthcare are promising. To create more sophisticated sensor platforms, we are investigating the potential of nanotechnology for the future. This chapter aims to provide a comprehensive overview of the current state of nanotechnology in sensor technology while highlighting important areas for further study and progress.

S. Kowsalya · T. Parthasarathi · S. Firdous (✉)
VIT School of Agricultural Innovations and Advanced Learning, Vellore Institute of Technology (VIT), Vellore, India
e-mail: saiyyeda.firdous@vit.ac.in

R. Divya
School of Advanced Sciences, Vellore Institute of Technology (VIT), Vellore, India

R. Divya · R. Sreena · M. Saravanakumar
Centre for Biomaterials, Cellular and Molecular Theranostics, Vellore Institute of Technology (VIT), Vellore, India

R. Sreena · M. Saravanakumar
School of Bio Sciences and Technology, Vellore Institute of Technology (VIT), Vellore, India

I. Uddin (ed.), *Smart Nanosensors*, Smart Nanomaterials Technology,
https://doi.org/10.1007/978-981-96-3878-9_8

Keywords Precision sensing · Carbon nanotubes · Magnetic nanomaterials · Strategic applications · Environmental monitoring · Medical diagnostics

1 Introduction

Nanotechnology is an area dedicated to the development of nanoparticles measuring between 1 and 100 nm, along with modifying the geometry and size of these particles using different synthesis methods. This complex interdisciplinary field includes nanoscience, nanomaterials, nano-chemistry, nanophysics, nanoelectronics, and nano-bionics. Nanomaterials serve as crucial tools for easy engineering and precisely adjusting extraordinary sensing configurations based on recognition events occurring at the nanoscale [1]. Consequently, a variety of nanomaterials are frequently utilized in sensor design across different platforms. Various forms of nanomaterials, including nanoparticles, nanotubes, nanofibers, nanowires, nanorods, nanocomposites, nano polymers, nanofilms, and nanoplates, are employed [2, 3]. Numerous sensors that rely on nanomaterials have been created for a wide range of applications utilizing carbon nanotubes, gold nanomaterials, graphene, nanomotors, nano cables, carbon spherical shells, and carbon nanoparticles [4, 5].

A sensor is an instrument that identifies a measurable quantity, typically in an electronic format, and translates this measurement into specific signals. Key requirements for sensors include diversity, sensitivity, the accuracy of the information obtained, selectivity, and stability [6, 7]. These features allow for a wide array of applications, such as quantum-effect devices, lasers, biological cells, solar cells, transistors, catalysts, molecular electronic systems, photonic band gap devices, photocatalysts, nano fuel cells, surface-enhanced Raman spectroscopy (SERS), nano drug delivery systems, nanoactuators, advanced energy storage solutions, and Nanosensors [8]. At present, numerous advanced nanomaterials are being utilized to create a variety of sensors for the detection of harmful toxins in the environment [9, 10].

In recent years, the advancement of Nanosensors utilizing nanomaterials has attracted considerable interest for a variety of applications, such as fundamental biological investigations, health monitoring, clinical diagnostics, pharmaceutical assessments, food safety, and environmental surveillance. Owing to their outstanding physicochemical properties, increased surface-to-volume ratio, substantial adsorption, and reactivity, in addition to other benefits absent in larger materials, nanomaterials present exciting prospects as analytical probes, providing improved sensitivity and facilitating a groundbreaking method for analyzing single molecules [11, 12]. Nanomaterials endowed with distinctive functionalities facilitate the promising advancement of Nanosensors, which can be easily activated upon encountering emerging chemical pollutants, as well as allow for real-time monitoring of the environment in air, soil, water, and food safety, underscoring the necessity to protect our environment. Sensors that incorporate nanostructured materials have been widely utilized as effective analytical tools, capitalizing on their ease of handling, cost-efficiency, high sensitivity and specificity, rapid response time, straightforward fabrication, and portability [13, 14].

Nanosensors enable the development of advanced technologies. They can interact at the nanoscale, such as with proteins, allowing them to identify unique phenomena not observable at larger scales. Currently, various techniques have been proposed for fabricating Nanosensors, including top-down lithography, bottom-up assembly, and molecular self-assembly. The nanoscale is a peculiar domain where numerous processes occur, significantly different from those at the macro dimension [15]. The fundamental necessity of Nanosensors aids in conducting multiple everyday tasks, leading to their successful application across diverse sectors to meet ongoing demands. In public areas, Nanosensors are utilized to monitor the spread of viruses and diseases [16]. Some practical uses comprise real-time testing of blood, disease, and breath within in-body networks. Moreover, these technologies can be seamlessly integrated into wearable health and environmental monitoring devices. Incorporating nanotechnology and biotechnology into the Internet of Things (IoT) may offer increased control over various aspects of our lives [17]. A key limitation of this sensing mechanism is the requirement for an external light source and a device capable of measuring and comparing the different resonance frequencies of particles. Coordination and communication among Nanosensors can address this challenge. For example, Nanosensors can emit identical particles with significantly reduced power and detect the re-radiated energy at different frequencies. This information can then be processed and transmitted to a central data sink [18].

In this scenario, it is noteworthy that the arrangement of nanomaterials and biological materials plays a crucial role in the design of hybrid nanostructured analytical devices "with enhanced detection performance in terms of response time, higher storage/operational stability, resistance toward environmental conditions, improved selectivity, reduced sample volumes, and easy sampling" [19].

2 Nanomaterial-Based Sensor

The field of nanomaterial-based sensors has experienced remarkable advancements in recent years, with researchers harnessing the unique properties of nanoscale materials to develop highly sensitive, selective, and efficient sensing devices for a wide range of applications [20, 21]. Nanomaterials possess exceptional physicochemical properties, such as high surface-to-volume ratios, high adsorption and reactive capacities, and other beneficial characteristics not found in their bulk counterparts, making them ideal candidates for sensor development. These properties have enabled the design of advanced sensor platforms that offer enhanced sensitivity, faster response times, and reduced size and cost compared to traditional sensors [14]. Developing analytical methods based on nanomaterials has garnered growing attention for numerous applications, including fundamental biological research, health monitoring, clinical diagnostics, pharmaceutical analysis, food safety, and environmental monitoring.

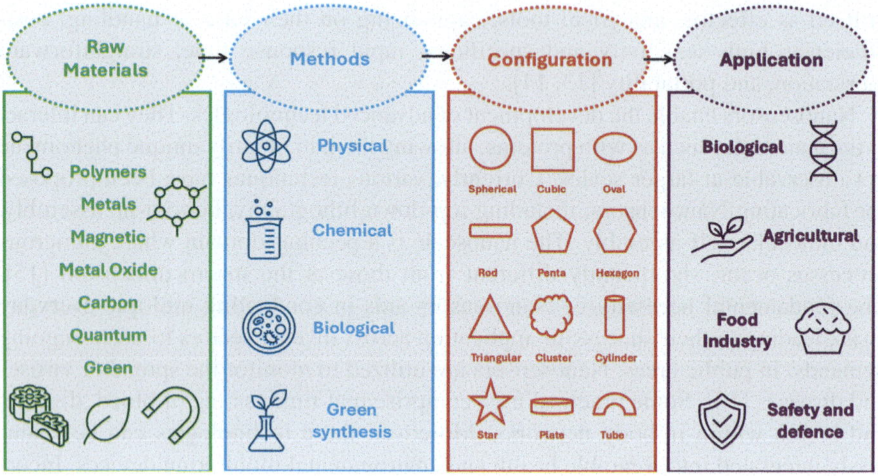

Fig. 1 Graphical representation of nanomaterials types, methods of fabrication and application

Nanomaterials have been successfully integrated into screen-printed electrochemical sensors for the detection of heavy metal ions in agricultural environments and food samples [20, 21] (Fig. 1).

2.1 Carbon Nanotubes

Carbon nanotube (CNT) based sensors leverage the unique properties of CNTs, such as high conductivity and large surface area, for various applications. They are used in electrochemical sensors for food safety, optical sensors for heavy metal detection, and field-effect devices for virus detection, including COVID-19 diagnostic [22, 23]. CNTs are used in gas sensors to detect various gases, including toxic and volatile organic compounds. Their high surface area and sensitivity to gas adsorption allow for the detection of gases like NO_2, NH_3, CO, and hydrocarbons at low concentrations [24–27]. CNTs serve as active elements in electrochemical sensors, where they can enhance the detection of biomolecules and chemical species. Their high electrical conductivity and catalytic activity make them suitable for applications in biosensing, such as glucose sensors and the detection of heavy metals [28]. CNT sensors can be designed to detect multiple health parameters simultaneously, for example, a combination of CNTs and other materials, like silk yarns with silver nanowires, has been used to create sensors capable of measuring both temperature and pressure, showcasing their versatility in health diagnostic. They are known for their excellent electrical conductivity and high surface area. These properties enable them to respond quickly to changes in their environment, such as temperature, pressure, or the presence of specific biomolecules. Their conductivity can change significantly in response to external stimuli, making them highly sensitive sensors. The specific mention of a temperature sensitivity of up to 1.23%/°C indicates that for every degree Celsius change in temperature, the electrical response of the CNT sensor changes by 1.23%. This level of sensitivity is advantageous for accurately monitoring temperature variations, which is crucial in many health diagnostics applications. Traditional temperature sensors often require additional components like digital converters or amplifiers to process the signals they generate. However, due to the high sensitivity of CNT-based sensors, they can provide a direct electrical signal that is already suitable for interpretation without the need for these extra components. This simplification can lead to more compact and efficient sensor designs. By reducing the number of components needed, CNT sensors can be made smaller, lighter, and more integrated into wearable technologies. This is particularly important for applications in health monitoring, where comfort and convenience are paramount. A simpler design also often translates to lower production costs and easier manufacturing processes [29]. The integration of CNTs into wearable sensors has enabled continuous health monitoring. This continuous monitoring allows for the early detection of potential health issues and provides users with valuable insights into their health status as in Fig. 2. These sensors can effectively track various vital parameters, including:

Heart Rate: By measuring the electrical signals associated with heartbeats, CNT sensors can provide accurate heart rate readings.

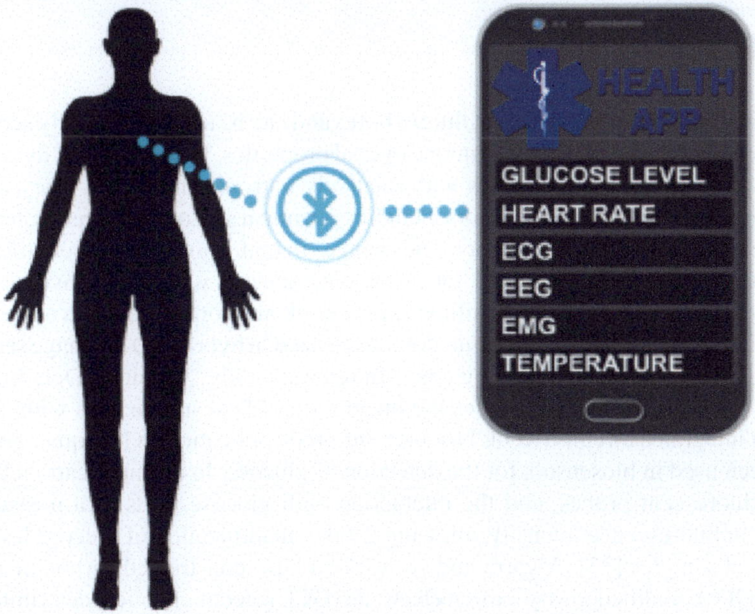

Fig. 2 Possible application of CNT sensors in smart textiles for health monitoring reproduced from [30] and reprinted from Creative Commons Attribution 4.0 international license (CC BY 4.0)

Blood Oxygen Saturation: CNTs can be used in optical sensors to measure the absorption of light by oxygenated and deoxygenated blood, allowing for non-invasive monitoring of blood oxygen levels.

Body Temperature: The high-temperature sensitivity of CNTs enables precise measurement of body temperature, which is essential for monitoring fever or other health conditions.

CNTs are known for their exceptional strength-to-weight ratio, making them lightweight. This characteristic is particularly important for wearable devices, as users prefer gadgets that are comfortable to wear for extended periods without causing discomfort. CNT sensors typically require low power to operate, which is a significant advantage for wearable technology. This low power consumption extends the battery life of devices, allowing them to function for longer periods without frequent recharging. Many wearable devices equipped with CNT sensors can connect to smartphones or other devices, enabling users to track their health metrics over time and share them with healthcare providers if needed [30]. Carbon nanotubes (CNT) can be used as biosensors in EEG applications due to their non-toxic nature, excellent electrical conductivity, and superior mechanical properties. They can be integrated into electrodes, providing stretchability and the ability to capture signals effectively, even though hair. This innovation represents a significant advancement in the development of EEG sensors [31].

2.2 *Quantum Dot*

Quantum dots (QDs) are excellent fluorophores and can be utilized as highly sensitive and selective sensors due to their unique optical properties. Unlike organic dyes, QDs have symmetric emission profiles with relatively narrow bandwidths. They can be employed as various types of sensors including chemical sensors, biosensors, temperature sensors, and humidity sensors. The changes in both photoluminescent lifetimes and spectral response are crucial for developing sensors with high sensitivity and selectivity, even at the parts per million (ppm) level, and sometimes even at the parts per billion (ppb) level [32]. Quantum dots are utilized in hybrid QD-Si image sensors, enhancing resolution and reducing costs. In relevant study "Quantum Dots Revolutionizing Image Sensors" discusses the integration of lead sulfide QDs with silicon readout integrated circuits (ROIC) for near-infrared applications. Gold quantum dots have been used in biosensors for the detection of glucose. In this application, GQDs act as fluorescent probes, and the interaction with glucose leads to a measurable change in fluorescence intensity, allowing for the quantification of glucose levels in biological samples [33]. Algarra and co-workers have constructed carbon quantum dots (CQDs) modified glassy carbon electrode (GCE) electrochemical determination of dopamine and acid uric, the proposed sensor exhibited high sensitivity. The lower limits of detection were found to be 1.3 μM and 2.7 μM for uric acid and dopamine, respectively [34]. Another group developed an electrochemical method employing cadmium telluride quantum dots (CdTe) in Printex 6L Carbon (P6LC) and within a poly (3,4-ethylene dioxythiophene) polystyrene sulfonate (PEDOT: PSS) film modified glassy carbon electrode (QDs-P6LC-PEDOT: PSS/GCE) was developed for the detection of amoxicillin. Under the optimum conditions, the obtained sensor exhibited good sensitivity, high selectivity, and stability. No significant interference was noticed from drugs and potential biological interferences such as paracetamol, ascorbic acid, uric acid, and caffeine. The proposed sensor could be used to simultaneously determine amoxicillin in tablets, urine, and milk samples [35]. Furthermore, QDs-based biosensors have been applied in toxicological applications. Elmizadeh et al. [36] designed a biosensor for the detection of digitalis toxin (digoxin) in biological fluids. $CuInS_2$ QDs have been used to detect lead ions (Pb^{2+}) in environmental samples. The presence of Pb^{2+} ions lead to a quenching of the photoluminescence of the QDs, allowing for sensitive detection of this toxic metal in the water sample [37]. QD barcode assays have been established for the point-of-care diagnosis of infectious diseases such as HIV and hepatitis B. These assays utilize microfluidics and magnetic separation to enhance the detection process, providing rapid and reliable results. And also, been used in the immunochromatographic detection of aflatoxin B_1 in maize. Luminescent QD beads were prepared to detect this food contaminant, showcasing the application of QDs in ensuring food safety. ZnS: Cu QD has been applied for the visualization of fingerprints. This method offers a brighter and more sensitive means of detecting (blood) fingerprints compared to conventional techniques, enhancing forensic investigations [38]. Another team of researchers has developed an ultra-thin, high-performance photosensor based on eco-friendly quantum dots, that can stably measure light signals without any external power source, due to the photovoltaic effect. The joint research team has also produced a *skin-attachable ultra-thin pulse sensor* based on this photosensor and introduced a wearable pulse sensor that can stably acquire pulse signals despite different deformations [39].

2.3 Gold

Gold nanoparticles (AuNPs) are effectively used in colorimetric sensors for detecting heavy metals, such as lead (Pb^{2+}). When AuNPs are mixed with a sample containing lead, the ions cause the nanoparticles to aggregate, resulting in a visible color change from red to blue due to shifts in surface plasmon resonance. This simple and rapid method allows for the detection of low lead concentrations, making it ideal for applications in environmental monitoring and food safety [38]. AuNPs can be functionalized with biomolecules such as antibodies or nucleic acids to bind to specific targets. The presence of a target analyte can result in a color change in the solution, indicating the presence of the analyte [40]. AuNPs are utilized in optical Nanosensors for screening food safety, monitoring disease biomarkers, and detecting pathogens. These sensors can rely on local surface plasmon resonance (LSPR), fluorescence resonance energy transfer (FRET), surface-enhanced Raman scattering (SERS), and other phenomena [41, 42]. By using different functionalization methods, gold nanoparticle-based sensors can detect multiple analytes simultaneously. This is particularly useful in applications where the identification of several targets is crucial, which can provide quick results, often within minutes, which is essential for applications in medical diagnostics and real-time monitoring [43]. AuNPs have been effectively used to detect various pathogens, including Salmonella, with detection limits ranging from 10 to 100 colony-forming units per millilitre (CFU/mL). This demonstrates their high sensitivity and specificity in evaluating microbial contamination [44].

Advancements in biosensors utilizing AuNPs for detecting diseases such as HIV and cancer biomarkers (e.g., prostate-specific antigen (PSA) and carcinoembryonic antigen (CEA)). For instance, AuNP-based biosensors for HIV detection have shown remarkable sensitivity, with detection limits as low as 5 femtomolar (fM) [45, 46]. Gold nanoparticles (AuNPs) play a crucial role in environmental monitoring, particularly in detecting air pollutants such as volatile organic compounds (VOCs) and gases. Their unique properties, including high surface area and the ability to undergo surface plasmon resonance, enable sensitive detection of harmful gases. For instance, AuNP-based sensors can detect nitrogen dioxide (NO_2) at concentrations as low as 50 parts per billion (ppb) and sulfur dioxide (SO_2) at an impressive 1 ppb. This high sensitivity is essential for assessing air quality and identifying pollution sources, as both NO_2 and SO_2 are significant contributors to respiratory problems and environmental degradation. The detection mechanism typically involves the interaction of these gases with functionalized AuNPs, leading to measurable changes in their optical or electrical properties, thereby providing reliable indicators of pollutant presence and concentration [47–49]. Gold nanoparticle (AuNP)-based biosensors have shown potential in detecting soil contaminants. Previous studies have provided evidence of their ability to identify pesticides, such as atrazine and 2,4-dichlorophenoxyacetic acid (2,4-D), in soil samples. It also plays a crucial role in environmental monitoring and ensuring the safety of agricultural products [50].

2.4 *Silver*

Silver nanoparticles (AgNPs) are well known for their superior optoelectronic and physicochemical properties, and have, therefore, attracted a great deal of interest in the sensor arena. The introduction of AgNPs onto the surface of two-dimensional (2D) structures, incorporation into conductive polymers, or within three-dimensional (3D) nanohybrid architectures is a common strategy to fabricate novel platforms with improved chemical and physical properties for analyte sensing [51]. AgNP-modified electrodes have proven effective in detecting various organic pollutants, including methylene blue, melamine, trifluralin, and a range of pesticides and dyes. Methylene blue, a synthetic dye, is monitored due to its toxicity, while melamine, used in plastics, raises health concerns when it contaminates food. Trifluralin, an herbicide, poses risks to aquatic life, making its detection crucial. The catalytic properties of AgNPs enhance the electrochemical signals for these compounds, allowing for sensitive and selective detection. This capability is essential for environmental monitoring and ensuring water safety, highlighting the value of AgNP-modified electrodes in addressing pollution challenges [52]. AgNPs are used to modify electrode surfaces, enhancing the electrochemical oxidation of glucose. This results in improved sensitivity and lower detection limits, making it suitable for diabetes monitoring and also for the detection of heavy metals like lead and mercury. The high conductivity of AgNPs allows for sensitive voltammetric detection methods, such as anodic stripping voltammetry [53]. AgNPs can enhance the fluorescence signal of labels in biosensors. For example, in the detection of prostate-specific antigen (PSA), AgNPs amplify the fluorescence signal, allowing for detection at very low concentrations (e.g., 15.66 pg/mL) [54]. which exhibit strong SPR, which can be utilized in biosensors for detecting biomolecular interactions. This technique is particularly useful for real-time monitoring of binding events [53]. The team developed a multichannel colorimetric sensor array utilizing silver nanoparticles (AgNPs) to detect SARS-CoV-2 and differentiate it from influenza. This sensor array employs five specific peptides that interact with viral proteases, leading to a color change indicative of the presence of the virus. The system achieves a limit of detection of 500,000 viral copies/mL and is adaptable for use in various clinical settings, providing rapid and accurate diagnostics without the need for complex chemical modifications [55]. AgNPs can be employed in assays for DNA detection, where the aggregation of nanoparticles leads to a color change that indicates the presence of specific DNA sequences, also used in MEF to enhance the fluorescence of nearby dyes. This technique has been applied in the detection of various biomolecules, including proteins and nucleic acids, by significantly increasing the sensitivity of the assay [53, 56]. Silver nanoparticles (AgNPs) are increasingly utilized as colorimetric sensors for detecting water pollutants, particularly heavy metals and pesticides. For instance, AgNPs can detect mercury ions (Hg^{2+}) through a redox reaction that forms a nano-alloy, altering the plasmonic properties of the nanoparticles, which can be monitored via colorimetric methods [2]. Additionally, a study developed a chemiluminescent sensor array based on luminol-functionalized AgNPs, capable of discriminating between organophosphate and carbamate pesticides, achieving an accuracy of 95% in identifying five specific pesticides at a concentration of 24 μg/mL [57]. The high extinction coefficients of AgNPs allow for significant changes in their optical properties based on size and distance, making them promising candidates for visual readout sensors for various target analytes [58]. A group of scientists have developed a silver nanoparticle sensor that utilizes Surface-Enhanced Raman Spectroscopy (SERS) for rapid and precise detection of gene mutations, particularly those causing hearing loss. This biosensor is noted for its stability, sensitivity, and potential applications in diagnosing various health issues, including cancer and Alzheimer's. The innovation aims to transform medical diagnostics by integrating these sensors into portable devices for real-time applications [59].

2.5 Magnetic Nanomaterials

Magnetic nanoparticles (MNPs) are increasingly used in sensor technology because of their unique properties, such as high surface area, low toxicity, and the ability to enrich magnetic signals. They can be functionalized with specific antibodies to create biosensors for detecting disease biomarkers in samples like blood or saliva, enhancing sensitivity and enabling the detection of low concentrations. MNPs help in detecting biomolecules by acting as labels that amplify signals in various assays, including enzyme-linked immunosorbent assays (ELISA) and DNA detection methods. They improve biosensor efficiency by generating detectable magnetic signals when they interact with target biomolecules, making them suitable for applications like point-of-care testing and environmental monitoring as in Fig. 3 [60, 61]. Recent advancements include the integration of MNPs with magneto resistive sensors, which improve detection limits and enable real-time monitoring of analytes like proteins and pathogens [62].

In environmental monitoring, MNPs can capture pollutants in water by binding to heavy metals or organic compounds, allowing for rapid quantification through magnetic separation. They are also effective in food safety, where they can detect pathogens such as *E. coli* by conjugating with specific antibodies, facilitating quick and reliable testing. Furthermore, MNPs are employed in advanced techniques for single molecule detection, monitoring electrical signals as individual molecules pass through nanopores, which is crucial for early disease diagnosis. Additionally, MNPs

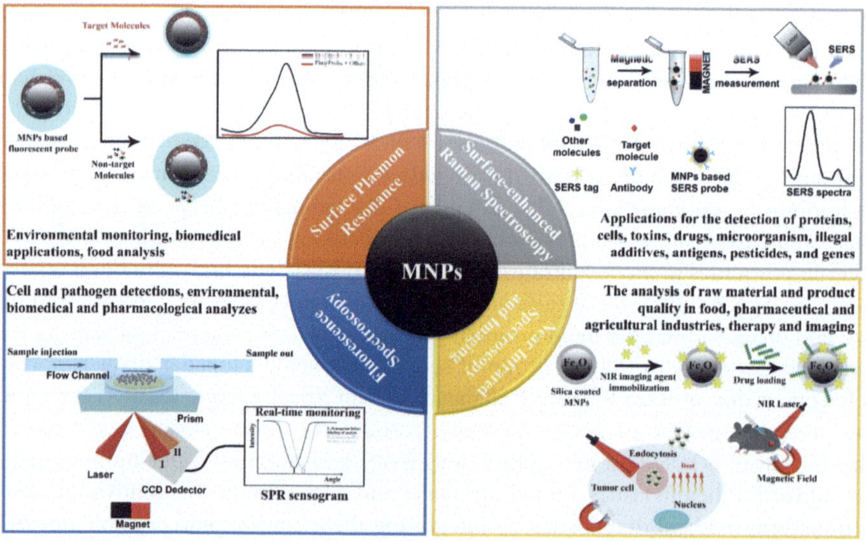

Fig. 3 Application of MNPs in sensing system reproduced from [63] and reprinted from the Creative Commons Attribution 4.0 license (CC BY 4.0)

enhance imaging techniques like magnetic resonance imaging (MRI) by improving contrast, aiding in the visualization of tumors and abnormalities [61].

One notable example is the use of superparamagnetic iron oxide nanoparticles (SPIONs) for the electrochemical detection of biomarkers. For instance, a study by Vargas et al. [64] demonstrated the use of magnetic beads-based sensors tailored for the rapid and single-step amperometric determination of microRNAs (miRNAs). The integration of MNPs allowed for efficient capture and isolation of target miRNAs from complex biological samples, significantly improving the sensor's sensitivity and detection limits. Additionally, Chiang et al. [65] synthesized trastuzumab-functionalized doxorubicin-encapsulated magnetic nanoparticles for targeted drug delivery in cancer treatment. This approach not only illustrates the application of MNPs in drug delivery systems but also highlights their potential in diagnostic sensors by enabling the detection of specific cancer biomarkers through targeted interaction. MNPs are also highlighted for their dual role in diagnostics, where they can be used for imaging purposes while simultaneously serving as components in electrochemical sensors. Their magnetic properties allow for tracking in vivo, which can be integrated with sensor technology for real-time monitoring of therapeutic agents [66]. In recent research, a magnetic biosensor (MBCPE/Fe_3O_4NPs@OA/anti-TET) was developed for the sensitive and selective detection of tetracycline. This sensor achieved a remarkable detection limit of 3.8×10^{-15} M, demonstrating excellent reproducibility and stability. The findings suggest that this aptasensor is capable of accurately measuring tetracycline levels in tablet solutions, even in the presence of other substances [67]. Magnetic nanoparticles (MNPs) are effective tools for detecting environmental contaminants such as bisphenol A (BPA) in water due to their high surface area, which allows for functionalization with specific receptors that selectively bind to BPA. Once bound, an external magnetic field can be applied to facilitate the magnetic separation of the MNPs along with the captured BPA from the water sample, simplifying the isolation process. This method not only enhances sensitivity and specificity in detection but also enables rapid and cost-effective testing, making it suitable for real-time monitoring of water quality. The integration of MNPs into portable sensing devices further supports their utility in environmental monitoring, ensuring the detection of contaminants in various water sources [68]. Also, MNPs can alter their magnetic properties upon interaction with gases, allowing for the detection of volatile organic compounds (VOCs). An innovative gas sensor based on MNPs demonstrated high sensitivity to harmful VOCs like benzene and toluene, utilizing frequency shifts in a magnetostatic spin wave oscillator [69].

2.6 Polymer-Based Nanomaterial

Polymer-based nanomaterials are used in sensors to detect target analytes by changing their physical or chemical properties. These can be used to detect heavy metals, pesticides, and antibiotics in various mediums. For example, Prussian blue (PB) and graphitic carbon nitride-based nanostructures can remove environmental pollutants

[70]. Conducting polymers include polypyrrole (PPy), polyaniline (PANI), polythiophene (PTh), and Poly(3,4-ethylenedioxythiophene) (PEDOT). Each of these polymers has specific properties that are beneficial for sensor applications [71]. Bio-based polymers can be derived from curcumin, polycurcumin acrylate (PCUA), and polycurcumin methacrylate (PCUMA). These polymers can detect nitroaromatic chemicals [72]. Natural polymers, such as cellulose and chitosan, are used due to their biodegradability and biocompatibility. In one study they demonstrated the use of desferrioxamine B immobilized on cellulose for the selective visual determination of vanadium (V) ions in highly acidic solutions [73]. A nanostructured chitosan-based electrode was developed for salivary glucose sensing. This sensor utilized layer-by-layer self-assembly of single-walled carbon nanotubes, chitosan, gold nanoparticles, and glucose oxidase on a screen-printed platinum electrode. The sensor demonstrated effective amperometric analysis for quantifying glucose concentrations in synthetic solutions and saliva samples [74]. Biesuz and colleagues developed low-cost colorimetric sensors based on the Colour Catcher® (CC), a commercial cellulose-based product used to prevent color running in the washing machine. The CC substrate was functionalized with different dyes to develop colorimetric devices for some metal ions, sulfur/thiols, and food spoilage detection [75–78]. Another research team developed a new polymer-based sensor that can quickly and accurately detect the pesticide pirimicarb. This sensor uses a special material made from magnetic nanoparticles and tiny particles called graphene quantum dots. It can detect very low levels of pirimicarb, with a limit of detection of 1.79 mg/L. The sensor was tested with real water samples and showed it could remove 99.92% of pirimicarb, proving it works well in real-life situations. Overall, this polymer-based sensor could be very useful for monitoring pesticides in the environment [79]. Polymer-based sensors are increasingly recognized for their effectiveness in various sensing applications, particularly in detecting gases. One notable example is the use of polyaniline nanofibers, which have been engineered to create highly sensitive chemo resistive sensors capable of detecting gases like ammonia and nitrogen dioxide at room temperature. This innovation highlights the potential of conducting polymers in environmental monitoring and safety applications [80, 81].

3 Synthesis of Nanomaterial for Sensor

Nanomaterials can be synthesized using two primary approaches like bottom-up approach and the top-down approach as in Fig. 4. The bottom-up approach includes methods such as chemical synthesis, electrochemical processes, sol–gel techniques, and solvothermal methods. In this approach, materials are built up atom by atom. In contrast, the top-down approach involves techniques like ball milling, laser ablation, and lithography, where materials are created from bulk substances. According to the literature survey it indicates that various researchers have demonstrated the use of different nanomaterials incorporating various dopants, catalysts, adhesives, binders,

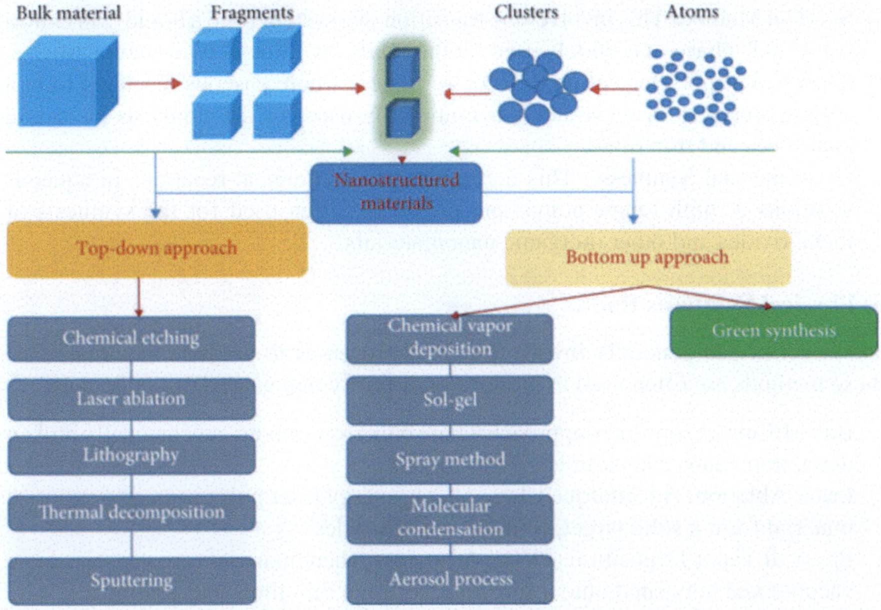

Fig. 4 A schematic representation of top-down and bottom-up approach for the synthesis of nanomaterial reproduced from [83] and reprinted from Creative Commons Attribution 4.0 international license (CC BY 4.0)

and surfactants to enhance the sensing characteristics of sensors made from these materials [82].

Synthesis methods described as random or chaotic are characterized by parameters that are challenging to control, resulting in a chaotic state. However, by manipulating certain conditions—such as reaction time, temperature, pressure, and limiting agents—during the process, it is possible to achieve a more favorable distribution of particle sizes or shapes. Examples of such methods include hydrothermal processes, pyrolysis, and various chemical synthesis techniques [84].

Synthesis methods can also be classified based on the routes involved in achieving the desired materials.

(i) Chemical routes
(ii) Physical routes
(iii) Biological routes

(i) Chemical Synthesis Routes

Chemical synthesis methods involve chemical reactions to produce nanomaterials. These methods are versatile and can produce a wide range of nanomaterials with controlled sizes and shapes. Common chemical synthesis methods include:

1. Sol–Gel Method: This involves the transition of a system from a liquid "sol" into a solid "gel" phase. It is widely used for the synthesis of metal oxide nanoparticles.
2. Chemical Vapor Deposition (CVD): A process where gaseous reactants form a solid material on a substrate. It is commonly used for the synthesis of carbon nanotubes and thin films.
3. Hydrothermal Synthesis: This method involves chemical reactions in aqueous solutions at high temperatures and pressures, often used for the synthesis of metal oxides and other inorganic nanomaterials.

(ii) Physical Synthesis Routes

Physical methods generally involve physical processes to produce nanomaterials. These methods are often used to produce high-purity nanomaterials and include:

1. Ball Milling: A top-down approach where bulk materials are mechanically broken down into nanoscale particles.
2. Laser Ablation: A technique where a high-energy laser pulse is used to vaporize material from a solid target, forming nanoparticles.
3. Physical Vapor Deposition (PVD): A process where materials are vaporized in a vacuum and deposited onto a substrate, forming thin films and coatings.

(iii) Biological Synthesis Routes

Biological synthesis methods use biological organisms or biomolecules to produce nanomaterials. These methods are environmentally friendly and often produce biocompatible nanomaterials. Examples include:

1. Microbial Synthesis: Using bacteria, fungi, or algae to produce nanoparticles. For instance, certain bacteria can reduce metal ions to form metal nanoparticles.
2. Plant Extracts: Plant extracts can act as reducing and stabilizing agents to synthesize nanoparticles. This method is often used for the synthesis of gold and silver nanoparticles.
3. Enzyme-Mediated Synthesis: Enzymes can catalyse the formation of nanoparticles, offering precise control over the size and shape of the particles [83–85] (Fig. 5).

4 Functional Mechanism of Nanosensors

Nanosensors are state-of-the-art devices that use nanoparticles to detect and quantify analytes in a range of samples. Their functioning is based on the interaction between the analyte and a bioreceptor, which is often installed on a nanoscale electrode. This interaction results in an electrical signal that matches the analyte concentration, making the sensor extremely effective at detecting minute quantities of substances. The addition of nanostructures, such as nanowires, nanotubes, and nanoparticles, greatly expands the sensor's surface area and boosts the number of bioreceptors that

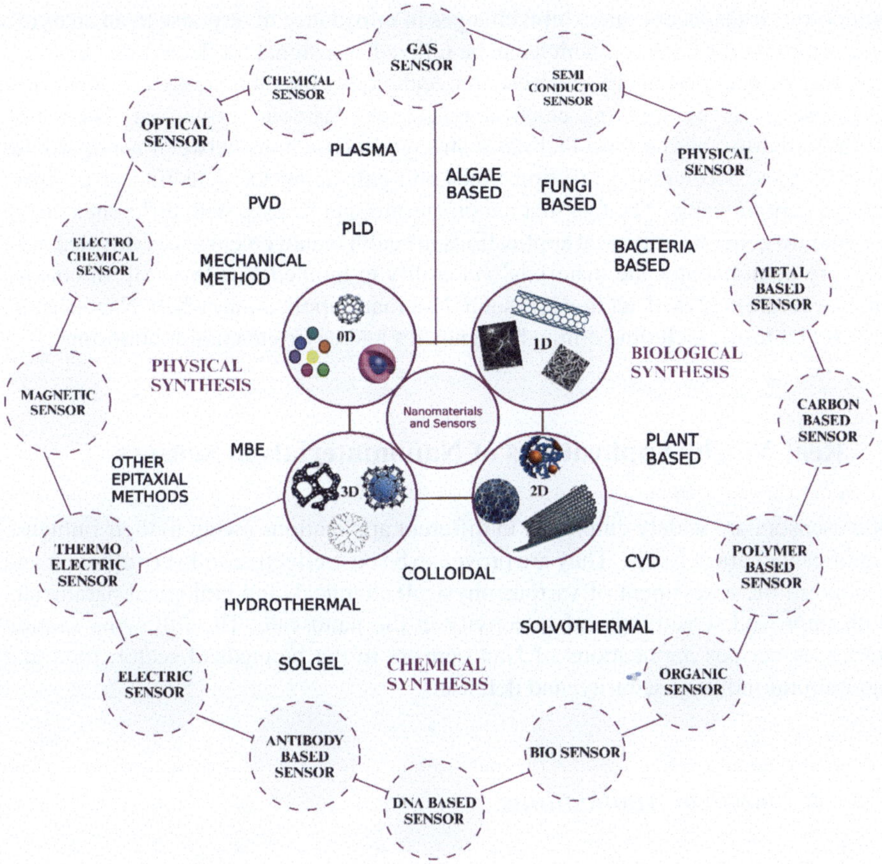

Fig. 5 Nanomaterials for sensors technology

can be accommodated [6]. The enhanced sensitivity that results from this increase in surface area allows for the detection of analytes at incredibly low concentrations, even at the femtomolar level. The electrocatalytic effect is enhanced and the sensor's reaction time is shortened by the rapid electron transfer made possible by nanostructures, such as metal nanoparticles, allowing for accurate and prompt results. Nanosensors can further improve selectivity by carefully changing their surfaces to allow targeted interactions with the selected analyte while lowering interference from other chemicals in the sample [86, 87].

Nanosensors use a range of methods to convert sensing events into quantifiable signals by taking advantage of the unique properties of nanomaterials. These sensors selectively detect bound analytes by measuring changes in resistance that occur when an analyte binds to a nanomaterial using electrochemical techniques. Biological Nanosensors, which combine bio-receptors and transducers, commonly employ fluorescence techniques due to their high sensitivity and ease of testing [60, 88].

Nonporous transducers can exhibit changes in impedance in response to an analyte's interaction with a receptor, which can be utilized as a signal for detection. Nanosensors can be operated using spectroscopic techniques like surface-enhanced Raman spectroscopy or by applying electromagnetic or plasmonic principles. Their high surface area-to-volume ratio facilitates stronger interactions between nanoparticles and analytes, accelerating reaction times and enhancing the detection of environmental contaminants. The fact that nanomaterials can be used with different sensing methods to generate improved applications in health management and controlled drug delivery demonstrates the materials' versatility in biomedicine [89, 90]. In conclusion, accurate analyte detection is essential for many applications where Nanosensors are crucial tools, including clinical diagnostics and environmental monitoring.

5 Real-World Applications of Nanomaterials in Sensors

Nanosensors are widely employed in different applications owing to their immense significant characteristics. They are proven to be cost-effective and very efficient and can aid in the assessment of various physical, chemical, and biological parameters with improved sensitivity and selectivity in the nanoscale. The following section discusses various applications of Nanosensors in the biomedical sector, food and agricultural industry, security, and defense.

5.1 Biomedical Applications

Nano-based sensors are versatile in nature and are employed in a variety of different biomedical applications. It aids in quick diagnosis and the use of nanomaterials has aided in the reduction of the usage of expensive enzymes employed for detection. Since nanotechnology is used, an improvised sensitivity and selectivity are observed and current research focuses on improving the limit of detection and also reducing the time and cost at the same point. In the field of fabrication of advanced diagnostic kits, Nanosensors have been employed to detect cancer, various other metabolic disorders, and also allergic responses. Blood glucose detection has been present for a long time now and is one of the most important breakthroughs in the domain of sensors. Single-walled carbon nanotubes have been modified to fabricate a non-enzymatic glucose sensor and were able to get a limit of detection up to 4.6 mM [91, 92]. Similarly, sensors using nanomaterials such as quantum dots, carbon dots, metal oxide nanoparticles, and polymeric nanoparticles have been used to detect cancer and other viral diseases [93, 94]. Carbon nanotube-based sensors have also been employed to identify various biomolecules produced by the cell and aid in diagnostic fabrication [95]. The identification of various pathogenic bacteria has been done using nanomaterial-based sensors and has shown accurate, fast, and sensitive results. Iron oxide nanoparticles have been employed in the detection of *Mycobacterium* species. Gold and silicon-based nanorods have been employed for the detection of *Salmonella* species [96, 97]. The metabolic activity of the pathogens can also be indirectly measured by employing various nanomaterials. Nano-mediated sensors have also been used in the detection of contaminants in the water, especially harm-causing bacteria such as *Escherichia coli*, *Staphylococcus aureus*, and various other enterotoxins [98, 99]

Biomolecular interactions are studied on a higher scale using Nanosensors with improved detection sensitivity and selectivity. Silver nanotubes have been made into a composite along with chitosan to detect IgG mouse antibodies [100, 101]. A higher level of telomerase activity is seen in cancer cells and a magnetic nanoparticle-based sensor has been employed and studied [102].

5.2 Food and Agricultural Applications

Food safety is one of the most critical challenges that requires adequate attention. One of the major causes concerning food safety is the food-borne illnesses caused by food-borne pathogens. The food gets infected first, affecting the humans who ingest it. *Listeria monocytogenes* is one of the most pathogenic organisms that cause severe diarrhoea and severe vomiting. It is of severe threat that it is very much necessary that it is identified with extensive sensitivity [103–105]. A lot of Nanosensors have been designed for an improved limit of detection of this dreadful pathogen [106]. Food safety has also led to a surge in research in the field of intelligent food packaging to have continuous monitoring and checking on the condition of the food inside the containers. Advanced research has been done where Nanosensors are inserted in the form of a code in the packaging containers and are completely invisible. They keep sending real-time data giving the status and the present condition of the food [107]. Nanomaterial-based gas sensors are employed to detect contamination in the food because of the gases they release when they spoil or any changes take place. Graphene-based nanomaterials, because of their ability to facilitate wireless transmission [108]. The most common use of nanomaterials in food-based Nanosensors is as transducers (conductive nanotubes or fluorescent nanoparticles). They can also be used as signal amplifiers (nanopatterned surface-enhanced Raman spectroscopy (SERS) substrates), recognition elements (molecularly imprinted polymers, or MIPs), capture agents (magnetic nanoparticles), and much more [109]. Food production in the agricultural ecosystem has to be continuously monitored to avoid the presence of pesticides or any other heavy metals and various other toxic particles. Their presence may negatively influence the environment and at the same time disrupt the food chain. It is very important to completely monitor all these factors using advanced systems such as Nanosensors rather than depending on the traditional expensive strategies such as mass spectroscopy or gas chromatography and also requires trained and skilled personnel to handle the equipment which makes it very complex. The Nanosensors can be used in the detection of molecules such as dyes, toxic elements, pesticides, heavy metals, gases such as oxygen or carbon dioxide, presence of organisms such as bacteria, fungi or viruses and also can help in monitoring the basic parameters of the food crops such as temperature, pH, light and humidity [109]. With the ability of nanomaterials for multiplex and real-time sensing capabilities advancements in the development of electrochemical, optical, electronic tongue and nose, nano-barcode, and wireless Nanosensors are on the rise [110]. Due to the unstable nature of enzymes, nanomaterials have been employed to fabricate nonenzymatic-based electrochemical sensors for the detection of pesticides using various nanomaterials such as carbon nanotubes, copper-molybdenum-based composites, and much more. Smart sensors are also being designed using various nanomaterials to achieve agricultural sustainability in the upcoming years [111]. Gold nanoparticles have been used on Nanosensors platforms and they significantly increase the sensitivity of surface-enhanced Raman scattering by 10^6–10^9 times and localized surface plasmon resonance by 2–10 times which is commendable. They have also been employed for the detection of various antibiotics such as streptomycin, kanamycin, or neomycin with an easy step of color change [110, 112]. The field is yet to grow and each day there are new advancements in employing nanomaterials for monitoring food and agricultural sectors.

5.3 Military and Defence Applications

The world has seemed to be overrun with threats in recent years, emanating from all perspectives. These threats include physical harm from fire or water, natural disasters like earthquakes or floods, interruptions to vital infrastructure like communication or electricity, and loss or compromise of information. Traditional techniques are very time-consuming and require skilled personnel [107]. Nanosensors have been widely employed for assessing global climatic change and security. Nano Teflon-based sensors have been used in smart fabrics in the defense sector for bullet as well as waterproof jackets. The nanomaterial-based sensors have also been employed to detect nuclear and life-threatening substances such as phosgene, sarin, or chlorine and also to detect biological warfare toxins such as botulinum and spores such as the Ebola virus. Carbon nanotube-based sensors have been employed as nano-fingerprints-based barcodes for the identification of passports and various other unique information and have been a breakthrough in the security aspects of the nation. Traces of explosives can be identified by using concepts such as electronic nose, nanotube, or nanowire-based sensor systems that can show improved selectivity and sensitivity and aid in the elimination of false positive results [15].

6 Challenges and Limitations of Nanomaterial-Based Sensors

The creation of nanoscale-based sensors is becoming a very popular area of study since it offers significant benefits like lower costs for routine health monitoring and early diagnosis and has applications beyond health, food, agriculture, and the defence sector. Nevertheless, response time and detection limit are two crucial metrics and parameters that are impacted by the reduction in vital dimensions of the sensing elements.

Other critical aspects to consider are non-specific binding, biofunctionalization simplicity, and manufacturing difficulty for large-scale production. The detection limit varies for each of the biomarkers and the physiochemical properties of the nanomaterial involved and the mechanism of interaction with the biomarker must be properly understood. The non-specific binding of the nanomaterial with the sensing element can also lead to a reduction in sensitivity. The cost of the sensor development is very high and can also show batch-to-batch variations because of their miniaturized version when compared to the bulk counterpart [107]. Many previous literatures have claimed difficulty in incorporating nanowires to the sensor systems. The selectivity parameter plays a key role and appears to be one of the main challenges owing to match with a real-time complex system [91]. The level of toxicity in the nanomaterials is sometimes of concern and has to be well understood before it is commercialized and used on a higher scale [95]. The stability of Nanosensors in terms of food preservation has to be stable for a long time to monitor its freshness or quality. Potential developments on the field can aid in looking into these challenges and pave way for future nano sensor research.

7 Prospects and Emerging Trends

Concerning the employment of Nanosensors in the health sector, they have been showing remarkable advancements, especially in the recent pandemic Covid-19, wherein researchers employed different kinds of Nanosensors combinations to specifically detect the viral pathogen. So, in the future, we will be able to detect the pathogens that are a biological threat to the world population. Also, researchers will fabricate wearable Nanosensors that will aid in continuous health monitoring of vital parameters of the body. They can also be employed to have personalized treatment strategies for specific screening and diagnostic purposes. Advanced technologies such as the fabrication of DNA Nanosensors are present for the identification of genetically modified organisms and various other contaminants. Nanosensors will also aid the farmers in detecting any particular pesticide or any other critical parameters concerning the growth of crops [107, 110]. Future research will also focus on the development of new combinations of nanomaterials with improved biocompatibility, stability, and sustainability. It will also be integrated into advanced technologies such as IoT and blockchain technology to develop advanced real-time sensing systems. Improved fabrication strategies or techniques will also be developed to produce high-quality Nanosensors and to minimize batch-to-batch variation [15].

8 Conclusion

This chapter discussed the various uses and synthesis techniques of nanomaterials, with a focus on their importance in sensor technology. Several kinds of nanomaterials, such as nanoparticles, nanotubes, and nanocomposites, are necessary to create sophisticated sensors that can identify dangerous environmental pollutants. The chapter focused on technologies like laser ablation and ball milling that make it easier to create nanoparticles, while also highlighting important synthesis routes like chemical, physical, and biological approaches. Accurate measurements also depend on the critical requirements for functional sensors, such as stability, selectivity, and sensitivity. From quantum devices to nanomedicine delivery systems, the wide range of uses for nanomaterials was also discussed, demonstrating its adaptability and potential influence on numerous sectors. In summary, this chapter highlights how nanoparticles are revolutionizing sensor technology and hold promise for future advancements in environmental monitoring and medical diagnostics.

References

1. Ghosh T, Raj GVSB, Kumar K (2022) A comprehensive review on nanotechnology-based sensors for monitoring quality and shelf life of food products. Measurement: Food 7(June):100049. https://doi.org/10.1016/j.meafoo.2022.100049
2. Chen L, Chan L, Fu X, Lu W (2013) Highly sensitive and selective colorimetric sensing of Hg^{2+} based on the morphology transition of silver nanoprisms. ACS Appl Mater Interfaces 5(2):284–290. https://doi.org/10.1021/am3020857
3. Willner MR, Vikesland PJ (2018) Nanomaterial enabled sensors for environmental contaminants. J Nanobiotechnol 1–16. https://doi.org/10.1186/s12951-018-0419-1
4. Aliyeva N, Eroglu S, Denizli A (2024) Nanomaterial-based sensors for coumarin detection. https://doi.org/10.1021/acsomega.4c01945
5. Kumar A (2019) Nanosensors: applications and challenges. Int J Sci Res 8(7):2018–2020
6. Malik S, Singh J, Goyat R, Saharan Y, Ameen S, Baskoutas S (2023) Heliyon nanomaterials-based biosensor and their applications: a review. Heliyon 9(9):e19929. https://doi.org/10.1016/j.heliyon.2023.e19929
7. Shimada T (2024) Nanomaterial-based sensors in analytical chemistry. Anal Sci 40(3):357–358. https://doi.org/10.1007/s44211-023-00498-z
8. Halicka K, Meloni F, Czok M, Spychalska K, Baluta S, Malecha K, Pilo MI, Cabaj J (2022) New trends in fluorescent nanomaterials-based bio/chemical sensors for neurohormones detection a review. https://doi.org/10.1021/acsomega.2c04134
9. Rohilla D, Chaudhary S, Umar A (2021) An overview of advanced nanomaterials for sensor applications, pp 47–70

10. Reda Y (2020) Review—nanostructured materials-based nanosensors. https://doi.org/10.1149/1945-7111/ab67aa
11. Sharma P, Pandey V, Mukut M, Sharma M, Patra A, Singh B (2021) A review on biosensors and nanosensors application in agroecosystems. Nanoscale Res Lett. https://doi.org/10.1186/s11671-021-03593-0
12. Tarannum N, Gautam A, Chauhan T, Kumar D (2024) Nanomaterial based sensors for detection of food contaminants: a prospect. Sens Technol 2(1). https://doi.org/10.1080/28361466.2024.2373196
13. Wang X, Li F, Guo Y, Wang X (2020) Recent trends in nanomaterial-based biosensors for point-of-care testing. Front Chem 8(October):1–6. https://doi.org/10.3389/fchem.2020.586702
14. Maduraiveeran G, Jin W (2017) Trends in environmental analytical chemistry nanomaterials based electrochemical sensor and biosensor platforms for environmental applications. Biochem Pharmacol 13:10–23. https://doi.org/10.1016/j.teac.2017.02.001
15. Darwish MA, Abd-elaziem W, Elsheikh A (2024) Nanoscale advances advancements in nanomaterials for nanosensors, pp 4015–4046. https://doi.org/10.1039/d4na00214h
16. Kang J (2021) Applications of nanotechnology in virus detection, tracking, and infection mechanisms, pp 1–11. https://doi.org/10.1002/wnan.1700
17. Akyildiz IF, Jornet JM (2010) Nano communication networks electromagnetic wireless nanosensor networks. Nano Commun Netw 1(1):3–19. https://doi.org/10.1016/j.nancom.2010.04.001
18. Luo X, Han Y, Chen X, Tang W, Yue T, Li Z (2020) Carbon dots derived fluorescent nanosensors as versatile tools for food quality and safety assessment: a review. Trends Food Sci Technol 95(March 2019):149–161. https://doi.org/10.1016/j.tifs.2019.11.017
19. Javaid M, Haleem A, Pratap R, Rab S, Suman R (2021) Exploring the potential of nanosensors: a brief overview. Sens Int 2(October):100130. https://doi.org/10.1016/j.sintl.2021.100130
20. Liu X, Yao Y, Ying Y, Ping J (2019) Recent advances in nanomaterial-enabled screen-printed electrochemical sensors for heavy metal detection. TrAC Trends Anal Chem 115:187–202. Elsevier B.V. https://doi.org/10.1016/j.trac.2019.03.021
21. Ullah N, Mansha M, Khan I, Qurashi A (2018) Nanomaterial-based optical chemical sensors for the detection of heavy metals in water: recent advances and challenges. TrAC Trends Anal Chem 100:155–166. Elsevier B.V. https://doi.org/10.1016/j.trac.2018.01.002
22. Ferrier DC, Honeychurch KC (2021) Carbon nanotube (CNT)-based biosensors. Biosensors 11(12). MDPI. https://doi.org/10.3390/bios11120486
23. Meskher H, Mustansar HC, Thakur AK, Sathyamurthy R, Lynch I, Singh P, Han TK, Saidur R (2022) Recent trends in carbon nanotube (CNT)-based biosensors for the fast and sensitive detection of human viruses: a critical review. Nanoscale Adv 5(4):992–1010. Royal Society of Chemistry. https://doi.org/10.1039/d2na00236a
24. Chopra S, McGuire K, Gothard N, Rao AM, Pham A (2003) Selective gas detection using a carbon nanotube sensor. Appl Phys Lett 83(11):2280–2282. https://doi.org/10.1063/1.1610251
25. Chopra S, Pham A, Gaillard J, Parker A, Rao AM (2002) Carbon-nanotube-based resonant-circuit sensor for ammonia. Appl Phys Lett 80(24):4632–4634. https://doi.org/10.1063/1.1486481
26. Snow ES, Perkins FK, Houser EJ, Badescu SC, Reinecke TL (2005) Chemical detection with a single-walled carbon nanotube capacitor. Science 307(5717):1942–1945. https://doi.org/10.1126/science.1109128
27. Yeow JTW, She JPM (2006) Carbon nanotube-enhanced capillary condensation for a capacitive humidity sensor. Nanotechnology 17(21):5441–5448. https://doi.org/10.1088/0957-4484/17/21/026
28. Zaporotskova IV, Boroznina NP, Parkhomenko YN, Kozhitov LV (2016) Carbon nanotubes: sensor properties. A review. Mod Electron Mater 2(4):95–105. https://doi.org/10.1016/j.moem.2017.02.002

29. Wu R, Ma L, Hou C, Meng Z, Guo W, Yu W, Yu R, Hu F, Liu XY (2019) Silk composite electronic textile sensor for high space precision 2D combo temperature–pressure sensing. Small 15(31). https://doi.org/10.1002/smll.201901558
30. Rdest M, Janas D (2021) Carbon nanotube wearable sensors for health diagnostics. Sensors 21(17). https://doi.org/10.3390/s21175847
31. Hajare R, Kadam S (2021) Comparative study analysis of practical EEG sensors in medical diagnoses. Global Trans Proc 2(2):467–475. https://doi.org/10.1016/j.gltp.2021.08.009
32. Kunwar N, Zala H, Parveen K (2023) Application of quantum dots in sensors. In: Quantum dots: fundamentals, synthesis and applications, pp 313–338. https://doi.org/10.1016/B978-0-12-824153-0.00013-6
33. Kumar YR, Deshmukh K, Sadasivuni KK, Pasha SKK (2020) Graphene quantum dot based materials for sensing, bio-imaging and energy storage applications: a review. RSC Adv 10(40):23861–23898. Royal Society of Chemistry. https://doi.org/10.1039/d0ra03938a
34. Algarra M, González-Calabuig A, Radotić K, Mutavdzic D, Ania CO, Lázaro-Martínez JM, Jiménez-Jiménez J, Rodríguez-Castellón E, del Valle M (2018) Enhanced electrochemical response of carbon quantum dot modified electrodes. Talanta 178:679–685. https://doi.org/10.1016/j.talanta.2017.09.082
35. Wong A, Santos AM, Cincotto FH, Moraes FC, Fatibello-Filho O, Sotomayor MDPT (2020) A new electrochemical platform based on low cost nanomaterials for sensitive detection of the amoxicillin antibiotic in different matrices. Talanta 206. https://doi.org/10.1016/j.talanta.2019.120252
36. Elmizadeh H, Faridbod F, Soleimani M, Reza M (2020) Fluorescent apta-nanobiosensors for fast and sensitive detection of digoxin in biological fluids using rGQDs: comparison of two approaches for immobilization of aptamer. Sens Actuators: B. Chem 302(September 2019):127133. https://doi.org/10.1016/j.snb.2019.127133
37. Efremenko E, Lyagin I, Senko O, Maslova O, Stepanov N (2021) Bioluminescent nano- and micro-biosensing elements for detection of organophosphorus compounds. In: Macro, micro, and nano-biosensors: potential applications and possible limitations, pp 239–261. https://doi.org/10.1007/978-3-030-55490-3_14
38. Liu G, Lu M, Huang X, Li T, Xu D (2018) Application of gold-nanoparticle colorimetric sensing to rapid food safety screening. Sensors (Switzerland) 18(12). MDPI AG. https://doi.org/10.3390/s18124166
39. Successful development of the world's highest-performance eco-friendly quantum dot photosensor with no external power source required | Asia Research News. (n.d.). https://www.asiaresearchnews.com/content/successful-development-worlds-highest-performance-eco-friendly-quantum-dot-photosensor. Accessed 15 Oct 2024
40. Ferrari E (2023) Gold nanoparticle-based plasmonic biosensors. Biosensors 13(3). Multidisciplinary Digital Publishing Institute (MDPI). https://doi.org/10.3390/bios13030411
41. Anh NH, Doan MQ, Dinh NX, Huy TQ, Tri DQ, Ngoc Loan LT, Van Hao B, Le AT (2022) Gold nanoparticle-based optical nanosensors for food and health safety monitoring: recent advances and future perspectives. RSC Adv 12(18):10950–10988. Royal Society of Chemistry. https://doi.org/10.1039/d1ra08311b
42. Chen H, Zhou K, Zhao G (2018) Gold nanoparticles: from synthesis, properties to their potential application as colorimetric sensors in food safety screening. Trends Food Sci Technol 78:83–94. Elsevier Ltd. https://doi.org/10.1016/j.tifs.2018.05.027
43. Karnwal A, Kumar Sachan RS, Devgon I, Devgon J, Pant G, Panchpuri M, Ahmad A, Alshammari MB, Hossain K, Kumar G (2024) Gold nanoparticles in nanobiotechnology: from synthesis to biosensing applications. ACS Omega 9(28):29966–29982. American Chemical Society. https://doi.org/10.1021/acsomega.3c10352
44. Behoftadeh F, Faezi Ghasemi M, Mojtahedi A, Issazadeh K, Golshekan M, Alaei S (2023) Development of a newly designed biosensor using multi-walled carbon nanotubes (MWCNTs) with gold nanoparticles (AuNPs) in the presence of acetaminophen for detection of Escherichia coli. Arch Microbiol 205(2). https://doi.org/10.1007/s00203-023-03418-x

45. Farzin L, Shamsipur M, Samandari L, Sheibani S (2020) HIV biosensors for early diagnosis of infection: the intertwine of nanotechnology with sensing strategies. Talanta 206:120201. https://doi.org/10.1016/J.TALANTA.2019.120201
46. Shayesteh OH, Ghavami R (2020) A novel label-free colorimetric aptasensor for sensitive determination of PSA biomarkers using gold nanoparticles and a cationic polymer in human serum. Spectrochim Acta Part A Mol Biomol Spectrosc 226:117644. https://doi.org/10.1016/J.SAA.2019.117644
47. Gan Z, Zhou Q, Zheng C, Wang J (2023) Challenges and applications of volatile organic compounds monitoring technology in plant disease diagnosis. Biosens Bioelectron 237:115540. https://doi.org/10.1016/J.BIOS.2023.115540
48. Heo JH, Sung M, Trung TQ, Lee Y, Jung DH, Kim H, Kaushal S, Lee NE, Kim JW, Lee JH, Cho SY (2023) Sensor design strategy for environmental and biological monitoring. EcoMat 5(5). Wiley. https://doi.org/10.1002/eom2.12332
49. Mi K, Tong L, Yu M, Zhao Y, Dong H, Hou S (2023) Fabrication and application of a 3D-rGO electrochemical sensor for SO_2 detection in ionic liquids. Anal Methods 15(29):3522–3531. https://doi.org/10.1039/D3AY00805C
50. Du H, Xie Y, Wang J (2021) Nanomaterial-sensors for herbicides detection using electrochemical techniques and prospect applications. TrAC Trends Anal Chem 135:116178. https://doi.org/10.1016/J.TRAC.2020.116178
51. Ivanišević I (2023) The role of silver nanoparticles in electrochemical sensors for aquatic environmental analysis. Sensors 23(7). MDPI. https://doi.org/10.3390/s23073692
52. Zahran M, Khalifa Z, Zahran MAH, Abdel Azzem M (2021) Recent advances in silver nanoparticle-based electrochemical sensors for determining organic pollutants in water: a review. Mater Adv 2(22):7350–7365. Royal Society of Chemistry. https://doi.org/10.1039/d1ma00769f
53. Beck F, Loessl M, Baeumner AJ (2023) Signaling strategies of silver nanoparticles in optical and electrochemical biosensors: considering their potential for the point-of-care. Microchimica Acta 190(3). Springer. https://doi.org/10.1007/s00604-023-05666-6
54. Peng W, Qin Y, Li W, Chen M, Zhou D, Li H, Cui J, Chang J, Xie S, Gong X, Tang B (2020) Nonenzyme cascaded amplification biosensor based on effective aggregation luminescence caused by disintegration of silver nanoparticles. ACS Sens 5(7):1912–1920. https://doi.org/10.1021/acssensors.9b02355
55. Lam B, Retout M, Clark AE, Garretson AF, Carlin AF, Jokerst JV (2024) Silver nanoparticle sensor array for the detection of SARS-CoV-2. ACS Appl Nano Mater 7(8):9136–9146. https://doi.org/10.1021/acsanm.4c00654
56. Jeong Y, Kook YM, Lee K, Koh WG (2018) Metal enhanced fluorescence (MEF) for biosensors: general approaches and a review of recent developments. Biosens Bioelectron 111:102–116. https://doi.org/10.1016/J.BIOS.2018.04.007
57. He Y, Xu B, Li W, Yu H (2015) Silver nanoparticle-based chemiluminescent sensor array for pesticide discrimination. J Agric Food Chem 63(11):2930–2934. https://doi.org/10.1021/acs.jafc.5b00671
58. Prosposito P, Burratti L, Venditti I (2020) Silver nanoparticles as colorimetric sensors for water pollutants. Chemosensors 8(2). Multidisciplinary Digital Publishing Institute (MDPI). https://doi.org/10.3390/CHEMOSENSORS8020026
59. AlSafadi I, Ramachandran K, Columbus S, Tlili A, Daoudi K, Gaidi M (2024) Highly efficient, label free, ultrafast plasmonic SERS biosensor (silver nanoarrays/Si) to detect GJB2 gene expressed deafness mutations in real time validated with PCR studies. Int J Biol Macromol 259. https://doi.org/10.1016/j.ijbiomac.2024.129381
60. Wang M, Jin L, Hang-Mei Leung P, Wang-Ngai Chow F, Zhao X, Chen H, Pan W, Liu H, Li S (2024) Advancements in magnetic nanoparticle-based biosensors for point-of-care testing. Front Bioeng Biotechnol 12. Frontiers Media SA. https://doi.org/10.3389/fbioe.2024.1393789
61. Gloag L, Mehdipour M, Chen D, Tilley RD, Gooding JJ (2019) Advances in the application of magnetic nanoparticles for sensing. Adv Mater 31(48). Wiley-VCH Verlag. https://doi.org/10.1002/adma.201904385

62. Koh I, Josephson L (2009) Magnetic nanoparticle sensors. Sensors 9(10):8130–8145. https://doi.org/10.3390/s91008130
63. Üzek R, Sari E, Merkoçi A (2019) Optical-based (bio) sensing systems using magnetic nanoparticles. Magnetochemistry 5(4):59. https://doi.org/10.3390/MAGNETOCHEMISTRY5040059
64. Vargas E, Torrente-Rodríguez RM, Montiel VRV, Povedano E, Pedrero M, Montoya JJ, Campuzano S, Pingarrón JM (2017) Magnetic beads-based sensor with tailored sensitivity for rapid and single-step amperometric determination of miRNAs. Int J Mol Sci 18(11). https://doi.org/10.3390/ijms18112151
65. Chiang CS, Shen YS, Liu JJ, Shyu WC, Chen SY (2016) Synergistic combination of multistage magnetic guidance and optimized ligand density in targeting a nanoplatform for enhanced cancer therapy. Adv Healthcare Mater 5(16):2131–2141. https://doi.org/10.1002/ADHM.201600479
66. Mollarasouli F, Zor E, Ozcelikay G, Ozkan SA (2021) Magnetic nanoparticles in developing electrochemical sensors for pharmaceutical and biomedical applications. Talanta 226. https://doi.org/10.1016/j.talanta.2021.122108
67. Jahanbani S, Benvidi A (2016) Comparison of two fabricated aptasensors based on modified carbon paste/oleic acid and magnetic bar carbon paste/Fe_3O_4@oleic acid nanoparticle electrodes for tetracycline detection. Biosens Bioelectron 85:553–562. https://doi.org/10.1016/j.bios.2016.05.052
68. Huang L, Wang X (2021) Rapid and sensitive detection of Bisphenol A in water by LF-NMR based on magnetic relaxation switch sensor. Microchem J 163:105911. https://doi.org/10.1016/J.MICROC.2020.105911
69. Matatagui D, Kolokoltsev OV, Qureshi N, Mejía-Uriarte EV, Saniger JM (2015) A magnonic gas sensor based on magnetic nanoparticles. Nanoscale 7(21):9607–9613. https://doi.org/10.1039/C5NR01499A
70. Shakeel A, Rizwan K, Farooq U, Iqbal S, Iqbal T, Awwad NS, Ibrahium HA (2022) Polymer based nanocomposites: a strategic tool for detection of toxic pollutants in environmental matrices. Chemosphere 303. https://doi.org/10.1016/j.chemosphere.2022.134923
71. Yoon H (2013) Current trends in sensors based on conducting polymer nanomaterials. Nanomaterials 3(3):524–549. https://doi.org/10.3390/NANO3030524
72. Alam MW, Bhat SI, Al Qahtani HS, Aamir M, Amin MN, Farhan M, Aldabal S, Khan MS, Jeelani I, Nawaz A, Souayeh B (2022) Recent progress, challenges, and trends in polymer-based sensors: a review. Polymers 14(11). MDPI. https://doi.org/10.3390/polym14112164
73. Takagai Y, Yamaguchi H, Kubota T, Igarashi S (2007) Selective visual determination of vanadium(V) ion in highly acidic solution using desferrioxamine B immobilization cellulose. Chem Lett 36(1):136–137. https://doi.org/10.1246/CL.2007.136
74. Du Y, Zhang W, Wang ML (2016) Sensing of salivary glucose using nano-structured biosensors. Biosensors 6(1):10. https://doi.org/10.3390/BIOS6010010
75. Alberti G, Nurchi VM, Magnaghi LR, Biesuz R (2019) A portable, disposable, and low-cost optode for sulphide and thiol detection. Anal Methods 11(35):4464–4470. https://doi.org/10.1039/C9AY01538H
76. Alberti G, Zanoni C, Magnaghi LR, Biesuz R (2020) Low-cost, disposable colourimetric sensors for metal ions detection. J Anal Sci Technol 11(1):1–12. https://doi.org/10.1186/S40543-020-00221-X/TABLES/5
77. Biesuz R, Nurchi VM, Magnaghi LR, Alberti G (2019) Inexpensive alizarin red S-based optical device for the simultaneous detection of Fe(III) and Al(III). Microchem J 149:104036. https://doi.org/10.1016/J.MICROC.2019.104036
78. Magnaghi LR, Capone F, Zanoni C, Alberti G, Quadrelli P, Biesuz R (2020) Colorimetric sensor array for monitoring, modelling and comparing spoilage processes of different meat and fish foods. Foods 9(5):684. https://doi.org/10.3390/FOODS9050684
79. Saadatidizaji Z, Sohrabi N, Mohammadi R (2024) Development of a simple polymer-based sensor for detection of the Pirimicarb pesticide. Sci Rep 14(1). https://doi.org/10.1038/s41598-024-60748-6

80. Verma A, Gupta R, Verma AS, Kumar T (2023) A review of composite conducting polymer-based sensors for detection of industrial waste gases. In: Sensors and actuators reports, vol 5. Elsevier B.V. https://doi.org/10.1016/j.snr.2023.100143
81. Zhang Y, Kim JJ, Chen D, Tuller HL, Rutledge GC (2014) Electrospun polyaniline fibers as highly sensitive room temperature chemiresistive sensors for ammonia and nitrogen dioxide gases. Adv Func Mater 24(25):4005–4014. https://doi.org/10.1002/ADFM.201400185
82. MedCrave online (n.d.) https://medcraveonline.com/JNMR/nanomaterials-for-sensing-applications.html?utm_source=perplexity. Accessed 19 Oct 2024
83. Noah NM (2020) Design and synthesis of nanostructured materials for sensor applications. J Nanomater 2020. https://doi.org/10.1155/2020/8855321
84. Nandre V, Jadhav Y, Das DK, Ahire R, Ghosh S, Jadkar S, Kodam K, Waghmode S (2022) Nanomaterials for sensors: synthesis and applications. In: Advanced nanomaterials for point of care diagnosis and therapy. Elsevier, pp 477–492. https://doi.org/10.1016/B978-0-323-85725-3.00011-8
85. Mohamad Nor N, Ramli NH, Zakaria ND, Abu Bakar AH, Abdul Razak K (2024) Preparation of nanomaterials-based sensors. Handb Nanosens 39–67. https://doi.org/10.1007/978-3-031-47180-3_3
86. Liu R, Zhang Y, Ali S, Haruna SA, He P, Li H, Ouyang Q, Chen Q (2021) Development of a fluorescence aptasensor for rapid and sensitive detection of Listeria monocytogenes in food. Food Control 122:107808. https://doi.org/10.1016/j.foodcont.2020.107808
87. Dhahi TS, Kamal A, Dafhalla Y, Tayfour OE, Mubarakali A, Alqahtani AS, Elsir A, Ahmed T, Elobaid ME, Adam T (2024) Advances in nano sensors for monitoring and optimal performance enhancement in photovoltaic cells. iScience 27(4):109347. https://doi.org/10.1016/j.isci.2024.109347
88. Mendonsa AA, Sodia TZ, Cash KJ (2024) The impact of zwitterionic surfactants on optode-based nanosensors via different fabrication approaches and sensing mechanisms. Analyst. 149(18):4615–4622. https://doi.org/10.1039/d4an00687a
89. Lin Y (2021) Mechanism of operation of a nanosensor and its applications. J Biosens Bioelectron 12(December):37421. https://doi.org/10.37421/jbsbe.2021.12.306
90. Huang X, Zhu Y, Kianfar E (2021) Nano biosensors: properties, applications and electrochemical techniques. J Market Res 12:1649–1672. https://doi.org/10.1016/j.jmrt.2021.03.048
91. Rabbani M, Hoque ME, Mahbub ZB (2020) Chapter 7—Nanosensors in biomedical and environmental applications: perspectives and prospects. In: Pal K, Gomes F (eds) Nanofabrication for smart nanosensor applications. Elsevier, pp 163–186. https://doi.org/10.1016/B978-0-12-820702-4.00007-6
92. Safarkhani M, Aldhaher A, Heidari G, Zare EN, Warkiani ME, Akhavan O, Huh Y, Rabiee N (2024) Nanomaterial-assisted wearable glucose biosensors for noninvasive real-time monitoring: pioneering point-of-care and beyond. Nano Mater Sci 6(3):263–283. https://doi.org/10.1016/j.nanoms.2023.11.009
93. Shaheen S, Fatima B, Hussain D, Najam-ul-Haq M (n.d.) Carbon dots in cancer detection and therapy. Springer International Publishing, pp 1–25. https://doi.org/10.1007/16833_2024_431
94. Xue Y, Liu C, Andrews G, Wang J, Ge Y (2022) Recent advances in carbon quantum dots for virus detection, as well as inhibition and treatment of viral infection. Nano Converg 9(1):15. https://doi.org/10.1186/s40580-022-00307-9
95. Banigo A, Azeez T, Ejeta K, Lateef A, Ajuogu E (2020) Nanobiosensors: applications in biomedical technology. IOP Conf Ser: Mater Sci Eng 805(1):012028. https://doi.org/10.1088/1757-899X/805/1/012028
96. Kaittanis C, Naser SA, Perez JM (2007) One-step, nanoparticle-mediated bacterial detection with magnetic relaxation. Nano Lett 7(2):380–383. https://doi.org/10.1021/nl062553z
97. Naseer B, Srivastava G, Qadri OS, Faridi SA, Islam RU, Younis K (2018) Importance and health hazards of nanoparticles used in the food industry. Nanotechnol Rev 7(6):623–641. https://doi.org/10.1515/ntrev-2018-0076

98. Rai M, Gade A, Gaikwad S, Marcato PD, Durán N (2012) Biomedical applications of nanobiosensors: the state-of-the-art. J Braz Chem Soc 23:14–24. https://doi.org/10.1590/S0103-50532012000100004
99. Vinayaka AC, Thakur MS (2010) Focus on quantum dots as potential fluorescent probes for monitoring food toxicants and foodborne pathogens. Anal Bioanal Chem 397(4):1445–1455. https://doi.org/10.1007/s00216-010-3683-y
100. Noori R, Ahmad R, Sardar M (2020) Nanobiosensor in health sector: the milestones achieved and future prospects. In: Mohsin M, Naz R, Ahmad A (eds) Nanobiosensors for agricultural, medical and environmental applications. Springer, pp 63–90. https://doi.org/10.1007/978-981-15-8346-9_4
101. Zhang D, Sun Y, Wu Q, Ma P, Zhang H, Wang Y, Song D (2016) Enhancing sensitivity of surface plasmon resonance biosensor by Ag nanocubes/chitosan composite for the detection of mouse IgG. Talanta 146:364–368. https://doi.org/10.1016/j.talanta.2015.08.050
102. Dzyadevych SV, Arkhypova VN, Martelet C, Jaffrezic-Renault N, Chovelon J-M, El'Skaya AV, Soldatkin AP (2006) Potentiometric biosensors based on ISFETs and immobilised cholinesterases. Int J Appl Electromagnet Mech 23(3–4):229–244
103. Fu Y, Liu T, Wang H, Wang Z, Hou L, Jiang J, Xu T (2024) Applications of nanomaterial technology in biosensing. J Sci: Adv Mater Devices 9(2):100694. https://doi.org/10.1016/j.jsamd.2024.100694
104. Huayhongthong S, Khuntayaporn P, Thirapanmethee K, Wanapaisan P, Chomnawang MT (2019) Raman spectroscopic analysis of food-borne microorganisms. LWT 114:108419. https://doi.org/10.1016/j.lwt.2019.108419
105. Liu X, Huang L, Qian K (2021) Nanomaterial-based electrochemical sensors: mechanism, preparation, and application in biomedicine. Adv NanoBiomed Res 1(6):2000104. https://doi.org/10.1002/anbr.202000104
106. Servarayan KL, Krishnamoorthy G, Sundaram E, Karuppusamy M, Murugan M, Piraman S, Vasantha VS (2023) Optical immunosensor for the detection of Listeria monocytogenes in food matrixes. ACS Omega 8(18):15979–15989. https://doi.org/10.1021/acsomega.2c07848
107. Adam T, Gopinath CB (2022) Nanosensors: recent perspectives on attainments and future promise of downstream applications. Process Biochem 117:153–173. https://doi.org/10.1016/j.procbio.2022.03.024
108. Lallas E (2019) Key roles of plasmonics in wireless THz nanocommunications—a survey. Appl Sci 9(24), Article 24. https://doi.org/10.3390/app9245488
109. Yang T, Duncan TV (2021) Challenges and potential solutions for nanosensors intended for use with foods. Nat Nanotechnol 16(3):251–265. https://doi.org/10.1038/s41565-021-00867-7
110. Srivastava AK, Dev A, Karmakar S (2018) Nanosensors and nanobiosensors in food and agriculture. Environ Chem Lett 16(1):161–182. https://doi.org/10.1007/s10311-017-0674-7
111. Sharma P, Pandey V, Sharma MMM, Patra A, Singh B, Mehta S, Husen A (2021) A review on biosensors and nanosensors application in agroecosystems. Nanoscale Res Lett 16(1):136. https://doi.org/10.1186/s11671-021-03593-0
112. Jain PK, Huang X, El-Sayed IH, El-Sayed MA (2007) Review of some interesting surface plasmon resonance-enhanced properties of noble metal nanoparticles and their applications to biosystems. Plasmonics 2(3):107–118. https://doi.org/10.1007/s11468-007-9031-1

Biosensors for Monitoring Climatic Stressors in Agricultural Crops

Oluwatoyin Adenike Fabiyi and Ayorinde Victor Ogundele

Abstract Agriculture is regarded as substructure of nations for the reason that, generality of population counts on it to acquire livelihood and subsist. Environmental effects, including variabilities in weather are central components of food insecurity. Severe alterations in the environment are caused by instability of climatic factors. This dynamic state of the weather influences development of crops and reduction in yield. There is thus a need for sustainable production in the face of weather variables. To manage ecosystem and maintain the crucial role of agriculture, resilient management of environment and natural resources are necessary. This will safeguard environment, while strengthening productivity in the absence of ecological disorder. Consequently, uncommon cultivation operations are essential to enlarge harvest. Application of nanosensors to address climatic challenges in crop production is important to overcome yield loss due to climatic stressors. The efficiency, design and functionality of nanosensors in plant health monitoring is considered and reviewed in this chapter.

Keywords Climate change · Nanosensors · Plant stress · Nanoparticles

1 Introduction

Sensors are intended to monitor environmental variables, such as temperature, humidity, or light intensity, and then relay this data to other electronic circuits or instruments for measurement or analysis Fig. 1 depicts representation of biosensor

O. A. Fabiyi (✉)
Department of Crop Protection, Faculty of Agriculture, University of Ilorin, Ilorin, Nigeria
e-mail: fabiyitoyinike@hotmail.com; fabiyi.oa@unilorin.edu.ng

A. V. Ogundele
Department of Chemistry, College of Pure and Applied Sciences, Kwara State University, Malete, Nigeria

Laboratory of Natural Products & Drug Discovery, Centre CEBIM, Department of Basic Sciences, Faculty of Medicine, Universidad de La Frontera, 4780000 Temuco, Chile

I. Uddin (ed.), *Smart Nanosensors*, Smart Nanomaterials Technology,
https://doi.org/10.1007/978-981-96-3878-9_9

design. Biosensors use biological elements, such as cells, enzymes, or antibodies, coupled to tiny transducers that change the signal shape so that it can be measured with ease. They provide important information for a variety of applications, including food safety, disease analysis, and environmental monitoring (Fig. 2). They can detect changes in cells and molecules even at low concentrations of tested chemicals [1]. Biosensors have become sophisticated detection instruments utilized in many scientific domains, such as illness diagnosis, environmental monitoring, food and beverage quality control, and pathogen detection. A thorough grasp of plant biology is necessary to address enduring global issues like securing sustainable food sources and creating renewable energies. The selectivity and specificity of biological recognition systems built into the sensing platform play a critical role in the efficacy of biosensors. A typical biosensor consists of two primary parts: a physical part (like a transducer or amplifier) and a biological part (like enzymes or cells) [2]. The biological component recognizes the target analyte and initiates a physical alteration (a signal) that the transducer may pick up on. Specific interactions, such as antibody-antigen binding, DNA and RNA hybridization, receptor-hormone binding, or enzyme–substrate reactions, are the basis for the majority of biosensors' functions (Fig. 3). Transducers are able to identify changes in form or composition that arise from the interaction between the biological component and the analyte. They transform these alterations into electrical signals, which are further processed and analyzed to ascertain the analyte's concentration with the aid of a processor and amplifier before being shown on a screen [3].

Depending on the properties of the target, biosensors use a variety of methods, including optical, electrochemical, thermometric, and magnetic transductions as illustrated in Fig. 4, as well as bio electrochemical signals, to specifically determine the presence or amounts of a specific biological component. They enable the quantitative examination of a wide range of substances by converting biological activity

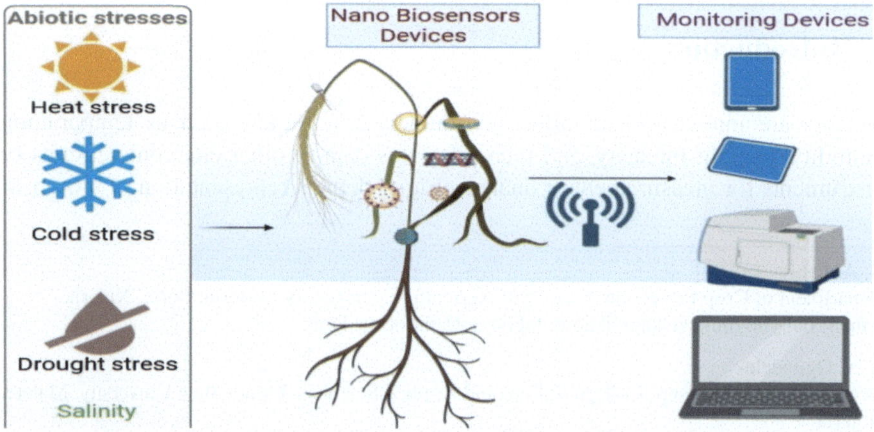

Fig. 1 Representation of biosensor for monitoring abiotic stressors on crop

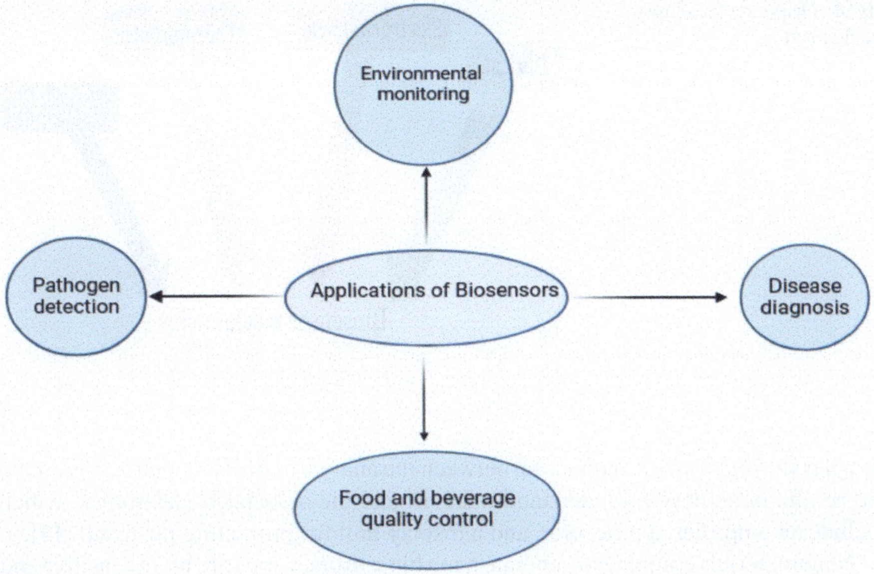

Fig. 2 The diversity of biosensor applications

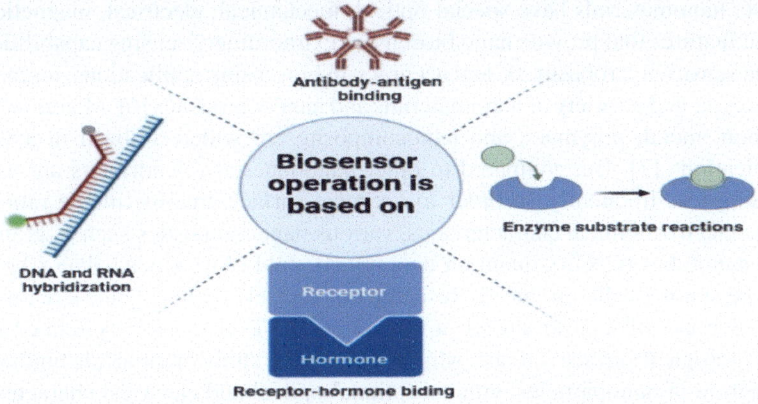

Fig. 3 Biosensor operation is premised on DNA, hormone, enzyme and antibodies

into quantifiable signals. Most biosensors use immobilized enzymes, and the biological reaction's stability, specificity, and sensitivity are essential to their functionality. Nanomaterials have the potential to greatly improve the efficacy of current detection methods in biosensing applications. A biosensor's schematic typically consists of four main parts: the analyte, which is the substance being detected; the bioreceptor, which is a biological element that recognizes and binds to the analyte, such as cells, antibodies, or enzymes; the transducer, which detects the bio-recognition

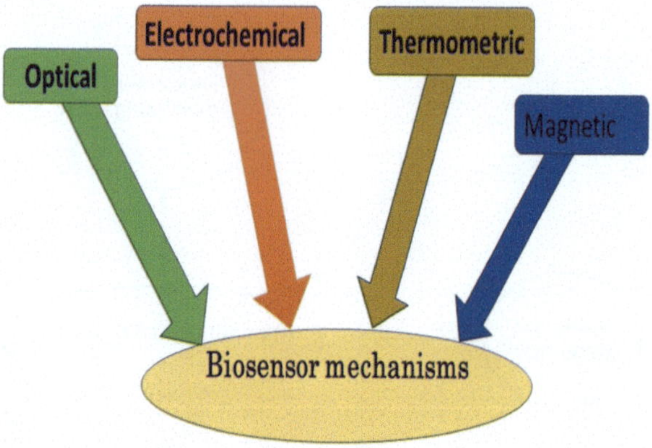

Fig. 4 Disparate biosensor mechanisms

signal resulting from the interaction between the analyte and bioreceptor and presents the results in an easy-to-understand manner; and the associated electronics, which include an amplifier, a processor, and a display unit for presenting the results [4].

Nanomaterials combine to generate nano biosensors, which are more sensitive and capable of precisely identifying atomic-level changes. Nanobiosensors use materials that are nanoscale in size as its receptors and transducers, in contrast to typical biosensors, which use materials that are micro- or macroscale in size [5]. Because of their tiny size, nanomaterials have special optical, mechanical, electrical, magnetic, and physical features that provide nano biosensors extraordinary sensing capabilities. To improve sensitivity, robustness, and accuracy in biosensing applications, researchers have investigated a variety of nanomaterials and nanocomposites [6]. Materials based on carbon, metal, polymers, and nanocomposite are widely utilized in a variety of applications [7]. But compared to other nanomaterials, dendrimers are used in biosensing less frequently. In order to optimize surface area-to-volume ratios and enhance electrochemical characteristics, various nanocomposites such as graphene, carbon nanotubes (CNTs), quantum dots (QDs), gold (Au), silica, silver (Ag), and others are synthetically produced. Research on precisely managing the structure and surface features of nanostructures at the nanoscale level is strongly related to the field of biological surface science, which includes the study of molecule binding [8]. The creation of nanoparticles with exceptional optical and electrical characteristics appropriate for uses in electronics and sensing has been made easier by developments in nanotechnology [9]. By increasing the detection limit and overall performance of biosensors, the use of nanomaterials can improve their efficiency. Because of their large surface area and electrical conductivity, nanomaterials provide an excellent substrate for the assembly of bio-recognition components, increasing the detection limit [10].

2 Overview of Nanomaterials and Their Potential for Abiotic Stress Detection

Stressors are unfavourable conditions or substances that interfere with a plant's growth, development, or metabolic functions. It stands for a crucial area where the agriculture industry has to develop. Stress exposure alters the metabolic balance of the plant, causing several reactions that serve as signaling systems [11]. There are two types of stress: high-level, long-term stress and low-level, short-term stress. Many times, stress mechanisms or adaptations that take place at different morphological and molecular levels might reduce short-term stress. But extended stress may be harmful, perhaps killing the plant or creating lasting damage. All of these challenges have a major effect on how plants use their metabolism [12]. There are two primary categories of plant stresses: biotic and abiotic.

Plant illnesses and subsequent crop loss are caused by biotic stressors, which include assaults from a variety of pathogens such as bacteria, viruses, nematodes, and fungus [13]. Unfavourable environmental factors that negatively affect plant growth and viability include droughts, floods, high temperatures, salt, soil deterioration, nutritional imbalances, chemical pollutants, and oxidative stress. These factors are sometimes referred to as abiotic stressors. Plant growth, development, and production can be directly hindered by these variables, which can cause physiological and biochemical changes in plants [14]. The plant responds to abiotic stress by initiating a series of molecular processes that culminate in physiological, metabolic, and developmental responses. Gaining understanding of plant resilience requires tracking the processes behind abiotic stress responses [15].

Nano biosensors have been essential in identifying phytohormone imbalances and allowing prompt intervention to reduce abiotic stress-induced plant damage and yield loss. These sensors play a major role in accomplishing the objectives of precision agriculture. For instance, utilizing a porous nanogold screen, a label-free electrochemical impedance immunosensor has been created to detect abscisic acid (ABA) [16]. Gibberellic acid (GA) may also be found in hybrid rice grain samples using nano biosensors that use electrochemical impedance spectroscopy [17]. Furthermore, the quantities of auxin and cytokinin in tomato xylem sap exudates may be measured in real time using nano biosensors [18]. These developments in nano biosensing technologies assisted in avoiding major crop losses and produce quality crops that are maintained by early detection of abiotic stress situations.

3 Applications of Engineered Nanomaterials in Plant Stress Control

One cutting-edge approach to enhancing sustainable farming methods is nanomaterial engineering. Benefits from using nanomaterials into agriculture include lower labour and cost, increased productivity, and the encouragement of ecologically friendly methods. In order to maintain soil and plant health, product quality, and overall sustainability, the development of nanosensors offers promise for monitoring a variety of crop growth, soil conditions, nutrient levels, diseases, and environmental consequences of agrochemicals [19]. It may be possible to increase the selectivity, sensitivity, and responsiveness of sensors by fusing biological organisms with nanomaterials [20]. To increase agricultural output, nanobiotechnology presents a viable method for developing intelligent bio-nanosensors that communicate with and manage electrical equipment [21]. Crops are negatively impacted by abiotic stress, which interferes with their regular development cycles and ultimately kills the plants. Abiotic stress in agricultural plants can thus be identified by a variety of visible or invisible changes, thermal analysis, changes in protein, reduced functions, the formation of novel receptors, and the presence of reactive oxygen species.

Alova et al. [22] developed oxygen nanosensors that measure oxygen levels in *Chara corallina* internodal cells electrochemically. These sensors are made of carbon-filled quartz micropipettes with platinum-coated tips. The investigation of cellular oxygen metabolism is made easier by the use of platinum nanoelectrodes. After micro damage, the study found that the concentration of oxygen in *Chara corallina* cells significantly decreased [22].

In a similar investigation, Wang and Li [23] used the sol–gel method to create ZrO_2/Au (zirconium dioxide/gold) nanocomposite films. These films were used to detect organophosphate insecticides by voltametric sensing. In wave voltammetry, the ZrO_2/Au nano sensor film electrode produced signals due to its remarkable sensitivity and selectivity towards the phosphate group on parathion molecules [23]. The liposome bio-nanosensor that Vamvakaki and Chaniotakis [24] developed to offer potential for monitoring organophosphorus pesticides, such as dichlorvos and paraoxon, was used in a related study to help with pollution control efforts. It allowed for the detection of specific toxins present at very low concentrations in a variety of environmental sources, including sewage sludge, herbicides, and pollutants in urban and industrial effluents. Moreover, the detection of hazardous cadmium ions (Cd^{2+}) and acetylcholinesterase (AChE) activity in water samples has been made easier by recent developments in nanosensor creation that employ the FRET configuration [25].

Through processes like transpiration and evaporation, soil temperature and moisture play critical roles in controlling temperature dynamics and water exchange between the soil surface and the surrounding environment. The ability of a crop plant to withstand heat stress and continue to grow, develop, and yield at levels that are economically feasible is known as thermotolerance. According to Kordrostami et al. [26] plants that are thought to be thermotolerant may have anatomical characteristics that allow them to maintain a steady internal temperature just below a critical point at which heat-induced damage becomes permanent. Heat stress causes a substantial reduction in photosynthetic activity, which in turn affects plant growth and yield. Photosynthesis is a very sensitive biological process [27]. By decreasing photosynthesis and respiration rates, high temperatures indirectly cause stomatal closure by raising the concentration of carbon dioxide under the leaf surface [26]. Heat stress has a significant impact on plants since they are stationary creatures. It causes changes in physiological and developmental processes as well as decreased growth and yield [28]. One major effect of both high and low temperatures is oxidative stress, which is caused by an excess of reactive oxygen species (ROS) [29]. Researchers have developed nanosensors based on micro electromechanical systems (MEMS) that function based on shear stress. According to Jackson et al. [30], these sensors combine a Wheatstone bridge piezoresistor circuit with a microsensor chip that has a proprietary polymer sensing element. This allows for the production of two DC output voltages that are linearly associated with temperature and moisture content. Similar alterations in membrane characteristics take place during cold stress, resulting in a modified metabolism and the production of toxic compounds, both of which cause secondary damage to the plant [31]. The rate of energy transmission to Photosystem II's (PSII) core reduces with decreasing temperatures [32]. In order to detect H_2O_2 in Arabidopsis, explain its dynamics with a high temporal precision, and demonstrate insensitivity to pH variations during elicitor-induced oxidative bursts, Nietzel et al. [33] developed biosensors. Exposito-Rodriguez et al. [34] showed that when cells are subjected to variations in light intensity, H_2O_2 signaling helps to promote communication between chloroplasts and nuclei. Plants that experience temperature stress exhibit elevated levels of phytohormones such as abscisic acid (ABA) and indole-3-acetic acid (IAA). Furthermore, in reaction to unfavourable circumstances, salicylic acid (SA), ethylene (ET), and jasmonates (JAs) each play unique functions [35]. As a result, scientists developed an electrochemical biosensor that can identify the phytohormones SA and IAA, allowing for the observation of a plant's response to abiotic stress [36]. Amreen et al. [37] created a pH-stable nanosensor (FLIP-H_2O_2) that is highly selective for H_2O_2 and insensitive to other oxidants such nitric oxide, superoxide anions, and peroxynitrite by genetic engineering. Although oxidative stress caused by high H_2O_2 levels in cells can stunt plant and animal growth, H_2O_2 is essential for regulating many metabolic activities, especially when under stress [38]. Additionally, FLIP-H_2O_2 may be directed towards cellular organelles like mitochondria and chloroplasts, making it easier to monitor subcellular signals in living things [37]. A microsensor chip that combines a proprietary polymer sensing element and a piezoresistive circuit inside a Wheatstone bridge to transmit two DC output voltages that are directly proportional to moisture and temperature has been developed using the shear stress principle [30].

A major obstacle to agricultural output is drought, which affects plants because of decreased water availability or higher transpiration rates, especially in semi-arid and dry areas [39]. According to projections, by the early twenty-first century, a significant section of the world's population would experience water scarcity [40]. Complex alterations in the physical, chemical, and molecular properties of plants are brought about by drought; these alterations begin with growth inhibition and metabolic modifications [41]. In order to withstand drought stress, plants utilize a variety of defence mechanisms, including changes to their metabolic pathways, stomatal conductance, anthocyanin accumulation, osmoprotectant participation, carotenoid degradation, and antioxidant enzyme activity as shown in Fig. 5 [42]. According to Nakashima and Suenaga [43], transcription factors that modulate plant responses to drought stress include abscisic acid-responsive element-binding factor (ABRE), dehydration-responsive element-binding protein (DREB), and no apical meristem (NAM). In water-limited areas, agricultural profitability may be maintained or increased by the use of ecologically friendly nanoproducts to increase crop drought resistance [44]. Many plant species have shown to be capable of withstanding drought thanks to nanotechnology [45]. For example, it has been demonstrated that sorghum's resistance to drought is improved by cerium oxide nanoparticles, or nanoceria [46]. It has been suggested that carbon nanotubes might increase water intake by interacting with aquaporins to lessen the negative consequences of drought stress [47]. According to research on *Hyoscyamus niger*, by triggering the plant's antioxidant machinery, seed priming with single-walled carbon nanotubes (SWCNTs) might reduce oxidative damage brought on by drought stress [48]. 16 genes, sensitive to drought were found among 351 genes exhibiting differential regulation when *A. thaliana* was treated with TiO_2NPs, AgNPs, and multi-walled carbon nanotubes, according to research by Garcia-Sanchez and Lujan-Garcia [49]. Furthermore, Borišev et al. [50] showed that plants under drought stress may change their intracellular water metabolism in response to foliar delivery of nanofullerenol. Si-NPs, or silicon nanoparticles, have great potential for use in agriculture and may help reduce a variety of abiotic stressors. Si-NPs may be used in conjunction with nanozeolite and nanosensors to detect plant toxins. Rastogi et al. [51] report that Si-NPs are also efficient transporters of proteins and nucleotides. Moreover, the use of Si-NPs as nanosensors presents significant benefits due to their core–shell structure, which amplifies sensor stability, sensitivity, and accuracy [52]. Si-NPs have the ability to improve soil water retention, especially when there is drought. Liu et al. [53] successfully demonstrated the use of carbon dot encapsulation onto rhodamine B-doped Si-NPs to construct a dual-emission nanosensor for the detection of hazardous copper ions (Cu^{2+}) in water.

For ages, the environment has been exposed to non-biodegradable heavy metals. They mostly build up in the sediments of lakes, seas, and estuaries. Eventually, they make their way up the food chain and end up at dangerously high concentrations in the bodies of apex predators [54]. Significant worldwide risk is posed by the high quantities of heavy metals in the environment [55]. Heavy metal levels in the environment are rising quickly due to a variety of activities, including mining and industrial processes, fuel emissions, inappropriate treatment of industrial waste, and the use of agrochemicals [56, 57]. Sensor performance has been greatly improved by using Quantum Dots (QDs)-based nanomaterials, which have overcome the drawbacks of conventional techniques such as slowness, low sensitivity, high cost, and complexity [58]. With the growing significance of heavy metal detection, electrochemical approaches have proven to be effective in enhancing detection and outperforming conventional analytical methods. Biological receptors, carbon nanotube

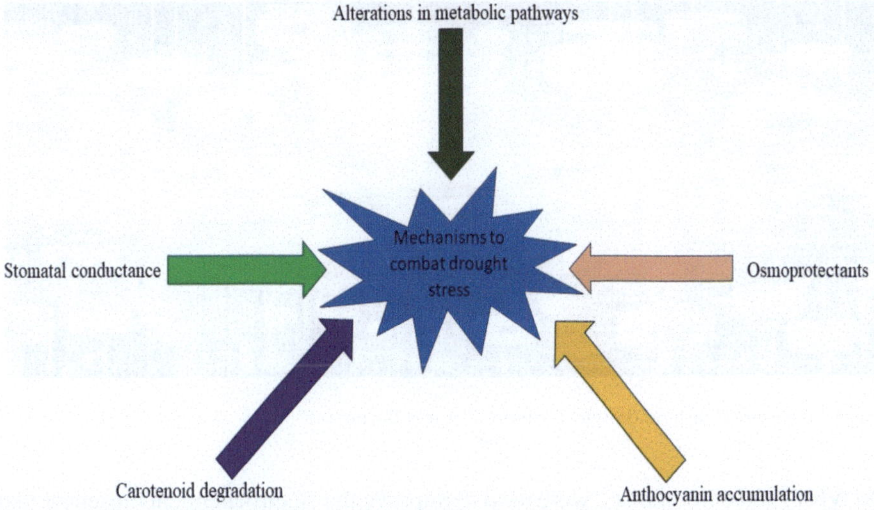

Fig. 5 Plant defense mechanisms to combat drought

(CNT) networks, and multi-nanoparticle (MNP) assemblies have all been widely used to increase the adaptability of nanomaterial-based electrochemical sensing [54]. Xiang et al. [59] have created biosensors that use changes in electrochemical impedance to detect the production of vitronectin-like (VN) proteins as a biomarker of plant cell damage. This helps identify cadmium and lead stress in plant species, including soybeans and Arabidopsis. A smartphone-enabled detection apparatus that can detect Hg(II) at concentrations as low as 2.53 nm was developed by Guo et al. [60]. An alternative method for mercury and cadmium detection involved utilizing a lateral flow assay with Up converting Nanoparticles (UCNPs), Gold Nanoparticles (AuNPs), aptamers, DNA probes, and antibody probes, achieving detection limits at 5 ppb and 0.18 ng/mL for mercury and cadmium, respectively. Lew et al. [61] engineered a nanosensor with near-infrared fluorescent capabilities to enable continuous, real-time detection of arsenic. Formaldehyde represents a prevalent environmental pollutant commonly encountered in industrial settings. Its presence poses abiotic stress on plants, impacting various factors such as chlorophyll content, ascorbate peroxidase (APX) activity, carotenoid content, peroxidase (POD) activity, soluble protein and sugar levels, superoxide anion production, and lipid peroxidation rate [62]. In order to detect formaldehyde using Surface-Enhanced Raman Spectroscopy (SERS) analysis, silver and gold nanoparticles (AgNPs and AuNPs) have been developed as surface-enhanced Raman-active substrate sensors [63].

One of the main causes of crop failure in arid and semiarid regions is the uneven accumulation of salt. The harmful effects of salinity primarily depend on the concentration and source of the salt [64]. Salinity causes plants' water relations to be disrupted because of the adverse water potential brought on by the toxic effects of sodium and chloride ions in the soil solution [65]. Multiwall carbon nanotubes

Fig. 6 Illustration of field designs for biosensor monitoring (a)

(MWNT) and carbon black were used to improve the electrochemical biosensor that Hu et al. [36] developed for soybean seedlings under saline stress. The biosensor had a detection limit of 1.99 M for IAA and 3.30 M for SA [56]. According to Baz et al. [66], seed priming using materials based on carbon nanoparticles greatly increases lettuce seed vigor and seedling development under high salt stress conditions by preventing primary root elongation. Single-wall carbon nanotube (SWCNT) nanosensors have been developed and implemented by MIT engineers within plant leaves. These sensors use near-infrared (nIR) optical signals to detect and transmit information about different stress conditions, which can then be received via emails or text messages on smartphones and laptops. Air, soil and water contamination could be seamlessly detected through mounting of sensors on the field. Pollen grains and other air borne pathogens are captured via sensing (Figs. 6, 7 and 8). Furthermore, pivotal information regarding requirements of crops is transmitted via sensing techniques to enable timely provision vis a vis resource conservation. This is premised on the fact that, inclusion of fertilizers, water and pesticides are done optimally without excesses leading to wastage. Health of crops are equally properly maintained via signals before symptom appearance. Specific point of disease is corrected without the injuring the crop and issues of residuals from pesticides are removed and avoided (Figs. 6, 7 and 8).

4 Challenges and Opportunities

Biosensor technology will likely progress in the near future to include embedded systems, a wider variety of analytes, microfluidics, a stronger focus on entire cell and tissue biosensors, telecommunication features, and wearable capabilities. Farmers and other end users will soon be able to use their smartphones and other portable devices to conduct real-time, on-the-spot testing for quality and safety evaluations

Fig. 7 Illustration of field designs for biosensor monitoring (b)

Fig. 8 Illustration of field designs for biosensor monitoring (c)

in a variety of locations, including residences, market places, factories, and fields. Although there are sensors on the market right now for some uses, they are not enough to handle all of the problems that face agriculture. Regarding detection limit, sensitivity, specificity, electrode design, matrix production, prices, and field usage viability, there is still much space for development.

While conventional matrices have given way to nanoparticles as the preferred matrix for biosensor fabrication, there are a number of obstacles that frequently stand in the way of their commercialization. Because sample matrices have complicated chemical compositions, biosensors-which can operate as offline, online, or in vivo devices in agriculture—are susceptible to interference. This interference is a major problem, especially in agricultural samples where metabolites, proteins, macromolecules, and cells in samples coming from both plants and animals frequently produce electrochemical disturbances [67]. These samples' complex natural biological interactions have the potential to change their chemical composition, which can then affect biomolecule activities and biosensor performance. Sample pretreatment is often used to lessen the impact of interferents. The sensitivity and accuracy of biosensors are improved by methods including extraction, derivatization, pre-concentration, and filtering. However, the direct use of biosensors in the field is necessary for many agricultural applications. Certain chemicals in agricultural samples can cause passivation and biofouling, which adds another layer of difficulty [59]. Implantable biosensors have a major challenge from biofouling, yet this problem may be lessened with careful consideration of an ideal working voltage. Since the stability of biomolecules plays a major role in biosensor development, shelf life and operational stability are both important factors to take into account. Deactivation and denaturation of proteins are frequent reasons why biosensors have shorter shelf lives. Furthermore, agricultural biosensors frequently use methods utilizing polymeric films, including nafion [68]. The stability and endurance of biosensors are largely dependent on the immobilization of biomolecules. Biomolecules are often more stable when they are dry than when they are in soluble solution form. They also show better stability after immobilization. Biomolecule stabilization may be accomplished efficiently by combining many techniques.

There are several ways to implement these sensors: you may put them in the soil around the plant, affix them to the plant's surface, or integrate them right into the plant itself. All things considered, wearable plant sensors and plant nanosensors are exciting new technologies that might revolutionize farming methods. These sensors can help increase agricultural yields, cut water usage, and lessen dependency on potentially hazardous pesticides by giving real-time data on crop conditions. Recent agricultural research has focused more on developing wearable sensors and biosensors specifically for plant applications, as our evaluation of several publications demonstrates. These sensors are designed to encourage the use of fewer chemicals while simultaneously monitoring a variety of plant growth and development-related characteristics [69].

These sensors are also important for the detection of other components such calcium ions, molecular oxygen, ROS, and ATP. Farmers now have a sophisticated tool at their disposal to maximize crop yields and harvesting by using nanosensors to make educated decisions. They may be incorporated into plant tissues to monitor plant development, spot nutritional deficits, and make the most use of fertilizer. Another use for nanosensors is in smart irrigation systems. These systems determine the optimal quantity of water to apply to crops based on evapotranspiration rates, soil moisture sensors, and meteorological data.

Plant nanosensors have many beneficial qualities, but they also have certain drawbacks. Since these technologies are still in their early stages of development, there are a number of obstacles that need to be overcomed, such as scalability and cost concerns. To improve these technologies' accuracy and efficacy, a great deal of research is needed. Their lack of cost-effectiveness is a significant disadvantage, since the complex production process might result in expensive costs. Furthermore, the operation of nanosensors, like electrochemical sensors, requires an on-site power source. Photobleaching and phototoxicity can cause reduced stability in some types of nanosensors, such as FRET sensors, which can impact fluorescence readings. Another challenge is the labour-intensive nature of nanosensors, which calls for specialized tools and knowledgeable operators. Because of their small size, plant sensors are very sensitive, and there are also health and environmental issues with nanosensors [70]. It is imperative to acknowledge the possible advantages that these sensors may provide, notwithstanding these disadvantages. These technologies have a bright future ahead of them and are expected to be crucial in transforming agriculture. All things considered, the use of plant nanosensors has the potential to transform agricultural operations through the facilitation of real-time plant development monitoring, the reduction of dependency on pesticides and other chemicals, and the promotion of sustainable farming practices. We may expect even more advanced sensors to be developed as technology develops further to improve the sustainability and efficiency of agricultural activities [71].

5 Conclusions

The widespread application of nanosensors and bionanosensors advances smart agricultural systems and improves crop yield and quality. Various diagnostic procedures have been created to identify environmental pollutants in the agriculture sector and analyze illnesses in medical contexts by utilizing biosensors inspired by nanotechnology. These mechanisms include carbon nanotubes, liposome-based bio nanosensors, genetically modified nanosensors like FLIP-H_2O_2, and nanoparticle-based sensors. These biosensors handle a variety of agricultural problems including both biotic and abiotic stressors and provide real-time detection capabilities. They are proficient in recognizing issues including temperature fluctuations, heavy metal and chemical pollution of soil, and phytopathology of plants.

Future nano-inspired biosensors could combine robotics and GPS to produce intelligent sensors that can identify and address plant problems before they become serious or show symptoms. Furthermore, complex tolerance mechanisms have been created to assist plants to resist stress conditions. Even though these studies have demonstrated potential in lab settings, more investigation is required to convert these findings into useful agricultural applications that would allow for the reduction of abiotic stresses for enhanced plant development and growth. Proper care is essential for use of bio-nanosensors on agricultural fields.

References

1. Fan X, White IM, Shopova SI, Zhu H, Suter JD, Sun Y (2008) Sensitive optical biosensors for unlabeled targets: a review. Analytica Chimica Acta 620(1):8–26. https://doi.org/10.1016/j.aca.2008.05.022
2. Chepyala R (2020) Chapter 4—Applications and success of MIPs in optical-based nanosensors. In: Pal K, Gomes SNA (eds) Micro and nano technologies, pp 89–121. https://doi.org/10.1016/B978-0-12-820702-4.00004-0
3. Ramesh M, Janani R, Deepa C, Rajeshkumar L (2023) Nanotechnology-enabled biosensors: a review of fundamentals, design principles, materials, and applications. Biosensors 13. https://doi.org/10.3390/bios13010040

4. Sawant SN (2017) 13—Development of biosensors from biopolymer composites. In: Sadasivuni KK, Ponnamma D, Kim J, Cabibihan J-J, AlMaadeed E (eds). https://doi.org/10.1016/B978-0-12-809261-3.00013-9
5. Doria G, Conde J, Veigas B, Giestas L, Almeida C, Assunção M, Baptista PV (2012) Noble metal nanoparticles for biosensing applications. Sensors 12:1657–1687. https://doi.org/10.3390/s120201657
6. Maduraiveeran G, Jin W (2017) Nanomaterials based electrochemical sensor and biosensor platforms for environmental applications. Trends Environ Anal Chem 13:10–23. https://doi.org/10.1016/j.teac.2017.02.001
7. Khan I, Saeed K, Khan I (2019) Nanoparticles: properties, applications and toxicities. Arabian J Chem 12(7):908–931. https://doi.org/10.1016/j.arabjc.2017.05.011
8. Lim T-C, Ramakrishna S (2006) A conceptual review of nanosensors. Zeitschrift für Naturforschung A 61(7–8):402–412. https://doi.org/10.1515/zna-2006-7-815
9. Shipway AN, Katz E, Willner I (2000) Nanoparticle arrays on surfaces for electronic, optical, and sensor applications. ChemPhysChem 1(1):18–52. https://doi.org/10.1002/1439-7641(20000804)1:1<18::AID-CPHC18>3.0.CO;2-L
10. Kumar P, Rajan R, Upadhyaya K, Behl G, Xiang X-X, Huo P, Liu B (2024) Metal oxide nanomaterials based electrochemical and optical biosensors for biomedical applications: recent advances and future prospectives. Environ Res 247:118002. https://doi.org/10.1016/j.envres.2023.118002
11. Shulaev V, Cortes D, Miller G, Mittler R (2008) Metabolomics for plant stress response. Physiologia Plantarum 132(2):199–208. https://doi.org/10.1111/j.1399-3054.2007.01025.x
12. Jan N, Rather AM-U-D, John R, Chaturvedi P, Ghatak A, Weckwerth W, Mir RR (2023) Proteomics for abiotic stresses in legumes: present status and future directions. Crit Rev Biotechnol 43(2):171–190. https://doi.org/10.1080/07388551.2021.2025033
13. He Z, Webster S, He SY (2022) Growth–defense trade-offs in plants. Curr Biol 32(12):R634–R639. https://doi.org/10.1016/j.cub.2022.04.070
14. Yaqoob U, Jan N, Raman PV, Siddique KHM, John R (2022) Crosstalk between brassinosteroid signaling, ROS signaling and phenylpropanoid pathway during abiotic stress in plants: does it exist? Plant Stress 4:100075. https://doi.org/10.1016/j.stress.2022.100075
15. dos Santos TB, Ribas AF, de Souza SGH, Budzinski IGF, Domingues DS (2022) Physiological responses to drought, salinity, and heat stress in plants: a review. Stresses 2:113–135. https://doi.org/10.3390/stresses2010009
16. Li Y-W, Xia K, Wang R-Z, Jiang J-H, Xiao L-T (2008) An impedance immunosensor for the detection of the phytohormone abscisic acid. Anal Bioanal Chem 391(8):2869–2874. https://doi.org/10.1007/s00216-008-2214-6
17. Li J, Xiao L-T, Zeng G-M, Huang G-H, Shen G-L, Yu R-Q (2005) Immunosensor for rapid detection of gibberellin acid in the rice grain. J Agric Food Chem 53(5):1348–1353. https://doi.org/10.1021/jf0487553
18. Wend S, Bosco CD, Kämpf MM, Ren F, Palme K, Weber W, Zurbriggen MD (2013) A quantitative ratiometric sensor for time-resolved analysis of auxin dynamics. Sci Rep 3(1):2052. https://doi.org/10.1038/srep02052
19. Singh H, Sharma A, Bhardwaj SK, Arya SK, Bhardwaj N, Khatri M (2021) Recent advances in the applications of nano-agrochemicals for sustainable agricultural development. Environ Sci Process Impacts 23(2):213–239. https://doi.org/10.1039/D0EM00404A
20. Dubey A, Mailapalli DR (2016) Nanofertilisers, nanopesticides, nanosensors of pest and nanotoxicity in agriculture. Sustain Agric Rev 19:307–330. https://doi.org/10.1007/978-3-319-26777-7_7
21. Giraldo JP, Wu H, Newkirk GM, Kruss S (2019) Nanobiotechnology approaches for engineering smart plant sensors. Nat Nanotechnol 14(6):541–553. https://doi.org/10.1038/s41565-019-0470-6
22. Alova A, Erofeev A, Gorelkin P, Bibikova T, Korchev Y, Majouga A, Bulychev A (2020) Prolonged oxygen depletion in microwounded cells of Chara corallina detected with novel oxygen nanosensors. J Exp Bot 71(1):386–398. https://doi.org/10.1093/jxb/erz433

23. Wang M, Li Z (2008) Nano-composite ZrO_2/Au film electrode for voltammetric detection of parathion. Sens Actuators B: Chem 133(2):607–612. https://doi.org/10.1016/j.snb.2008.03.023
24. Vamvakaki V, Chaniotakis NA (2007) Pesticide detection with a liposome-based nano-biosensor. Biosens Bioelectron 22(12):2848–2853. https://doi.org/10.1016/j.bios.2006.11.024
25. Giardi MT (2006) Biotechnological applications of photosynthetic proteins: biochips, biosensors and biodevices. In: Biotechnology intelligence unit. Springer, New York
26. Kordrostami M, Mafakheri M, Al-Khayri JM (2021) Contributions of nano biosensors in managing environmental plant stress under climatic changing era. Nanobiotechnol Mitig Abiotic Stress Plants. https://doi.org/10.1007/978-3-030-73606-4_6
27. Fahad S, Bajwa AA, Nazir U, Anjum SA, Farooq A, Zohaib A, Huang J (2017) Crop production under drought and heat stress: plant responses and management options. Front Plant Sci 8. https://doi.org/10.3389/fpls.2017.01147
28. Lobell DB, Field CB (2007) Global scale climate–crop yield relationships and the impacts of recent warming. Environ Res Lett 2(1):14002. https://doi.org/10.1088/1748-9326/2/1/014002
29. Hasanuzzaman M, Nahar K, Fujita M (2013) Extreme temperature responses, oxidative stress and antioxidant defense in plants. In: Vahdati K, Leslie C (eds). https://doi.org/10.5772/54833
30. Jackson T, Mansfield K, Saafi M, Colman T, Romine P (2008) Measuring soil temperature and moisture using wireless MEMS sensors. Measurement 41(4):381–390. https://doi.org/10.1016/j.measurement.2007.02.009
31. Cai X, Magwanga RO, Xu Y, Zhou Z, Wang X, Hou Y, Wang K (2019) Comparative transcriptome, physiological and biochemical analyses reveal response mechanism mediated by CBF4 and ICE2 in enhancing cold stress tolerance in Gossypium thurberi. AoB Plants 11(6):plz045. https://doi.org/10.1093/aobpla/plz045
32. Su L, Dai Z, Li S, Xin H (2015) A novel system for evaluating drought–cold tolerance of grapevines using chlorophyll fluorescence. BMC Plant Biol 15(1):82. https://doi.org/10.1186/s12870-015-0459-8
33. Nietzel T, Elsässer M, Ruberti C, Steinbeck J, Ugalde JM, Fuchs P, Schwarzländer M (2019) The fluorescent protein sensor roGFP2-Orp1 monitors in vivo H_2O_2 and thiol redox integration and elucidates intracellular H_2O_2 dynamics during elicitor-induced oxidative burst in Arabidopsis. New Phytol 221(3):1649–1664. https://doi.org/10.1111/nph.15550
34. Exposito-Rodriguez M, Laissue PP, Yvon-Durocher G, Smirnoff N, Mullineaux PM (2017) Photosynthesis-dependent H_2O_2 transfer from chloroplasts to nuclei provides a high-light signalling mechanism. Nat Commun 8(1):49. https://doi.org/10.1038/s41467-017-00074-w
35. Siddiqi KS, Husen A (2019) Plant response to jasmonates: current developments and their role in changing environment. Bull Natl Res Centre 43(1):153. https://doi.org/10.1186/s42269-019-0195-6
36. Hu Y, Wang X, Wang C, Hou P, Dong H, Luo B, Li A (2020) A multifunctional ratiometric electrochemical sensor for combined determination of indole-3-acetic acid and salicylic acid. RSC Adv 10(6):3115–3121. https://doi.org/10.1039/C9RA09951D
37. Amreen, Ali HM, Ahmad M, Salem MZM, Ahmad A (2020) Construction of a nanosensor for non-invasive imaging of hydrogen peroxide levels in living cells. Biology 9. https://doi.org/10.3390/biology9120430
38. Amjad S, Serajuddin M (2022) Role of nanosensors and bionanosensors in crop abiotic stress. In: Ansari SA, Ansari MI, Husen A (eds) Augmenting crop productivity in stress environment. https://doi.org/10.1007/978-981-16-6361-1_1
39. Ghaffar N, Farrukh MA, Naz S (2020) Applications of nanobiosensors in agriculture. In: Javad S (ed) Nanoagronomy. https://doi.org/10.1007/978-3-030-41275-3_10
40. Nezhadahmadi A, Prodhan ZH, Faruq G (2013) Drought tolerance in wheat. Sci World J 2013:610721. https://doi.org/10.1155/2013/610721
41. Tiwari S, Lata C, Chauhan PS, Nautiyal CS (2016) Pseudomonas putida attunes morphophysiological, biochemical and molecular responses in Cicer arietinum L. during drought stress and recovery. Plant Physiol Biochem 99:108–117. https://doi.org/10.1016/j.plaphy.2015.11.001
42. Varshney RK, Tuberosa R, Tardieu F (2018) Progress in understanding drought tolerance: from alleles to cropping systems. J Exp Bot 69(13):3175–3179. https://doi.org/10.1093/jxb/ery187

43. Nakashima K, Suenaga K (2017) Toward the genetic improvement of drought tolerance in crops. Jpn Agric Res Q JARQ 51(1):1–10. https://doi.org/10.6090/jarq.51.1
44. Wu H, Li Z (2022) Recent advances in nano-enabled agriculture for improving plant performance. Crop J 10(1):1–12. https://doi.org/10.1016/j.cj.2021.06.002
45. Sun L, Song F, Zhu X, Liu S, Liu F, Wang Y, Li X (2021) Nano-ZnO alleviates drought stress via modulating the plant water use and carbohydrate metabolism in maize. Archiv Agron Soil Sci 67(2):245–259. https://doi.org/10.1080/03650340.2020.1723003
46. Djanaguiraman M, Nair R, Giraldo JP, Prasad PVV (2018) Cerium oxide nanoparticles decrease drought-induced oxidative damage in sorghum leading to higher photosynthesis and grain yield. ACS Omega 3(10):14406–14416. https://doi.org/10.1021/acsomega.8b01894
47. Martínez-Ballesta MC, Zapata L, Chalbi N, Carvajal M (2016) Multiwalled carbon nanotubes enter broccoli cells enhancing growth and water uptake of plants exposed to salinity. J Nanobiotechnol 14(1):42. https://doi.org/10.1186/s12951-016-0199-4
48. Chandrashekar HK, Singh G, Kaniyassery A, Thorat SA, Nayak R, Murali TS, Muthusamy A (2023) Nanoparticle-mediated amelioration of drought stress in plants: a systematic review. 3 Biotech 13(10):336. https://doi.org/10.1007/s13205-023-03751-4
49. Garcia-Sanchez S, Lujan-Garcia C (2015) M-learning and U-learning environments to enhance EFL communicative competence. In: Zhang (Aimee) (ed) Handbook of mobile teaching and learning. https://doi.org/10.1007/978-3-642-41981-2_74-1
50. Borišev M, Borišev I, Župunski M, Arsenov D, Pajević S, Ćurčić Ž, Djordjevic A (2016) Drought impact is alleviated in sugar beets (Beta vulgaris L.) by foliar application of fullerenol nanoparticles. PLOS ONE 11(11):e0166248. https://doi.org/10.1371/journal.pone.0166248
51. Rastogi A, Tripathi DK, Yadav S, Chauhan DK, Živčák M, Ghorbanpour M, Brestic M (2019) Application of silicon nanoparticles in agriculture. 3 Biotech 9(3):90. https://doi.org/10.1007/s13205-019-1626-7
52. Sun D, Hussain HI, Yi Z, Rookes JE, Kong L, Cahill DM (2016) Mesoporous silica nanoparticles enhance seedling growth and photosynthesis in wheat and lupin. Chemosphere 152:81–91. https://doi.org/10.1016/j.chemosphere.2016.02.096
53. Liu X, Zhang N, Bing T, Shangguan D (2014) Carbon dots based dual-emission silica nanoparticles as a ratiometric nanosensor for Cu^{2+}. Anal Chem 86(5):2289–2296. https://doi.org/10.1021/ac404236y
54. Numan A, Gill AAS, Rafique S, Guduri M, Zhan Y, Maddiboyina B, Nguyen Dang N (2021) Rationally engineered nanosensors: a novel strategy for the detection of heavy metal ions in the environment. J Hazard Mater 409:124493. https://doi.org/10.1016/j.jhazmat.2020.124493
55. Mitra D, Saritha B, Janeeshma E, Gusain P, Khoshru B, Abo Nouh FA et al (2022) Arbuscular mycorrhizal fungal association boosted the arsenic resistance in crops with special responsiveness to rice plant. Environ Exp Botany 193:104681. https://doi.org/10.1016/j.envexpbot.2021.104681
56. Mondal R, Dam P, Chakraborty J, Paret ML, Katı A, Altuntas S, Husen A (2022) Potential of nanobiosensor in sustainable agriculture: the state-of-art. Heliyon 8(12):e12207. https://doi.org/10.1016/j.heliyon.2022.e12207
57. Wawata IG, Fabiyi OA (2024) Sustainable application of nanomaterials in the removal of heavy metals from water. In: Uddin I (ed) Sustainable nanomaterials. Sustainable materials and technology. Springer, Singapore, pp 21–44. https://doi.org/10.1007/978-981-97-2761-2_2
58. Xinyi W, Kong L, Zhou S, Ma C, Lin W, Sun X, Wang P (2022) Development of QDs-based nanosensors for heavy metal detection: a review on transducer principles and in-situ detection. Talanta 239:122903. https://doi.org/10.1016/j.talanta.2021.122903
59. Xiang Y, Camarada MB, Wen Y, Wu H, Chen J, Li M, Liao X (2018) Simple voltammetric analyses of ochratoxin A in food samples using highly-stable and anti-fouling black phosphorene nanosensor. Electrochimica Acta 282:490–498. https://doi.org/10.1016/j.electacta.2018.06.055
60. Guo Z, Kang Y, Liang S, Zhang J (2020) Detection of Hg(II) in adsorption experiment by a lateral flow biosensor based on streptavidin-biotinylated DNA probes modified gold nanoparticles and smartphone reader. Environ Pollut 266:115389. https://doi.org/10.1016/j.envpol.2020.115389

61. Lew TTS, Park M, Cui J, Strano MS (2021) Plant nanobionic sensors for arsenic detection. Adv Mater 33(1):2005683. https://doi.org/10.1002/adma.202005683
62. Erofeeva EA (2018) Hormesis and paradoxical effects of pea (Pisum sativum L.) parameters upon exposure to formaldehyde in a wide range of doses. Ecotoxicology 27(5):569–577. https://doi.org/10.1007/s10646-018-1928-2
63. Zhang Z, Zhao C, Ma Y, Li G (2014) Rapid analysis of trace volatile formaldehyde in aquatic products by derivatization reaction-based surface enhanced Raman spectroscopy. Analyst 139(14):3614–3621. https://doi.org/10.1039/C4AN00200H
64. Litalien A, Zeeb B (2020) Curing the earth: a review of anthropogenic soil salinization and plant-based strategies for sustainable mitigation. Sci Total Environ 698:134235. https://doi.org/10.1016/j.scitotenv.2019.134235
65. Munns R (2005) Genes and salt tolerance: bringing them together. New Phytol 167(3):645–663. https://doi.org/10.1111/j.1469-8137.2005.01487.x
66. Baz H, Creech M, Chen J, Gong H, Bradford K, Huo H (2020) Water-soluble carbon nanoparticles improve seed germination and post-germination growth of lettuce under salinity stress. Agronomy 10. https://doi.org/10.3390/agronomy10081192
67. Noor Aini B, Siddiquee S, Ampon K (2016) Development of formaldehyde biosensor for determination of formalin in fish samples; malabar red snapper (Lutjanus malabaricus) and longtail tuna (Thunnus tonggol). Biosensors 6. https://doi.org/10.3390/bios6030032
68. Wei Q, Zhao Y, Du B, Wu D, Li H, Yang M (2012) Ultrasensitive detection of kanamycin in animal derived foods by label-free electrochemical immunosensor. Food Chem 134(3):1601–1606. https://doi.org/10.1016/j.foodchem.2012.02.126
69. Shaw DS, Honeychurch KC (2022) Nanosensor applications in plant science. Biosensors 12. https://doi.org/10.3390/bios12090675
70. Sharma P, Pandey V, Sharma MMM, Patra A, Singh B, Mehta S, Husen A (2021) A review on biosensors and nanosensors application in agroecosystems. Nanoscale Res Lett 16(1):136. https://doi.org/10.1186/s11671-021-03593-0
71. Li Z, Yu T, Paul R, Fan J, Yang Y, Wei Q (2020) Agricultural nanodiagnostics for plant diseases: recent advances and challenges. Nanoscale Adv 2(8):3083–3094. https://doi.org/10.1039/C9NA00724E

Smart Nanosensors: A New Approach in Crop Protection and Pest Management

Oluwatoyin Adenike Fabiyi, Ibrahim Garba Wawata, Oyesolape Basirat Akinsipo, and Ayorinde Victor Ogundele

Abstract The world is facing a pressing challenge of feeding a growing population with limited food resources, exacerbated by crop diseases and pests. To address this, the agricultural sector must adopt smart and innovative solutions. Nanosensors are designed to detect and analyse crop diseases and pest infestations. The signalling is in real time. This facilitates effective and timely management of pest and diseases. In the face of high variability in climate, crops are confronted with great risk of infection and infestation, nanosensors are promising methods to minimise losses and reduce impact of diseases. These innovative tools offer a beacon of hope in minimizing losses and mitigating diseases' impact, showcasing nanotechnology's vast potential in ensuring global food security. This chapter explores the transformative potential of nanotechnology in enhancing crop protection and pest management. Also, the chapter raises awareness of the economic and environmental impacts of implementing nanosensor technology in sustainable farming as a substitute for synthetic pesticides, which threatens human health and environment.

O. A. Fabiyi (✉)
Department of Crop Protection, Faculty of Agriculture, University of Ilorin, Ilorin, Nigeria
e-mail: fabiyitoyinike@hotmail.com; fabiyi.oa@unilorin.edu.ng

I. G. Wawata
Department of Pharmaceutics and Pharmaceutical Technology, School of Pharmacy, Kampala International University Western Campus, Ishaka Bushenyi District, Uganda

Institute of Biomedical Research, Kampala International University Western Campus, Ishaka Bushenyi District, Uganda

Department of Pure and Applied Chemistry, Kebbi State University of Science and Technology Aliero, Aliero, Nigeria

O. B. Akinsipo
Department of Chemical Sciences, College of Science and Information Technology, Tai Solarin University of Education Ijagun, Ijebu-Ode, Nigeria

A. V. Ogundele
Department of Chemistry, College of Pure and Applied Sciences, Kwara State University, Malete, Nigeria

Laboratory of Natural Products & Drug Discovery, Centre CEBIM, Department of Basic Sciences, Faculty of Medicine, Universidad de La Frontera, Temuco, Chile

I. Uddin (ed.), *Smart Nanosensors*, Smart Nanomaterials Technology,
https://doi.org/10.1007/978-981-96-3878-9_10

Keywords Crop protection · Nanosensors · Plant diseases · Environmental factors

1 Introduction

Agriculture is an important sector owing to the direct link it has with several industries for the supply of raw materials, hence maintaining and balancing the economy of countries worldwide. The sector has the highest number of workforces globally [1]. Several cultivable lands have gone into industrialisation, hence supply of food is significantly low relative to quality and quantity. There is a worldwide increase in food insecurity on yearly basis, while global human population is expected to rise [2]. Owing to the present and projected population, farmers have stepped up production of crops with increasing application of pesticides. This aims to increase productivity to support growing population in order to manage malnutrition.

Similarly, traditional biodiversity is lost coupled with recent global warming and this has brought about new pest populations which has called for incessant pesticide use in order to maintain productivity. In some cases, the pesticides are ineffective thus aggravating the problem. The overuse of these pesticides has contaminated the cultivable lands. Furthermore, sundry other contaminants such as heavy metals are equally present in the soil and are transferred to consumers after harvest [3]. For safety assurance to consumers, it is necessary to identify and detect synthetic pesticide residues, heavy metal accumulation and infected crops to ensure food security. Several available methods for detecting contaminants are tedious and expensive, while some requires expert monitoring. Though they are clearly accurate and efficient techniques, which necessitates laboratories to be equipped with state-of-the-art equipment plus trained personnel. Farmers find this difficult and unaffordable. This then calls for an excellent method with swift onsite analysis, which can provide regular crop health check to minimise losses, hence technology must advance.

Nanotechnology provides avenue to scaling such obstacles [4]. It involves sustainable practices which primarily protects the environment. Soil physical properties, pesticide toxicity, pest and pathogen symptoms could be detected seamlessly using nanotechnology. Several approaches are involved with energy use efficiency technology. Sensors are used extensively for agricultural pathogen detection, they function rapidly, highly sensitive, efficient, reliable with minimal sample and are very easy to handle [5]. The potential use of nano biosensors is wide and tremendous [6]. Biosensors are devices which use receptors of bio-origin in order to detect biomolecules which are then presented inform of analytical data. Transducers which may be physical or chemical in nature are joined with detection or recognition elements in order to identify products of biological origin [7]. Transducers convert meaningful information from the interaction of biological elements for onward output through the screen. Sensors could be based on quantum-dots, dendrimers, nanorods or nanowires (Fig. 1).

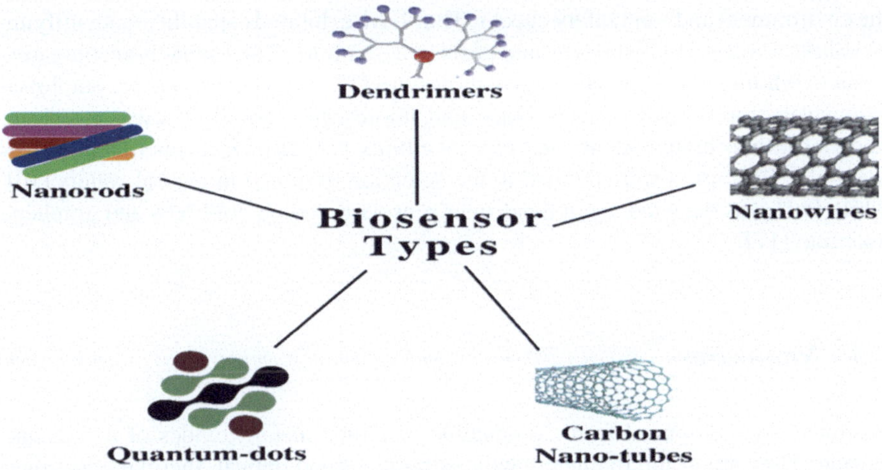

Fig. 1 Different types of sensors used in crop protection and pest management

2 Biosensor Types

2.1 Nanorods

These are rods with dimensions between 1–100 nm which originates from dissimilar materials, from oxides of many metals to graphenes including materials with conducting characteristics [8]. Nanorods are reliable materials for detecting metal ions, carbohydrates and acids. Phosphate contaminants were detected using nano rods from zinc oxide on the field [9]. Sensors for lead ion detection were developed with gold nanorods with high efficiency [10]. In the environment, organic compounds with volatile nature abound. Swargiary et al. [11] applied zinc oxide nanorods based on optical fibre in the detection of volatile organic compounds. Vapour was detected with very high accuracy.

2.2 Dendrimers

These are macromolecules with hyper-branched nature and of three-dimensional shape which combines specific properties like dense branching, distinct structures and turntable surface. Due to these functionalities, they are extremely stable, highly sensitive and selective [12]. This increases target analyte interactions. They possess several functional groups and wide area surface; this has assisted intricate amplification of signals which widens the limit of detection. Thus, the sensors are suitable accurately for analytes of low concentration, while enhancing good signal output. Application of dendrimer sensors are diverse; they are a good choice in monitoring of the environment and food safety checks. They have exhibited capability in identifying biomarkers related to pathogens and pests in agriculture. Their wide branching give room to binding of biologically active molecules [12, 13]. Dendrimers are employed extensively for sensors of impedance and fluorescence based. These sensors are known for their high diversity and reproducibility [14, 15]. Dendrimer based sensitive gold NP sensors were applied in the detection of uranyl in natural waters [16], while Pb^{2+} was detected using dendrimer sensors based on gold NPs and graphene quantum [17].

2.3 Nanowires

Nanowires are produced from conductible nanotube metals, oxides of metals and carbon. They are of nm diameters with extremely good optical, thermal, electronic and mechanical characteristics that are not available in materials of bulk nature [18, 19]. Polysilicon nanowires and optical silicon nanowires have been extensively used for detection of sensitive signals of biomolecules in analytes. Wu et al. [20] applied AuPt nanowires for detecting organophosphorus pesticide. A good sensitivity for detection was achieved.

2.4 Carbon-Nano Tubes

These possess carbon structure with hollow and nm diameters. They are stiff, strong hardened materials owing to the linkage of well-arranged carbon atoms through sp2 bond. They are frequently employed as biosensors in different fields of science. Tang et al. [21] developed carbon nano single wall sensors that are highly responsive and sensitive. Similarly, Li et al. [22] detected dopamine at a very low limit level with SW carbon nano tubes. In the same vein, toluene, ammonia gas, N-nitrosodialkylamines were detected using carbon nano tube sensor devices at room temperature. Low detection limit was achieved for all elements [23–26]. Nano tubes of hydrogel hybridized carbon were equally effective in the monitoring of activity of *Aspergillus niger.* Signals of growth of microbes were detected at various developmental stages. Species of fungi were detected relative to their individual signals. Metabolites were produced by the fungi, thus allowing for the detection of their various activities thorough change in conductance of the CNT [27].

2.5 Quantum-Dots

Quantum dots have nanometers between 2.0–10 nm and are made up of nanocrystalline conductible materials [28]. QDs have varying properties which is determined by their sizes, structures and shapes. For example, 5.0–6.0 nm QDs come with bright colours such as red and QDs with diameters ranging between 2.0–3.0 nm are characterized by darker shades like blue colours. QDs are employed in place of fluorophores in the manufacturing of biosensors for macromolecules and organic compound detection [29]. QD based sensors have been extensively applied in assays. QDs doped with nitrogen effectively identified mercury ions at a lower limit of detection (0.23 μM) [30]. Detection of arsenite from water sample was facilitated by the use of QD chemo sensors premised on functionalized thiol fluorescent, a wide limit of detection range was achieved (5–100 ppb) [31]. QDs combined with nitrogen and sulphur to manufacture biosensor of chemiluminescence nature was employed in the determination of indomethacin, 65 μg L^{-1} was the limit of detection, and a range of concentration between 1.0–1.5 mg [32].

3 Environmental Factors

Developments have been achieved in the production of agricultural sensors and are tailored to match the agricultural environment. Crop productivity relies on several factors in the environment. In the last decades, agriculture has developed and made substantial progress in electronics [33]. Environmental considerations are important for good yield attainment. This is because the process of farming and the output products are influenced and impacted by changes in climate [34]. For evaluation of crop growth, it is highly essential to measure environmental factors which are supportive to development of crops. Important factors like irradiation, velocity of wind, direction of wind, temperature of air and humidity of the air all require precise measurement for good quality yield (Fig. 2).

3.1 Irradiation

The photosynthesis activity of crop is controlled by radiation from sunlight. The radiation comes to the earth as an electromagnetic wave. Sensors for monitoring and measuring intensity of the sun are highly efficient and cost effective even in the measurement of tree canopies in the forest [35]. The photosynthetic sensor is made with groups of silicon photodiodes in a casing which is protected with infrared protection filter. The sensor consists of data logger in which agricultural data are integrated. Stored data could be displayed on the screen if there is any need for it. Accuracy including details of date and time data as recorded is seen on the displayed screen.

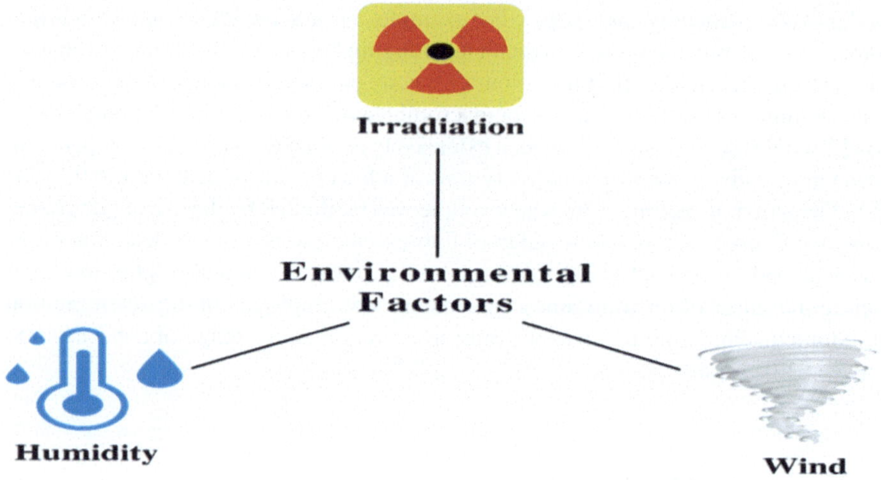

Fig. 2 Environmental factors for measurement of crop productivity

3.2 Wind

Advancement in modern agricultural technology has brought about sensitive measurements in agriculture. Wind speeds on the fields are measured with MEMS sensors based on cantilever, these instruments tilt in response and relative to velocity of wind. The miniature sensors consume less power and are developed with high precision and accuracy [36]. New in the series are the 2-D octagon-like MEMS sensors. They exhibit error level of $\pm 1\%$ at high wind speed and their capability of high wind speed record is at 360°. The octagon shape ensures accuracy, Data are taken separately and the average is presented, thus further improving accuracy.

3.3 Humidity

Information about air humidity is considered important in pest and diseases management, moisture quantity in the surroundings has a relationship with pathogen build up. Resistive and capacitive sensors are combined and premised on organic nickel phthalocyanine. It is known to be highly effective even when relative humidity is around 35–95 and 35–75%. The sensor functions on organic semiconductor principle with many associated benefits like eco friendliness, cost effective and easy of manufacture [37]. The device swings into action when water vapour settles on the inbuilt sensor, the vapours are absorbed into the device and alters the electrical sensitivity of the organic film. The alteration of the film is calculated to arrive at the level of humidity in the surrounding environment [37].

4 Biosensors in Detection of Plant Diseases

Globally, pathogens are the key limiting factors which reduces crop yields tremendously. The annual monetary loss to pests and pathogens destruction is calculated to be over two billion dollars. Total yields of important crops such as soybean, maize, potatoes and peanuts are lost up to the tune of 25% to field and storage pathogens and pests [38, 39]. This has resulted in staple food shortage and reduction in necessary nutritional requirements. Thus, posing serious food security issues.

Repeated use of pesticides which is the main method of control has significantly polluted the environment, for efficient sustainability in agriculture and food security, swift detection of plant pathogens is necessary [40, 41]. Incorporation of nanosensors in plant disease detection facilitates rapid identification of disease, which in turn assists swift response. In disease epidemiology and plant disease detection, early onsite identification of pathogens could be achieved through nano biosensors. Sensors could be distributed on to farm fields and connected to GPS for accurate and immediate monitoring of crop health, quality of soil and plant diseases [42], with this approach, plant diseases are detected long before appearance of symptoms. This gives the exact and accurate data, enhances precision farming, good decisions are made and output is improved [43].

Platforms of nano sensing with nanomaterials such as fibre, carbon, tubes and oxides of metals employed in pathogen detection minimizes the time of pathogen identification. Sensitivity also increases in nanostructures premised on microfluids [44, 45]. Many types of structured nanomaterials are used in the detection of fungi, nematode and bacteria (Fig. 3). Indicators of food pathogens are in built into nanomaterials to sense pathogens. Nano silicon oxide and chitosan embedded in immunoelectrode was identified to be efficient in detection of *Aspergillus ochraceus*. *Aspergillus flavus* toxins at very low concentrations in food was equally detected through electrochemical immunosensor [46, 47]. Diverse approaches have been used to detect zearalenone produced by *Fusarium* spp. Hervas et al. [48] in their research came up with a highly sensitive and selective prototype for zearalenone detection, while Panini et al. [49] developed multiwalled carbon nanotubes in a continuous flow technology to form an ultra-sensitive sensor for fast and exact identification of zearalenone in samples of maize silage.

Plants under pathogen infection are known to exude volatile organic compounds. These compounds serve as markers that a host plant is under biotic or abiotic stress. The volatile substances could serve as finger prints in such diseased plants. Multilayered gas sensors are applied in identification of volatile organic stress biomarkers in conditions like, insect invasion, pathogen/pest infection or plant injury [50, 51]. Nano plasmonic sensor arrays premised on gold nanoparticles solgel has been employed in the detection of terpenes selectively. Ethylene and phytohormones could equally be identified with colorimetry procedures. Plasmonic nanoparticles built with smart phones could sense infections swiftly. Li et al. [52] in their research differentiated tomato late blight from other tomato infections under 48 h, irrespective of similar symptoms displayed on tomato leaves with smartphone premised on plasmonic

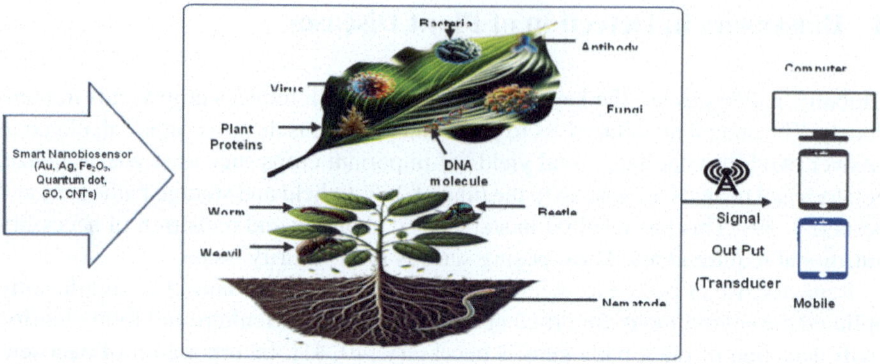

Fig. 3 Detection of plant pathogens and pest using smart nano biosensors

nanoparticles. This sensor type could equally serve in detection of plant nutrition deficiency, drought and various plant pathogens and pests. Several mini handheld sensor devices for onsite identifications of plant diseases abound. Some portable vapour analysers are incorporated with gas chromatography. They have been reported to be very efficient devices for volatile organic compound detection.

Inorganic nanomaterials such as metal oxides of silicon, silver and gold are used in some nanosensor. These categories of sensors work with transducers and ligands for signal and recognition respectively in order to detect and evaluate the biological molecule of interest. The sensors work in accordance with bio-analytes that are specifically related to the plant disease. Sensor responses are recorded using surface plasmon resonance of nanoparticles.

Bacterial species have been reported to cause yield reductions in crops. The most common ones include *Proteobacteria*, *Xanthomonas Pseudomonas*, *Agrobacterium* and *Erwinia.* The infection of these organisms results in symptoms like blights, cankers, soft rots, galls, wilts, specks, scabs, leaf spots, galls and over growths. A number of nano sensing techniques have been used in detection of bacterial diseases in plants [53–56]. For example, silver nanorods are applied regularly for identification of sundry bacterial diseases. Sensitivity of Au nanorods was enhanced with addition of platinum nanosensors enabled with IgG for detection of *Bacillus thuringiensis and B. subtilis* around carrot roots and the rhizosphere soil. A new enzyme nanosensor with non-covalent centre was developed by Sun et al. [57]. The urease censor works with 90.7% accuracy and can detect bacterial load up to about 102 CFT/LL under 30 min. In like manner, Wilson et al. [58] created biosensor arrays containing silver interdigitated microelectrode for *Escherichia coli* detection of up to about 1 CFU mL1. Nightshade plant disease caused by *Xanthomonas campestris* has been effectively identified using nanoparticles fluorescent silica connected with antibody and doped using Ru (II) complex [59, 60]. An accurate detection of *Pseudomonas syringe* using electrochemical biosensor premised on AuNPs was carried out by Lau and Botella [61]. Rapid isothermal amplification was applied to increase the targeted sequence of the pathogen DNA for accuracy of identification. Likewise, an important organism of quarantine, *Ralstonia solanacearum* which impacts potato tubers with brown rot was identified using test strips premised on lateral flow immunoassay. The procedure also involves, enlargement of gold nanoparticles, amplification and development of colour [55, 62]. Lately, infection of an endophytic bacterial was done using electrochemical method. H_2O_2 exuded by leaf tissues of infected plant was detected electrochemically.

Optical DNA hybridization biosensors have been reported to offer high sensitivity of detection by providing visible nucleic acid identification without the use of PCR. Gold nanoparticles surface plasmon resonance (SPR), using DNA based diagnostic biosensor was employed to identify tomato yellow leaf curl virus (TYLCV) genome in infected plants, through the use of colorimetric nano-biosensing system. Here, alterations around nanoparticles in the face of the pathogenic virus was mirrored and evaluated through alterations in the local surface resonance of gold nanoparticles. Visual colour changes and UV visible calculations were used for recording [63]. This approach removes the use of PCR equipment. Identification of cucumber green mottle mosaic virus, known to cause serious reduction in production owing to its mechanical transmission via seeds was reported by Wang et al. [64]. Enzyme AcII was use sequentially, while terminal deoxynucleotidyl transferase was connected with DNAzymes amplification. Mg^{2+} based DNAzyme and hemin/G-quadruplex DNAzyme was equally involved. The assay was visual colour alteration dependent. Lower detection limit was allowed under short observation time. A similar method was applied in the identification of water melon mosaic virus. Polymerase activity of T4 DNA polymerase and exonuclease, with Mg^{2+} based DNAzyme and hemin/G-Quadruplex DNAzyme reactions was involved, this was done via voltametric measurements of electro-chemical differential pulse. Another virus, which causes great loss of grape quality and production, the grapevine leafroll-associated virus 3 (GLRAV-3) was detected through the use of immunochromatographic assay, using sandwich immunoassay procedure. The antibody of the virus in gold nanoparticles was detected within minutes [65]. Similarly, potato virus Y was identified through sandwich lateral flow immunoassay. Gold nanoparticles and polyvalent antigen were involved, while cross verification was achieved with surface plasmon resonance [66]. Furthermore, in order to recognize citrus virus, gold nanoparticle premised on impedimetric biosensor was made by Khater et al. [67]. The citrus tristeza virus was detected through the technique which works on real time. The level of detection was 0.1–10 μM.

Infection of plants by fungal pathogens is enormous. They act as decomposers in some cases and can maintain several relationships with plants, ranging from pathogenic, symbiotic or saprophytic. It is crucial to detect fungal infection because they inflict tremendous damage on crops. Banana plants are highly susceptible to leaf streak disease, widely known as black sigatoka. The causative organism, *Pseudocerocospora fijiensis* was accurately detected in leaf samples with immunosensor assay based on surface plasmon resonance. Antibody of the cell wall protein of the organism was put on gold chip and a linear response was received with detection limit of 11.7 μg per mL [68]. Furthermore, potato production plays a significant role in sustainable food security, however, 40% of cultivated potatoes are lost to sundry factors on yearly basis. Fungal infection of potatoes is brought about by *Alternaria solani*, *Collectorichum coccodes*, *Rhizocotonia solani*, *Spongospora subterranean*, *Fusarium* spp. and *Snchytrium endobioticum*. Multiplex field diagnostics of these fungi were conducted in open microreactors with qPCR-based microarray. DNA of target species was identified between 0.6 and 43.5 detection limit. The fungus *Puccinia striiformis* which causes yellow rust in wheat, with 100% production and quality damage requires timely identification to minimise the degree of damage. The bio physical variation in crop age, that is, early to mid-growth stage was captured using hyperspectral analysis. This approach was also used in the detection of bacterial leaf blight of rice and rust diseases of sugar beet [69–71].

Visual detection of *Phytophthora infestans* was achieved by using gold nanoparticle lateral flow biosensor in connection with universal primer mediated asymmetric PCR. The primer accomplished generation of several amounts of single-stranded DNA through amplification of a part of *P. infestans*. Visible detection was done with the lateral flow biosensor with sandwich hybridization [72]. Infections of fungi have been detected through the application of gold nanoparticles. It is crucial to use a sensitive method for fungal pathogen detection owing to their virulence. Seeds, soil and weeds are usually infected, while their spores' spreads through water and wind. The lateral flow assay is in combination with chromatography. The procedures are simple and swift on-site analysis premised on visual evaluation. Additionally, *Sclerotinia sclerotiorum* infection of oil seeds was accurately detected by Dubey and Mailapalli [73] by measuring level of salicylic acid using gold nano electrode connected with copper nanoparticles.

Nanosensor remote sensing is a good technique for identifying, predicting and controlling insects. Owing to their high sensitivity and accuracy, sensors have the capacity and solution to insect control. Gas sensors with micro-electro-mechanical systems were developed by Brezolin et al. [74] for detection of pheromones in insects. Sex pheromones of *Euschistus heros* were detected at 0.005–0.03 μg mL. Pheromones of insects are generally employed in their observation and detection. This is in order to facilitate a timely application of pesticide [75]. To perceive sex pheromones of female Lepidopteran insects that are pest to many important crops like, chickpea and rice, a micro-electro-mechanical-system premised on silicon dioxide for productive recognition was created by Moitra et al. [76]. Further to this, sensors for viriods, nematodes and phytoplasmas have also been built for easy identification of their presence on the field [77, 78].

5 Agriculture and Nano Biosensors

In order to transform agriculture, nano biosensors have been a part of the farming process. Viable farm practices, crop safety, seed viability, swift application of water and plant nutrient have been evaluated with biosensors. Impact of environmental factors on crop yields are observed ahead with sensors. Fertilizer and pesticide usage are also controlled via the sensing operation (Fig. 4). For prompt action, crop physiological conditions, residual pesticides, moisture and heavy metals detection in the agroecosystem have been reported to be monitored [79–81]. Biosensors are channelled to monitor and evaluate crop root rhizosphere and zones. This has availed the observation and detection of root oxygen variation, carbon levels, growth hormones and enzymes. In food manufacturing, sensors are again employed to monitor the production line and packaging for detection of pathogens for food safety assurance. The suitability of soil for cultivation, soil fertility and health for maximum yield are incorporated under field sensors for maximum yield [82–84]. Precise pesticide, water and fertilizer use are comprehensively delivered by sensors on time, thus eliminating water stress and excessive application of pesticides and fertilizers to crops. The importance of water for crop growth cannot be over emphasized, it is required for photosynthesis, micro elements and other nutrient movement for good crop growth. This then necessitates the need to observe water quantity flow. Chong et al. [85] developed a fluorescence optical biosensor for water stress monitoring. If there are any alterations in moisture conditions of the plant, the water current is equally altered, with this, it can be foreseen if the crop is under water stress. Similarly, Subraya et al. [86] created a biosensor to quantify the level of indole 3-acetic acid, using an amperometric technique. This was tested on corn plants, and it acted chemically as a pointer to water stress in corn plants. The researchers discovered a good interconnectedness between the electrochemical response evaluated and alterations in the water content of the corn plants.

For good post-harvest standard, freshness retention and extension of product shelf life, maturity sensing is crucial. This is necessary in order to check under ripe and over ripe fruits and vegetables for consumer acceptance. Maturity and harvesting of vegetables and fruits are checked using portable field sensors. The sensors are driven towards parameters like pH and titrable acidity content of fruits. Lupetti et al. [87] and colleagues applied jack fruit-capric acid biosensor for determination of total phenolic, while Vermeir et al. [88] determined level of ascorbic acid using differential calorimetric microplate biosensor. Several issues in agriculture could be resolved with multidisciplinary approach by expertise from industry and academia (Table 1).

6 Economic and Environmental Impact

A number of prepared smart nano sensors have the ability to dramatically transform the economic and ecological aspects of crop protection and pest control. This has led to change from normal extensive practices which are less accurate to intensive practices plus accurate and efficient farming methods.

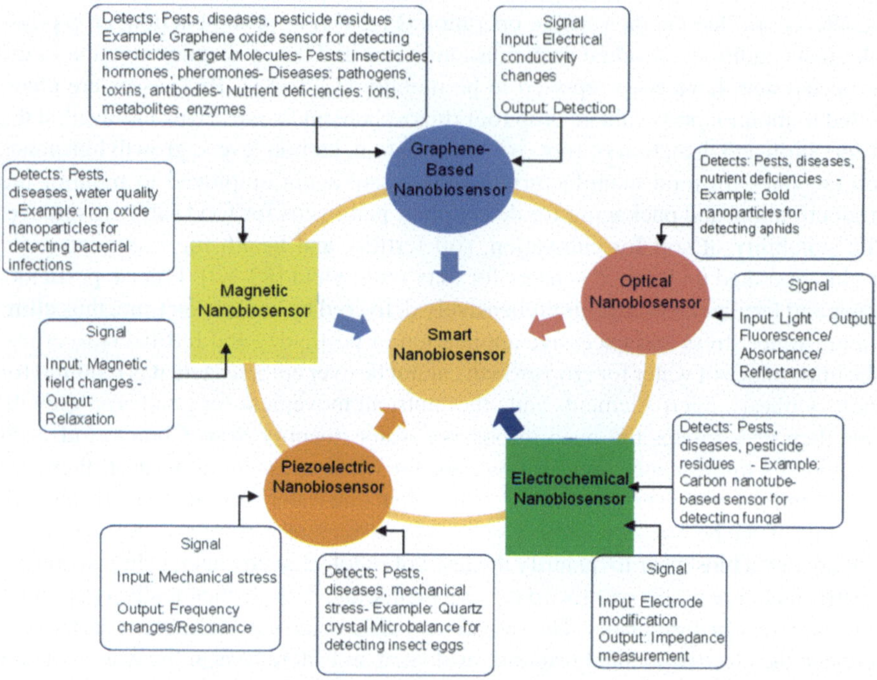

Fig. 4 Inputs, outputs and detection of smart nano biosensors

6.1 Economic Impact

The use of nansensors in agriculture is most rewarding in terms of cost efficiency, which is one of the most important determining factors of economic performance. Until recent time, farmers have a tendency of using large quantity of pesticides and fertilizers since they merely lack sufficient details of the health of their crops or the level of pest infestation. Nanosensors are however unique in their nature since they enable the farmers to monitor pests, diseases, and nutrient deficiencies in real time. When tackled professionally and at the right time, farmers can work towards minimizing their input cost especially on pesticides and fertilizers. There is typically up to 30% saving of input costs by incorporating precision farming systems such as sensors in farming [89]. Besides, nanosensors can also minimize crop losses because they give timely information that halts the pest and diseases from spreading. Some of the biotic stress factors which can easily be detected include pathogens and pests which, if early detected, cause minimal harm to the produce both in yield and quality. This translates into higher economic returns for farmers because better market prices are usually accorded to healthier crops [94]. Increased connectiveness of nanosensors with Internet of things (IoT) and data analytics increases profitability of the farm. These smart systems can control the use of water, nutrients and chemicals by reducing their application to only what is required. It is good to note that such

Table 1 Key applications of smart nano sensors in crop protection and their impacts

Application area	Description	Example of nanosensors	Impact	References
Early detection of biotic stress	Self-powered nano sensors are used to monitor and identify in advance the presence of pathogens such as bacteria, fungi and virus. This makes it possible for intervention, to be done in a timely manner and reduce on crop damage	Biosensors based pathogen for detecting bacterial blight in rice (e.g. carbon nanotube-based)	This way the spread of diseases is minimized hence a reduced use of broad spectrum pesticides	[89]
Pest detection	Using some form of bio or chemical signals given off by pests, nano sensors track the movement of pests in real time and point out the type of pest	Nanosensors based on surface plasmon resonance for the detection of pheromones of insect pests like bollworm	Reduces environmental destruction by minimizing pesticide use through specific targeting of pests	[90]
Soil nutrient monitoring	Farmers can be assisted in optimizing fertilizer use by considering the actual conditions of soil. With nanosensors, soil nutrient levels is analysed (e.g., nitrogen, phosphorus, potassium)		Efficient nutrient usage reduces fertilizer overuse, preserving soil fertility and enhancing crop productivity	[91]
Monitoring of Soil moisture	Nano sensors continuously monitor soil moisture levels, optimizing irrigation and conserving water resources by ensuring that crops receive adequate water based on soil conditions	Graphene oxide-based sensors for soil moisture monitoring	Ensures water use efficiency in agriculture, critical for regions with limited water availability	[92]

(continued)

Table 1 (continued)

Application area	Description	Example of nanosensors	Impact	References
Environmental condition	Nano sensors measure environmental factors such as temperature, humidity, and sunlight intensity, allowing farmers to adjust practices to match real-time conditions	Nano-scale temperature and humidity sensors deployed in agricultural fields	Improves crop resilience by adapting quickly to weather fluctuations, thereby reducing crop loss	[93]
Detection of environmental stressors	These sensors detect harmful elements, such as toxic metals, or unfavorable environmental conditions that may negatively impact plant growth	Quantum dot sensors for detecting heavy metals like lead in soil	Prevents soil contamination and ensures safe food production by identifying pollutants early	[94]

optimization does not only help cut costs but it also improves the practicability of farming. Smart agricultural technologies such as nanosensors are becoming more popular over time, therefore the market for nanosensors is expected to expand in the future since farmers are inclined towards adopting precision farming in order to obtain better economic yields [93].

6.2 Environmental Impact

The advantages of nanosensors in crop protection and pest control are apparent in terms of environmental aspects. The conventional extension approaches of seed, fertilizers, and chemical pesticides are efficient but they offer negative impacts on crop yield, the environment, and other particulate beneficial organisms and wildlife. Nanosensors decrease the use of these chemical inputs since they offer targeted solutions where they are required. For instance, it has been estimated that nanosensors can help identify certain pests or pathogens at a molecular level, thereby enabling spraying of pesticides in specific areas or at times when they are required most rather than applying them over large pieces of farming land [91]. This leads to decrease in chemical wash off on water bodies, and improves soil structure which is important for long term agricultural productivity.

Integrated pest management (IPM) is the other environmental benefit that comes with smart nanosensors as they help in minimizing the use of pesticide. IPM involves the utilization of biological, cultural, mechanical as well as chemical processes on control of pests. Nanosensors improve on an already effective IPM strategy through offering specific data that can help in tracking the behaviour of pests, as well as the efficiency of the implemented control measures. It also assists the government to reduce the application of chemical pesticides and increase the use of IPM which includes use of biological control agents and organic farming methods [90]. In addition, with the use of nanosensors, the use of water is also preserved. These sensors also aid farmers to apply water when necessary and in the right measure by portraying to them, real time soil moisture levels. This especially takes place in areas that experience water deficit. While irrigation is important to prevent crop failure, doing this in excess results in soil salinization and reduction in the groundwater reserves [92].

Lastly, smart nanosensors also help to decrease the level of pesticides and fertilizers usage which, in turn, reduces the carbon emission of agricultural organisations. These are chemicals whose production, transportation and use in their various applications require energy, hence, emitting greenhouse gases. The latter approach with the help of nanosensors used in the agricultural industry will generally assist in the decrease of the sector's negative effects on the environment, therefore meeting the global climate objectives [95].

All in all, the application of smart nanosensors in crop protection and pest control involves significant economic and environmental values. From an economic point of view, these sensors save inputs and increase yields and profitability because of precision agriculture. Environmentally, they are involved in sustainable practices such as reducing the use of environmentally unfriendly chemicals, water conservation and extension of services on integrated pest management. These factors are expected to make nanosensor technology even more relevant as new technology for the future of farming in response to the world's demand for sustainable and efficient agricultural practices.

7 Challenges in the Use of Nanobiosensors

The rate of global population increase is a crucial challenge to agriculture. Seed use, and other inputs of agriculture are to be incorporated into smart agriculture in order to have fast increase in out-put for the teeming populace. Advancement in nanotechnology is exceptional and has expanded into various facets of science and technology with sundry materials to choose from.

Although nanosensors presents enormous potential to agriculture owing to the great precision it offers in symptom detection, pathogen identification and water use efficiency. However, these attributes of nanosensor are not yet within the reach of peasant farmers who dominates production in developing nations. Commercialization of nanosensors is a primary challenge. Nanosensors are not yet across the counter. The marketability of nanosensors will be determine by safety of use, ethical concerns of the general public and enactment of standards.

8 Conclusion

This chapter presents report on the successful application of biosensors in agricultural fields for crop pest, pathogens and sundry other uses. Nanosensors are important versatile tool in smart agriculture, their superior characteristics points to their performance. The commercialisation of nanosensors may be driven by artificial intelligence (AI) and internet of things (IoT). Incorporation of sensors into agricultural management aims to remodel agriculture with immense increase in crop production. Finally, researchers are encouraged to focus on plant growth monitoring considering the significance of biosensors.

References

1. Food and Agriculture Organization of the United States (FAO) statistical yearbook 2021. Food and Agricultural Organization. https://www.fao.org/3/cb4477en/online/cb4477en.html
2. Duro JA, Lauk C, Kastner T, Erb KH, Haberl H (2020) Global inequalities in food consumption, cropland demand and land-use efficiency: a decomposition analysis. Global Environ Change 64:102124
3. Schieber A (2018) Introduction to food authentication. In: Modern techniques for food authentication. Academic Press, pp 1–21
4. Konur O (2021) Nanobiosensors in agriculture and foods: a scientometric review. In: Biosensors in agriculture: recent trends and future perspectives. Springer, pp 365–384
5. Fraceto LF, Grillo R, de Medeiros GA, Scognamiglio V, Rea G, Bartolucci C (2016) Nanotechnology in agriculture: which innovation potential does it have? Front Environ Sci 4:20
6. Nadporozhskaya M, Kovsh N, Paolesse R, Lvova L (2022) Recent advances in chemical sensors for soil analysis: a review. Chemosensors 10:35
7. Bhalla N, Jolly P, Formisano N, Estrela P (2016) Introduction to biosensors. Essays. Biochem 60:1–8
8. Ibupoto ZH, Ali SMU, Khun L, Chey CO, Nur O, Willander M (2011) ZnO nanorods based enzymatic biosensor for selective determination of penicillin. Biosensors 1:153–163
9. Ahmad R, Ahn MS, Hahn YB (2017) ZnO nanorods array-based field-effect transistor biosensor for phosphate detection. J Colloid Interface Sci 498:292–297
10. Liu G, Feng DQ, Qian Y, Wang W, Zhu JJ (2019) Construction of FRET biosensor for off-on detection of lead ions based on carbon dots and gold nanorods. Talanta 201(2019):90–95
11. Swargiary K, Metem P, Kulatumyotin C, Thaneerat S, Ajchareeyasoontorn N, Jitpratak P, Bora T, Mohammed WS, Dutta J, Viphavakit C (2022) ZnO nanorods coated single-mode–multimode–single-mode optical fiber sensor for VOC biomarker detection. Sensors 22:6273
12. Malik S, Singh J, Goyat R, Saharan Y, Chaudhry V, Umar A, Ibrahim AA, Akbar S, Ameen S, Baskoutas S (2023) Nanomaterials-based biosensor and their applications: a review. Heliyon 9:e19929
13. Abbasi E, Aval SF, Akbarzadeh A, Milani M, Nasrabadi HT, Joo SW, Hanifehpour Y, Koshki KN, Asl RP (2014) Dendrimers, synthesis applications, and properties. Nanoscale Res Lett 9:247
14. Satija J, Sai VVR, Muherji S (2011) Dendrimers in biosensors: concept and applications. J Mater Chem 21:14367–14386
15. Caminade AM, Turrin CO (2014) Dendrimers for drug delivery. J Mater Chem B 2:4055–4066
16. Shi S, Wu H, Zhang L, Wang S, Xiong P, Qin Z, Chu M, Liao J (2021) Gold nanoparticles based electrochemical sensor for sensitive detection of uranyl in natural water. J Electroanal Chem 880:114884

17. Niu XF, Zhong Y, Chen R, Wang F, Liu Y, Luo D (2018) A "turn-on" fluorescence sensor for Pb^{2+} detection based on graphene quantum dots and gold nanoparticles. Sens Actuators B Chem 255(2018):1577–1581
18. Ramanathan K, Bangar MA, Yun M, Chen W, Myung NV, Mulchandani A (2005) Bioaffinity sensing using biologically functionalized conducting polymer nanowire. J Am Chem Soc 127:496–497
19. Patolsky F, Zheng G, Lieber CM (2006) Nanowire-based biosensors. Anal Chem 78:4260–4269
20. Wu Y, Jiao L, Xu W, Gu W, Zhu C, Du D, Lin Y (2019) Polydopamine-capped bimetallic AuPt hydrogels enable robust biosensor for organophosphorus pesticide detection. Small 15(17):1900632
21. Tang XX, Bansaruntip S, Nakayama N, Yenilmez E, Chang YL, Wang Q (2006) Carbon nanotube DNA sensor and sensing mechanism. Nano Lett 6:1632–1636
22. Li WS, Hou PX, Liu C, Sun DM, Yuan J, Zhao SY, Yin LC, Cong H, Cheng HM (2013) High-quality, highly concentrated semiconducting single-wall carbon nanotubes for use in field effect transistors and biosensors. ACS Nano 7(8):6831–6839
23. Panes-Ruiz LA, Shaygan M, Fu Y, Liu Y, Khavrus V, Oswald S, Gemming T, Baraban L, Bezugly V, Cuniberti G (2018) Toward highly sensitive and energy efficient ammonia gas detection with modified single-walled carbon nanotubes at room temperature. ACS Sens 3(1):79–86
24. Aniello Falco FC, Loghin M, Becherer P, Lugli Jose FS, Almudena R (2019) Low-cost gas sensing: dynamic self-compensation of humidity in CNT-based devices. ACS Sens 4(12):3141–3146
25. He M, Croy RG, Essigmann JM, Swager TM (2019) Chemiresistive carbon nanotube sensors for N-nitrosodialkylamines. ACS Sens 4(10):2819–2824
26. Seekaew Y, Wisitsoraat A, Phokharatkul D, Wongchoosuk C (2019) Room temperature toluene gas sensor based on TiO_2 nanoparticles decorated 3D graphene carbon nanotube nanostructures. Sens Actuator B Chem 279:69–78
27. Park M, Kim HS, Kim T, Kim J, Seo S, Lee BY (2018) Real-time monitoring of microbial activity using hydrogel-hybridized carbon nanotube transistors. Sens Actuator B Chem 263:486–492
28. Ramesh M, Janani R, Deepa C, Rajeshkumar L (2022) Nanotechnology-enabled biosensors: a review of fundamentals, design principles, materials, and applications. Biosensors 13(1):40
29. Ma F, Li CC, Zhang CY (2018) Development of quantum dot-based biosensors: principles and applications. J Mater Chem B 6:6173–6190
30. Zhang R, Chen W (2014) Nitrogen-doped carbon quantum dots: facile synthesis and application as a "turn-off" fluorescent probe for detection of Hg^{2+} ions. Biosens Bioelectron 55(2014):83–90
31. Pooja D, Saini S, Thakur A, Kumar B, Tyagi S, Nayak MK (2017) "Turn-On" thiol functionalized fluorescent carbon quantum dot based chemosensory system for arsenite detection. J Hazard Mater 328:117–126
32. Hallaj T, Amjadi M, Manzoori JL, Azizi N (2017) A novel chemiluminescence sensor for the determination of indomethacin based on sulfur and nitrogen co-doped carbon quantum dot-KMnO4 reaction. Luminescence 32(7):1174–1179
33. Council NR (2007) A strategy for assessing science: behavioural and social research on aging. National Academies Press
34. Bandara JS, Cai Y (2014) The impact of climate change on food crop productivity, food prices and food security in South Asia. Econ Anal Policy 44:451–465
35. Caya MVC, Alcantara JT, Carlos JS, Cereno SSB (2018) Photosynthetically active radiation (PAR) sensor using an array of light sensors with the integration of data logging for agricultural application. In: 2018 3rd international conference on computer and communication systems (ICCCS). IEEE, pp 377–381
36. Ye Y, Yi Z, Gao S, Qin M, Huang Q-A (2018) Octagon-shaped 2-D micro machined thermal wind sensor for high-accurate applications. J Microelectromech Syst 27:739–747

37. Harrey P, Ramsey B, Evans P, Harrison D (2002) Capacitive-type humidity sensors fabricated using the offset lithographic printing process. Sens Actuators B 87:226–232
38. Kaminski J, Christiaensen L (2014) Post-harvest loss in sub-Saharan Africa-what do farmers say? Glob Food Secur 3(3–4):149–158. https://doi.org/10.1016/j.gfs.2014.10.002
39. Kumar D, Kalita P (2017) Reducing postharvest losses during storage of grain crops to strengthen food security in developing countries. Foods 6(1):8. https://doi.org/10.3390/foods6010008
40. Strange RN, Scott PR (2005) Plant disease: a threat to global food security. Annu Rev Phytopathol 43(1):83–116. https://doi.org/10.1146/annurev.phyto.43.113004.133839
41. Shivashakarappaa K, Reddy V, Tupakulaa VK, Farniana A, Vuppulac A, Gunnaiah R (2022) Nanotechnology for the detection of plant pathogens. Plant Nano Biol 100018
42. Nezhad AS (2014) Future of portable devices for plant pathogen diagnosis. Lab Chip 14(16):2887–2904. https://doi.org/10.1039/C4LC00487
43. Rai M, Ingle A (2012) Role of nanotechnology in agriculture with special reference to management of insect pests. Appl Microbiol Biotechnol 94(2):287–293. https://doi.org/10.1007/s00253-012-3969-4
44. Baeumner A (2004) Nanosensors identify pathogens in food. Food Technol (Chic) 58(8):51–55
45. Malhotra BD, Srivastava S, Ali M, Singh C (2014) Nanomaterial-based biosensors for food toxin detection. Appl Biochem Biotechnol 174(3):880–896. https://doi.org/10.1007/s12010-014-0993-0
46. Kaushik A, Solanki PR, Ansari AA, Ahmad S, Malhotra BD (2009) A nanostructured cerium oxide film-based immunosensor for mycotoxin detection
47. Paniel N, Radoi A, Marty JL (2010) Development of an electrochemical biosensor for the detection of aflatoxin M1 in milk. Sensors 10(10):9439–9448. https://doi.org/10.3390/s101009439
48. Hervás M, López MA, Escarpa A (2011) Integrated electrokinetic magnetic bead-based electrochemical immunoassay on microfluidic chips for reliable control of permitted levels of zearalenone in infant foods. Analyst 136(10):2131–2138. https://doi.org/10.1039/C1AN15081B
49. Panini NV, Bertolino FA, Salinas E, Messina GA, Raba J (2010) Zearalenone determination in corn silage samples using an immunosensor in a continuous-flow/stopped-flow systems. Biochem Eng J 51(1–2).7–13. https://doi.org/10.1016/j.bej.2010.04.005
50. Cellini A, Buriani G, Rocchi L, Rondelli E, Savioli S, Rodriguez Estrada MT, Spinelli F (2018) Biological relevance of volatile organic compounds emitted during the pathogenic interactions between apple plants and *Erwinia amylovora*. Mol Plant Pathol 19(1):158–168. https://doi.org/10.1111/mpp.12509
51. Mahmud MM, Constantino N, Seok C, Yamaner FY, Dean RA, Oralkan O (2018) ACMUT-based electronic nose for real-time monitoring of volatiles emitted by plants: preliminary results. IEEE Sens 2018:1–4. https://doi.org/10.1109/icsens.2018.8589740
52. Li Z, Paul R, Ba Tis T, Saville AC, Hansel JC, Yu T (2019) Non-invasive plant disease diagnostics enabled by smartphone-based fingerprinting of leaf volatiles. Nat Plants 5(8):856–866. https://doi.org/10.1038/s41477-019-0476-y
53. Jackson RW (ed) (2009) Plant pathogenic bacteria: genomics and molecular biology. Caister Academic Press. ISBN 978-1-904455-37-0
54. Williams SD, Boehm MJ, Hand FP (2017) Department of Plant Pathology, College of Food Agriculture and Environmental Sciences, Ohio State University extension. https://ohioline.osu.edu/factsheet/plpath-gen-6. Agriculture and Natural Resources
55. Borse V, Srivastava R (2019) Process parameter optimization for lateral flow immunosensing. Mater Sci Energy Technol 2:434–441
56. Lertanantawong B, Krissanaprasit A, Chaibun T, Gothelf KV, Surareungchai W (2019) Multiplexed DNA detection with DNA tweezers in a one-pot reaction. Mater Sci Energy Technol 2:503–508
57. Sun J, Gan T, Zhai R, Fu W, Zhang M (2018) Sensitive and selective electrochemical sensor of diuron against indole-3-acetic acid based on core-shell structured SiO_2@Au particles. Ionics (Kiel) 24(2018):2465–2472

58. Wilson D et al (2019) Electrical detection of pathogenic bacteria in food samples using information visualization methods with a sensor based on magnetic nanoparticles functionalized with antimicrobial peptides. Talanta 194:611–618
59. Yao KS, Li SJ, Tzeng KC, Cheng TC, Chang CY, Chiu CY, Lin ZP (2009) Fluorescence silica nanoprobe as a biomarker for rapid detection of plant pathogens. Adv Mater Res 79–82:513–516. https://doi.org/10.4028/www.scientific.net/amr.79-82.513
60. Ahmad F, Siddiqui MA, Babalola OO, Wu HF (2012) Biofunctionalization of nanoparticle assisted mass spectrometry as biosensors for rapid detection of plant associated bacteria. Biosens Bioelectron 35(1):235–242. https://doi.org/10.1016/j.bios.2012.02.055
61. Lau HY, Botella JR (2017) Advanced DNA-based point-of-care diagnostic methods for plant disease detection. Front Plant Sci 8. https://doi.org/10.3389/fpls.2017.0
62. Razo SC, Panferova NA, Panferov VG IV, Drenova Safenkova NV, Varitsev YA, Zherdev AV, Pakina EN, Dzantiev BB (2019) Enlargement of goldnanoparticles for sensitive immunochromatographic diagnostics of potato brown rot. Sensors (Basel) 19(2019):E153
63. Razmi A, Golestanipour A, Nikkhah M, Bagheri A, Shamsbakhsh M, Malekzadeh-Shafaroudi S (2019) Localized surface plasmon resonance biosensing of tomato yellow leaf curl virus. J Virol Methods 267:1
64. Wang Y, Liu J, Zhou H (2019) Visual detection of cucumber green mottle mosaic virus based on terminal deoxynucleotidyl transferase coupled with DNAzymes amplification. Sensors (Basel) 14(1–12):E1298
65. Byzova NA, Vinogradova SV, Porotikova EV, Terekhova UD, Zherdev AV, Dzantiev BB (2018) Lateral flow immunoassay for rapid detection of grapevine leaf roll-associated virus. Biosensors (Basel) 15:E11
66. Razo SC, Panferov IV VC, Varitsev Safenkova YA, Zherdev AV, Pakina EN, Dzantiev BB (2018) How to improve sensitivity of sandwich lateral flow immunoassay for corpuscular antigens on the example of Potato virus Y? Sensors (Basel) 15:E3975
67. Khater M et al (2019) Electrochemical detection of plant virus using gold nanoparticle modified electrodes. Anal Chim Acta 1046:123–131
68. Luna-Moreno D, Sánchez-Álvarez A, Islas-Flores I, Canto-Canche B, Carrillo-Pech M, Villarreal-Chiu JF, Rodríguez-Delgado M (2019) Early detection of the fungal banana black sigatoka pathogen *Pseudocercospora fijiensis* by an SPR immunosensor method. Sensors (Basel) 23:E465
69. Mahlein AK, Rumpf T, Welke P, Dehne HW, Plumer L, Steiner U, Oerke EC (2013) Development of spectral indices for detecting and identifying plant diseases. Remote Sens Environ 128:21–30
70. Das PK, Laxman B, Rao SVCK, Seshasai MVR, Dadwal VK (2015) Monitoring of bacterial late blight in rice using ground based hyperspectral and LISS IV satellite dated in Kurnool, Andhra Pradesh, India. Int J Pest Manag 61:2593–2668
71. Zheng Q, Huang W, Cui X, Dong Y, Shi Y, Ma H, Liu L (2018) Identification of wheat yellow rust using optimal three-band spectral indices in different growth stages. Sensors (Basel) 19:E35
72. Zhan F, Wang T, Iradukunda L, Zhan J (2018) A gold nanoparticle-based lateral flow biosensor for sensitive visual detection of the potato late blight pathogen, Phytophthora infestans. Anal Chim Acta 1036:153–161
73. Dubey A, Mailapalli DR (2016) Nanofertilisers, nanopesticides, nanosensors of pest and nanotoxicity in agriculture. Sustain Agric Rev 19:307–330
74. Brezolin AN et al (2020) Nanostructured cantilever sensor using with Pani/MWCNT-COOH nanocomposites applied in the detection of pheromone. J Mater Sci Mater Electron 31(8):6008–6016
75. Martinazzo J et al (2022) Sensing of pheromones from *Euschistus heros* (F.) stink bugs by nanosensors. Sens Actuators Rep 4:100071
76. Moitra P et al (2016) A novel bio-engineering approach to generate an eminent surface-functionalized template for selective detection of female sex pheromone of *Helicoverpa armigera*. Sci Rep 6(1):37355

77. Sharma S, Dhadly DK (2023) Nano-inspired biosensors and plant diseases: recent advances and challenges. Nanoparticles and plant-microbe interactions. Academic Press, pp 135–162. ISBN 978-0-323-90619-7
78. Dutta P, Bhattacharyya A, Kumari A (2023) Innovative integrated pest management paradigm for sustainable crop production with special reference to North East India. Integrated pest management in diverse cropping systems. Apple Academic Press, pp 61–90
79. Dar FA, Qazi G, Pirzadah TB (2020) Nano-biosensors: nextGen diagnostic tools in agriculture. In: Hakeem K, Pirzadah T (eds) Nanobiotechnology in agriculture. Nanotechnology in the life sciences. Springer, Cham. https://doi.org/10.1007/978-3-030-39978-8_7
80. Bhagat R, Ingle AP, Chen H (2023) Nanosensors and nanobiosensors for sustainable agriculture. Nanotechnology in agriculture and agroecosystems. Elsevier, pp 93–112
81. Gangwar J, Kadanthottu Sebastian J, Puthukulangara Jaison J, Kurian JT (2023) Nano-technological interventions in crop production—a review. Physiol Mol Biol Plants 1–15
82. Negrete JC (2020) Nanotechnology an option in Mexican agriculture. J Biotechnol Bioinform Res 3
83. Dimri A, Pathak N, Sharma S (2020) Nanosensors for root zone parameters influencing plant growth. In: Nanomaterials for agriculture and forestry applications. Elsevier, pp 387–406
84. Thakur M, Wang B, Verma ML (2022) Development and applications of nano biosensors for sustainable agricultural and food industries: recent developments, challenges and perspectives. Environ Technol Innov 102371
85. Chong JPC, Liew OW, Li BQ, Asundi AK (2007) Optical fluorescence biosensor for plant water stress detection. In: Saratov fall meeting 2006: optical technologies in biophysics and medicine: volume III. International Society for Optics and Photonics, pp 65350–65358
86. Subraya KK, Diggs A, Porterfield DM (2013) Amperometric biosensor approaches for quantification of indole 3-acetic acid in plant stress responses. Commun Soil Sci Plant Anal 44:1749–1763
87. Lupetti KO, Vieira IC, Fatibello-Filho O (2004) Jack fruit-capric acid biosensor for total phenols determination in wastewaters. Anal Lett 37:1833–1846
88. Vermeir S, Nicolai BM, Verboven P, Van Gerwen P, Baeten B, Hoflack L et al (2007) Microplate differential calorimetric biosensor for ascorbic acid analysis in food and pharmaceuticals. Anal Chem 79:6119–6127
89. Prasad S, Bhattacharyya S, Roy G (2019) Smart farming using IoT and nanotechnology: a systematic review. IEEE Access 7:120431–120452
90. Norton BG, De Vries F (2021) Integrated pest management and smart technologies. Environ Manag 65(52–66):2021
91. Singh RP (2020) Nanotechnology applications in sustainable agriculture. J Nanobiotechnol 18:134–144
92. Zhang J, Liu L (2019) Real-time soil moisture monitoring using nano sensors. IEEE Sens J 19:2764–2772
93. Sharma M, Singh K (2020) Growth of smart agriculture market: trends, drivers, and future. IEEE J Emerg Sel Top Power Electron 8:2365–2376
94. Hu D, Khan MA, Taylor RW (2020) Economic benefits of precision agriculture technologies. J Agric Econ 71(678–694):2020
95. Jones RL (2021) Sustainable agriculture: the role of smart technologies in reducing the carbon footprint. Sustainability 13:2521–2536

Nanosensors: How Nanotechnology Is Being Used to Detect and Monitor Food Quality, Safety, and Authenticity

Ayaz Mukarram Shaikh, Tejas R. Suthar, Shivangi Srivastava, Vinay Kumar Pandey, Béla Kovács, and József Prokisch

Abstract This book chapter explores the application of nanotechnology in developing nano sensors for the detection and monitoring of food quality, safety, and authenticity. The chapter discusses the use of diverse types of nano sensors, including optical, electrochemical, and mechanical sensors, and their potential in detecting foodborne pathogens, contaminants, and adulterants. It also highlights the advantages and limitations of using nano sensors in food analysis and the challenges in commercializing these technologies. Overall, the chapter provides insights into the potential of nanotechnology in improving food safety and quality control.

Keywords Nanotechnology · Nano sensors · Food quality · Food safety · Authenticity

A. M. Shaikh (✉) · B. Kovács
Faculty of Agriculture, Food Science & Environmental Management, Institute of Food Science, University of Debrecen, Debrecen 4032, Hungary
e-mail: Ayaz.shaikh@agr.unideb.hu

T. R. Suthar
Chicago, IL 60616, USA

S. Srivastava
Department of Food Technology, Harcourt Butler Technical University, Kanpur 208002, Uttar Pradesh, India

V. K. Pandey
Research and Development cell, Biotechnology Department, Manav Rachna International Institute of Research and Studies, Faridabad 121002, Haryana, India

J. Prokisch
Institute of Animal Science, Biotechnology and Nature Conservation, Faculty of Agricultural and Food Sciences and Environmental Management, University of Debrecen, Debrecen 4032, Hungary

I. Uddin (ed.), *Smart Nanosensors*, Smart Nanomaterials Technology,
https://doi.org/10.1007/978-981-96-3878-9_11

1 Introduction

Nanotechnology is considered as one of prominent technology that is making various implications in fields like chemical, biological, electronics, mechanical etc. According to Greek terms, nanotechnology describe as the technology which involves studies, construction, and manufacture of nanometer-sized materials [36]. Nanomaterials are specified as a material having size between 1 and 100 nm [18]. Nanotechnology has been used in various experiments specified with the issues related to food processing, food preservation, food processing, and food packaging [4]. As per the historical background, nanotechnology comes long way when an American physicist Richard Feynman showed the concept of nanotechnology in 1959. During this period, Feynman explained a method for channelizing individual atoms and molecules to operate nanomaterials importance using set of specific tools. Still, at present time the idea of nanotechnology considers new in reference to experimental theories, but its evaluability comes in life few years ago with some secret information. Since, life came into existence, attention to details regarding working of nano-sized proteins on much smaller level in DNA matrix and depicting its various functionality in number of identified/unidentified organism are of unique capabilities (Fig. 1) [21].

In order to enhance physicochemical properties and increase nutritional stability and bioavailability, different methods of food processing have been introduced using nanotechnology [26]. The technology advancement for solving issues in food industry has been revolutionized in some or other way. Nanotechnology is one of them. The potential application of nanotechnology in the food sector can be covered in wide range. As per the various research, the four defined specific that comes withing the food industry are Food Processing, Food Packaging, Food Safety and

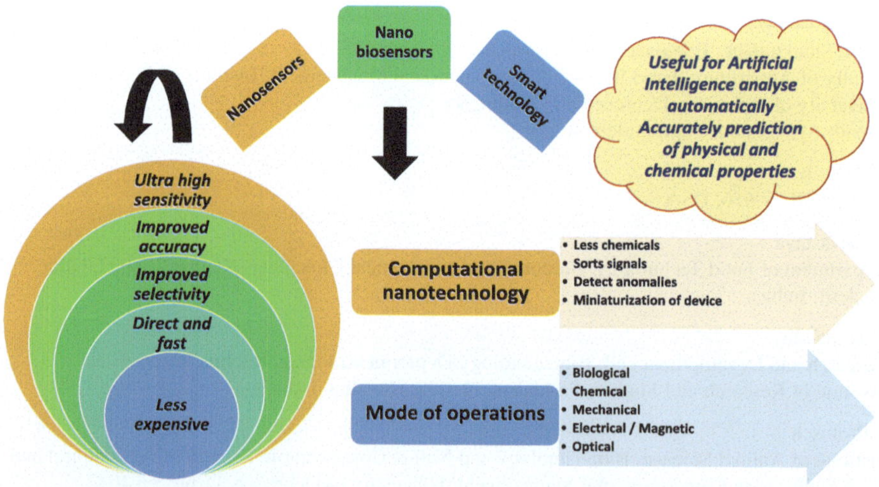

Fig. 1 Smart nanosensors for the detection of heavy metals and pesticides in foods

Food Preservation. Further, some of the examples of this specified hierarchy have been identifies in Table 1.

Safety of food is one of significant considerable factor that is associated with variety of food commodities ranging from confectionery, beverage to meat products. Crucial factors in the global food industry that directly affect customer health, trust, and satisfaction are food quality, safety, and authenticity. Growing worries about foodborne illnesses, dishonest business practices, and adulteration in recent years have underlined the necessity of strict regulations to protect the integrity of the food supply chain. The qualities, traits, and aspects of food products that determine their acceptability and suitability for ingestion are referred to as food quality. In addition to offering vital nutrients, high-quality food also enhances consumers sensory enjoyment and overall happiness. Food quality is determined by several elements, including nutritional content, flavour, freshness, and appearance. Food producers may provide consumers with safe, wholesome, and delectable products by upholding quality standards [38].

Food safety is of utmost importance since tainted or improperly handled food can result in foodborne infections, which can have serious health effects. The World Health Organization (WHO) estimates that each year, tainted food causes six hundred

Table 1 Pillars of nanotechnology in food industry

Food processing	Food preservation	Food packaging	Food safety
Anticaking agent—improvement in consistency and prevent crystallization	Antimicrobial agent—work in accordance with reducing microbial activity in foods	Improved packaging—use of nanoparticles or nanomaterials to improve physical performance of food	Detection of food borne pathogen and works with analysing microbial activity in food systems
Nutraceuticals—nutritional value increment in various foods	Nano encapsulation—for successful delivery of bioactive compounds and flavors	Active packaging—nanoparticle as antimicrobial agent	Works in against heavy metal reductions
Gelatine agent—used for providing stability in food matrix and helps in providing texture	Improving food's physical attributes, such as texture, appearance, and stability	Smart packaging—nano biosensors for pathogen detection such as bacteria and viruses	Against pesticides and drugs and helpful in toxic detection
Nanocarriers—to protect organoleptic properties in foods	Protection of foods against chemical deterioration	Intelligent packaging—to warn food reacted with environmental changes	Preventing the growth and development of biofilms on surfaces or in organisms

million illnesses and 420,000 deaths worldwide. To avoid microbiological contamination, chemical risks, and physical contaminants, proper food safety standards cover every step of the food chain, from manufacturing to consumption. Governments and regulatory organizations have established standards and laws, such as Hazard Analysis Critical Control Points (HACCP) and Good Manufacturing Practices (GMP), to assure the safety of food. Following these recommendations lowers the possibility of foodborne illnesses and safeguards the public's health by assisting in the identification and control of possible dangers [15].

The accurate representation of a food product in terms of its origin, content, and labelling is referred to as food authenticity. For the food business to continue to enjoy consumer confidence and trust, authenticity must be upheld. Food fraud, which includes actions like adulteration, incorrect labelling, and counterfeiting, not only misleads customers but also puts their health at danger and damages the reputation of food companies. Reliable testing procedures, strong traceability systems, and stringent regulatory enforcement are all necessary to ensure the authenticity of food products. More and more, traceability is being improved through the use of technologies like blockchain and DNA analysis, giving customers the capacity to confirm the provenance and authenticity of food products. Food authenticity can be protected by promoting openness and accountability in the supply chain, fostering consumer trust and loyalty [9, 29].

2 Types of Nano Sensors

One of the essential components of nanotechnology are nano sensors, which is one of the most amazing developments that have helped with problems relating to food analysis, appropriateness for spotting pathogens in food, and their ability to regulate food quality. These sensors, which are further discussed below, include optical, electrochemical, and mechanical ones (Fig. 2).

2.1 *Optical Sensors*

Optical sensors are used for tracking better output from food analysis perspective. These sensors are based on nanomaterials and nano-effects which give valuable result in terms of rapid detection, low-key operation, low cost, and high sensitivity. Optical instruments or visual colorimetric detection is the working principal of nano sensors. At present, food analysis is done with using high-performance liquid chromatography (HPLC), gas chromatography (GC), near-infrared (NIR) and ultraviolet–visible (UV–vis) spectroscopies. These methods offer excellent detection results but at the same time they suffer from time variations and associate with prohibitive cost. The traditional chromatographic and spectroscopy methods consist of various problems such as time-consuming, low accuracy and peak overlap. As compared to this,

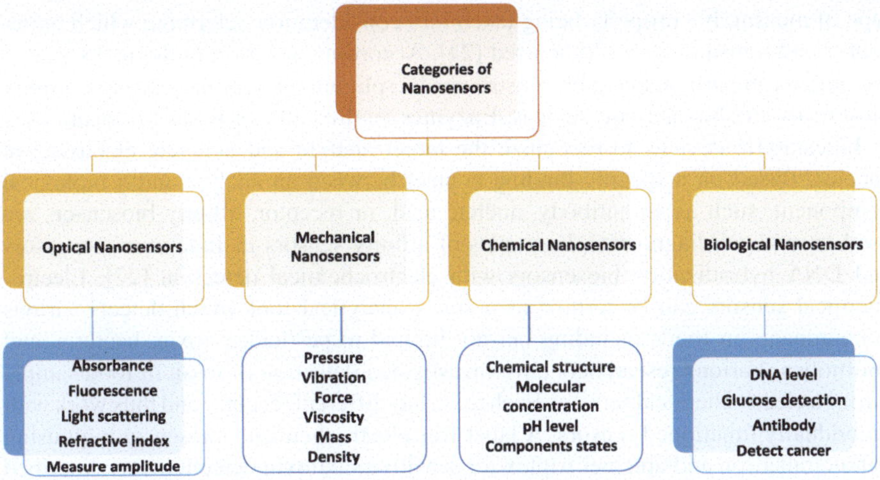

Fig. 2 Nanosensors classification according to the sensing event category

the spectroscopy that consist of nano sensors can detect data with high accuracy and in shorter time which fulfil the requirement of rapid and point to the mark detection in food safety [10]. Apart from this, nano sensors materials have higher thermal and chemical stability. For detection of pesticides in food, fluorescence quenching is often used. The nanomaterial (such as Quantum Dots) interacts with pesticides result in fluorescent sensing, which further can be converted into pesticide content. According to research data provide by Y. Fan et al., it is stated that Quantum Dots can detect multicomponent pesticides with high sensitivity [14]. With one such example is QDs compositor used for detection of glyphosate in food matrix [20].

2.2 Electrochemical Sensors

Conventionally, food analysis being conducted in end of production process which led to passing of contaminating agents through entire production chain and its even be distributed in markets before contamination is noticed. At convectional level, chromatography, mass spectrometry or fluorescence techniques is used, taking longer time for accurate detection, and providing positive results. On the other hand, technological advancement with respect to electrochemical biosensors can been seen in making its mark for fast, rapid, and accurate detection in food analysis [31].

Principal working of electrochemical sensors is based on the determination of amount of molecule in a sample. Generally, it is demonstrated that bioreceptor connected with a transducer. A physicochemical or biological signal is produced by the precise interaction between the target molecule and the biocomponent, and the transducer transforms it into a measured property. The features of the sample and the type of measurable property being taken into consideration determine which bioreceptor and transducer should be used [24]. According to how a biological signal is recognized, electrochemical biosensors can be split into two primary groups: affinity biosensors and biocatalytic devices. Enzymes, entire cells, or tissue slices are used in biocatalytic devices to recognize the target analyte and generate electroactive species. Based on a specific binding contact between an analyte and a biological component, such as an antibody, nucleic acid, or receptor, affinity biosensors are used in medical diagnostics. Examples of affinity sensors include immunosensors and DNA hybridization biosensors with electrochemical detection [27]. Electrochemical sensors can be termed as a smart analytical tool which detects various contaminants in foods including but not limited to pesticides, toxin, bacteria, and antibiotics. Various researchers have investigated detection of toxin in food sample through electrochemical sensors. Each toxin has different receptor and thus work with individually functioned sensors. A label-free electrochemical aptasensor built using nanotetrahedron and aptamer triplex for sensitive saxitoxin detection was described by Wang et al. For the initial occasion, a label-free electrochemical aptasensor was constructed to achieve highly sensitive detection of small compounds by merging various techniques such as aptamer technology, DNA nanotetrahedron, DNA triplex, and electrochemistry. Saxitoxin, a typical tiny chemical with high toxicity and a low molecular weight, was selected as a model target. One of the main toxins in paralytic shellfish poison (PSP), STX, can kill persons and fisheries by causing shock and suffocation [28].

2.3 Mechanical Sensors

For detecting the presence of chemical species in food and water, mechanical nano sensors have been used widely in recent area of food technological advancement. Unique characteristics of mechanical nanosensors make them ideally suited for applications involving the monitoring of food quality. First off, because to their small size, non-destructive and non-invasive measurements are possible, which reduces the need for sample preparation and waste. Second, because of their high sensitivity, they can detect even the smallest changes in food's physical and chemical characteristics, including those related to moisture, pressure, temperature, and gas composition. This sensitivity enables prompt intervention and stops the distribution of subpar food products by enabling early identification of spoilage, contamination, or deterioration [1].

Various food packaging components, including films, coatings, and labels, can incorporate in mechanical nanosensors. These sensors can keep an eye on important factors like humidity, oxygen levels, and temperature changes while being transported and stored. They support the preservation of ideal storage conditions and the safety and quality of food by giving real-time data. Additionally, crucial food production processes including fermentation, drying, and frying can be monitored and controlled by mechanical nano sensors, improving the uniformity and quality of the finished goods. Numerous food quality control activities involve the use of mechanical nano sensors. The monitoring of freshness and spoilage is one notable example. This technology enables the early identification of microbiological contamination or food degradation by detecting changes in volatile organic molecules released by spoiled food. By measuring variables like headspace gas composition or pressure, they may also evaluate the quality of packed food, giving vital details regarding the strength of the seal and potential spoiling. The ability to identify allergens and pollutants in food is another application. Even at very low concentrations, hazardous compounds like pesticides, heavy metals, or allergenic proteins can be detected by mechanical nano sensors. This capacity is especially important for guaranteeing food safety and avoiding health hazards brought on by tainted or allergic components [12].

2.4 Advantages and Limitations of Nano Sensors in Food Analysis

In today's era of technological advancement, nanotechnology has emerged as a game changer in various fields, including medicine, environmental observing, and food quality control. As per the various data, it is stated that nano sensors offer number of benefits as compared to traditional sensors. From perspective of manufacturing capabilities, nano sensors are made with a material having minute particle size. And due to this development, the sensors are of high sensitivity, which enables the detection of environmental changes as well as minuscule quantities of substances. At nanoscale, sensor exhibit unique physical, chemical, and electrical properties that enhance sensitivity. And due to concept, nano sensor is even capable of detecting ultra-low concentrations of multiple analytes and pathogens simultaneously [39].

The selectivity of nano sensors, or their capacity to distinguish between various chemicals or analytes, is another important benefit. Nano sensors can be designed to interact only with targeted molecules or chemicals and not be influenced by other substances in the surrounding environment. Applications like the detection of pollutants in food and water depend on this selectivity. By detecting specific diseases, poisons, or chemical residues, nano sensors can ensure food safety and reduce health hazards. Additionally, nano sensors can detect and measure target substances in industrial processes, improving quality control and lowering the possibility of product contamination. Nano sensors have an exceptional advantage in terms of quick detection, providing data in real-time or very close to real-time. Nano sensors have quick reaction times because of their small size and special characteristics, which facilitate quick analysis and decision-making. Nano sensors' quick detecting capabilities are used in a variety of situations. Nano sensors in the food industry can deliver quick diagnostic findings, enabling signals if the food product no longer suitable for consumption. Nano sensors in security and defines can quickly spot chemical or biological agents, improving emergency response. In addition, nano sensors can also continually check the quality of the air or water in environmental monitoring, swiftly alerting authorities to any dangers or safety standards violations (Table 2).

Having various benefits associated with nano sensors being used in food industry comes with challenges too. At present, most research on nano sensor and their capabilities being tested in laboratory environment which many be varies from those real conditions. In manufacturing industry, sensors should be capable of working under extreme pressure, temperature, pH, and other unstable parameters. Sensors also have to operate with n number of microorganisms and other substance in food matrix which could interferes with output data.

Regarding the cost of nano sensor devices, it is important to consider that, until the technology becomes more widely adopted and popularized, there may be significant infrastructure expenses associated with their implementation. As with any emerging technology, the initial stages of development and deployment often incur higher costs due to factors such as research, design, production, and distribution. With a larger market and improved production capabilities, the costs of manufacturing and infrastructure are expected to decrease. Other challenges nano sensor may have been the stability of sensors when other species present. When the analytes interact with

Table 2 Types of nanosensors in food analysis, applications, advantages, and limitations

Nanosensors type	Applications in food analysis	Advantages	Limitations	References
Nano biosensors	Detection of pathogens, toxins, and allergens in food products	High sensitivity and selectivity, rapid detection, real-time monitoring	Complex fabrication process, potential for cross-contamination	Anh et al. 3
Nano electrochemical sensors	Determination of food quality parameters such as pH, antioxidants, and heavy metals	High sensitivity, low cost, miniaturization potential	Limited applicability for some analytes, need for electrode modification	Sohrabi et al. 34
Nano plasmonic sensors	Detection of food spoilage markers, contaminants, and allergens	Label-free detection, high sensitivity, real-time monitoring	Expensive instrumentation, limited compatibility with complex food matrices	Bhagat et al. 7
Nano photonic sensors	Identification of food adulteration, traceability of food products	Ultra-high sensitivity, label-free detection, multiplexing capability	Limited robustness, interference from environmental factors	Sharma et al. 32
Nano mechanical sensors	Monitoring of food packaging integrity, detection of volatile compounds	High sensitivity, small sample volume requirement	Susceptible to mechanical noise, limited stability	Das et al. 11

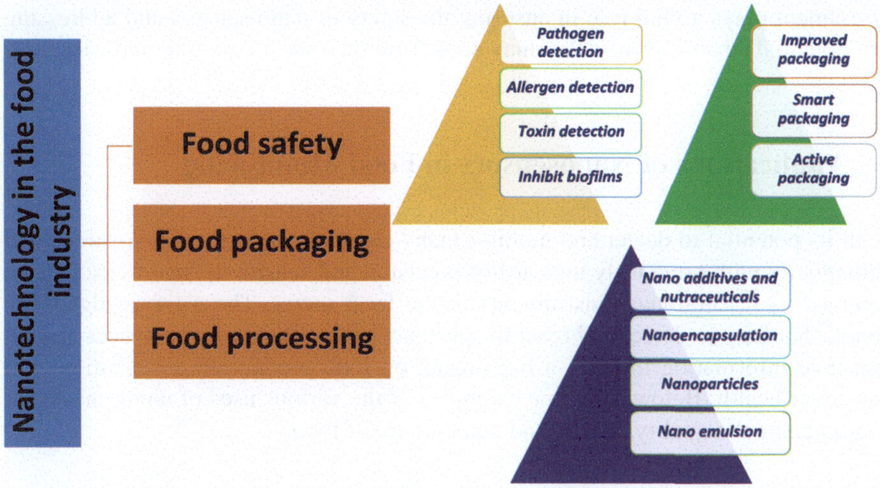

Fig. 3 Application of nanotechnology in different aspects of the food industry

several unknow species, changes in the fingerprint signature may occur, making detection and identification more difficult (Fig. 3).

There are significant concerns regarding sampling in both internal and external nano sensors. One of the main challenges is related to the minuscule size of these sensors in comparison to the products or conditions they are designed to measure. Consequently, the sensors can only capture data from a small portion of the food or environment being examined. To address this limitation, one potential solution that is frequently discussed is the implementation of autonomous wireless sensor networks in distinct locations. These networks could help overcome the issue of limited sampling by providing a more comprehensive and distributed coverage, enhancing the accuracy and effectiveness of data collection [23].

The introduction of nanotechnology in the food industry faces significant resistance in terms of raising awareness and gaining acceptance. Like any modern technology, there is a need to overcome skepticism and apprehension surrounding its use. Currently, there are no specific regulations in place to control or limit the production of nano sensors specifically in the food industry. This is primarily due to a lack of comprehensive understanding regarding the potential risks associated with these nano sensors. In essence, the field of nano sensor research and applications in the food sector is still in its early stages, with significant advancements yet to be made. In comparison to medical applications, which have seen more extensive exploration, the use of nano sensors in food is relatively new. Consequently, the safety implications of utilizing nano sensors present a challenge for both the government and industry. On an industrial scale, it becomes crucial to assess the release of nanomaterials into the environment and estimate the subsequent levels of exposure to these materials. It is imperative to identify and quantify the risks associated with such exposure. This assessment plays a vital role in ensuring the safety of nano sensors and addressing any potential adverse effects that may arise from their use in the food industry [30].

3 Applications of Nanosensors in Food Monitoring

With its potential to detect and monitor many elements of food quality, safety, and authenticity with previously unheard-of precision and sensitivity, nanosensors have emerged as revolutionary instruments in the food sector. These tiny gadgets use nanotechnology to detect and react to substances, viruses, or circumstances, giving real-time information that has a big impact on food production, distribution, and consumer health. Below are some examples of the various uses of nanosensors for guaranteeing the quality, safety, and authenticity of food.

3.1 Detection of Contaminants

By acting as extremely sensitive and effective tools for detecting a variety of contaminants that pose threats to consumer health, nanosensors have transformed the subject of food safety. To detect infections, allergies, pesticides, and heavy metals in food goods, these tiny gadgets use nanotechnology. In order to improve food safety and keep potential risks from reaching consumers, their capacity to identify these compounds at extremely low quantities is crucial. Nanosensors' ability to identify dangerous germs quickly and precisely like E. coli and Salmonella is one of their most important uses [37]. These infections have the potential to produce serious foodborne illnesses that can result in fatalities or serious health problems. Even in complicated matrices like raw materials or processed meals, nanosensors can detect these pathogens at very low concentrations. As a result of this early discovery, it is possible to take prompt corrective action to stop tainted products from entering the market and reaching customers. One significant benefit of nanosensor-based detection is its speed. Traditional pathogen identification techniques frequently call for time-consuming laboratory testing, which could have happened while tainted goods were still being sold and consumed. Real-time data from nanosensors enables early responses and lowers the likelihood of epidemics and widespread disease. Allergens, insecticides, and heavy metals can be easily and effectively detected with nanosensors. Nanosensors can detect even minute levels of allergens, protecting those with allergies and ensuring the safety of food products. Heavy metal and pesticide residue detection is essential to preventing chronic health problems brought on by prolonged exposure to these toxins. By identifying pollutants at incredibly low quantities, nanosensors are at the forefront of improving food safety. Their quick, delicate, and in-the-moment detection abilities enable preventive actions to be performed, averting any health problems and protecting the wellbeing of customers. The food sector is more prepared than ever to uphold the highest standards of safety and offer customers products they can rely on as these technologies develop [34].

3.2 Freshness and Shelf-Life Assessment

A revolutionary method of determining and maintaining food freshness and shelf life in storage conditions has been made possible by nanosensors. Nanosensors play a vital role in decreasing food waste and preserving product quality across the supply chain by providing real-time monitoring of critical parameters including temperature, humidity, and gas levels. The rate of food deterioration and rotting is significantly influenced by temperature and humidity. Nanosensors offer ongoing monitoring of these parameters due to their great sensitivity and quick reaction times. This is especially important for businesses that deal with perishable goods like produce, dairy, meat, and other dairy products. Premature spoiling can be caused by microbial growth, chemical reactions, and enzyme activity that are accelerated by variations from the ideal temperature and humidity ranges. The benefit of fast detection provided by nanosensors enables stakeholders to immediately remedy problems by changing storage conditions or speeding up distribution [17]. Oxygen and ethylene gas concentrations are important for maintaining food quality. The buildup of gases caused by ripening or spoiling processes can be detected by nanosensors. For instance, increased ethylene levels may shorten the shelf life of fruits and vegetables by causing them to mature more quickly. Nanosensors enable suppliers, distributors, and retailers to proactively manage their inventory and optimise the timing of distribution by spotting such changes in real-time. Reducing food waste is one of the most beneficial effects of nanosensor-based monitoring. Nanosensors enable stakeholders to make knowledgeable decisions about product distribution by delivering precise and timely information. This helps to ensure that items are delivered to the right markets before they expire. By preventing premature product disposal, both consumer and industrial waste are decreased. Nanosensors promote accountability and transparency across the supply chain. By employing nanosensors to verify the conditions under which their items have been stored and transported, suppliers can prove their dedication to supplying high-quality products. Retailers may easily reassure customers that the goods they buy have undergone rigorous observation and upkeep [25].

3.3 Authentication and Traceability

The authenticity and traceability of food products are being revolutionised by the innovative equipment known as nanosensors. They play a crucial role in preserving the integrity of the global food supply chain thanks to their unmatched capacity to determine the origin and authenticity of foodstuffs. Nanosensors are successfully preventing dishonest activities that undermine consumer confidence and degrade manufacturer credibility by accurately identifying fake additions, contaminated components, and misleading labelling [6]. In the context of high-end products, such as organic or expensive things, the significance of nanosensors becomes particularly apparent. Due to their high calibre and distinctive qualities, these goods frequently attract a premium price. Instances of fraud, where inferior replacements are misrepresented as real premium items, are prevalent in the market. Nanosensors take on this problem head-on by offering a discrete yet effective layer of verification. They examine complex molecular and elemental fingerprints to find any variations from the composition and structure that should be expected. Customers are guaranteed to get precisely what they pay for, whether it's the real organic food they want or a high-end good they want to indulge in. This fosters customer and producer loyalty and trust. The reputation of manufacturers is strengthened because these cutting-edge technologies validate their dedication to supplying genuine and pure items. As a result, both customers and regulatory organisations commend the commitment to quality, creating the conditions for increased market share and longevity [5].

3.4 Quality Control in Production

In the field of food production, nanosensors have become essential tools, significantly advancing quality control procedures. Nanosensors play a vital role in guaranteeing constant and exceptional product quality throughout the manufacturing pipeline because to their extraordinary capacity to monitor and evaluate important aspects like moisture content, pH levels, and texture variations. The ability of nanosensors to offer in-the-moment insights into many parameters is one of their key features. Nanosensors assist avoid unfavourable consequences like bacterial growth, mould development, or textural degeneration that might result from too much or not enough moisture by continuously monitoring moisture levels [13]. Similar to this, keeping an eye on pH levels is essential to maintaining the quality, security, and shelf life of food goods. Precision pH measurement made possible by nanosensors enables manufacturers to quickly rectify any discrepancies that could jeopardise the quality of their products. Nanosensors are also quite useful for identifying minute differences in product texture. Any sudden changes could be a sign of a process deviation or an issue with the equipment. Nanosensors enable operators to take corrective action before minor problems turn into major production issues by quickly spotting these variances. This proactive strategy minimises the chance of batch rejections while also ensuring product consistency, cutting down on waste and related costs. The application of nanosensors to quality control also encourages adherence to SOPs. These sensors may be designed to rapidly detect any deviations by comparing current data to established benchmarks. These characteristic supports maintaining product homogeneity and observing legal regulations. The capacity to continuously track multiple variables enables producers to decide intelligently on changes to production settings, resulting in streamlined operations and reduced batch-to-batch fluctuations [35].

3.5 Allergen Detection

For those who suffer from allergies, accurate allergen identification is crucial since even minute levels of allergenic chemicals can result in life-threatening reactions. Nanosensors have become essential tools in this critical area, completely changing the way allergy management is done. Due to its extraordinary sensitivity and specificity, allergists can be protected from injury and have a better quality of life by being able to identify even very little amounts of allergens. For people who have allergies, even a small amount of exposure to allergenic chemicals can result in potentially fatal reactions. Nanosensors offer a strong barrier against contamination and unintentional exposure since they can identify allergens at exceedingly low concentrations [19]. Nanosensors help both producers and consumers make wise consumption decisions by quickly detecting the presence of allergenic substances in food products [33]. Preventing the mixing of chemicals that could endanger sensitive people is especially important in surroundings where various allergens may be present. Nanosensors also provide a proactive method of allergy detection. From the inspection of the raw materials through the evaluation of the finished product, they can be incorporated into many stages of food production. This thorough coverage guarantees that potential allergy sources are found at every stage, lowering the possibility of unintentional contamination. Nanosensors allow for prompt interventions and remedial measures to stop allergen-related problems from entering the market by quickly identifying any deviations from the norm. Nanosensors' effect on allergen detection goes beyond personal safety to address greater public health issues. Nanosensors help to lessen the stress placed on healthcare systems and resources, as well as the psychological and financial costs experienced by allergic people and their families, by reducing the frequency of allergen-induced occurrences [8].

3.6 Monitoring Food Packaging

A new era of intelligent and responsive packaging has arrived as a result of the incorporation of nanosensors into packaging materials, revolutionising how food is transported and preserved. Real-time information on important environmental variables like temperature, humidity, and gas permeability is provided by these nanosensors [2, 22]. This development has broad ramifications since it guarantees that packaged goods stay under ideal conditions throughout the whole supply chain, reducing the danger of rotting and maintaining product quality. In order to maintain the freshness and security of perishable items, temperature regulation is essential. Nanosensors built into packaging can track temperature differences over time and promptly notify stakeholders of any changes that would jeopardise the quality of the items. The preservation of the products' sensory qualities and the prevention of the growth of dangerous germs depend heavily on this functionality. Another important element that might affect the quality and shelf life of packaged goods is humidity. Real-time feedback provided by nanosensors that can measure humidity levels enables prompt corrective action when humidity levels surpass permissible thresholds. These sensors assist prevent mould growth, texture changes, and other types of damage that might result from excess moisture by regulating moisture levels. Often disregarded, gas permeability has a big impact on how effective packaging is. With the help of nanosensors, packaging may be monitored for variations in gas concentrations that could indicate leaks or oxygen infiltration. Nanosensors stop oxidative reactions that lead to product degradation and loss of quality by quickly detecting such anomalies. The use of nanosensors into packaging materials satisfies today's desire for sustainability and openness. By making sure that products are stored and transported in the best conditions possible, real-time monitoring enables supply chain stakeholders to minimise waste. Cost savings and a smaller environmental impact follow from this [16].

4 Conclusion

From a scientific perspective, nanotechnology is a progressive technique that is evolving to address and benefit major issues in the food industry. In this article, we have demonstrated capabilities including detection of pathogens, heavy metals, chemical residues. These nano sensors having unique properties helping in real-time monitoring and early detection of potential issue in food matrix. Along with this, due to its microscopic structure, nano sensors have high sensitivity, selectivity, and rapid response. However, despite the promising potential, various challenges need to be stated. Some include, understanding possible toxicity of nanomaterials and their unintentional impact on overall human health. Also, regulatory framework must be addresses for continuing reliable performance, parallelly further research regarding technical limitations, reproducibility, scalability of nano sensors and consumer acceptance has to be taken into considerations. Furthermore, prior to their broad acceptance, it is essential to carefully assess these materials' safety features throughout the sectors including preservation, packaging and ultimately food safety.

References

1. Agüero-Chapin G, Cortés-Rodríguez M, del Real A (2021) Nano and microsensors for food safety and quality control: state of the art and challenges. Foods 10(3):638
2. Alfei S, Marengo B, Zuccari G (2020) Nanotechnology application in food packaging: a plethora of opportunities versus pending risks assessment and public concerns. Food Res Int 137:109664. https://doi.org/10.1016/j.foodres.2020.109664

3. Anh NH, Doan MQ, Dinh NX, Huy TQ, Tri DQ, Ngoc Loan LT, Van Hao B, Le AT (2022) Gold nanoparticle-based optical nanosensors for food and health safety monitoring: recent advances and future perspectives. RSC Adv 12(18):10950–10988. https://doi.org/10.1039/d1ra08311b
4. Anu K, Bhanu M, Harsha K (2015) Advancement of nanotechnology in food packaging. World J Pharm Pharmaceut Sci 4(04):1054–1057
5. Ba LP (2022) Study of the quantum efficiency of the excitonic optical amplification of nanosensors by linear and rainbow nonlinear optical spectroscopy
6. Baksa R (2015) Continuous monitoring of enterprise risks: a Delphi feasibility study
7. Bhagat B, Baruah P, Mukherjee K (2022) Application of nanosensors in food inspection. In: Nanosensors for smart agriculture, pp 705–735
8. Bonlokke JH, Bang B, Aasmoe L, Rahman AMA, Syron LN, Andersson E, Dahlman-Höglund A, Lopata AL, Jeebhay M (2019) Exposures and health effects of bioaerosols in seafood processing workers—a position statement. J Agromed 24(4):441–448. https://doi.org/10.1080/1059924X.2019.1646685
9. Brereton P (2020) Food authenticity and traceability: overview and challenges. In: Food authenticity and traceability. Woodhead Publishing, pp 1–13
10. Chen H, Zhang L, Hu Y, Zhou C, Lan W, Fu H, She Y (2021) Nanomaterials as optical sensors for application in rapid detection of food contaminants, quality, and authenticity. Sens Actuators Part B 329:129135. https://doi.org/10.1016/j.snb.2020.129135
11. Das PP, Dhara S, Purkait MK (2023) Nanosensors in food science and technology. In: Advances in smart nanomaterials and their applications. Elsevier, pp 247–272
12. Dey P, Barfidokht A (2020) Nanomaterials as sensors and biosensors for food quality assurance. In: Advanced sensing techniques for food safety and quality assurance. Elsevier, pp 147–168
13. Dubey N, Chitranshi S, Dwivedi SK, Sharma A (2023) Postharvest physiology, value chain advancement, and nanotechnology in fresh-cut fruits and vegetables. In: Nanotechnology horizons in food process engineering, vol 3: Trends, nanomaterials, and food delivery, p 99
14. Fan Y, Liu L, Sun D, Lan H, Fu H, Yang T, She Y, Ni C (2016) 'Turn-off' fluorescent data array sensor based on double quantum dots coupled with chemometrics for highly sensitive and selective detection of multicomponent pesticides. Anal Chim Acta 916:84–91. https://doi.org/10.1016/j.aca.2016.02.021
15. Food and Agriculture Organization (FAO) (2023) Food quality and safety systems—a training manual on food hygiene and the hazard analysis and critical control point (HACCP) system. http://www.fao.org/3/y1576e/y1576e00.htm. Accessed 25 June 2023
16. Hari N, Radhakrishnan S, Nair AJ (2020) Nanosensors as potential multisensor systems to ensure safe and quality food. In: Biotechnological approaches in food adulterants. CRC Press, pp 204–238
17. Hofmann T, Lowry GV, Ghoshal S, Tufenkji N, Brambilla D, Dutcher JR, Gilbertson LM, Giraldo JP, Kinsella JM, Landry MP, Lovell W, Naccache R, Paret M, Pedersen JA, Unrine JM, White JC, Wilkinson KJ (2020) Technology readiness and overcoming barriers to sustainably implement nanotechnology-enabled plant agriculture. Nat Food 1(7):416–425. https://doi.org/10.1038/s43016-020-0110-1
18. Hyunjong Y, Jun-Young P, Chang Woo K, Sung-Chul H, Kyung-Min P, Pahn-Shick C (2018) An overview of nanotechnology in food science: preparative methods, practical applications, and safety. Chemistry 2018:10, article ID 5427978. https://doi.org/10.1155/2018/5427978
19. Jagtiani E (2022) Advancements in nanotechnology for food science and industry. Food Front 3(1):56–82. https://doi.org/10.1002/fft2.104
20. Jiménez-López J, Llorent-Martínez EJ, Ortega-Barrales P, Ruiz-Medina A (2020) Graphene quantum dots-silver nanoparticles as a novel sensitive and selective luminescence probe for the detection of glyphosate in food samples. Talanta 207:120344. https://doi.org/10.1016/j.talanta.2019.120344
21. Khan RA, Khan NM, Nisar A (2015) Nanotechnology and food. Is nanotechnology for food applications safe?: A review. Nanosci Technol Open Access 2(3):1–7. https://doi.org/10.15226/2374-8141/2/3/00133

22. Koo YEL, Cao Y, Kopelman R, Koo SM, Brasuel M, Philbert MA (2004) Real-time measurements of dissolved oxygen inside live cells by organically modified silicate fluorescent nanosensors. Anal Chem 76(9):2498–2505. https://doi.org/10.1021/ac035493f
23. Kowalczyk JW (2013) Tracing emerging irreversibilities and development pathways in the field of nanosensor technologies. [Dissertation], Utrecht University
24. Moro G, De Wael K, Moretto LM (2019) Challenges in the electrochemical (bio)sensing of nonelectroactive food and environmental contaminants. Curr Opin Electrochem 16:57–65. https://doi.org/10.1016/j.coelec.2019.04.019
25. Osmólska E, Stoma M, Starek-Wójcicka A (2022) Application of biosensors, sensors, and tags in intelligent packaging used for food products—a review. Sensors 22(24):9956. https://doi.org/10.3390/s22249956
26. Patra JK, Shin HS, Paramithiotis S (2018) Editorial: application of nanotechnology in food science and food microbiology. Front Microbiol 9:714. https://doi.org/10.3389/fmicb.2018.00714
27. Patris S, Vandeput M, Kauffmann JM (2016) Antibodies as target for affinity biosensors. TrAC Trends Anal Chem 79:239–246. https://doi.org/10.1016/j.trac.2015.12.005
28. Qi X, Yan X, Zhao L, Huang Y, Wang S, Liang X (2020) A facile label-free electrochemical aptasensor constructed with nanotetrahedron and aptamer-triplex for sensitive detection of small molecule: saxitoxin. J Electroanal Chem 858:113805. https://doi.org/10.1016/j.jelechem.2019.113805
29. Ragaert P, Smits M, Van Poucke C (2013) Food fraud and the technology solutions available to tackle it. Trends Food Sci Technol 31(1):87–99
30. Ranjan S, Dasgupta N, Chakraborty AR, Melvin Samuel S, Ramalingam C, Shanker R, Kumar A (2014) Nanoscience and nanotechnologies in food industries: opportunities and research trends. J Nanopart Res 16(6):1–23. https://doi.org/10.1007/s11051-014-2464-5
31. Ronkainen NJ, Halsall HB, Heineman WR (2010) Electrochemical biosensors. Chem Soc Rev 39(5):1747–1763. https://doi.org/10.1039/b714449k
32. Sharma A, Ranjit R, Pratibha N, Kumar N, Kumar M, Giri BS (2023) Nanoparticles based nanosensors: principles and their applications in active packaging for food quality and safety detection. Biochem Eng J 193:108861. https://doi.org/10.1016/j.bej.2023.108861
33. Shawon ZBZ, Hoque MF, Chowdhury SR (2020) Nanosensors and nanobiosensors: agricultural and food technology aspects. In: Nanofabrication for smart nanosensor applications. Elsevier, pp 135–161
34. Sohrabi H, Salahshour Sani P, Zolfaghari R, Majidi MR, Yoon Y, Khataee A (2022) MOF-based mycotoxin nanosensors for food quality and safety assessment through electrochemical and optical methods. Molecules 27(21):7511. https://doi.org/10.3390/molecules27217511
35. Sun X, Zhang Y, Ma C, Yuan Q, Wang X, Wan H, Wang P (2021) A review of recent advances in flexible wearable sensors for wound detection based on optical and electrical sensing. Biosensors 12(1):10. https://doi.org/10.3390/bios12010010
36. Suthar TR, Gavane AB, Shah AY, Devkatte AN (2021) Nanotechnology, a revolutionary technique in the food industry: systematic review. Sustain Agric Food Environ Res 9(1):127–141. https://doi.org/10.7770/safer-V0N0-art2256
37. Vanegas DC, Gomes CL, Cavallaro ND, Giraldo-Escobar D, McLamore ES (2017) Emerging biorecognition and transduction schemes for rapid detection of pathogenic bacteria in food. Compr Rev Food Sci Food Saf 16(6):1188–1205. https://doi.org/10.1111/1541-4337.12294
38. World Health Organization (WHO) Food safety. https://www.who.int/health-topics/food-safety#tab=tab_1. Accessed 25 June 2023
39. Xu K, Huang J, Ye Z, Ying Y, Li Y (2009) Recent development of nanomaterials used in DNA biosensors. Sensors 9(7):5534–5557. https://doi.org/10.3390/s90705534

Development and Applications of Nanobiosensors for Precision Agriculture

Nizamul Haque Ansari, Shumaila Shahid, Mohd Shoeb Khan, Navaid Zafar Rizvi, Nafis Ahmad, and Mangey Ram Nagar

Abstract Nanobiosensors, leveraging many nanomaterials such as carbon nanotubes, quantum dots, nanowires, nanoparticles, and nanorods, are revolutionizing agriculture by enhancing the sensitivity and efficiency of monitoring systems. Nanobiosensors are extremely reliable and highly sensitive in biosensing approaches as compared to the conventional sensors due to the unique properties of nanomaterials such as good conductivity as well as physicochemical, optical, electrochemical, mechanical and magnetic properties. These sensors are very useful in agriculture and play a critical role in detecting plant pathogens, optimizing soil quality, understanding plant-soil-microbe interactions and enhancing the quality of agri-food products. Nanotechnology-based biosensors enable precise detection of pesticide residues, nutrient deficiencies, and environmental stress, promoting sustainable and precision agriculture by minimizing chemical inputs and improving crop management. Future advancements, including the integration of artificial intelligence and wireless monitoring, will enable real-time, remote sensing, offering predictive analytics for crop health and environmental parameters. These developments hold great promise for sustainable farming practices, enhancing food security, and ensuring environmental safety in response to global agricultural challenges.

N. H. Ansari
Department of Physical Sciences (Chemistry), Sant Baba Bhag Singh University, Jalandhar, Punjab, India

S. Shahid (✉)
Division of Plant Pathology, ICAR-Indian Agricultural Research Institute, New Delhi, India
e-mail: shumaila24amu@gmail.com; shumaila@iari.res.in

M. S. Khan · N. Z. Rizvi
Department of Applied Sciences and Humanities, Faculty of Engineering and Technology, Jamia Millia Islamia, New Delhi, India

N. Ahmad
Faculty of Dentistry, Jamia Millia Islamia, New Delhi, India

M. R. Nagar
Department of Electronics and Communication Engineering, Noida Institute of Engineering and Technology, Noida, Uttar Pradesh, India

I. Uddin (ed.), *Smart Nanosensors*, Smart Nanomaterials Technology,
https://doi.org/10.1007/978-981-96-3878-9_12

Keywords Nanobiosensors · Precision agriculture · Nanomaterials · Agri-food industry · Plant pathogen detection · Agricultural sustainability

1 Introduction

In recent decades, several issues such as the growing pressure from the world's population, the constantly changing climate, and increased competition for resources have posed serious threats and created an urgent need to ensure food guarantee for the growing population [1]. Present agricultural practices aimed at meeting food demands rely heavily on resource exploitation, advanced machinery, and excessive use of agrochemicals. These action have led to substantial degradation of soil, air, and water assets, significantly increasing pollution in agricultural environments and negatively impacting the health of both humans and animals. The scientific community is under increasing pressure to develop cutting-edge farming techniques and technologies to address the rising levels of environmental contamination. Similarly, farmers are facing growing demands to reduce their use of agrochemicals by adopting alternative farming methods, driven by increased global awareness of this issue [2].

Precision agriculture presents a viable alternative, as it minimizes agrochemical use and provides crop-specific, tailored treatments to enhance profitability. Its primary goal is to increase crop yields while reducing the need for pesticides, herbicides, and fertilizers. One example of precision agriculture is the use of biosensors. Biosensors, a unique type of bioelectronic device, have practical applications across various sectors, including biosecurity, environmental monitoring, agriculture, pharmaceuticals, and food and animal industries (Fig. 1). The term "biosensors" was coined by Clark and Lyons in 1962 following the development of an enzyme-based electrode system [3]. Biosensors are compact, highly specialized analytical devices that detect and measure chemical and biological data by combining an analyte with biorecognition materials and signaling transducers. These devices convert biological signals into electrical outputs, which are then amplified, processed, and presented using computer-aided readouts, enabling users to interpret the interactions between ligands and target materials effectively.

2 Use of Nanotechnology for the Development of Nanobiosensors

The study and control of matter at the nanoscale ranging from 1 to 100 nm reveals physical, chemical, and biological properties that can differ significantly from those at larger scales. For example, certain nanomaterials can be stronger, more flexible, or more efficient at conducting heat or electricity than other forms of the same material. Nanotechnology encompasses the creation, manufacturing, and utilization of various

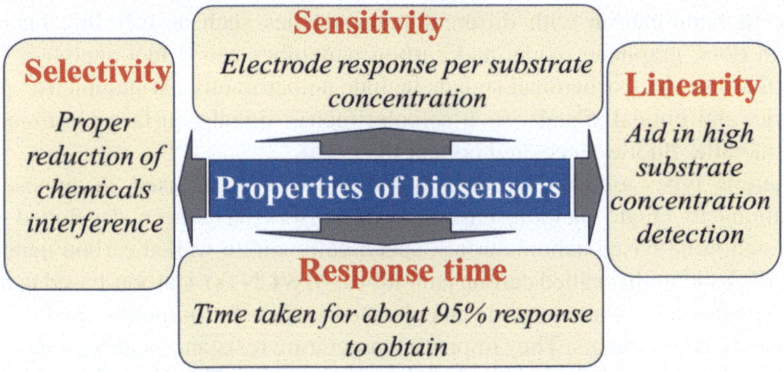

Fig. 1 A diagram showing characteristics and description of biosensors

structures, devices, and systems. It integrates various fields, including medicine, electronics, biomaterials, chemical engineering, and biotechnology, to synthesize materials that benefit humanity. Advancements in this rapidly evolving field have led to the introduction of nanoscale biosensors for the agricultural industry. Other function includes environmental monitoring, where nanotechnology is used to develop device for pollution removal. Biosensors, known for their ability to analyze even the smallest biological interactions with exceptional precision, adaptability, and sensitivity, have recently been reported [4].

A nanobiosensor is an analytical tool on the nanometer scale used to detect or measure biochemical material. It generally includes a sensing part, or "bio-receptor," that collaborate with the target analyte to produce an assessable physical signal. This signal is then converted by a transducer component, enabling the quantification of biological and biochemical signals through optical, electronic, thermal, or magnetic methods.

A nanobiosensor consists of a transducer that transform biological signals into digital form, a data writing unit for storing and transferring information, and a biological probe equipped with affinity-based materials. These materials include interactions such as antibody-antigen, enzyme–substrate, nucleic acid, and cell-based interactions [5], Fig. 2.

Examples of biological components include DNA, enzymes, antibodies, biomimetic substances, and aptamers. Synthetic receptors have replaced natural biological components to provide faster and more precise detection. Biological

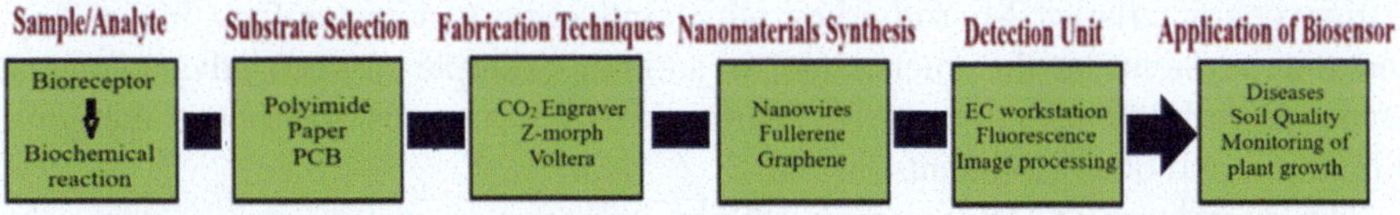

Fig. 2 Diagram showing nanobiosensors (from analyte to application)

probes, in combination with different nanoparticles such as metallic, magnetic, quantum dots, graphene oxide, and carbon nanotubes can detect analytes. Transducers that use electrochemical signals include amperometric, voltammetric, potentiometric, and optical signals, such as colorimetric signals, surface plasmon resonance, metallic fluorescence, and optical fibers [6].

Different types of cutting-edge materials have been used in nanosensing applications to create nanobiosensors. Nanosensors have been developed using carbon nanotube-based nanomaterials, specifically single-walled carbon nanotubes (SWCNTs) and multi-walled carbon nanotubes (MWCNTs). Carbon-based nanomaterials have been shown to provide an enhanced surface for immobilizing biological components in biosensors. They improve temperature resistance and conductivity by imparting desirable mechanical and physical properties. However, the main drawback is that these nanoparticles have low solubility in water. Other challenges include toxicity, the fabrication of nanobiosensors, the sustainability of nanostructures, and a lack of selectivity. Silicon oxides, titanium, and fullerenes are additional materials that are used. For encapsulation, polymers such as zeolites, chitosan, and polyacrylic acid can be employed. Due to their safer properties, certain metallic nanoparticles, often made of gold and silver, can be utilized in nanobiosensors.

The most common materials used in food nanobiosensors to detect pollutants and pathogens in food and water are gold- and silver-based nanoparticles. Nanobiosensors based on metallic components offer safer alternatives for detecting harmful pollutants in food products. Another widely used material for nanobiosensors is developed based on fluorescence measurement. Fluorescence-based nanobiosensors employ two sensing methodologies: one relies on quantum dot sensors with semiconducting properties, and the other uses fluorescence resonance energy transfer (FRET) and intermolecular charge transfer (ICT). These biosensors offer simpler alternatives for detection. Additional strategies, such as single fluorophore, dual fluorophore, and modular designs, have also been utilized in their development. However, a considerable disadvantage of fluorescence-based biosensors is their high cost, which can be mitigated by using the metal-enhanced fluorescence (MEF) strategy.

Zinc Oxide (ZnO) is another important metallic component used in nanobiosensor fabrication, particularly in light-emitting diodes (LEDs) for detecting free radicals in food packaging materials. High catalytic power, high isoelectric point, and strong adsorption capacity are among the notable characteristics of ZnO. Magnetoelastic materials, made from gold nanoparticles and magnets, represent a new advancement. Magnetic nanomaterials can reduce the signal-to-noise ratio by preventing non-specific binding of biological molecules and enabling easy separation. These materials are categorized into three groups metals, alloys, and oxides each with distinct characteristics. Microfluidics, which integrates nanotechnology, biosensing, and microsystems at the microscale, is another example of such advancements. These microfluidic techniques have been used to detect toxins, foodborne allergens, infections, and other pollutants.

The integration of DNA into microfluidic systems is one way to overcome the challenges of miniaturized fabrication, along with the use of electrokinetic and hydrodynamic flows. A different strategy is to amalgamate microfluidic approaches with cell morphology to enhance food safety. Biological constituents, such as bilayer lipid membranes, have been used to replace metals and other materials in the fabrication of nanobiosensors. Most lipid membrane-based nanobiosensors have been developed and employed for detecting metals, toxins, and microorganisms; however, technical challenges include reduced strength and sensitivity to damage outside of electrolyte solutions. Their stability can be improved by using glass, ZnO and graphene to strengthen the stability of membranes.

Agricultural waste materials offer a promising, cost-effective resource for creating nanobiosensors due to their availability in the environment. Nanocellulose fibers have been developed and can serve a variety of purposes. Rice husks, a byproduct of rice milling, can also be utilized in nanobiosensor production. When combined with silica, rice husks can enhance functionality for diverse applications. Such materials are often used in agricultural practices. Glass, ZnO, and graphene are materials that can enhance the stability of membranes. Since agricultural waste is abundant in the environment and offers a low-cost alternative, it can be used as an innovative material for nanobiosensors. Nanocellulose fibers have been developed and can be applied in various ways. Rice husks, a waste or by-product of rice milling, can also be utilized to create nanobiosensors. Silica extracted from rice husks can be repurposed for more beneficial uses. These materials have been widely used in agricultural practices and have proven effective for various applications. However, a significant challenge is the high cost associated with their design and processing. To address this, different types of nanomaterials, from expensive options to more affordable organic waste sources, have been integrated to design a robust nanobiosensor.

3 Different Materials Used in Production of Nanobiosensors

(a) **Carbon Nanotubes (CNTs)**: CNTs are excellent candidates for chemical and biological sensing applications due to their distinct structure and nanoscale size. Their characteristics include high electrical conductivity, strong chemical stability, luminescent properties, bigger surface area/volume ratios, and high aspect ratios. CNTs have been primarily used in electrochemical [7], photonic [8], and field-effect device [9] biosensors. Although CNTs offer several advantages such as enhanced mechanical and physical properties that improve conductivity and temperature resistance along with surface features that prevent the movement of biological components, they exhibit low solubility in aqueous environments.

(b) **Quantum Dots (QDs)**: It is possible to create biosensors using QDs. QDs are a type of novel luminescent nanomaterial, consisting of an inorganic core with organic molecules bonded to its surface. They possess exceptional optical and chemical properties, including wide excitation spectra, tunable particle sizes, constrained emission spectra, the emission of multiple fluorescence colors, superior signal brightness, and longer fluorescence lifetimes. Due to the large number of working groups on QDs, it is comfortable to create hybrid nanomaterials with excellent analytical performance. By hybridizing QDs with other nanomaterials, functionalized sensing systems enable the sensitive and selective detection of metal ions, biomarkers, and antibiotics. In agriculture, quantum dot (QD) biosensors can be used to assess plant health, identify plant diseases, and enhance agricultural yields. QD biosensors can detect and eradicate plant diseases caused by bacteria and fungi without harming the plant. However, QD nanobiosensor have high toxicity, which make them unsuitable for in vitro systems [10].

(c) **Nanowires (NWs)**: The electrical, mechanical, and chemical properties of C-nanotubes are exceptional. Nanotubes can be either metallic or semiconducting, depending on their precise diameter and the arrangement of their C atoms. An ideal nanotube exhibits low heat dissipation and ballistic (negligible scattering) electrical conduction. Because of this, a wire made from a nanotube, or nanowire, can conduct significantly more current than a regular metal wire of the same size. Many different materials, including silicon, germanium, carbon, and various conductive metals like Cu and Au, can be utilized to make nanowires. Nanowires have unique electrical, optical, and magnetic properties. For instance, the creation of an optical photonic switch with specialized cadmium sulfide nanowires, which enables photons to travel through the wire and function as binary signals (0 and 1), could significantly boost computer processing speeds. Numerous techniques, such as lithography, milling, thermal oxidation, suspension, electrochemical deposition, and vapor deposition, can be used to create nanowires.

The physical characteristics of nanowires are influenced by factors such as their shape, the diameter-dependent band gap, and the carrier density of the material's state. NWs are often referred to as one-dimensional (1-D) materials because they typically exhibit aspect ratios (length/width) of 10^3 or greater. Neither bulk materials nor their larger three-dimensional structures display the conductive properties that nanowires possess. Because they are laterally quantum-confined, electrons in nanowires occupy energy levels distinct from the conventional continuum of bands or energy levels seen in bulk materials. Semiconductor and super lattice nanowires can be used to create simple electronic devices, including logic gates, junction diodes, transistors, and field-effect transistors (FETs). NWs are also being developed for bio-electromechanical and biomedical sensor utilization. The use of nanowire sensors can help ensure food safety in agriculture [11].

(d) **Nanoparticles**: Materials with external dimensions ranging from 1 to 100 nm are called nanoparticles. The compact surface area provides greater accessibility for improved surface functionality within a given volume. Additionally, nanoparticles have a higher surface area/mass ratio compared to their bulk counterparts. This affects the energy of the surface atoms, which, in turn, influences the material properties of nanoparticles. The incorporation of nanoparticles in biosensor development has enhanced their sensitivity and performance, while also enabling the combination of several novel signal transduction technologies. Advances in instruments and techniques for fabricating, measuring, and imaging nanoscale objects have led to the development of sensors capable of interconnecting with acutely small molecules that require analysis. Nanoparticles, such as gold nanoparticles and magnetic nanoparticles, are actively being examined for their use in biosensors, which have become a new interdisciplinary frontier between biological detection and material science [12].

(e) **Nanorods (NDs)**: NDs are 1-D structures that provide a directed way for electrical transport and are utilized to control the bandgap by changing the radius of the rods and leveraging the quantum magnitude effect. The efficiency of quantum dot-conjugated polymer solar cells can be enhanced through the quantum confinement effect, which influences the length and width of nanorods, resulting in thinner devices for optimal absorption of incident light. Nanorods offer advantages over nanotubes and nanowires, as they can be synthesized from a broad range of materials, including both metallic and non-metallic substances like carbon, zinc oxide, and gold, using various flexible methods. The diversity of nanoprocessing techniques has expanded their applications across multiple fields. For instance, they are widely used in the fabrication of magnetic, electronic, optical, and micromechanical devices. C, Au, ZnO, and magnetic NDs are among the most common types. Additionally, NDs are employed in biosensors to enhance sensitivity, increase stability, and improve detection capabilities [13].

4 Advantages of Nanobiosensors in Agriculture

Compared to the conventional biosensors, nanomaterial-based biosensors currently demonstrate exciting potential. Nanobiosensors offer significant benefits, such as improved sensitivity and detection specificity, and hold great promise for use in various fields, including biodefense, agriculture, food quality control, and environmental and bioprocess management. Our focus here, however, is on the role of nanobiosensors in agriculture and agro-products. These sensors are highly advanced and enhanced with improved techniques to detect pesticide and herbicide residues, regulate fertilizer and herbicide applications, identify nutrient deficiencies, assess soil quality, and optimize soil health.

Agriculture is becoming more sustainable due to the use of nanomaterials and biosensors, which enable optimal and productive yields with fewer inputs. Researchers have developed a carbon nanotube-based nanobiosensor to monitor how plants respond to stress, damage, and infection. These nanobiosensors, which detect and report hydrogen peroxide signaling waves, can be implanted in plant leaves. Nanyang Nonyoung and Tematec Life Sciences Laboratory have also developed rapid-testing nanobiosensors for plant hormones.

A highly sensitive, self-driving, and culture-free DNA nanosensor has been developed by researchers from Taiwan and Japan to identify lethal superbugs. Deoxyribonucleic acid (DNA), which encodes the instructions required for the synthesis of genes and proteins, is a vital component of all living organisms, including plants. The development of DNA biosensors and microarrays has been facilitated by the relative simplicity of DNA nanofabrication and the ease of managing complementary base pairing. These technologies, also known as DNA chips, have demonstrated potential in agriculture and have been successfully applied in medical diagnostics.

DNA-based detection techniques play a crucial role in forensic investigations, identifying biological contaminants, and analyzing gene expression. Conventional DNA chip systems function by attaching specific DNA sequences to complementary probes, such as single-stranded DNA (ssDNA) or synthetic oligonucleotides, which are fixed onto a solid substrate or suspended in buffer solutions to facilitate hybridization with target DNA [14]. Masrie et al. [15] developed optical transducers along with LED and photodiode-based biosensors for monitoring nitrogen, phosphorus, and potassium levels in soil. Fiber-optic biosensors have also been suggested for tracking nutrient availability and water stress in plants. Additionally, a dual-probe heat pulse (DPHP) technique has been introduced as a low-power, cost-effective solution for real-time soil moisture monitoring. This sensor consists of two probes: one for temperature measurement and another that functions as a heater, powered by a 3.3 V AC supply. It was tested with both red and white clay soil samples to measure volumetric water content.

5 Applications of Nanobiosensors in Agriculture

Nanobiosensors have numerous applications in agriculture including plant pathogen detection, assessment of soil quality, precision farming etc. (Fig. 3). These may be of great value for effective monitoring of several parameters and thus assist as an efficient tool to improve agricultural productivity.

Nanobiosensors have immense potential in agriculture as well as food industry and its applications has been discussed in detail as below.

5.1 Nanobiosensors for Plant Pathogen Detection

Plant pathogens are significant factors that limit crop yields and severely impact agricultural productivity worldwide, creating challenges for food security [16]. While chemical treatments remain the primary method for controlling plant diseases, repeated applications can lead to reduced pathogen susceptibility and environmental contamination, disrupting soil microbiota. To promote sustainable agriculture and secure food production, the early diagnosis of plant pathogens is crucial. Although molecular techniques for pathogen detection are available, they are often

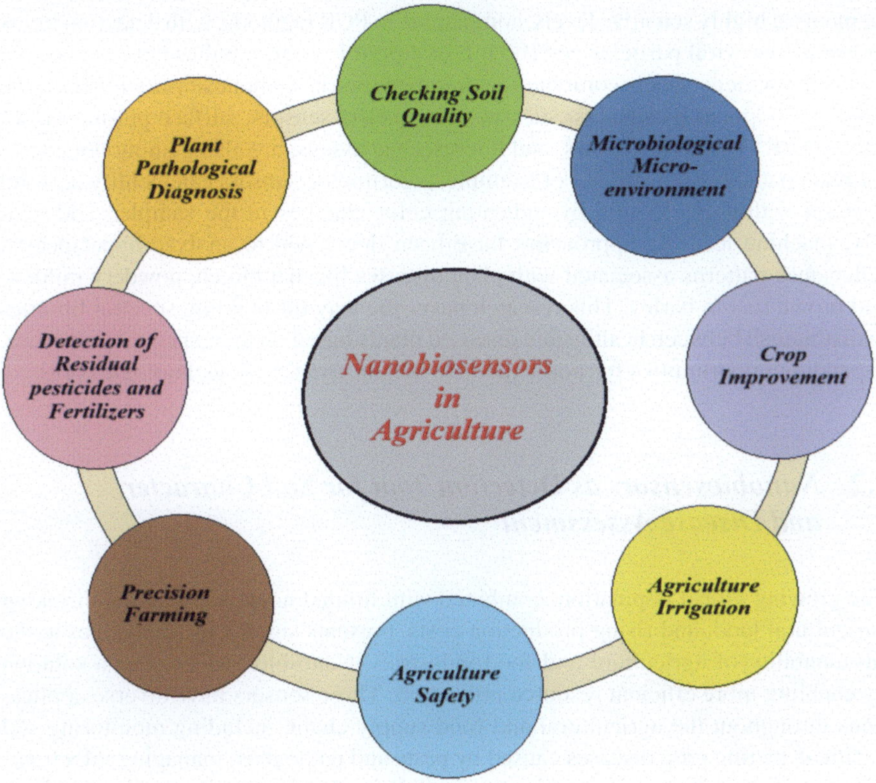

Fig. 3 Applications of nanobiosensors in agriculture

expensive, time-consuming, and require skilled operators, making them impractical for on-site analysis. Nanobiosensors, including microneedle patches, nanopore sequencing, quantum dots, nano-barcoding systems, and miRNA-based diagnostic kits, have emerged as promising tools to enhance the precision, speed, and efficiency of pathogen detection [17]. These sensors have various agricultural applications, such as identifying plant viruses, monitoring soil nutrients, and detecting crop infections, which contribute to improved crop management, environmental safety, and food quality. Electrochemical nanobiosensors can be categorized based on their biorecognition components into biocatalytic and biocomplexing types. Enzymes are frequently used in biosensors as labels for pathogen detection in a secondary binding process, facilitating the interaction between analytes and macromolecules.

The performance of these biosensors rely on factors such as assay design, material properties, sample composition, and the specific detection method. Functionalized electrodes are useful in electrochemical techniques like amperometric, conductometric, potentiometric, and impedimetric methods, with signals being measured either continuously or intermittently to classify the detection technique. Recent advancements have shown that electrochemical sensors can detect virus-associated elements at highly sensitive levels, comparable to PCR methods, with detection limits as low as two viral particles per 100 mL or 4 pg/mL. Some studies have introduced reduced graphene sheets combined with impedimetric immunosensors for bacterial detection. Optical biosensors, such as colorimetric sensors, surface plasmon resonance (SPR)-based biosensors, and fluorescence assays, are also gaining popularity for plant pathogen detection. For example, colorimetric sensors can identify harmful bacteria within 10–15 min by indicating color changes in the sample. Additionally, machine learning approaches have been developed to analyze hyperspectral reflectance patterns associated with plant diseases like net blotch, powdery mildew, and brown rust in barley. This research paves the way for building spectral libraries to distinguish between healthy and diseased plants based on their spectral properties, expanding opportunities for non-experts to benefit from these technologies [18].

5.2 Nanobiosensors as Detection Tool for Soil Character and Disease Assessment

The growing global population, combined with limited natural resources, shrinking agricultural land, and rising production costs, presents significant challenges to the sustainability of agriculture and food industries. Nano-biosensors offer a solution by enabling more efficient resource utilization. These sensors have diverse applications throughout the agricultural and food supply chain, including monitoring soil health, detecting crop diseases caused by pests and pathogens, managing infections, and ensuring product quality during storage. Nano-biosensors play a crucial role in assessing soil properties, such as fertility, pH levels, moisture content, and mineral concentrations, while also detecting nutrient deficiencies and pest presence. They help identify soil-borne diseases by measuring differences in oxygen consumption between beneficial and harmful microbes in the soil. This process involves immersing two separate sensors, one containing beneficial microbes and the other containing harmful ones into a soil suspension in a buffer solution. By comparing the oxygen consumption of the two microbial groups, it becomes possible to determine which type of microbe dominates the soil environment. Additionally, this method can predict the likelihood of disease outbreaks, providing early warnings and enabling proactive measures. Nano-biosensors thus offer an innovative, semi-quantitative approach to soil diagnostics, helping farmers optimize crop management and prevent potential losses [19–22].

5.3 Nanobiosensor to Study Plant-Soil-Microbe Interaction

Nano-inspired biosensors play a crucial role in enhancing standard of life by supporting different clinical, environmental, and quality-control applications globally. These biosensors are used for rapid diagnosis of plant infections such as fungal, viral, and bacterial pathogens as well as abiotic stress, metabolic compounds, phytohormones, microRNAs, genetically modified (GM) plants, and transcriptional activity. Advances in nano-technology have enabled the customization or "nano-tuning" of these sensors, leveraging the unique properties of nanomaterials to overcome challenges posed by conventional methods. As a result, these next-generation nanobiosensors are capable of detecting extremely low concentrations of target molecules for both in vitro and in vivo applications.

The interactions between plants, soil, and microbes are highly complex, ranging from symbiosis to mutualism, depending on the type of microbial association with plants. A deeper knowledge of these interactions is essential for meeting the rising global demand for food, as agriculture must undergo significant transformations to remain sustainable. Reducing the use of pesticides and fertilizers is essential to protect the health of humans, animals, and plants. Plants benefit from both above-ground and below-ground interactions, which contribute to their development. Microbial communities living on plant surfaces and in the root zone, known as rhizospheric microbes, play a particularly critical role in supporting plant growth. Rhizospheric associations are dynamic and have a significant influence on plant nutrition and development.

Beneficial microorganisms such as plant growth-promoting rhizobacteria (PGPR) and other plant growth-promoting microbes (PGPMs) assist in nutrient acquisition, produce antioxidants, enhance stress resistance, and help plants adapt to environmental challenges. These microbes not only improve plant health but also contribute to maintaining soil quality. During colonization of the rhizosphere, microbial communities influence root system development and play a key role in plant growth. Monitoring microbial diversity and interactions in the soil has traditionally relied on conventional methods, but advanced sensing technologies are now making these processes more efficient [23–25].

5.4 Nanobiosensors for Agri-food Industry

The population of the world is estimated to reach 8.5 billion by the year 2030 which will put extra pressure on natural resources and food production units [26]. Besides this, another problem that is raising concern globally is the contamination of food and food products with several microorganisms during storage as well as processing. Therefore, there is an urgent and crucial requirement to develop innovative sustainable techniques/approaches that can be utilized by agri-food industries to make advancement in the agricultural and food products in the form of improved quality.

In this regard, nanotechnology has a huge ability to suggest robust and keener biosensors which can quickly recognize the contaminants, pathogens, additives and toxins present in the food. Hence, nanobiosensors with advanced augmented nanomaterials are gaining attention to check and maintain the quality standards of food materials. Uses of nanobiosensors in the agri-food industry is expanding very fast and these food industries use nanobiosensors to ensure food safety during production as well as packaging of the food/food products [27, 28] and detect pathogens and/or allergens (Fig. 4).

Besides safety, several challenges also exist in case of food industry, such as food security, availability as well as utilization which is provided by nanobiosensors giving it advantage over the conventional sensors [29, 30]. Nanobiosensors can be applied to monitor the quality as well as safety of the food products through detection of contaminants thus ensuring that only superior quality product reaches the market. Nanobiosensor-based detection is becoming progressively important in food safety monitoring (Fig. 5).

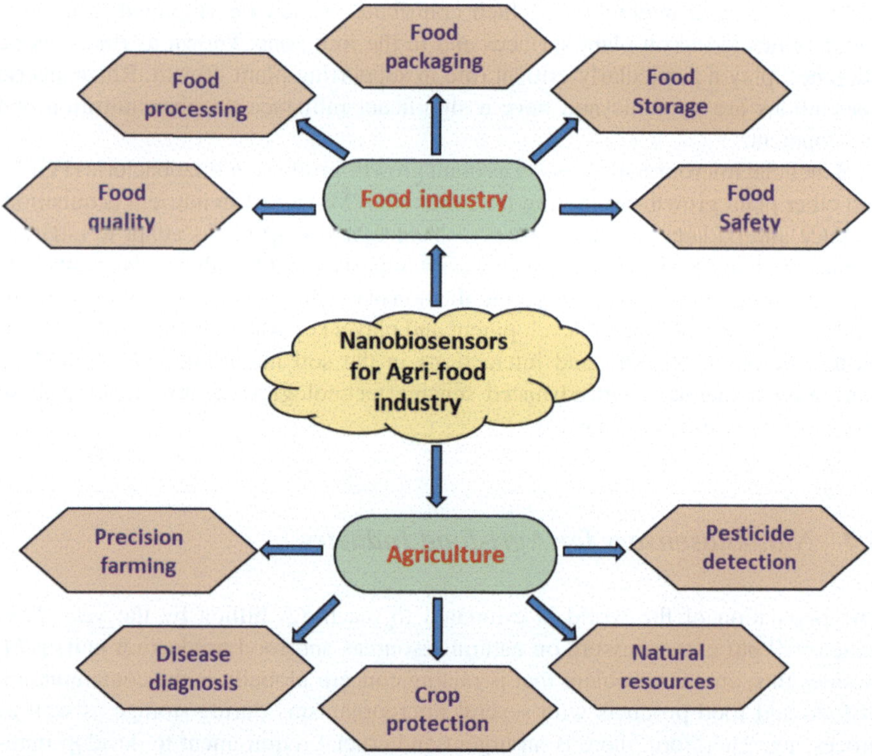

Fig. 4 Nanobiosensors for agri-food industry

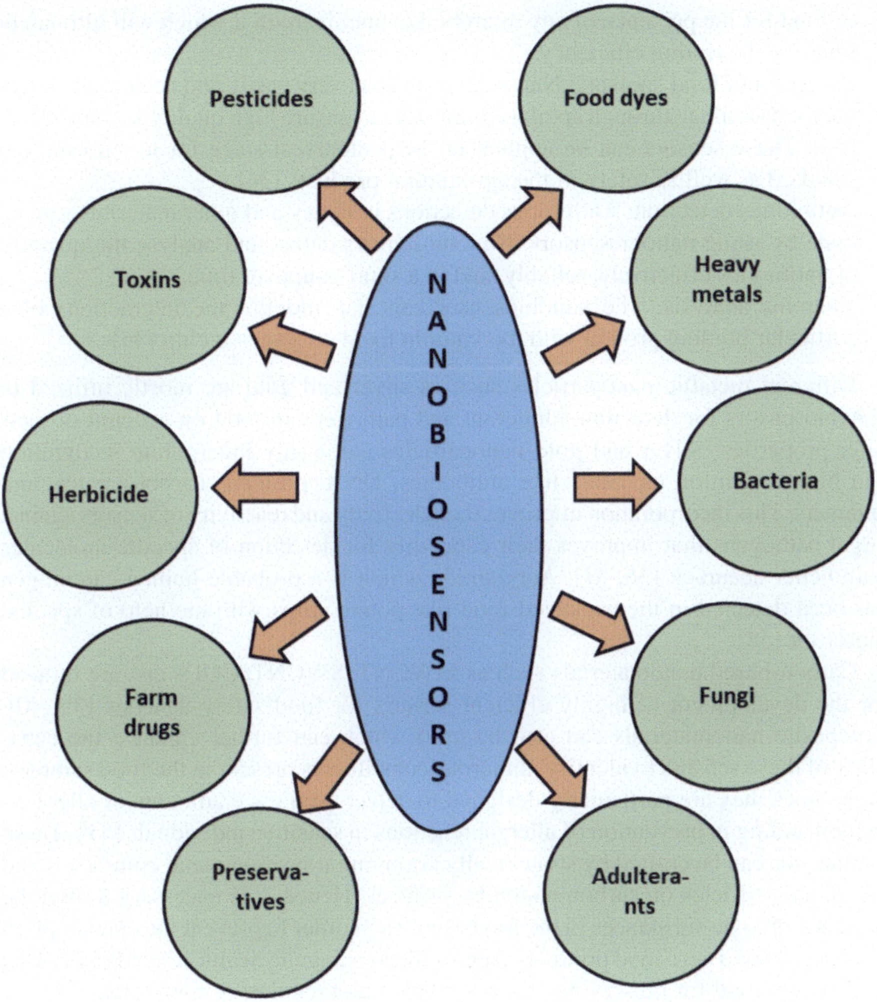

Fig. 5 Role of nanobiosensors in food safety monitoring

Nanobiosensors has numerous uses in enhancing the quality of food and food products through detection of various contaminants and pathogens which is discussed as below-

- Detection of microbial contamination: Microbial contaminants can be detected with metallic components based nanobiosensors [31]. Nanosensors are designed to concurrently detect and quantify the existence of several pathogenic microbes such as bacteria (e.g. *Escherichia coli*, *Salmonella*) as well as some viruses in food products [32]. This multiplex ability allows the complete screening of samples of food for the presence of any microbial contaminant in it which will ultimately improve the testing efficiency.

- Detection of food spoilage: Nanobiosensors can very easily and accurately detect the food spoilage through spoilage indicators to ensure high quality food products [33]. These sensors can be applied in the postharvest stage for monitoring the standard as well as safety of the agricultural products [34].
- Antibiotics detection: Antibiotics detections in honey and other materials, can be done by using nanobiosensors. Nanobiosensors detect and analyse the quantity of antibiotics effectively, reliably, and in a short course of time.
- Vitamins analysis: The nanobiosensor can also monitor the interactions of a particular binding protein with the vitamin fixed on a sensor chip [35].

Different metallic nanoparticles such as silver and gold are mostly utilized in nanobiosensors for detecting adulterant and pathogens in food on account of their safer properties. Silver and gold nanoparticles can easily incorporate recognition and bio-recognition elements like antibodies, DNA, proteins, carbohydrates, and aptamers. This incorporation improves the selectivity and reactivity of sensors against target pathogens that improves their capability for detection of specific molecules with better accuracy [36, 37]. Acrylamide which is a probable human carcinogen has been detected in the packaged food like potato chips with the help of specific biosensor [38].

Carbon-based nanomaterials such as MWCNT, SWCNT, CQDs etc. are utilized for the development of highly efficient sensors for food safety analysis [39, 40]. Composite nanomaterials can also be used which can further enhance the capability of these sensors to identify numerous constituents present in the food samples. Sometimes they are particularly designed to detect even trace amounts of allergens in food, aiding in prevention of allergic reactions in sensitive individuals [41]. These composites can be crafted by strategically arranging transition metal complexes and metal nanoparticles on carbon nanotube surfaces. Hence, it is necessary to find the presence of toxic substances in the food samples in order to prevent the consumption of contaminated agri-food products thereby, decreasing the health-related risks. This is also important for preserving consumer belief and regulatory compliance.

6 Conclusion

Nanobiosensors have potential to contribute to smart and precision agriculture. The integration of nanobiosensors with the farm systems holds great promise for improving the farming practices. The application of nanobiosensors in the present modern agri-food industries is expanding to increase production through utilization of more natural resources as well as plant health monitoring, that supports the sustainability of both agricultural and food sectors. Nanobiosensors can be used for detection of pathogenic organisms, detection of chemicals and adulterants, disease management, evaluation of moisture, pH of soil, natural resources, soil assessment etc. Nowadays, various nanomaterials are being used for the development of nanobiosensors resulting into portable broad-spectrum nanobiosensors and more commercialisation of these nanobiosensors may result based on bioassays. As new technologies evolve, nanobiosensors are poised to play a crucial role in promoting sustainable agriculture by reducing chemical inputs, improving environmental safety, and ensuring food quality, ultimately contributing to a resilient and efficient global food system.

7 Future Perspectives

The future of nanobiosensors in agriculture holds immense potential for transforming crop management and sustainable farming practices. As agricultural systems face increasing challenges from climate change, pests, soil degradation, and food security demands, nanobiosensors are expected to become integral tools for precision agriculture. These sensors will enable real-time monitoring of soil conditions, plant health, and environmental parameters, allowing farmers to make data-driven decisions to optimize crop yield and resource efficiency. Advancements in wireless nanobiosensors will further facilitate remote sensing, enabling continuous monitoring with minimal human intervention. Future innovations may also focus on developing multifunctional nanobiosensors capable of detecting multiple parameters such as nutrient levels, pathogen presence, and abiotic stress within a single platform. Additionally, the combination of artificial intelligence (AI) and machine learning with nanobiosensors will enhance predictive analytics, helping to prevent crop diseases and identify emerging threats early. In the coming years, nanobiosensors can be potentially associated with the GPS system which will enable the farmers to make better decisions on pest control, fertilization, irrigation, as well as harvesting which will help in optimizing natural resource utilization enhancing the agricultural productivity.

Acknowledgements The authors would like to thank to the Division of Plant Pathology, Indian Agricultural Research Institute, New Delhi-110012, India and Department of Applied sciences and Humanities, Faculty of Engineering and Technology, Jamia Millia Islamia, New Delhi, India.

Declarations

Conflict of Interest The authors declare that they have no conflict of interest.

References

1. Shahid S, Sharma BB, Khan AA, Ansari NH (2023) Int J Trop Agri 41:121–127
2. Shahid S, Khan MR (2019) Indian Phytopathol 72(1):89–98

3. Clark LC, Lyons C (1962) Electrode systems for continuous monitoring in cardiovascular surgery. Ann N Y Acad Sci 102:29–45
4. Touhami A (2014) Biosensors and nanobiosensors: design and applications. Nanomed 15:374–403
5. Verma ML, Rani V (2021) Biosensors for toxic metals, polychlorinated biphenyls, biological oxygen demand, endocrine disruptors, hormones, dioxin, phenolic and organophosphorus compounds: a review. Environ Chem Lett 19:1657–1666
6. Shawon, ZBZ, Hoque ME, Chowdhury SR (2020) Nanosensors and nanobiosensors: agricultural and food technology aspects. In: Nanofabrication for smart nanosensor applications. Elsevier, pp 135–161
7. Ahammad AJS, Lee JJ, Rahman MA (2009) Electrochemical sensors based on carbon nanotubes. Sensors 9:2289–2319
8. Chen Z, Zhang X, Yang R, Zhu Z, Chen Y, Tan W (2011) Single-walled carbon nanotubes as optical materials for biosensing. Nanoscale 3:1949–1956
9. Park RS, Hills G, Sohn J, Mitra S, Shulaker MM, Wong HSP (2017) Hysteresis-free carbon nanotube field-effect transistors. ACS Nano 11:4785–4791
10. Guan J, Sagar LK, Li R, Wang D, Bappi G, Wang W, Watkins N, Bourgeois MR, Levina L, Fan F (2020) Quantum dot plasmon lasing with controlled polarization patterns. ACS Nano 14:3426–3433
11. Jiang X, Herricks T, Xia Y (2002) CuO nanowires can be synthesized by heating copper substrates in air. Nano Lett 2(12):1333–1338
12. Sonker RK, Yadav BC, Gupta V, Tomar M (2020) Synthesis of CdS nanoparticle by sol-gel method as low temperature NO_2 sensor. Mater Chem Phys 239:121975
13. Krahne R, Morello G, Figuerola A, George C, Deka S, Manna L (2011) Physical properties of elongated inorganic nanoparticles. Phys Rep 501(3–5):75–221
14. Hahn S, Mergenthaler S, Zimmermann B, Holzgreve W (2005) Nucleic acid based biosensors: the desires of the user. Bioelectrochemistry 67(2):151–154
15. Masrie M, Rosman MSA, Sam R, Janin Z (2017) Detection of nitrogen, phosphorus and potassium (NPK) nutrients of soil using optical transducer. In: 2017 IEEE 4th international conference on smart instrumentation, measurement and application (ICSIMA). IEEE, pp 1–4
16. Shahid S, Khan MR (2016) Plant Pathol J 15:27–39
17. El-Khawaga MA, Mukhtar S, Shahid S (2024) Sustainable nanomaterials as promising antibacterial agents. In: Uddin I, Ahmad I (eds) Sustainable nanomaterials: synthesis and environmental applications. Springer Nature Singapore, pp 203–225
18. Dyussembayev K, Sambasivam P, Bar I, Brownlie JC, Shiddiky MJA, Ford R (2021) Biosensor technologies for early detection and quantification of plant pathogens. Front Chem 9:2021
19. Miguel-Rojas C, Pérez-de-Luque A (2023) Nanobiosensors and nanoformulations in agriculture: new advances and challenges for sustainable agriculture. Emerg Top Life Sci 7(2):229–238
20. Jabran M, Ali MA, Muzammil S, Zahoor A, Ali F, Hussain S, Muhae-Ud-Din G, Ijaz M, Gao L (2024) Exploring the potential of nanomaterials (NMs) as diagnostic tools and disease resistance for crop pathogens. Chem Biol Technol Agric 11:75
21. Mahmood Q, Shaheen S, Azeem M (2024) Nanobiosensors: application in healthcare, environmental monitoring and food safety. Asian J Agric Biol 1:2023157
22. Darwish MA, Abd-Elaziem W, Elsheikh A, Zayed AA (2024) Advancements in nanomaterials for nanosensors: a comprehensive review. Nanoscale Adv 6:4015–4046
23. Khan ST, Adil SF, Shaik MR, Alkhathlan HZ, Khan M, Khan M (2022) Engineered nanomaterials in soil: their impact on soil microbiome and plant health. Plants 11:109
24. Renella G, Giagnoni L (2016) Light dazzles from the black box: whole-cell biosensors are ready to inform on fundamental soil biological processes. Chem Biol Technol Agric 3:8
25. Enespa CP (2022) Tool and techniques study to plant microbiome current understanding and future needs: an overview. Commun Integr Biol 15:499
26. Thakur M, Wang B, Verma ML (2022) Development and applications of nanobiosensors for sustainable agricultural and food industries: recent developments, challenges and perspectives. Environ Technol Innov 26:102371

27. Sridhar A, Ponnuchamy M, Kumar PS, Kapoor A (2021) Food preservation techniques and nanotechnology for increased shelf life of fruits, vegetables, beverages and spices: a review. Environ Chem Lett 19:1715–1735
28. Shahid S, Khan MS, Kumar A, Rahman S, Arshad M, Kaushik P, Saini P, El-Khawaga MA (2024) Role of nanomaterials in sustainable agriculture. In: Uddin U, Ahmad I (eds) Sustainable nanomaterials: synthesis and environmental applications. Springer Nature Singapore, pp 227–248
29. Calicioglu O, Flammini A, Bracco S, Bellu L, Sims R (2019) The future challenges of food and agriculture. An integrated analysis of trends and solutions. Sustainability 11:222
30. Dasgupta N, Ranjan S, Ramalingam C (2017) Applications of nanotechnology in agriculture and water quality management. Environ Chem Lett 15:591–605
31. Oluwaseun AC, Phazang P, Sarin N B (2018) Biosensors: a fast growing technology for pathogen detection in agriculture and food sector. In: Rinken T, Kivirand K (eds) Biosensing technologies for the detection of pathogens—a prospective way for rapid analysis. Intech Open. https://doi.org/10.5772/intechopen.74668
32. Aladhadh M (2023) A review of modern methods for the detection of foodborne pathogens. Microorganisms 11(15):111
33. Mohammad ZH, Ahmad F, Ibrahim SA, Zaidi S (2022) Application of nanotechnology in different aspects of the food industry. Discover Food 2(1):12
34. Kumar P, Chugh P, Ali SS, Chawla W, Sushmita S, Kumar R, Raval AV, Shamim S, Bhatia A, Kumar R (2024) Trends of nanobiosensors in modern agriculture systems. Appl Biochem Biotechnol. https://doi.org/10.1007/s12010-024-05039-6
35. Rai V, Acharya S, Dey N (2012) Implications of nanobiosensors in agriculture. J Biomater Nanobiotechnol 3:315–324
36. Malekzad H, Zangabad PS, Mirshekari H, Karimi M, Hamblin MR (2017) Noble metal nanoparticles in biosensors: recent studies and applications. Nanotechnol Rev 6(3):301–329
37. Hashem A, Hossain MAM, Marlinda AR, Mamun MA, Simarani K, Johan MR (2021) Nanomaterials based electrochemical nucleic acid biosensors for environmental monitoring: a review. Appl Surf Sci Adv 4:100064
38. Alvarez-Morezuelas A, Ortiz-Barredo A, Barandalla L, Ritter E, de Galarreta JIR (2021) Estimation of glucose content in raw potatoes with a biosensor as an indicator of acrylamide level in processed potatoes. Potato Res 64(4):601–609
39. Pan M, Yin Z, Liu K, Du X, Liu H, Wang S (2019) Carbon-based nanomaterials in sensors for food safety. Nanomaterials 9(9):1330
40. Tian J, An M, Zhao X, Wang Y, Hasan M (2023) Advances in fluorescent sensing carbon dots: an account of food analysis. ACS Omega 8(10):9031–9039
41. Fritea L, Banica F, Costea TO, Moldovan L, Dobjanschi L, Muresan M, Cavalu S (2021) Metal nanoparticles and carbon-based nanomaterials for improved performances of electrochemical (bio)sensors with biomedical applications. Materials 14(21):6319

Smart Materials as Nanosensors in Solar Cell Perspectives

Md Nazrul Islam and Arijit Bag

Abstract The chapter explores the integration of smart materials and nanosensors into solar cell technologies to enhance their efficiency, durability, and functionality. Smart materials, which can respond to external stimuli like temperature, light, and pressure, are used to optimize solar energy capture by dynamically adjusting to environmental changes. The chapter provides an overview of different generations of solar cells, focusing on first-generation silicon-based cells, second-generation thin-film cells, and third-generation technologies such as perovskite and quantum dot cells. Each generation offers improvements in cost, efficiency, and application versatility. The integration of nanosensors into solar cells enables real-time performance monitoring, defect detection, and environmental adaptation. Key smart materials like Shape Memory Alloys (SMAs), Piezoelectric Materials, and Thermochromic Materials are discussed in detail for their applications in solar panels, such as self-cleaning mechanisms, dynamic solar tracking, and temperature regulation. The use of these materials can significantly enhance solar cell performance by improving light absorption, maintaining optimal operating conditions, and enabling adaptive systems. The chapter also outlines challenges, including cost, durability, and material compatibility, but emphasizes the potential of these advanced materials to revolutionize the solar energy industry. The future directions and challenges of incorporating smart materials and nanosensors in solar technologies are discussed, highlighting their role in making solar energy more sustainable and efficient.

Keywords Solar cells · Nanosensors · Smart materials · Piezoelectric materials · Thermochromic materials

M. N. Islam
Chemistry Department, Jawahar Lal Nehru College, Chakradharpur, West Singhbhum, Jharkhand, India

A. Bag (✉)
Department of Applied Chemistry, Maulana Abul Kalam Azad University of Technology West Bengal, Simhat, Nadia, Haringhata, West Bengal, India
e-mail: bagarijit@gmail.com

I. Uddin (ed.), *Smart Nanosensors*, Smart Nanomaterials Technology,
https://doi.org/10.1007/978-981-96-3878-9_13

1 Introduction

The escalating global demand for sustainable and efficient energy sources has led to a significant surge in research and development in advanced solar cell technologies [1–9]. With the world's energy consumption continuously rising and the pressing need to mitigate climate change, renewable energy sources like solar power have become increasingly crucial. Being abundant and renewable, solar energy offers a viable solution to the energy crisis. However, enhancing the efficiency and reliability of solar cells remains a critical challenge. The present trends of research works are intensively motivated toward the innovation of metal-free and green technology for solar energy harvesting [10–16].

In this context, the advent of smart materials and nanosensors [17–25] marks a pivotal role in the invention of renewable energy harvester. Smart materials, defined by their ability to respond dynamically to external stimuli such as temperature, light, pressure, and electric fields, offer remarkable advantages in various applications. When smart sensors are integrated into solar cells, these components can significantly improve the performance of the solar cells by adapting to environmental changes, optimizing energy capture, and increasing overall efficiency [26–33].

2 Different Genres of Solar Cells

Solar cells work based on the photovoltaic effect and are continuously under research for their modification to enhance their efficiency, applicability, stability, and other aspects. These modifications include advancements in materials, design, and manufacturing processes, aiming to make solar cells more cost-effective and efficient in converting sunlight to electricity. Solar cells can be classified into three distinct generations based on their working principles, the type of materials employed, and the time of their invention.

2.1 First Generation Solar Cells

The first generation of the solar cell [34–39] was primarily silicon-based cells which were three different types; monocrystalline [40–42], polycrystalline silicon [43–45] and amorphous silicon cells [46–49]. These are the most established and widely used solar cells developed in the mid-twentieth century. In 1954, the first silicon-based solar cell was manufactured in the Bell Laboratories [50]. At its first invention, the efficiency was only 6%. However, this was the beginning of the solar energy era. This technology relies on the basic p–n junction principle to create an electric field, which separates and collects charge carriers generated by sunlight.

The earliest technology of solar cells, particularly monocrystalline silicon solar cells, remains crucial and valuable. These cells are manufactured from thin silicon wafers using the Czochralski process, which covers about 30% of the PV market. Monocrystalline solar cells, introduced commercially by Sharp Company in 1963, have maximum efficiencies of up to 26%, making them highly efficient for generating electricity from a specific area of panels. Despite their higher cost compared to polycrystalline silicon solar cells, which have a maximum efficiency of about 21%, monocrystalline cells are preferred for their high efficiency.

Silicon solar panels typically contain two layers: a p-type positive layer and an n-type negative layer. Pure crystalline silicon is a poor conductor of electricity as its electrons are not labile. By doping silicon with atoms such as phosphorous, which has an extra electron, the silicon becomes n-type and a better conductor. This process is known as doping, and it significantly improves the conductive properties of silicon. On the other hand, doping silicon with boron, which has fewer electrons, creates p-type silicon with free openings commonly known as holes that carry a positive charge.

2.1.1 Classification Based on the Material Used

First-generation solar cells are primarily silicon-based and are classified into three main types based on the form of silicon used: monocrystalline silicon (Mono-Si), polycrystalline silicon (Poly-Si), and amorphous silicon (a-Si). Each type has distinct characteristics, efficiencies, and applications.

Monocrystalline Silicon Solar Cells (Mono-Si)

Monocrystalline silicon is a material made from a single, continuous crystal structure, typically produced using the Czochralski process, where a seed crystal is drawn from molten silicon to form a cylindrical ingot, which is then sliced into wafers. Monocrystalline silicon cells are known for their high efficiency, ranging from 15 to 22%, due to the uniform crystal structure that enhances electron flow and reduces energy loss. They are also highly durable, with an operational lifespan often exceeding 25 years. Visually, these cells are typically dark black with a distinctive rounded-edge appearance, a result of the cylindrical shape of the silicon ingot. Because of their efficiency and longevity, monocrystalline silicon cells are ideal for residential and commercial installations where space is limited, and maximizing energy output is crucial.

Polycrystalline Silicon Solar Cells (Poly-Si)

Polycrystalline silicon, also known as multicrystalline silicon, is produced by melting silicon fragments together to form an ingot, which is then sliced into wafers, resulting in multiple small silicon crystals. Due to this structure, polycrystalline cells have moderate efficiency, typically ranging from 13 to 16%, as the multiple crystal boundaries lead to increased electron scattering, reducing energy efficiency compared to monocrystalline cells. However, they are more cost-effective because of the simpler manufacturing process and lower silicon purity requirements. Polycrystalline cells have a distinctive bluish hue and a grainy texture, with square edges reflecting the shape of the ingot. They are commonly used in large-scale solar power installations and residential applications where cost considerations outweigh the need for maximum efficiency. While less efficient, these cells offer a budget-friendly option for renewable energy projects, making them a popular choice in regions focused on affordability in solar energy deployment.

Amorphous Silicon Solar Cells (a-Si)

Amorphous silicon is a non-crystalline form of silicon created by depositing a thin layer of silicon onto substrates like glass, plastic, or metal. Due to its disordered atomic structure, amorphous silicon cells have relatively low efficiency, typically around 6–8%, as higher electron recombination rates result in energy loss. However, these cells offer significant advantages in terms of flexibility and weight, as they can be made lightweight and adaptable to various surfaces. This flexibility, combined with a lower production cost due to reduced silicon material usage, makes amorphous silicon cells a more cost-effective option for certain applications. They are commonly used in scenarios where efficiency is less critical, such as portable solar chargers, thin-film solar panels for rooftops, and building-integrated photovoltaics (BIPV), where flexibility, lightweight construction, and affordability are key factors in product design

2.1.2 Characteristics of First-Generation Solar Cell

Monocrystalline silicon cells are known for their high efficiency, often exceeding 20%, making them some of the most efficient solar cells available. Their uniform crystal structure allows for better electron flow, resulting in less energy loss and higher power output. In addition to their performance, monocrystalline cells are highly durable, with a long operational lifespan and proven reliability, often lasting more than 25 years with minimal degradation. However, this high efficiency comes at a cost, as the manufacturing process requires highly pure silicon and is energy-intensive, making monocrystalline cells more expensive compared to other types of solar cells. Despite the higher cost, they are widely used in residential, commercial, and industrial solar power systems, particularly in situations where space is limited and maximizing energy output is essential. Their combination of high performance and long-term reliability makes them a preferred choice for many solar installations.

2.2 *Second Generation Solar Cells*

Second-generation solar cells [51–54], often called thin-film solar cells, represent a significant advancement in photovoltaic technology aimed at reducing manufacturing costs and enhancing application versatility. Unlike traditional solar cells, which are known as first-generation silicon wafer-based solar cells, thin-film solar cells utilize semiconductor materials that are only a few micrometers thick. These materials include amorphous silicon (a-Si) [55, 56], cadmium telluride (CdTe) [57–59], and copper indium gallium selenide (CIGS) [53, 60–62].

The adoption of these materials offers several advantages. First, it significantly reduces production costs compared to the expensive silicon wafers used in first-generation cells. Thin-film solar cells are also more flexible and lightweight, making them suitable for a wider range of applications, such as integration into building materials and portable devices. This flexibility opens up new possibilities for innovative applications, including solar windows [63], curved surfaces, and even wearable solar technology [64–66].

In addition to cost benefits and versatility, the manufacturing processes for thin-film solar cells generally consume less energy and produce fewer emissions, thereby contributing to their environmental benefits. This aspect is crucial for making solar energy more sustainable and eco-friendly.

Despite generally lower efficiency rates compared to first-generation cells, ongoing research and developments are continuously enhancing the performance of thin-film solar cells. Innovations in material science and fabrication techniques are leading to improvements in their energy conversion efficiency and overall durability. These advancements make second-generation solar cells a promising and sustainable solution for expanding solar energy adoption on a global scale.

2.2.1 Classification Based on the Material Used

Second-generation solar cells are classified based on the type of material used in their construction. The primary categories include:

Amorphous Silicon (a-Si): These solar cells utilize non-crystalline silicon, which has a disordered structure compared to the highly ordered crystalline silicon used in first-generation cells. Amorphous silicon cells are cheaper to produce and use significantly less silicon, making them more cost-effective. They are flexible and lightweight, suitable for a variety of applications, including portable electronics and building-integrated photovoltaics (BIPV). However, they typically have lower efficiency rates, generally around 6–8%.

Cadmium Telluride (CdTe): CdTe is a semiconductor compound that is highly efficient at converting sunlight into electricity. CdTe solar cells are known for their cost-effectiveness and relatively high efficiency, often ranging from 9–11%. They are the second most common type of solar cell after silicon-based cells. The manufacturing process for CdTe cells is simpler and less expensive than for crystalline silicon cells, and they perform well under a variety of lighting conditions. However, concerns about the toxicity and availability of cadmium limit their widespread adoption.

Copper Indium Gallium Selenide (CIGS): CIGS solar cells use a compound of copper, indium, gallium, and selenium. They are known for their high efficiency, often achieving rates between 12–14%, and their potential for flexibility. CIGS cells can be deposited on various substrates, including glass, plastic, and metal, making them versatile for different applications. The technology allows for both rigid and flexible panels, suitable for residential, commercial, and portable uses. The main challenges with CIGS cells are the complexity of their manufacturing process and the availability of indium and gallium.

Gallium Arsenide (GaAs): While not as common as the other second-generation materials, GaAs solar cells [67–69] are known for their high efficiency and excellent performance in low-light conditions. They are used in specialized applications such as space missions and concentrator photovoltaics. GaAs cells have high electron mobility and a direct bandgap, which makes them highly efficient but also more expensive to produce. They are typically used where high performance is more critical than cost.

2.2.2 Working Principle

Second-generation solar cells, also known as thin-film solar cells, emerged in the latter half of the twentieth century, with significant developments occurring from the 1970s through the early 2000s. This period marked the transition from the traditional first-generation crystalline silicon solar cells to more advanced and cost-effective photovoltaic technologies. The working principle of second-generation solar cells, or thin-film solar cells, is based on the photovoltaic effect, similar to first-generation silicon-based solar cells. However, the key difference lies in the materials and structure used. Here is an overview of the working principles for the main types of second-generation solar cells.

General Working Principle

When sunlight strikes a solar cell, photons are absorbed by the semiconductor material, which in the case of thin-film cells, is only a few micrometers thick. This absorption of energy excites electrons within the semiconductor, leading to the formation of electron–hole pairs, or excitons. These excitons are then separated by an internal electric field generated at the junction between different material layers, such as the p-type and n-type layers. This separation is crucial for preventing recombination and allowing the free movement of charges. Once separated, the electrons and holes are collected by electrodes, generating an electric current that can be harnessed for external use. This process of light absorption, exciton generation, charge separation, and charge collection underpins the fundamental operation of solar cells, converting solar energy into usable electricity. Thin-film cells, with their reduced thickness, offer advantages in flexibility and material efficiency, although their overall efficiency can be lower than traditional crystalline cells.

Specific Working Principles for Different Materials

Amorphous Silicon (a-Si) Solar Cells

Amorphous silicon (a-Si) cells are composed of a thin layer of non-crystalline silicon deposited on substrates like glass, plastic, or metal, making them flexible and lightweight. However, the disordered atomic structure of a-Si cells leads to a high number of dangling bonds, which can trap charge carriers and reduce efficiency by increasing electron–hole recombination rates. To address this issue, a buffer layer or hydrogenation process is typically employed to passivate these dangling bonds, enhancing charge separation and improving overall cell performance. Despite these improvements, a-Si cells have lower efficiency compared to crystalline silicon cells, usually around 6–8%, primarily due to the inherent disordered structure that causes higher recombination rates. Nonetheless, their low cost, flexibility, and adaptability to various applications make them a viable option for specific uses like portable electronics and building-integrated photovoltaics.

Cadmium Telluride (CdTe) Solar Cells

CdTe (cadmium telluride) solar cells consist of a thin layer of cadmium telluride deposited on a substrate, typically glass, with a cadmium sulfide (CdS) buffer layer that forms a junction essential for charge separation. When sunlight hits the CdTe layer, it absorbs photons and generates electron–hole pairs, which are efficiently separated by the built-in electric field at the junction with the CdS layer, preventing recombination and allowing charge collection. CdTe cells generally achieve efficiencies in the range of 9–11%, performing well not only in standard sunlight conditions but also in low-light environments and high-temperature conditions, which makes them particularly advantageous for large-scale solar installations in diverse climates. In addition to their solid performance, CdTe cells are known for their cost-effectiveness, as they require less material and energy to produce compared to other types of photovoltaic cells, contributing to their widespread use in utility-scale solar farms.

Copper Indium Gallium Selenide (CIGS) Solar Cells

CIGS (copper indium gallium selenide) solar cells consist of a thin layer of copper, indium, gallium, and selenium, typically deposited on substrates such as glass or plastic, with additional buffer layers like cadmium sulfide or zinc oxide to enhance performance. The CIGS layer absorbs sunlight, creating electron–hole pairs, and its multi-layered structure, with materials that have different bandgaps, facilitates efficient charge separation and collection. This design helps reduce recombination losses and improves overall cell performance. CIGS cells are known for their relatively high efficiency, typically ranging between 12–14%, and can also be manufactured in flexible formats, making them suitable for a variety of applications, including portable electronics, building-integrated photovoltaics, and lightweight solar panels for specialized uses. Their combination of efficiency, flexibility, and adaptability has made them an attractive option in both conventional and innovative solar technologies.

Gallium Arsenide (GaAs) Solar Cells

GaAs (gallium arsenide) solar cells use gallium arsenide as the primary semiconductor material, often incorporating additional layers to form multi-junction cells, which further enhance performance. GaAs has a direct bandgap, enabling it to absorb light very efficiently and generate electron–hole pairs quickly, resulting in highly effective charge separation and minimal recombination losses. This property contributes to GaAs cells achieving some of the highest energy conversion efficiencies among thin-film technologies, often exceeding 20%. Due to their superior performance, particularly in converting sunlight across a broad spectrum and their ability to maintain efficiency in high-radiation environments, GaAs cells are used in specialized applications such as space exploration and concentrator photovoltaics, where performance is critical despite higher production costs. Their durability and high-efficiency make them ideal for high-stress, high-performance environments.

2.2.3 Characteristics of Second-Generation Solar Cell

Thin-film solar cells, such as those made from materials like cadmium telluride (CdTe) or copper indium gallium selenide (CIGS), require significantly less semiconductor material compared to traditional silicon wafers, leading to lower material costs. They can also be manufactured on flexible substrates, which enhances their versatility and suitability for various applications, including integration into building materials and portable devices. Although their efficiency is generally lower than that of first-generation silicon cells, typically ranging from 10 to 15%, ongoing improvements are gradually increasing their performance. The simpler production processes and reduced material usage contribute to lower manufacturing costs, making thin-film cells cost-effective. They are particularly well-suited for large-scale utility installations, building-integrated photovoltaics (BIPV), and portable solar products, offering a balance of flexibility, cost-efficiency, and versatility in diverse solar energy applications.

2.2.4 Advantages and Challenges to Second-Generation Solar Cell

Thin-film solar technologies offer significant cost-effectiveness due to their lower material and energy requirements during production, making them cheaper than first-generation silicon-based solar cells. Their flexibility and lightweight nature allow them to be deposited on various substrates, enabling innovative applications such as solar windows, portable chargers, and wearable technology. Additionally, certain thin-film materials, like cadmium telluride (CdTe) and copper indium gallium selenide (CIGS), exhibit improved performance in low-light conditions compared to traditional silicon cells. However, second-generation solar cells generally have lower efficiency rates, and the availability and environmental impact of some materials, such as cadmium and indium, raise concerns. Furthermore, the production processes for technologies like CIGS can be complex, requiring precise control over material composition and deposition. In conclusion, while second-generation solar cells present advancements in cost, versatility, and environmental considerations, they also come with their own set of challenges, reflecting a trade-off between efficiency and innovative application potential in the quest for more cost-effective and sustainable solar energy solutions.

2.3 Third Generation Solar Cells

Third-generation solar cells [70–76] encompass a variety of emerging technologies that promise to revolutionize the photovoltaic industry with their innovative materials and sophisticated working principles. These include Perovskite solar cells [71, 77–80], Organic Photovoltaics (OPVs) [81–85] Quantum Dot Solar Cells [86–90] and Multi-Junction (Tandem) Solar Cells [75, 91–93] Each of these technologies leverages advanced materials and novel concepts to address and surpass the limitations inherent in traditional semiconductor materials, such as silicon (Table 1).

2.3.1 Classification Based on the Material Used

Perovskite Solar Cells: These utilize perovskite-structured compounds that are highly efficient at light absorption and charge-carrier mobilization. Perovskites have rapidly garnered attention due to their high efficiency and relatively low production costs compared to traditional silicon cells.

Organic Photovoltaics (OPVs): OPVs employ organic molecules or polymers to absorb light and convert it into electricity. Their flexibility, lightweight nature, and potential for low-cost production via printing technologies make them highly attractive for a range of applications, including portable and wearable electronics.

Quantum Dot Solar Cells: These use semiconductor nanocrystals known as quantum dots to create multiple excitons from a single photon of sunlight, potentially leading to higher efficiency rates. Quantum dots can be tuned to absorb different wavelengths of light, enhancing their ability to capture more of the solar spectrum.

Table 1 Comparison between 1st generation and 2nd generation solar cells

Aspect	1st Generation solar cells	2nd Generation solar cells
Material type	Primarily crystalline silicon (c-Si)	Thin-film materials (e.g., amorphous silicon, CdTe, CIGS)
Structure	Thick silicon wafers	Thin layers of semiconductor materials
Efficiency	High efficiency (15–26%)	Moderate efficiency (10–20%)
Manufacturing cost	High due to expensive silicon processing	Lower due to less material usage and simpler processes
Production process	Czochralski process or float-zone process	Various methods including sputtering, vapor deposition
Flexibility	Rigid and heavy	Flexible and lightweight
Thickness	Typically around 150–200 μm	1–2 μm
Market share	Dominates the market (~70–80%)	Growing but less dominant (~20–30%)
Environmental impact	Higher energy consumption in production	Lower energy consumption and material use
Durability and lifetime	Long lifespan (25–30 years)	Slightly shorter lifespan (10–20 years)
Applications	Large-scale solar farms, rooftops	Integrated into building materials, portable devices
Temperature sensitivity	Less sensitive to temperature variations	More sensitive to temperature and light variations
Research and development	Development mature technology with incremental improvements	Active research for improving efficiency and stability
Energy payback time	Longer due to energy intensive manufacturing	Shorter due to lower material and energy requirements

Multi-junction (Tandem) Solar Cells: These cells stack multiple layers of different semiconductor materials, each designed to capture a different part of the solar spectrum. By optimizing the use of available sunlight, multi-junction cells can achieve higher efficiencies than single-junction cells.

2.3.2 Working Principle

Third-generation solar cells, emerging predominantly in the early twenty-first century, represent the forefront of solar technology, driven by rapid advancements in material science, nanotechnology, and fabrication techniques. These cells utilize cutting-edge materials and innovative concepts to address the limitations of traditional semiconductor technologies. By employing techniques such as multi-junction cells, which stack several layers of different semiconductor materials to capture a broader spectrum of sunlight, and integrating nanostructures or quantum dots to enhance light absorption and conversion efficiency, third-generation solar cells aim to significantly surpass the performance of earlier generations. These advanced approaches seek to optimize energy capture and conversion, overcoming efficiency barriers associated with conventional solar cells and paving the way for highly efficient and versatile photovoltaic technologies.

Third-generation solar cells are engineered to capture a broader spectrum of sunlight, with multi-junction cells employing layers of materials with varying bandgaps to absorb different portions of the solar spectrum, thereby significantly enhancing overall efficiency. Quantum dot solar cells introduce the concept of multiple exciton generation (MEG), where more than one exciton can be produced from a single absorbed photon, potentially leading to higher efficiencies compared to traditional cells. Additionally, organic photovoltaic (OPV) materials offer flexibility and lightweight properties, enabling innovative applications such as solar-powered clothing and portable chargers. Perovskite solar cells further exemplify the advances of third-generation technology, combining high efficiency with low-cost production processes, positioning them as a highly promising solution for large-scale solar power generation and advancing the feasibility of widespread solar energy adoption.

2.3.3 Characteristics of Third-Generation Solar Cell

Third-generation solar cells hold high potential for efficiency, with multi-junction cells capable of exceeding 40% efficiency under concentrated sunlight and perovskite cells achieving over 25% efficiency in laboratory settings. These advancements are driven by the use of innovative materials and nanostructures that enhance light absorption and charge carrier mobility. Additionally, some third-generation cells, such as organic photovoltaics (OPVs) and perovskites, can be fabricated to be flexible and lightweight, opening up new possibilities for integration into a diverse array of products, including wearable devices, flexible electronics, and building materials. However, these technologies still face significant challenges, including issues related to stability, scalability, and long-term durability, which need to be addressed before they can be widely adopted. Despite these hurdles, the potential applications of third-generation solar cells are extensive, promising to revolutionize both traditional power generation and emerging fields in renewable energy technology.

Third-generation solar cells are characterized by their use of cutting-edge materials and advanced technological concepts to enhance the efficiency and versatility of solar energy harvesting. These technologies, developed predominantly in the early twenty-first century, represent the forefront of solar cell innovation, with the potential to significantly impact the future of sustainable energy.

3 Smart Materials and Their Use in Solar Cell

Smart materials, also known as intelligent or responsive materials, have properties that can be significantly altered in a controlled fashion by external stimuli, such as stress, temperature, moisture, pH, and electric or magnetic fields. These materials can respond to changes in their environment and can potentially return to their original state once the external influence is removed. The responsiveness and adaptability of smart materials make them ideal for a wide range of applications, including in the fields of aerospace [94–96], biomedical [97–102] automotive [103–105] and consumer electronics [106–109]. To improve the efficiency and durability of a solar cell, the integration of smart nanomaterials with the working cells is essential. Smart materials can be classified based on their response to various stimuli, such as light, temperature, pH, and electric or magnetic fields.

3.1 Different Kinds of Smart Materials Useful in Solar Cell

3.1.1 Shape Memory Alloys (SMAs)

Shape Memory Alloys (SMAs) [110–112] are a unique class of smart materials that have the ability to return to their original shape after being deformed when exposed to an appropriate thermal condition. This remarkable property is due to a reversible phase transformation between two different crystal structures: austenite (high-temperature phase) and martensite (low-temperature phase). SMAs are primarily composed of metals such as nickel-titanium (NiTi) [113] copper–aluminium–nickel (CuAlNi) [114] and copper–zinc–aluminum (CuZnAl) [115].

Characteristics of SMAs

Shape Memory Alloys (SMAs) exhibit unique properties such as the Shape Memory Effect (SME) and superelasticity. In the one-way SME, the material returns to its original shape upon heating above a certain temperature (austenite phase), while in the two-way SME, the material can remember two distinct shapes—one in the low-temperature martensite phase and another in the high-temperature austenite phase. Superelasticity, or pseudoelasticity, allows SMAs to endure large strains (up to 8%) and recover their original shape once the load is removed, all while maintaining a constant temperature. SMAs also show hysteresis, which refers to the temperature difference between heating and cooling cycles required for phase transformation, providing energy dissipation that enhances their damping capabilities. Additionally, SMAs possess a high damping capacity, absorbing significant energy during phase transformations, making them ideal for damping applications.

General Applications of SMAs

a. *Biomedical Devices*

Shape Memory Alloys (SMAs) have found widespread use in biomedical applications due to their unique properties, such as the ability to change shape in response to temperature or stress. In the field of cardiology, SMAs are used to manufacture stents, which are implanted in coronary arteries to prevent or alleviate blockages by expanding when heated to body temperature. In orthodontics, SMA-based archwires are used for teeth alignment, offering a consistent and gentle force that adjusts as the wire returns to its pre-programmed shape. Additionally, SMAs are utilized in surgical tools designed for minimally invasive procedures, where their ability to change shape within the body allows for greater precision and flexibility during operations. These applications highlight the versatility of SMAs in advancing medical technology and improving patient outcomes.

b. *Aerospace Industry*

Shape Memory Alloys (SMAs) find extensive applications in the aerospace industry due to their unique properties like the shape memory effect and superelasticity. One notable application is in adaptive wing structures, where SMAs enable wings to change shape in response to flight conditions, optimizing aerodynamics and improving fuel efficiency. This capability allows aircraft to adapt to varying speeds, altitudes, and weather conditions. Additionally, SMAs are used in vibration control systems, where their high damping capacity helps reduce vibrations in critical components, enhancing the structural integrity and performance of aircraft. These applications improve flight stability, reduce wear on components, and extend the lifespan of aerospace systems, contributing to the overall efficiency and safety of modern aviation technology.

c. *Automotive Industries and Robotics*

Shape Memory Alloys (SMAs) have diverse applications in industries like automotive, robotics, and consumer electronics. In the automotive sector, SMAs are used in actuators, enabling components such as variable geometry turbochargers to change shape as needed for optimal performance. Additionally, they are integrated into damping systems to absorb shocks, enhancing ride comfort. In robotics, SMAs act as artificial muscles, replicating natural muscle movements for soft robotics, and are also used as actuators for precise movement and control in robotic systems. In consumer electronics, SMAs enable flexible devices that can alter their shape and return to their original form, while wearable technology leverages SMAs for adaptive fits in products like smartwatches and fitness trackers, providing both flexibility and comfort.

Applications of SMAs in Solar Cell

Shape Memory Alloys (SMAs) have found a wide range of applications across various industries due to their unique ability to revert to their original shape after deformation when subjected to a thermal stimulus. In the field of solar energy, SMAs offer innovative solutions to enhance the efficiency, durability, and functionality of solar cells. Below are some of the key applications of SMAs in solar cells.

a. *Self-cleaning Mechanisms*

One of the challenges faced by solar panels is the accumulation of dust and debris on their surfaces, which can significantly reduce their efficiency. SMAs can be used to develop self-cleaning mechanisms for solar panels. When the temperature rises, SMA-based actuators can change shape to remove dust and debris from the surface of the panels. This ensures that the panels remain clean and maintain their maximum efficiency without the need for manual cleaning.

b. *Dynamic Solar Tracking Systems*

Solar tracking systems are used to orient solar panels towards the sun to maximize energy capture. SMAs can be integrated into these tracking systems to provide a dynamic and responsive mechanism. The SMAs can adjust the angle of the solar panels based on the temperature changes caused by sunlight, thereby optimizing the position of the panels throughout the day. This increases the overall efficiency of the solar energy system.

c. *Temperature Regulation*

Overheating of solar panels can lead to reduced efficiency and damage over time. SMAs can be employed in temperature regulation systems to control the cooling of solar panels. When the panels reach a certain temperature, SMA elements can change shape to open vents or activate cooling fans, thereby dissipating excess heat and maintaining the optimal operating temperature of the solar panels.

d. *Deployable Structures*

SMAs can be used to create deployable structures for solar panels, such as foldable or expandable arrays. These structures can be compactly stored during transport and automatically deploy to their functional state when exposed to sunlight or heat. This is particularly useful for portable solar panels and space applications where efficient use of space and weight is crucial.

e. *Adaptive Lenses and Covers*

SMAs can be utilized in the design of adaptive lenses or covers for solar panels that adjust their transparency or position based on the intensity of sunlight. For example, SMA-driven mechanisms can tilt or move lenses to focus or diffuse sunlight onto the solar cells, optimizing light capture under varying conditions. This adaptability can lead to improved energy conversion efficiency.

f. *Structural Health Monitoring*

In large solar installations, maintaining the structural integrity of the panels is important. SMAs can be integrated into the mounting systems of solar panels to monitor and adjust for structural deformations or misalignments. The shape memory effect can be used to correct any misalignments or deformations automatically, ensuring that the panels remain in their optimal position for maximum energy capture.

Advantages of Using SMAs in Solar Cells

The use of Shape Memory Alloys (SMAs) in solar cells offers several key advantages. SMAs enhance efficiency by optimizing the orientation and cleanliness of solar panels, ensuring that they consistently operate at peak performance. They also contribute to reduced maintenance through self-cleaning and self-regulating mechanisms, minimizing the need for manual intervention. Additionally, SMAs improve the durability of solar panels by enabling temperature regulation and structural health monitoring, which extend the lifespan and reliability of the systems. Furthermore, SMAs facilitate innovative designs, allowing for deployable and adaptive structures that make solar energy systems more versatile, space-efficient, and suitable for various applications, from rooftops to large-scale installations.

Challenges and Considerations of Use of SMAs in Solar Cell

While Shape Memory Alloys (SMAs) offer several benefits for solar cells, they also present some challenges and considerations. One major challenge is cost, as the integration of SMAs can be expensive, potentially increasing the overall price of solar energy systems. Additionally, the complexity of designing and implementing SMA-based systems requires precise engineering and sophisticated control mechanisms, adding further difficulty to their application. Material fatigue is another concern, as repeated phase transformations in SMAs can lead to degradation over time, negatively impacting the long-term performance and reliability of solar panels. Addressing these issues is crucial for the widespread adoption of SMA-enhanced solar technology.

Shape Memory Alloys offer promising applications in enhancing the functionality and efficiency of solar cells. By leveraging their unique properties, solar energy systems can become more adaptive, efficient, and low-maintenance, contributing to the broader adoption of renewable energy solutions. As research and development in SMAs continue, their integration into solar technologies is likely to expand, offering new opportunities for innovation in sustainable energy.

3.1.2 Piezoelectric Materials

Piezoelectric materials [116–118] are a class of materials that have the ability to generate an electric charge in response to mechanical stress and, conversely, to undergo mechanical deformation when an electric field is applied. This unique property, known as the piezoelectric effect, was first discovered in 1880 by Pierre and Jacques Curie [119]. Since then, piezoelectric materials have found widespread applications across various fields, including electronics [120], automotive [121], aerospace [122], healthcare [123], and energy harvesting [124, 125].

Characteristics

Piezoelectric materials possess unique characteristics that make them valuable in various applications. The primary feature is the piezoelectric effect, where they generate an electric charge in response to mechanical stress and undergo mechanical deformation when subjected to an electric field. This effect is divided into the direct piezoelectric effect, where mechanical stress produces voltage, used in sensors and energy harvesting, and the inverse effect, where applying an electric field causes mechanical deformation, used in actuators and precision devices. These materials are known for their high mechanical strength, allowing them to endure repetitive stress, and they maintain functionality across a wide operational temperature range, from cryogenic to high temperatures. Their high-frequency response enables use in applications requiring rapid mechanical or electrical changes, while their high sensitivity makes them ideal for detecting minute mechanical stress. The piezoelectric effect originates from their non-centrosymmetric crystal structure, which lacks a center of symmetry.

Common Types of Piezoelectric Materials

Piezoelectric materials come in various forms, each with distinct properties and applications. Natural piezoelectric materials include quartz (SiO_2), a stable and reproducible crystal, and Rochelle salt, which offers a high piezoelectric response but is sensitive to environmental conditions. Ceramic piezoelectric materials are widely used, with lead zirconate titanate (PZT) being the most common due to its high piezoelectric coefficients and versatility, while barium titanate ($BaTiO_3$) is an early ceramic material still used in specific applications, and potassium niobate ($KNbO_3$) is utilized for high-performance tasks. Polymeric piezoelectric materials are represented by polyvinylidene fluoride (PVDF), a flexible and easy-to-process polymer known for its practical use in flexible devices. Each type of material offers unique advantages based on its structure and properties.

Applications

a. *Sensors*

Piezoelectric materials are widely used in sensor applications due to their ability to convert mechanical stress into electrical signals. In pressure sensors, they measure changes in pressure by generating an electrical response proportional to the applied mechanical force, making them ideal for precise pressure monitoring. In accelerometers, piezoelectric materials detect acceleration forces, converting them into electrical signals, which are commonly employed in automotive systems and consumer electronics for motion detection, vibration analysis, and stabilization technologies. These applications highlight the versatility and responsiveness of piezoelectric materials in sensing devices.

b. *Actuators*

Piezoelectric materials are widely used in actuator applications where precision and control are essential. In precision actuators, they enable highly accurate movements, making them ideal for optical devices and nano-positioning systems that demand fine, controlled adjustments. Additionally, piezoelectric actuators are employed in ultrasonic transducers, which play a crucial role in medical imaging technologies like ultrasound and in industrial non-destructive testing, where they help detect structural flaws without causing damage. These applications leverage the material's ability to convert electrical signals into precise mechanical motion.

c. *Energy Harvesting*

Piezoelectric materials are widely used in energy harvesting applications due to their ability to convert mechanical energy into electrical energy. Vibration energy harvesters, for instance, capture ambient mechanical vibrations and convert them into electrical energy, which is particularly useful for powering small electronic devices and wireless sensors. In the medical field, piezoelectric materials are key components in ultrasound transducers, which are essential for both diagnostic imaging and therapeutic treatments. Additionally, piezoelectric motors, known for their precision and compact size, are used in medical implants and surgical instruments, offering highly controlled movements in delicate medical procedures. These diverse applications highlight the versatility of piezoelectric materials in both energy harvesting and advanced medical technologies.

d. *Consumer Electronics*

Piezoelectric materials are widely used in consumer electronics due to their ability to convert mechanical energy into electrical signals and vice versa. In microphones, piezoelectric elements are used to convert sound waves into electrical signals, enabling sound capture with high sensitivity and precision. In lighters, these materials generate sparks for ignition by producing a high voltage when subjected to mechanical pressure, offering a reliable and durable ignition mechanism without the need for external power sources. These applications highlight the versatility of piezoelectric materials in everyday electronic devices.

Application in Solar Cells

Piezoelectric materials can enhance solar cell efficiency and functionality through various innovative approaches.

a. *Hybrid Energy Systems*

Piezoelectric materials are increasingly being integrated into hybrid energy systems to complement solar cells by harvesting mechanical energy, such as vibrations and wind, alongside solar energy. This dual energy capture enhances overall energy output, especially in environments where solar irradiance fluctuates, such as during cloudy weather or at night. By converting mechanical energy into electricity, piezoelectric components provide a continuous energy source, improving system efficiency and reliability. This approach is particularly advantageous in renewable energy systems, enabling more consistent and sustainable energy production in diverse environmental conditions.

b. *Self-cleaning Mechanisms*

Piezoelectric materials play a crucial role in self-cleaning mechanisms, particularly in solar panels, where they are used to enhance efficiency through vibration-induced cleaning. By embedding piezoelectric materials within the panels, controlled vibrations can be generated when an electrical signal is applied, dislodging dust and debris from the surface. This self-cleaning action helps maintain optimal light absorption, ensuring that the panels operate at peak efficiency over time without the need for manual cleaning. This innovative application not only extends the lifespan of solar panels but also enhances their overall energy output.

c. *Enhanced Light Absorption (Through Surface Texturing)*

Piezoelectric materials are increasingly used to enhance light absorption in solar cells through surface texturing. By leveraging their precise actuation capabilities, piezoelectric materials can dynamically alter the surface texture of solar cells, creating micro- and nano-scale patterns that improve light trapping. These textured surfaces reduce reflection and increase the absorption of incident light, thereby boosting the overall efficiency of the solar cells. This innovative approach not only optimizes energy capture but also allows for adaptable and responsive surface designs in advanced photovoltaic systems.

d. *Structural Health Monitoring (Through Stress Sensors)*

Piezoelectric materials play a crucial role in structural health monitoring, particularly in solar panel systems, by acting as stress sensors embedded within the structure. These sensors can detect mechanical stress, deformation, or damage in real-time, providing valuable data on the integrity of the panels. By continuously monitoring the mechanical condition of solar installations, piezoelectric sensors enable early detection of potential issues, allowing for preventive maintenance. This proactive approach not only reduces the risk of costly failures but also extends the lifespan and efficiency of solar panels, contributing to more reliable and sustainable energy production.

e. *Dynamic Orientation*

Piezoelectric materials are effectively utilized in dynamic orientation applications, particularly in optimizing the positioning of solar panels. By employing piezoelectric actuators, these systems can adjust the angle of solar panels in real-time based on environmental conditions, such as the position of the sun. This dynamic adjustment ensures that the panels maintain optimal solar exposure throughout the day, maximizing energy capture and enhancing overall efficiency. The ability of piezoelectric actuators to provide precise and rapid movements makes them ideal for tracking systems, contributing to more effective harnessing of solar energy and improving the performance of solar installations.

Challenges and Considerations

The application of piezoelectric materials in solar cell technologies presents several challenges and considerations that must be addressed for optimal performance. One significant issue is material fatigue, as repeated mechanical deformation can lead to degradation over time, potentially reducing the lifespan and effectiveness of the devices. Additionally, the high cost of certain piezoelectric materials, particularly high-performance ceramics and polymers, can pose a barrier to widespread adoption and scalability in solar applications. Environmental sensitivity is another critical factor; for instance, materials like Rochelle salt are susceptible to changes in humidity and temperature, which can affect their piezoelectric properties and overall reliability. Addressing these challenges is essential for enhancing the viability and longevity of piezoelectric materials in solar energy applications.

Piezoelectric materials offer a wide range of characteristics that make them suitable for various applications, including sensors, actuators, energy harvesting devices, and consumer electronics. Their integration into solar cells presents innovative ways to enhance efficiency, maintain cleanliness, and ensure structural integrity, making them a valuable component in advancing solar energy technologies.

3.1.3 Thermochromic Materials

Thermochromic materials [126–128] are substances that change their color in response to temperature changes. This reversible change is due to alterations in their molecular structure or crystal phase, which affects their optical properties, particularly their absorption and reflection of light. The temperature at which the color change occurs can be finely tuned based on the material's composition.

Types of Thermochromic Materials

a. *Liquid Crystals:* These materials exhibit changes in their molecular arrangement with temperature, which affects their reflection of specific wavelengths of light, leading to visible color changes.
b. *Leuco Dyes:* These dyes undergo a chemical transformation between two forms, one of which is colored and the other is colorless. The transformation is triggered by temperature changes.

Applications

Thermochromic materials have a wide range of applications, particularly in temperature indication and smart packaging solutions. In temperature indicators, they are utilized in medical and industrial thermometers to provide a visual representation of temperature changes, allowing for easy and immediate readings. Additionally, these materials serve as battery indicators, changing color to signal the charge level of batteries, enhancing user awareness. In the realm of smart packaging, thermochromic materials play a crucial role in food packaging by indicating whether products have been exposed to unsafe temperature conditions, thereby ensuring food safety. Similarly, in pharmaceutical packaging, they help verify that medications are stored within the recommended temperature ranges, ensuring efficacy and safety for consumers. These applications highlight the significant benefits of thermochromic materials in enhancing safety and functionality across various industries.

Thermochromic materials find diverse applications across various industries, particularly in textiles, security printing, and building materials. In textiles, these materials are utilized in clothing that changes color in response to body temperature or ambient temperature fluctuations, making them popular for both fashion and safety apparel. Heat-sensitive fabrics are especially valuable in sportswear, as they can indicate overheating, enhancing user safety during physical activities. In the realm of security printing, thermochromic materials serve as anti-counterfeiting features in banknotes, tickets, and official documents, providing a temperature-sensitive mechanism that adds an extra layer of security. Furthermore, in building materials, thermochromic technologies are incorporated into windows and coatings, enabling them to regulate indoor temperatures by altering their color and reflectivity in response to external temperature changes, thereby improving energy efficiency and comfort in living spaces.

Application in Solar Cells

Thermochromic materials can enhance the efficiency and functionality of solar cells through several innovative approaches discussed below.

a. *Smart Windows (Dynamic Regulation)*

Thermochromic materials integrated into window coatings can adjust their transparency and heat absorption based on temperature. This reduces the cooling load on buildings and improves energy efficiency, indirectly enhancing the performance of solar panels by maintaining a more consistent temperature.

b. *Temperature Management (Self-regulating Panels)*

Incorporating thermochromic materials into solar panels can help manage temperature. As solar panels heat up, thermochromic layers can change color to reflect more sunlight, thereby reducing the operating temperature and preventing efficiency losses due to overheating.

c. *Light Absorption Optimization (Adaptive Coatings)*

Thermochromic coatings on solar panels can adapt to different light conditions, optimizing the absorption of sunlight throughout the day and across seasons. This can improve the overall energy yield of solar installations.

Challenges and Considerations

The application of thermochromic materials is accompanied by several challenges and considerations that impact their performance and longevity. One significant concern is durability and stability; these materials can degrade over time with repeated temperature cycling, which can result in a diminished ability to change color and ultimately loss of functionality. Additionally, prolonged exposure to ultraviolet (UV) light can lead to photodegradation, further compromising their color-changing properties. Another critical factor is the temperature range and operational limits of thermochromic materials, as their effectiveness may not align with all application environments, particularly in extreme conditions. This limitation can restrict their usefulness in certain applications, making it essential for developers to carefully assess the suitability of thermochromic materials for specific uses while considering these challenges.

Other significant concerns are the response time and speed of transition; the rate at which these materials change color can vary, often resulting in a lag in environments with rapidly fluctuating temperatures, which can limit their utility in certain applications. Additionally, cost and scalability pose substantial hurdles; the manufacturing costs associated with producing thermochromic materials and integrating them into products can be high, potentially hindering widespread adoption across industries. Moreover, achieving consistent quality and performance when scaling up production remains a challenge, as variations in material properties can affect the reliability and effectiveness of the thermochromic response. Addressing these issues is essential for enhancing the viability of thermochromic materials in commercial applications.

Some of these materials contain chemical compositions that may be harmful to the environment, which raises concerns about their lifecycle management and disposal methods. It is essential to carefully evaluate the ecological implications of using such substances, including their potential toxicity and persistence in the environment. Additionally, considerations around sustainability, regulatory compliance, and the development of eco-friendly alternatives are critical for mitigating any adverse effects associated with thermochromic materials. By addressing these challenges, industries can ensure that the benefits of thermochromic technologies do not come at the expense of environmental health and safety.

Thermochromic materials offer a unique capability to change color with temperature, providing numerous applications in temperature indicators, smart packaging, textiles, security printing, and building materials. In solar cells, they can enhance efficiency through smart windows, temperature management, and light absorption optimization. However, challenges such as durability, temperature range, response time, cost, scalability, and environmental impact need to be addressed to fully realize their potential in various applications.

3.1.4 Photochromic Materials

Photochromic materials [129–133] are substances that undergo a reversible change in color when exposed to light, typically ultraviolet (UV) or visible light. This change occurs due to a photochemical reaction that alters the molecular structure of the material, leading to a change in its absorption spectrum. When the light source is removed, the material gradually returns to its original color through a reverse reaction. Photochromic materials can be found in both organic compounds and inorganic compounds. They are widely used in various applications due to their ability to change color in response to light

Characteristics

Photochromic materials exhibit distinct characteristics that make them unique and functional in various applications. One of their primary features is reversibility, as these materials can undergo color changes in response to light exposure and then revert to their original state, allowing for multiple cycles of transformation without degradation. This reversibility is crucial for applications such as photochromic lenses in eyewear, where the material adapts to changing light conditions. Additionally, photochromic materials are sensitive to specific wavelengths of light, typically ultraviolet (UV) or visible light, which trigger the underlying photochemical reactions responsible for the color change. This light sensitivity enables precise control over their activation and makes them suitable for a variety of uses, from sunglasses to smart windows that adjust tint based on environmental lighting.

There are other notable features of photochromic materials which are also very important for present day technology. One important characteristic is the speed of response. While some photochromic materials can change color almost instantaneously, others may take several seconds to minutes to fully transition, impacting their suitability for different uses. Additionally, high-quality photochromic materials demonstrate excellent fatigue resistance, allowing them to undergo numerous cycles of color change without significant degradation in performance, which is vital for longevity in applications such as eyewear and coatings. Stability is another critical characteristic, as photochromic materials must maintain their performance under various environmental conditions, including fluctuations in temperature, humidity, and light exposure. This stability ensures reliable operation and effectiveness in real-world scenarios, making these materials valuable in a range of applications.

Applications

Photochromic materials are widely utilized in various applications, most notably in eyewear and smart windows. In eyewear, photochromic lenses are designed to darken in bright sunlight, providing enhanced protection against UV rays, and return to a clear state indoors or in lower light conditions. This functionality not only improves visual comfort but also eliminates the need for multiple pairs of glasses. Additionally, photochromic materials are employed in smart windows, which feature adaptive tinting capabilities that adjust their level of tint in response to sunlight. This technology effectively controls indoor lighting, reduces glare, and enhances energy efficiency by minimizing the need for artificial lighting and cooling. Together, these applications highlight the versatility and practicality of photochromic materials in improving everyday experiences and promoting energy conservation.

Photochromic materials are also adapted in a variety of applications that leverage their ability to change color in response to light exposure. In security printing, they are incorporated into documents, currency, and tickets, providing an additional layer of protection against counterfeiting through the use of photochromic inks that reveal hidden features under specific lighting conditions. In textiles, these materials are employed in fashion and sportswear to create dynamic patterns or colors that shift with changes in light, enhancing aesthetic appeal and functionality. Additionally, photochromic materials are increasingly used in advanced optical data storage systems, where information can be written, read, and erased using light, allowing for efficient and versatile data management. This versatility across multiple domains highlights the innovative potential of photochromic technologies in enhancing security, design, and data storage capabilities.

Application in Solar Cells

Photochromic materials offer innovative applications in solar cell technology, enhancing both performance and functionality. One significant application is the use of adaptive coatings on solar panels, which can dynamically regulate light absorption properties in response to sunlight intensity, optimizing energy capture throughout the day. Additionally, integrating photochromic materials into smart windows with solar cells can significantly improve energy efficiency in buildings by controlling light and heat entry while simultaneously generating electricity. Furthermore, certain photochromic materials possess photocatalytic properties that contribute to self-surface-cleaning, effectively minimizing dirt accumulation on solar panels and maintaining their efficiency. These advancements highlight the potential of photochromic materials to enhance the effectiveness of solar technologies and promote more sustainable energy solutions.

Challenges and Considerations

The application of photochromic materials comes with several challenges and considerations that must be addressed to ensure their effectiveness and longevity. One significant issue is durability; repeated exposure to light can lead to the degradation of these materials over time, diminishing their ability to change color and reducing their overall effectiveness. Additionally, temperature sensitivity poses another challenge, as fluctuations in temperature can affect the response time and efficiency of photochromic materials, potentially limiting their functionality in extreme environments. This sensitivity may restrict their use in applications where consistent performance is critical, highlighting the need for ongoing research and development to enhance the durability and resilience of photochromic technologies.

There are other challenges which also affect their commercial viability. One major concern is the manufacturing expense associated with producing high-quality photochromic materials and integrating them into various products, which can drive up costs and limit widespread adoption. Additionally, the speed of color change, or response time, can pose practical limitations for certain applications; if the material does not change color quickly enough, it may fail to meet the performance requirements necessary for effective use in specific contexts. Addressing these challenges is crucial for enhancing the functionality and market acceptance of photochromic technologies, ensuring that they can be effectively utilized in diverse applications.

The environmental impact of photochromic materials must be considered before their integration into existing technologies. Some photochromic materials may contain hazardous chemicals that pose risks to both environmental and human health, requiring careful handling and appropriate disposal methods to mitigate these risks. Furthermore, the integration of photochromic materials into established technologies, such as solar cells, can present technical challenges. Ensuring compatibility and maintaining long-term performance in conjunction with existing systems necessitates careful engineering and innovation. Addressing these challenges is crucial for maximizing the benefits of photochromic materials while safeguarding environmental and health standards.

Photochromic materials offer unique capabilities for reversible color change in response to light, finding applications in eyewear, smart windows, security printing, textiles, and data storage. In the context of solar cells, they provide innovative solutions for dynamic light regulation, smart window integration, and self-cleaning surfaces. However, challenges such as durability, temperature sensitivity, cost, speed of color change, environmental impact, and integration complexity must be addressed to fully harness their potential.

3.1.5 Electrochromic Materials

Electrochromic materials [134–138] are substances that change their color or opacity when an electrical voltage is applied. This reversible change is due to the redox (reduction–oxidation) reactions that occur within the material, altering its optical properties such as light absorption and reflectance. Electrochromic materials can switch between different states (e.g., transparent to opaque) multiple times without significant degradation.

Characteristics

Electrochromic materials exhibit distinctive characteristics that make them valuable in various applications, primarily due to their ability to undergo reversible color changes. This reversibility allows for multiple cycles of switching between different states, enabling consistent performance in applications such as smart windows and displays. Additionally, the optical properties of these materials are voltage-dependent, meaning that they can be precisely controlled by applying an electrical voltage. This feature allows for fine-tuning of their coloration and transparency, enhancing their functionality in dynamic environments where control over light transmission is essential. The combination of reversibility and voltage control positions electrochromic materials as versatile components in modern technological solutions.

Electrochromic materials are highly energy-efficient and versatile in optical applications. One of their most significant advantages is their ability to maintain a switched state whether transparent or opaque without requiring continuous power, leading to minimal energy consumption. This makes them ideal for energy-saving technologies such as smart windows and displays. Additionally, electrochromic materials possess variable optical properties, allowing them to modulate transparency, reflectance, and light absorption. This capability provides a broad range of optical states, making them highly adaptable for applications in dynamic glazing, electronic displays, and energy-efficient lighting, where precise control over light transmission and reflection is critical.

The ability to change color or become transparent in response to an applied electrical voltage is also an important characteristics of Electrochromic materials. One key factor is their response time, which can vary depending on the material and the application. The speed at which these changes occur typically ranges from a few seconds to several minutes. Faster response times are desirable in applications like smart windows or displays, where quick transitions are needed, while slower response times may be acceptable in other applications like energy-efficient building materials, where gradual changes in light transmission are sufficient. Factors such as material composition, layer thickness, and applied voltage all influence the response time, making it a critical consideration in the design and selection of electrochromic systems.

General Applications

Electrochromic materials are employed in a wide range of innovative applications due to their ability to dynamically change optical properties in response to electrical signals. In smart windows, these materials enable dynamic tinting, adjusting the window's transparency to regulate indoor lighting and temperature, significantly reducing energy consumption for heating, cooling, and lighting. They are also utilized in display screens, such as e-readers, where their low power consumption and high visibility in various lighting conditions make them ideal for electronic paper technologies. In automotive rearview mirrors, electrochromic materials are used to create anti-glare mirrors that automatically adjust reflectance, minimizing glare from headlights of trailing vehicles and improving driving safety. Building facades benefit from electrochromic technologies as adaptive envelopes that alter their optical properties to optimize energy efficiency and aesthetic appearance. Additionally, protective eyewear, such as variable tint goggles, leverages electrochromic materials to change tint based on light exposure, providing protection from glare and harmful UV radiation. These diverse applications highlight the adaptability and efficiency of electrochromic materials in enhancing energy efficiency, comfort, and safety.

Application in Solar Cells

Electrochromic materials offer significant potential to enhance the functionality and efficiency of solar cells through various applications. One key innovation is their integration into smart windows combined with solar cells, providing both energy generation and dynamic regulation of light and heat entering a building. This dual functionality not only generates renewable energy but also improves building energy efficiency by controlling indoor temperatures. Additionally, electrochromic coatings on solar panels enable adaptive light management, adjusting transparency based on sunlight intensity. This ensures optimal light absorption while minimizing overheating, leading to more consistent and efficient energy generation throughout the day. Furthermore, electrochromic materials can be incorporated into hybrid systems that integrate solar cells with energy storage, allowing for seamless management of light, electricity generation, and energy storage, offering a comprehensive solution for both power production and energy conservation. These innovations demonstrate the versatility of electrochromic materials in improving solar energy technologies.

Challenges and Considerations

The application of electrochromic materials faces several challenges and considerations that impact their performance, longevity, and commercial viability. One key concern is the durability and lifespan of these materials, as repeated cycling between states can lead to degradation, reducing their effectiveness over time. Response time is another factor, with some materials taking longer to change optical states, which may not be fast enough for certain applications, such as dynamic displays or automotive windows. Additionally, the production and integration of high-quality electrochromic materials can be expensive, posing barriers to widespread commercialization. Technical challenges also arise when integrating electrochromic materials with existing technologies, such as solar cells, where ensuring long-term compatibility and performance requires innovative solutions. Environmental impact is a further consideration, as some electrochromic materials may contain toxic substances, requiring careful handling and disposal to minimize health and ecological risks. Although maintaining the switched state is energy-efficient, the initial switching process requires power, making energy management a critical aspect of their design. Addressing these challenges is essential to unlocking the full potential of electrochromic materials across various industries.

Electrochromic materials offer unique capabilities for reversible color change in response to electrical voltage, finding applications in smart windows, displays, rearview mirrors, building facades, and protective eyewear. In the context of solar cells, they provide innovative solutions for smart windows, light management, and hybrid energy systems. However, challenges such as durability, response time, manufacturing costs, complex integration, environmental impact, and energy requirements must be addressed to fully harness their potential.

3.1.6 Magnetostrictive Materials

Magnetostrictive materials [139–142] are materials that change their shape or dimensions in response to a magnetic field. This phenomenon is known as magnetostriction, and it occurs due to the reorientation of magnetic domains within the material when exposed to a magnetic field. Conversely, these materials can generate a magnetic field when mechanically deformed. Common magnetostrictive materials include alloys such as Terfenol-D (an alloy of terbium, dysprosium, and iron) and other iron-based compounds. These materials are used in various applications due to their ability to convert magnetic energy into mechanical energy and vice versa.

Characteristics

Magnetostrictive materials are defined by their unique ability to change shape or dimension in response to an applied magnetic field, and conversely, generate a magnetic field when mechanically deformed. This property, known as magnetostriction, is the core characteristic that enables their wide range of applications. These materials also exhibit high energy density, capable of producing significant strains and forces, making them suitable for high-power applications like actuators and sensors. Their broad frequency response allows them to operate from static to ultrasonic frequencies, depending on the application, providing versatility in both low and high-frequency environments. Additionally, magnetostrictive materials are known for their durability and robustness, capable of withstanding harsh environmental conditions such as extreme temperatures and mechanical stress. Their high sensitivity to magnetic fields further enhances their suitability for precision actuation and sensing tasks, making them indispensable in fields requiring both strength and accuracy.

Applications

Magnetostrictive materials are utilized in a variety of advanced applications due to their ability to change shape in response to magnetic fields, offering precise control and functionality across multiple industries. In actuators, they provide high-precision movement control, making them ideal for applications in adaptive optics, robotics, and aerospace systems where accuracy is paramount. As magnetic field sensors, these materials detect subtle changes in magnetic fields, proving valuable in automotive, industrial, and medical applications. Magnetostrictive materials are also key components in ultrasonic transducers, which are used in medical imaging, non-destructive testing, and underwater sonar systems. In the field of energy harvesting, they convert ambient vibrations into electrical energy, enabling the development of efficient vibration energy harvesters. Additionally, they are employed in active damping systems for noise and vibration control, helping reduce unwanted vibrations in machinery and structures, improving performance and longevity. This versatility makes magnetostrictive materials crucial for enhancing precision, energy efficiency, and control across various technological domains.

Application in Solar Cells

Magnetostrictive materials can be integrated into solar cell technology to enhance performance and functionality in several ways.

a. *Self-cleaning Solar Panels (Vibration-Induced Cleaning)*: Magnetostrictive actuators can be used to create vibrations that shake off dust and debris from the surface of solar panels, maintaining their efficiency.
b. *Structural Health Monitoring (Embedded Sensors)*: Magnetostrictive sensors can be embedded in the support structures of solar panels to monitor stress and strain, ensuring structural integrity and optimizing maintenance.

c. *Energy Harvesting (Hybrid Systems)*: Incorporating magnetostrictive materials in solar panels to harvest energy from environmental vibrations, complementing the energy produced by the photovoltaic cells.

Challenges and Considerations

Magnetostrictive materials face several challenges and considerations that impact their widespread adoption and performance. One major issue is the high cost of advanced alloys, such as Terfenol-D, which makes production and processing expensive. Integration complexity also poses a challenge, as incorporating magnetostrictive materials into systems like solar cells requires precise engineering and can be technically demanding. Additionally, energy conversion efficiency is a concern, as the process of converting mechanical energy into electrical energy, and vice versa, may result in efficiency losses that could reduce the overall system's performance. Environmental stability is another factor, with the performance of magnetostrictive materials potentially affected by changes in temperature, humidity, and other external conditions. Moreover, generating the necessary magnetic fields to induce magnetostriction requires additional power, which must be managed efficiently to avoid excess energy consumption. Addressing these issues is crucial for enhancing the viability of magnetostrictive materials in advanced applications.

Conclusion

Magnetostrictive materials offer unique capabilities for converting magnetic energy into mechanical energy and vice versa, finding applications in actuators, sensors, transducers, energy harvesting, and noise and vibration control. In the context of solar cells, they provide innovative solutions for self-cleaning panels, structural health monitoring, and hybrid energy systems. However, challenges such as material cost, integration complexity, energy conversion efficiency, environmental stability, and power requirements must be addressed to fully leverage their potential.

4 Nanosensors and Their Application in Solar Cell

Nanosensors are devices that use nanoscale materials to detect and measure physical, chemical, or biological properties with high sensitivity and specificity. In solar cell technology, nanosensors can be utilized to monitor and optimize the performance of solar panels [143–148]. They can detect minute changes in environmental conditions, such as temperature, light intensity, and chemical composition, allowing for real-time adjustments to enhance efficiency and longevity. Nanosensors can also identify defects or degradation in the solar cells, enabling predictive maintenance and reducing downtime [149–156]. Additionally, they can be integrated into self-cleaning systems that detect and respond to the accumulation of dust or debris on solar panels, ensuring maximum light absorption and energy conversion. The incorporation of nanosensors into solar cells represents a significant advancement in smart solar technology, providing critical data that can lead to improved performance and sustainability.

4.1 Optical Nanosensors

Optical nanosensors [157–160] are a class of sensors that leverage the unique optical properties of nanoscale materials to detect and measure various physical, chemical, or biological parameters. These sensors operate based on changes in light absorption, emission, reflection, or scattering when interacting with the target analytes. The nanoscale dimensions of the sensing elements provide a high surface-to-volume ratio, enhancing sensitivity and enabling the detection of minute quantities of substances.

4.1.1 Key Characteristics

Optical nanosensors offer several advanced features that make them highly effective for various applications. One key advantage is their high sensitivity, as their nanoscale size and large surface area enable the detection of extremely low concentrations of analytes. This, coupled with their efficient interaction at the nanoscale, makes them ideal for precise measurements. These sensors also provide a fast response time, allowing for real-time monitoring and rapid detection. Additionally, through functionalization, the surface of optical nanosensors can be tailored with specific molecules or receptors, enabling highly selective detection of target analytes. Their miniaturized form factor allows them to be easily integrated into compact and portable devices, enhancing versatility. Furthermore, optical nanosensors often utilize non-invasive sensing techniques, permitting detection without altering or damaging the sample, making them suitable for sensitive environments or biological applications.

4.1.2 Applications

Biomedical Diagnostics

Optical nanosensors are revolutionizing biomedical diagnostics by enabling highly sensitive and accurate detection methods. In disease detection, these nanosensors play a critical role in early diagnosis by identifying specific biomarkers in blood or other bodily fluids, allowing for timely and more effective treatment of conditions such as cancer, infections, and metabolic disorders. In advanced imaging techniques, optical nanosensors significantly enhance resolution and contrast, providing clearer and more detailed images at the cellular or molecular level. This improved imaging capability is crucial for precision diagnostics, allowing for better visualization of disease progression and more targeted therapeutic interventions. The integration of optical nanosensors in biomedical diagnostics represents a leap forward in both detection accuracy and imaging clarity, offering immense potential in personalized medicine.

Environmental Monitoring

Optical nanosensors play a crucial role in environmental monitoring by providing advanced detection capabilities for various applications. In air and water quality assessment, these sensors are employed to identify pollutants and hazardous substances, enabling timely responses to environmental hazards. In the realm of food safety, optical nanosensors help detect pathogens, toxins, and chemical residues, ensuring that food products meet safety standards and protecting public health. Additionally, they are instrumental in industrial process control, where they monitor chemical processes and ensure product quality in manufacturing. By leveraging their high sensitivity and specificity, optical nanosensors contribute significantly to maintaining environmental integrity, ensuring food safety, and enhancing industrial efficiency.

Security and Defense

Optical nanosensors play a crucial role in enhancing security and defense applications, particularly in explosive detection. These advanced sensors utilize nanoscale materials to identify explosives and hazardous substances during security screenings, offering high sensitivity and specificity. By harnessing the unique optical properties of nanomaterials, such as surface plasmon resonance or fluorescence, these sensors can detect minute traces of explosive compounds, significantly improving the accuracy and speed of threat assessments. Their ability to operate in real-time allows for swift decision-making in critical situations, making optical nanosensors an invaluable tool in safeguarding public spaces and ensuring the safety of military operations. Furthermore, their compact size and potential for integration into portable devices enhance their utility in various security contexts, from airport screenings to battlefield monitoring.

4.1.3 Application in Solar Cells

Optical nanosensors play a significant role in enhancing the performance and reliability of solar cells.

Performance Monitoring

Optical nanosensors are increasingly utilized in solar cells for performance monitoring, significantly enhancing their efficiency and longevity. These nanosensors can accurately monitor light intensity and spectral distribution incident on solar panels, allowing for real-time adjustments that optimize energy conversion efficiency. By providing detailed insights into how varying light conditions affect solar cell performance, they enable more effective energy harvesting strategies. Additionally, optical nanosensors are instrumental in temperature sensing, detecting variations that could lead to overheating. This capability is crucial for improving the thermal management of solar panels, as it helps prevent performance degradation and extends the lifespan of the solar cells. Overall, the integration of optical nanosensors in solar technology not only enhances operational efficiency but also contributes to the sustainable performance of solar energy systems.

Defect Detection and Quality Control

Optical nanosensors are equally effective for defect detection, playing a pivotal role in quality control during both manufacturing and operational phases. These sensors enable the identification of defects, such as microcracks and other structural imperfections, that can significantly impact the performance and longevity of solar cells. By employing advanced optical techniques, such as light scattering or photoluminescence, optical nanosensors can detect even the smallest anomalies, ensuring that only high-quality products reach the market. This proactive approach to quality assurance not only minimizes the risk of failures but also enhances the overall efficiency and reliability of solar energy systems. By integrating these nanosensors into the production and monitoring processes, manufacturers can optimize performance, extend the lifespan of solar panels, and ultimately contribute to the growth and sustainability of renewable energy technologies.

Environmental Adaptation

Optical nanosensors may be integrated into solar cells to create adaptive systems that optimize performance in response to varying environmental conditions. These advanced sensors can detect factors such as dust accumulation, shading, and changes in light intensity, enabling real-time adjustments to the solar cell's operation. By monitoring the conditions surrounding the solar panels, the nanosensors facilitate automatic adjustments, ensuring that the cells maintain optimal energy output even in less-than-ideal situations. This capability not only enhances energy efficiency but also prolongs the lifespan of the solar panels by minimizing the negative impacts of environmental stressors. The integration of optical nanosensors in solar technology represents a significant advancement in the pursuit of more resilient and efficient renewable energy systems, making solar power generation more reliable and effective in diverse settings.

Energy Harvesting

Energy harvesting through advanced photon management techniques is possible employing optical nanosensors. By employing nanoscale optical manipulation, these sensors enhance light absorption and significantly reduce losses associated with photon transmission into the solar cell. This precision allows for better control over how photons interact with the solar material, improving the efficiency of light capture and conversion into electrical energy. For instance, by utilizing nanostructures to create anti-reflective coatings or photonic crystals, the overall absorption spectrum can be broadened, enabling the solar cells to harness a greater range of wavelengths. Additionally, these nanosensors can help in mitigating effects such as reflection and scattering, which traditionally hinder performance. As a result, the integration of optical nanosensors in solar technology not only increases the energy yield but also contributes to the development of more efficient and cost-effective renewable energy solutions.

4.1.4 Challenges and Considerations

The implementation of optical nanosensors presents several challenges and considerations, particularly when integrating them with solar cell systems. One major challenge is the complexity of integration, which requires advanced manufacturing techniques and precise engineering to ensure compatibility and optimal performance. Additionally, the long-term durability and stability of nanosensors must be ensured, as they need to perform reliably under varying environmental conditions, including fluctuations in temperature and humidity. Economic feasibility is another concern; the high costs associated with producing and integrating these nanosensors can impact the overall cost-effectiveness of solar cell systems. Furthermore, maintaining the accuracy and reliability of optical nanosensors necessitates regular calibration and standardization to ensure consistent performance. They are also susceptible to signal interference from ambient light and other environmental factors, making effective noise reduction techniques essential for accurate detection. Addressing these challenges is vital for fully harnessing the potential of optical nanosensors in solar energy applications and beyond.

4.1.5 Conclusion

Optical nanosensors offer significant advantages in terms of sensitivity, specificity, and miniaturization, making them invaluable in various applications, including biomedical diagnostics, environmental monitoring, food safety, industrial process control, and security. In the context of solar cells, they provide critical capabilities for performance monitoring, defect detection, environmental adaptation, and energy harvesting. However, challenges such as integration complexity, durability, cost, calibration, and signal interference must be addressed to fully realize their potential in enhancing solar cell technology.

4.2 Thermal Nanosensors

Thermal nanosensors [161–164] are devices that detect and measure temperature variations at the nanoscale. These sensors leverage the unique properties of nanomaterials, such as high surface area, quantum effects, and enhanced sensitivity, to provide precise thermal measurements. Common nanomaterials used in thermal nanosensors include carbon nanotubes, graphene, and nanowires of various metals and semiconductors.

4.2.1 Key Characteristics

Thermal nanosensors are characterized by several advantageous features that enhance their utility in various applications. One of their primary strengths is high sensitivity, allowing them to detect extremely low concentrations of analytes thanks to their large surface area and efficient interactions at the nanoscale. Additionally, these sensors typically provide fast response times, making them ideal for real-time monitoring of environmental or biological processes. Their specificity is further enhanced by the functionalization of the nanosensor surface with tailored molecules or receptors, enabling selective detection of target analytes amidst complex backgrounds. The miniaturization of these sensors facilitates their integration into compact and portable devices, enhancing their practicality in field applications. Moreover, many optical detection methods employed by thermal nanosensors are non-invasive, allowing for measurements to be conducted without altering the sample, which is particularly beneficial in sensitive contexts such as biomedical diagnostics. Together, these features position thermal nanosensors as powerful tools in diverse fields, including environmental monitoring, healthcare, and industrial applications.

4.2.2 Applications

Biomedical

Thermal nanosensors have significant applications in the biomedical field, providing valuable insights and enhancing various medical processes. One prominent use is in cellular thermography, where these sensors measure the temperature of individual cells or tissues, facilitating accurate medical diagnostics and research into cellular behavior. They are also instrumental in monitoring temperatures during hyperthermia treatments for cancer, ensuring that the heat applied to target tissues remains within optimal therapeutic ranges. Beyond biomedical applications, thermal nanosensors are utilized in electronics, specifically for monitoring the temperature of semiconductor devices and integrated circuits, helping to prevent overheating and optimize overall performance. Furthermore, these sensors contribute to the development of advanced thermal management systems for electronic devices, ensuring effective heat dissipation and enhancing device reliability. Collectively, the capabilities of thermal nanosensors play a crucial role in advancing both medical diagnostics and electronic device performance.

Material Science

Other significant application of thermal nanosensors is in phase transition studies, where these sensors enable researchers to investigate thermal characteristics such as phase transitions and thermal conductivity of materials with high precision. This understanding is crucial for developing new materials with tailored properties for specific applications. Additionally, thermal nanosensors are utilized in environmental monitoring and temperature mapping, where they track temperature variations across diverse ecosystems. By offering high sensitivity and rapid response times, these sensors help researchers understand environmental dynamics and assess the impact of climate change. Overall, the application of thermal nanosensors in material science not only advances fundamental research but also supports practical environmental management efforts.

Industrial

In industries, particularly for process control, thermal nanosensors play vital roles in monitoring temperature during manufacturing processes. Their high sensitivity and rapid response times enable precise temperature measurements, which are crucial for ensuring quality control and optimizing various production parameters. By providing real-time data on temperature fluctuations, these nanosensors help maintain optimal operating conditions, reducing the risk of defects and enhancing product consistency. Furthermore, the compact size and integration capabilities of thermal nanosensors allow them to be seamlessly incorporated into existing production lines, facilitating continuous monitoring without disrupting workflows. This application not only improves efficiency and product quality but also contributes to energy savings and waste reduction in manufacturing processes, making thermal nanosensors an invaluable asset in modern industrial operations.

4.2.3 Application in Solar Cells

Thermal nanosensors can significantly enhance the performance and reliability of solar cells through various mechanisms:

Temperature Monitoring

By providing continuous monitoring of solar cell temperatures, these sensors help prevent overheating, a condition that can significantly degrade the performance and reduce the lifespan of solar panels. Additionally, thermal nanosensors contribute to optimizing thermal management systems, ensuring uniform temperature distribution across the solar panel. This uniformity not only improves the overall efficiency of the solar energy system but also minimizes the risk of hot spots that can lead to performance losses. Through the integration of thermal nanosensors, solar energy systems can achieve more reliable operation, maximizing energy output and extending the operational life of the panels.

Performance Optimization

Integration of these sensors into solar cell systems, adaptive systems can be developed that dynamically adjust the operation of the solar cells based on real-time temperature data, thereby optimizing energy output and enhancing efficiency. This capability allows for precise monitoring of thermal conditions, ensuring that the solar cells operate at their most efficient temperatures. Additionally, thermal nanosensors serve as early warning systems for failure detection by identifying hotspots or abnormal temperature rises that may indicate potential defects or failures within the solar cells. This proactive monitoring enables timely maintenance and repairs, significantly reducing downtime and extending the lifespan of solar energy systems. Overall, the incorporation of thermal nanosensors enhances the reliability and efficiency of solar cells, contributing to more effective harnessing of solar energy.

Energy Harvesting

Thermal nanosensors play a pivotal role in enhancing energy harvesting capabilities within hybrid solar cells through thermoelectric generation. By effectively monitoring and managing the temperature of thermoelectric materials, these nanosensors enable the efficient conversion of waste heat into additional electrical energy. This capability is particularly valuable in solar cell applications, where excess heat generated during operation can be harnessed to improve overall energy efficiency. The high sensitivity and fast response times of thermal nanosensors allow for real-time monitoring of thermal conditions, facilitating optimized performance and management of thermoelectric materials. As a result, the integration of thermal nanosensors in hybrid solar cell systems not only boosts energy output but also contributes to more sustainable and efficient energy utilization, making them an essential component in the advancement of renewable energy technologies.

4.2.4 Challenges and Considerations

The integration of thermal nanosensors into solar cells presents several challenges and considerations that must be addressed for optimal functionality. One significant hurdle is integration complexity, as incorporating these nanosensors into solar cell systems demands advanced fabrication and precise integration techniques to ensure seamless operation. Additionally, the durability and stability of thermal nanosensors must be ensured, particularly in the harsh outdoor environments where solar cells typically operate, to guarantee long-term reliability. Maintaining calibration and accuracy is also critical, as precise thermal measurements require regular calibration and careful consideration of varying environmental factors that can affect sensor performance. Material compatibility poses another challenge; it is essential that the materials used in thermal nanosensors do not negatively interact with those in solar cells, potentially compromising system performance. Lastly, the economic viability of developing and integrating these sensors must be justified, requiring that the performance enhancements and cost savings they offer outweigh their associated costs. Addressing these challenges is crucial for harnessing the full potential of thermal nanosensors in enhancing the efficiency and effectiveness of solar energy systems.

4.2.5 Conclusion

Thermal nanosensors offer high sensitivity, rapid response, and high spatial resolution, making them ideal for a wide range of applications, including biomedical diagnostics, electronics, material science, environmental monitoring, and industrial process control. In solar cells, they provide crucial capabilities for temperature monitoring, performance optimization, failure detection, and energy harvesting. However, challenges such as integration complexity, durability, calibration, material compatibility, and cost must be addressed to fully realize their potential in enhancing solar cell technology.

4.3 Chemical Nanosensors

Chemical nanosensors [165–169] are devices that use nanomaterials to detect and quantify the presence of specific chemical substances with high sensitivity and selectivity. These sensors operate based on various detection mechanisms, including changes in electrical conductivity, optical properties, or mechanical behavior when interacting with target chemicals. The nanoscale dimensions of the sensing elements enhance their interaction with chemical molecules, leading to improved detection capabilities.

4.3.1 Key Characteristics

Chemical nanosensors are distinguished by their high sensitivity and rapid response capabilities, making them invaluable tools in various applications. Their high surface area-to-volume ratio significantly enhances the likelihood of interactions with target molecules, allowing for the detection of very low concentrations of chemicals. This sensitivity is further amplified through functionalization, where the surface of nanosensors is tailored with specific receptors or binding sites to selectively detect particular chemicals, ensuring accurate measurements in complex environments. Additionally, the small size and high reactivity of nanomaterials enable these sensors to provide fast detection times, facilitating timely responses in dynamic settings. Their compact design allows for seamless integration into portable and miniature devices, enhancing their practicality for on-the-go monitoring and analysis. Together, these features position chemical nanosensors as powerful tools for advancing fields such as environmental monitoring, healthcare diagnostics, and industrial applications. Chemical nanosensors can utilize various transduction methods, such as electrochemical, optical, or piezoelectric, to detect chemical changes.

4.3.2 Applications

Chemical nanosensors have various applications, particularly in environmental monitoring and healthcare diagnostics. In environmental monitoring, these sensors are utilized for pollution detection, allowing for the real-time assessment of air and water quality by identifying pollutants and toxic chemicals, which is crucial for safeguarding public health and the ecosystem. They are also employed in soil analysis, measuring the concentration of nutrients and contaminants to inform agricultural practices and environmental remediation efforts. In the healthcare sector, chemical nanosensors are invaluable for disease detection, as they can identify biomarkers in bodily fluids, facilitating early diagnosis and personalized treatment strategies. Additionally, they are used for drug monitoring, enabling the tracking of pharmaceutical concentrations in the bloodstream to ensure therapeutic effectiveness and minimize adverse effects. This versatility underscores the significant impact of chemical nanosensors in enhancing safety and health outcomes across various domains.

They are also instrumental across various sectors, including industrial process control, food safety, and security and defense. In industrial settings, these sensors play a crucial role in quality assurance by continuously monitoring chemical processes to ensure product quality and consistency. They are also vital for leak detection, swiftly identifying hazardous chemical leaks, thereby enhancing workplace safety and compliance with regulations. In the realm of food safety, chemical nanosensors are employed to detect contaminants, enabling the identification of pesticides, toxins, or spoilage indicators in food products, which helps maintain public health standards. Additionally, in security and defense applications, these sensors are essential for explosive detection, as they can identify specific chemical signatures of explosives and hazardous materials, aiding in threat assessment and prevention. Collectively, the versatility and precision of chemical nanosensors make them invaluable tools for enhancing safety and quality across diverse industries.

4.3.3 Application in Solar Cells

Chemical nanosensors can play a vital role in enhancing the efficiency and longevity of solar cells through several mechanisms. The manufacturing and maintenance of solar cells may be improved by ensuring material integrity and optimizing performance. One key application of chemical sensors is purity detection, where these sensors are used to identify contaminants and impurities during the manufacturing process, guaranteeing the high quality of materials used in solar cells. Additionally, they facilitate the monitoring of surface cleanliness, detecting the accumulation of dust, debris, or organic materials on solar panels. This timely information allows for efficient cleaning and maintenance, which is essential for maintaining optimal light absorption and overall efficiency. Furthermore, chemical nanosensors are employed for degradation monitoring, enabling the detection of chemical changes in solar cell materials that indicate wear or degradation. This capability allows for predictive maintenance and timely replacement of components before significant performance loss occurs. In terms of encapsulation optimization, these sensors can identify moisture ingress in encapsulated solar cells, which can lead to corrosion and further degradation of the materials. By addressing these factors, chemical nanosensors significantly enhance the reliability and longevity of solar energy systems. In hybrid solar systems, chemical nanosensors can optimize the interaction between photovoltaic and chemical energy storage or conversion systems.

4.3.4 Challenges and Considerations

The integration of chemical nanosensors into solar cell systems presents several challenges that must be carefully addressed to ensure optimal performance and longevity. One significant hurdle is integration complexity, as incorporating these advanced sensors requires sophisticated engineering and fabrication techniques to achieve seamless functionality within the solar cell framework. Additionally, the durability and stability of chemical nanosensors are critical, as they must withstand the harsh outdoor environments in which solar cells operate, including fluctuations in temperature, humidity, and exposure to various chemicals. Calibration and accuracy are also vital; regular calibration is necessary to maintain precise measurements and reliability, particularly in light of potential interference from other environmental factors. Furthermore, material compatibility is a crucial consideration, ensuring that the materials used in chemical nanosensors do not adversely interact with solar cell materials, which could detrimentally affect overall performance. Finally, the economic viability of developing and integrating these sensors must be assessed, balancing the costs involved against the performance enhancements and potential cost savings they can provide. Addressing these challenges is essential for maximizing the benefits of chemical nanosensors in solar energy applications.

4.3.5 Conclusion

Chemical nanosensors offer high sensitivity, selectivity, rapid response, and versatility, making them invaluable in a wide range of applications, including environmental monitoring, healthcare, industrial process control, food safety, and security. In solar cells, they provide critical capabilities for material quality monitoring, environmental adaptation, degradation monitoring, encapsulation optimization, and hybrid system optimization. However, challenges such as integration complexity, durability, calibration, material compatibility, and cost must be addressed to fully realize their potential in enhancing solar cell technology.

4.4 Mechanical Nanosensors

Mechanical nanosensors [170–173] are devices designed to detect and measure mechanical properties such as force, pressure, displacement, and acceleration at the nanoscale. These sensors typically employ nanomaterials like carbon nanotubes, graphene, and nanowires, which offer exceptional mechanical properties, including high strength, flexibility, and sensitivity. The small size and unique properties of these materials allow mechanical nanosensors to perform highly precise and localized measurements.

4.4.1 Key Characteristics

Nanomechanical sensors offer exceptional sensitivity and resolution, making them invaluable tools for detecting and measuring forces at the nanoscale. These sensors leverage nanomechanical detection, allowing them to detect extremely small forces, pressures, or displacements due to the heightened mechanical responsiveness of nanomaterials. This high sensitivity is further enhanced by piezoelectric and piezoresistive effects, where mechanical deformation induces changes in electrical resistance or voltage, enabling precise force measurements. Nanomechanical sensors also provide high resolution, offering nanoscale precision that allows for detailed mapping of mechanical properties with fine spatial accuracy. Their wide dynamic range ensures the measurement of both small and large forces without sacrificing accuracy, making them versatile across various applications. The compact design of these sensors, enabled by the small size of nanomaterials, allows for easy integration into micro- and nanoscale devices. Additionally, nanomaterials like carbon nanotubes and graphene contribute to the robustness and mechanical durability of these sensors, ensuring long-term reliability even under demanding conditions. Together, these characteristics make nanomechanical sensors a powerful tool for advanced sensing and measurement applications.

4.4.2 Applications

Nanomechanical sensors have transformative applications across various fields, particularly in biomedical, industrial, and aerospace sectors. In biomedical applications, these sensors play a crucial role in measuring cellular mechanics, allowing researchers to observe forces exerted by and on individual cells. This is invaluable for understanding disease progression, cell motility, and tissue mechanics. Additionally, nanomechanical sensors are used in medical diagnostics to detect subtle mechanical changes in biological tissues or fluids, offering a new dimension of precision for diagnostic tools. In the industrial domain, they enhance quality control by monitoring the mechanical properties of materials and components during manufacturing, ensuring product consistency and reliability. They are also vital in structural health monitoring, where they detect stress, strain, and potential damage in critical infrastructure such as bridges, buildings, and aircraft, thus preventing catastrophic failures.

In electronics and aerospace, nanomechanical sensors offer significant advancements. They are integrated into Microelectromechanical Systems (MEMS), where their sensitivity and precision contribute to sensing, actuation, and signal processing functions, crucial for modern devices. In the aerospace industry, these sensors monitor mechanical stresses and vibrations in flight systems, ensuring the safety, reliability, and performance of critical components. Furthermore, in consumer electronics, nanomechanical sensors are widely used in touch-sensitive devices, where their compact design and high sensitivity enhance the user interface's precision and responsiveness, improving the overall experience in smartphones, tablets, and other gadgets. These versatile applications demonstrate the far-reaching impact of nanomechanical sensors in advancing both technology and safety across multiple sectors.

4.4.3 Application in Solar Cells

Mechanical nanosensors hold significant potential for enhancing solar cell technology by improving structural monitoring and optimizing performance. These sensors can detect stress and strain in solar panels caused by environmental factors such as wind, thermal expansion, and temperature fluctuations, which is essential for maintaining the structural integrity and extending the lifespan of solar panels. They can also detect surface deformations that may reduce light absorption efficiency, enabling corrective actions to ensure optimal energy conversion. Additionally, mechanical nanosensors can be integrated with self-cleaning mechanisms by identifying the accumulation of dust or debris on the panel surface, preventing a decline in performance. Through their precise monitoring, these sensors offer a proactive approach to maintenance, ensuring that solar cells operate efficiently under various environmental conditions.

Moreover, mechanical nanosensors are crucial for damage detection, identifying micro-cracks or other defects in solar panels before they become significant issues, allowing for timely repair or replacement. This early detection can prevent major performance degradation and extend the operational life of solar cells. Furthermore, these sensors can provide feedback for adaptive systems, enabling solar panels to adjust their orientation or position in response to mechanical stress or environmental factors like wind or snow. Such adaptive systems improve energy capture while protecting the panels from mechanical damage. Overall, the integration of mechanical nanosensors enhances the durability, efficiency, and maintenance of solar cells, making them more resilient and reliable for long-term energy generation.

4.4.4 Challenges and Considerations

Integrating mechanical nanosensors into solar cells poses significant challenges, particularly in terms of fabrication and environmental durability. Advanced fabrication techniques are necessary for incorporating these nanosensors, as they require precision engineering to seamlessly integrate with the existing solar cell materials and systems. This complexity can drive up production costs, making the economic viability of widespread adoption a key consideration. Furthermore, the durability of mechanical nanosensors must be ensured in harsh outdoor environments. Solar cells are regularly exposed to fluctuating temperatures, humidity, and mechanical stresses such as wind or snow load, which can affect the long-term stability and performance of the nanosensors. Maintaining accurate measurements in such conditions, while minimizing signal interference from environmental factors or electronic noise, is crucial for reliable operation.

Material compatibility is another critical factor in integrating mechanical nanosensors into solar cells. The materials used in these nanosensors must be carefully selected to avoid any adverse interactions with the solar cell components, as such interactions could degrade performance or cause long-term damage. Additionally, economic viability plays a significant role in determining whether the cost of developing and integrating these sensors can be justified by the performance enhancements and cost savings they provide. Striking a balance between material compatibility, durability, and cost-effectiveness will be essential for the successful integration of mechanical nanosensors into next-generation solar energy systems.

4.4.5 Conclusion

Mechanical nanosensors offer high sensitivity, resolution, dynamic range, and robustness, making them invaluable in various applications, including biomedical research, industrial quality control, electronics, aerospace, and consumer electronics. In solar cells, they provide critical capabilities for structural monitoring, efficiency optimization, self-cleaning mechanisms, damage detection, and adaptive systems. However, challenges such as integration complexity, environmental durability, signal interference, material compatibility, and cost must be addressed to fully realize their potential in enhancing solar cell technology.

5 Integration of Smart Materials into Solar Cells

Integrating smart materials into solar cells represents a promising approach to enhance their efficiency, flexibility, and overall performance. Smart materials can adapt to environmental changes and offer advanced functionalities, making them suitable for advanced solar cell technologies. Key strategies for integration include surface functionalization, layer integration, and the creation of hybrid nanocomposites.

5.1 Surface Functionalization

Surface functionalization [174–178] involves modifying the surface properties of solar cell materials to improve their performance or add new functionalities. This process typically includes the following.

5.1.1 Chemical Modification

Coating or treating the surface of solar cell materials with specific chemicals to enhance their interaction with light or improve their stability. For example, adding hydrophobic or hydrophilic layers can help manage surface contamination or improve water resistance.

5.1.2 Nano-coatings

Applying thin layers of nanomaterials such as metal nanoparticles, graphene, or carbon nanotubes to enhance light absorption, reduce reflection, or increase conductivity. These coatings can improve the light trapping and overall efficiency of solar cells.

5.1.3 Functional Groups

Attaching functional groups to the surface of materials to enable selective interaction with light or environmental stimuli. For instance, integrating photoactive groups can lead to enhanced light absorption and energy conversion.

5.1.4 Applications and Benefits

Surface functionalization offers significant advantages in solar cell technology, enhancing performance and longevity. By modifying the surface properties of solar cells, light absorption and photocurrent generation can be improved, leading to more efficient energy conversion. Additionally, functionalized surfaces contribute to greater stability and durability, protecting the cells from environmental degradation and wear over time. An important benefit is the potential for self-cleaning or contaminant-repellent surfaces, which help maintain optimal performance by preventing the accumulation of dust, debris, or other pollutants that could reduce light penetration and efficiency. These innovations collectively contribute to higher efficiency and reduced maintenance costs for solar cells.

5.1.5 Challenges and Considerations

Surface functionalization presents challenges in ensuring uniform and stable coatings across large surface areas, which is critical for consistent performance in applications such as solar cells or sensors. Achieving homogeneity in the functional layers can be difficult, particularly when scaling up production. Additionally, potential interactions between the functionalized surface and the underlying materials must be carefully managed, as these interactions can affect the stability, adhesion, and overall performance of the functional layer. Balancing these factors is key to optimizing the effectiveness and durability of surface functionalization in advanced materials and devices.

5.2 Layer Integration

Layer integration refers to combining different layers of materials to create multifunctional solar cells [179–181]. This process can involve:

5.2.1 Layer Stacking

Assembling multiple layers of materials with different properties to capture a broader spectrum of sunlight and enhance overall efficiency. For instance, integrating perovskite layers with silicon or thin-film materials can improve light absorption and energy conversion.

5.2.2 Flexible Substrates

Using flexible materials as substrates to enable the integration of smart materials and facilitate the development of flexible and lightweight solar cells. Flexible substrates can accommodate the expansion and contraction of smart materials in response to environmental changes.

5.2.3 Functional Interfaces

Creating interfaces between different material layers that enhance charge transfer and reduce energy loss. For example, using interlayers with high electrical conductivity can improve charge extraction and overall efficiency.

5.2.4 Applications and Benefits

Layer integration in solar cells offers several key applications and benefits that significantly enhance overall performance. By incorporating multiple layers of different materials, it improves light absorption and energy conversion efficiency, allowing solar cells to capture a broader spectrum of sunlight. This approach also facilitates the development of flexible and lightweight solar cells, which are more adaptable to diverse applications, including portable and wearable devices. Furthermore, layer integration enables the synergy of different materials, optimizing the unique properties of each layer to enhance overall efficiency, durability, and functionality, ultimately leading to more efficient and cost-effective solar energy solutions.

5.2.5 Challenges and Considerations

Layer integration in solar cells presents challenges related to ensuring compatibility and strong adhesion between different material layers, which is crucial for the overall efficiency and longevity of the system. Poor adhesion can lead to delamination, reducing the effectiveness of energy transfer and potentially causing device failure. Additionally, managing the stability and durability of these layers over time is vital, as environmental factors such as temperature fluctuations, humidity, and mechanical stress can weaken layer bonds and degrade performance. Advanced material engineering and precise fabrication techniques are needed to address these challenges and ensure long-term reliability.

5.3 *Hybrid Nanocomposites*

Hybrid nanocomposites [182, 183] involve combining nanomaterials with other materials to create composite structures with enhanced properties. This approach includes the following.

5.3.1 Nanomaterial Integration

Incorporating nanomaterials such as quantum dots, nanowires, or carbon nanotubes into the solar cell matrix to improve light absorption, charge transport, and overall efficiency. These materials can provide unique optical and electrical properties that enhance solar cell performance.

5.3.2 Composite Materials

Creating composite materials by blending smart materials with traditional solar cell materials. For example, combining organic–inorganic perovskites with silicon or other semiconductor materials to develop high-efficiency hybrid solar cells.

5.3.3 Enhanced Performance

Utilizing the unique properties of nanocomposites, such as increased surface area and high electrical conductivity, to achieve better light absorption, charge separation, and energy conversion.

5.3.4 Applications and Benefits

Hybrid nanocomposites offer significant applications and benefits in solar cell technologies by combining the unique properties of different materials to enhance overall efficiency and performance. The synergy between these materials enables the development of advanced solar cells with improved functionalities, such as greater light absorption, higher energy conversion rates, and increased durability. These nanocomposites also provide opportunities for innovation in designing flexible, lightweight, and more efficient solar cells, paving the way for large-scale manufacturing and commercialization. Their versatility and potential for cost-effective production make hybrid nanocomposites a promising solution for next-generation solar energy systems.

5.3.5 Challenges and Considerations

Hybrid nanocomposites face significant challenges in achieving uniform dispersion and integration of nanomaterials within the composite matrix, as uneven distribution can lead to inconsistent properties and compromised performance. Additionally, the stability and durability of these hybrid materials must be carefully addressed, as interactions between the nanomaterials and the matrix can lead to degradation over time, particularly under varying environmental conditions such as temperature and humidity. Ensuring robust interfacial bonding between components is crucial for maintaining the mechanical and thermal properties of the nanocomposite. Advanced fabrication techniques and thorough characterization methods are essential to optimize the dispersion of nanomaterials and enhance the overall stability and longevity of hybrid nanocomposites in practical applications.

5.4 Considerations

Integrating smart materials into solar cells through surface functionalization, layer integration, and hybrid nanocomposites offers a pathway to enhance their efficiency, flexibility, and performance. Each approach brings unique benefits and challenges, necessitating careful consideration of material compatibility, stability, and manufacturing processes. The ongoing research and development in these areas hold the potential to advance solar cell technology and contribute to the widespread adoption of solar energy.

6 Enhancements in Solar Cell Performance Employing Smart Materials as Nanosensors

The integration of smart materials as nanosensors in solar cells has emerged as a cutting-edge approach to enhancing the performance, efficiency, and longevity of photovoltaic systems. These smart materials possess the ability to respond dynamically to environmental changes, thereby optimizing the operation of solar cells under varying conditions.

6.1 Real-Time Monitoring and Adaptive Response

Smart materials, when used as nanosensors, enable real-time monitoring of critical parameters such as temperature, light intensity, and electrical performance within solar cells. By incorporating these nanosensors, solar cells can dynamically adjust to changing environmental conditions. For instance, thermochromic or photochromic materials can alter their optical properties in response to temperature or light intensity, optimizing the absorption of sunlight and minimizing energy losses.

6.2 Enhanced Light Absorption

Nanosensors made from materials like quantum dots or plasmonic nanoparticles can be integrated into the solar cell structure to enhance light absorption. These materials can be tuned to absorb a broader spectrum of sunlight, including infrared and ultraviolet wavelengths, which are typically not absorbed efficiently by conventional materials. This broader absorption spectrum leads to higher energy conversion efficiencies.

6.3 Improved Charge Separation and Transport

Smart materials can also be employed to improve charge separation and transport within solar cells [184–187] For example, piezoelectric nanosensors can generate an electric field in response to mechanical stress, aiding in the separation of electron–hole pairs generated by sunlight. This can reduce recombination losses and increase the overall efficiency of the solar cell.

6.4 Self-healing and Longevity

Some smart materials, such as shape memory alloys or self-healing polymers, can be used to extend the lifespan of solar cells [188–197]. These materials can detect and respond to microcracks or other structural damages, initiating a repair process that restores the cell's functionality. This self-healing capability reduces maintenance costs and improves the durability of solar installations.

6.5 Energy Harvesting and Storage

Incorporating smart materials as nanosensors can also contribute to more efficient energy harvesting and storage. For instance, ferroelectric or magnetostrictive materials can convert mechanical energy (such as vibrations or wind) into electrical energy, providing an additional power source for the solar cell system. This multi-functional approach maximizes energy capture and utilization.

6.6 Environmental Sensing and Adaptation

Smart nanosensors can detect environmental pollutants or other harmful agents and adapt the solar cell's functioning to maintain optimal performance [198–200]. For example, chemical nanosensors can monitor air quality and adjust the cell's protective coatings or surface treatments to prevent degradation from pollutants.

6.7 Challenges and Considerations

The integration of nanosensors into solar cells presents several challenges, primarily related to engineering precision, cost, scalability, and long-term reliability. Achieving compatibility with existing materials and manufacturing processes necessitates meticulous design and engineering to ensure seamless integration without compromising performance. While smart materials and nanosensors offer significant advantages, their production costs and scalability for large-scale deployment pose considerable obstacles that must be addressed to facilitate widespread use. Furthermore, ensuring the long-term stability and reliability of these nanosensors under harsh environmental conditions is crucial for their successful adoption in solar technologies, as fluctuations in temperature, humidity, and mechanical stress can significantly affect their functionality and lifespan. Addressing these challenges will be key to unlocking the full potential of nanosensors in enhancing solar energy systems.

6.8 Conclusion

The use of smart materials as nanosensors in solar cells represents a significant advancement in photovoltaic technology. By enabling real-time monitoring, adaptive responses, and enhanced functionalities, these materials can lead to substantial improvements in solar cell efficiency, durability, and overall performance. Ongoing research and development in this area hold the promise of making solar energy more accessible and effective in meeting global energy demands.

7 Future Directions and Challenges

The integration of smart materials as nanosensors in solar cells is an exciting frontier in renewable energy technology. These innovations promise to enhance solar cell efficiency, durability, and adaptability, paving the way for more sustainable and reliable solar energy systems. However, the path to widespread adoption is lined with both opportunities and challenges.

7.1 Future Directions

7.1.1 Advanced Material Design

The future of solar technology is poised for transformative advancements through the development of new smart materials that exhibit enhanced functionalities. One promising area of research involves the creation of multi-responsive materials capable of simultaneously reacting to various stimuli, such as temperature, light, and mechanical stress. This innovation could lead to the design of solar cells that not only harness solar energy more effectively but also adapt dynamically to environmental changes, optimizing their performance and efficiency in real-time.

In parallel, there is a growing emphasis on the exploration of hybrid nanocomposites that integrate the beneficial properties of different materials, particularly those combining piezoelectric and thermoelectric characteristics. Such multifunctional nanosensors could significantly enhance the performance of solar cells by enabling efficient energy conversion and improved monitoring of environmental conditions. By harnessing the synergistic effects of these materials, researchers aim to develop solar technologies that are more efficient, durable, and responsive to their operating environments, paving the way for the next generation of renewable energy solutions.

7.1.2 Enhanced Integration Techniques

Future advancements in solar cell technology will significantly benefit from enhanced integration techniques, particularly through innovations like layer-by-layer assembly and advanced surface functionalization. Techniques such as atomic layer deposition (ALD) and molecular self-assembly will facilitate precise control over the placement and functionality of nanosensors within solar cells. This level of precision will enable the creation of more efficient and responsive solar energy systems, as nanosensors can be strategically positioned to optimize their performance in real-time monitoring and environmental adaptation. Such advancements are expected to improve the overall efficiency and lifespan of solar cells, making them more competitive in the renewable energy market.

In addition to layer integration, the development of sophisticated surface functionalization methods will play a crucial role in enhancing the interface between nanosensors and solar cell materials. Improved functionalization techniques will enable better adhesion and interaction between the nanosensors and the underlying materials, leading to enhanced performance, stability, and durability. This will not only boost the reliability of nanosensors in varying environmental conditions but also foster the development of hybrid materials that can capitalize on the unique properties of both nanosensors and traditional solar cell components. As these technologies advance, they will pave the way for more innovative and efficient solar energy solutions, contributing to a more sustainable future.

7.1.3 Improving Efficiency and Lifespan

The future of solar cell technology holds exciting prospects for improving the efficiency and lifespan of nanosensors through the development of dynamic response systems. These systems are designed to actively monitor environmental changes in real time, enabling solar cells to adjust their properties and optimize performance under varying conditions. By implementing advanced sensors and adaptive algorithms, these dynamic systems can enhance energy capture and utilization, ensuring that solar cells maintain peak efficiency even in fluctuating environmental scenarios. This capability not only maximizes energy output but also contributes to the long-term reliability and durability of the solar cells by mitigating potential damage from external factors such as temperature shifts, humidity, and mechanical stress.

Another promising avenue for extending the lifespan of solar cells is the integration of self-healing smart materials. These innovative materials possess the ability to autonomously repair minor damages, such as micro-cracks and surface abrasions, that can occur during operation. By incorporating self-healing mechanisms into the solar cell design, maintenance costs can be significantly reduced while improving overall reliability. This approach not only prolongs the operational life of solar cells but also enhances their performance, allowing for sustained energy production over time. As research and development in this field advance, the incorporation of dynamic response systems and self-healing materials could revolutionize solar technology, making it more efficient, resilient, and cost-effective for widespread adoption.

7.1.4 Scalability and Manufacturing

The future of nanosensor-based solar cells hinges on the development of cost-effective and scalable manufacturing processes that can facilitate the widespread integration of smart materials. As the demand for sustainable energy solutions continues to rise, creating economically viable methods for producing these advanced solar technologies will be paramount. Innovations in manufacturing techniques, such as roll-to-roll printing or advanced coating methods, can significantly reduce production costs while maintaining high-quality standards. By optimizing these processes, the industry can enhance the commercial viability of nanosensor-integrated solar cells, making them more accessible to a broader market and driving their adoption in renewable energy applications.

In addition to cost-effective manufacturing, establishing standardized protocols for the production and integration of nanosensors in solar cells is essential for ensuring consistent quality and performance across various applications. Developing industry standards will help address challenges related to material compatibility, stability, and long-term reliability, providing a framework for manufacturers to follow. Quality control measures will also play a critical role in maintaining the integrity of nanosensor technologies, enabling manufacturers to meet the performance expectations of consumers and industry stakeholders alike. As the field advances, a collaborative approach involving researchers, manufacturers, and regulatory bodies will be vital to drive innovation and ensure the successful scalability of nanosensor-based solar cells in the competitive energy market.

7.2 Challenges

7.2.1 Material Stability and Durability

A significant challenge in the deployment of smart materials within solar cells is ensuring their long-term stability and functionality, especially under harsh environmental conditions such as UV exposure, temperature fluctuations, and humidity variations. These factors can significantly impact the performance and reliability of nanosensors, potentially compromising the overall efficiency of the solar cell. Additionally, nanosensors and other smart materials may experience degradation and aging over time, which can lead to a decline in their effectiveness and accuracy. To address these issues, ongoing research is crucial to identify and develop more durable materials and innovative protective coatings that enhance the resilience of nanosensors. By focusing on material advancements and protective strategies, the industry can improve the longevity and reliability of smart materials in solar applications, ultimately contributing to the broader adoption of solar technologies in varying environments.

7.2.2 Integration Complexity

Compatibility with Existing Technologies

Integrating nanosensors into existing solar cell technologies poses a significant challenge, primarily because it requires the seamless incorporation of these advanced components without disrupting the current operation or efficiency of the solar cells. Achieving this integration demands meticulous engineering to ensure that the nanosensors complement rather than interfere with the solar cells' energy conversion processes. Additionally, the physical footprint of the nanosensors must be minimized to avoid overshadowing the active areas of the solar cells, which could negatively impact light absorption and overall performance. Furthermore, the materials used for the nanosensors must be compatible with the existing solar cell materials to maintain structural integrity and performance reliability. Overcoming these challenges is crucial for enhancing the functionality of solar cells through the added benefits of nanosensors, such as improved monitoring, efficiency optimization, and predictive maintenance.

Complex Engineering

Integrating smart materials as nanosensors into solar cells involves a complex design and engineering process that demands precise control over both material properties and interfacial interactions. Achieving optimal functionality requires a thorough understanding of the nanosensors' physical, chemical, and electrical characteristics, as well as their compatibility with the host materials of the solar cells. This complexity is further compounded by the need to manage the integration of different materials at the nanoscale, where even minor variations can significantly affect performance. Effective engineering solutions must ensure robust bonding between the nanosensors and the solar cell layers, promoting efficient energy conversion and reliable operation under various environmental conditions. Consequently, advancements in fabrication techniques, such as layer-by-layer assembly and precision deposition methods, are essential to realize the full potential of smart materials in enhancing solar cell technology.

Economic Viability

The economic viability of integrating smart materials into solar cells is a critical consideration in their adoption and widespread use. Although smart materials have the potential to significantly enhance solar cell performance through improved efficiency, durability, and functionality, they often come with higher production costs. Therefore, striking a balance between the costs associated with the integration of these advanced materials and the benefits they offer such as increased energy output and longer lifespan will be essential for encouraging manufacturers and consumers to embrace this technology. Careful analysis and cost–benefit assessments will be necessary to determine the long-term financial implications of adopting smart materials in solar energy applications.

Moreover, scalability poses another significant challenge in the economic viability of smart materials for solar cells. The successful scaling up of production methods must ensure that the integration of these materials can achieve economies of scale comparable to those of traditional solar technologies. This involves optimizing manufacturing processes, reducing waste, and improving efficiency to lower costs. By developing innovative strategies for large-scale production and integration, the industry can enhance the competitiveness of smart materials against conventional solar cells. Ultimately, addressing both cost and scalability will be pivotal in fostering a market environment where smart materials can thrive, driving forward the next generation of solar technology.

7.2.3 Environmental and Safety Concerns

The integration of smart materials into solar cells raises significant environmental concerns that must be carefully addressed to promote sustainable practices in the industry. One major issue is the recycling of these materials, particularly when they contain toxic or rare elements that can be harmful to the environment if not disposed of properly. As the demand for solar technology increases, developing eco-friendly alternatives and effective recycling methods becomes crucial to mitigate potential ecological harm. Research into sustainable materials and innovative recycling processes is essential to ensure that the environmental footprint of smart materials remains minimal, thereby fostering a more circular economy within the renewable energy sector.

In addition to recycling, the safety of smart materials throughout their lifecycle is a paramount concern. Ensuring that these materials are safe for workers during the manufacturing process and do not pose health risks during their operational lifespan is vital to prevent adverse effects on both human health and the environment. Comprehensive safety assessments, regulatory compliance, and adherence to best practices in handling and disposal are necessary to minimize potential risks associated with the use of these advanced materials. By prioritizing environmental impact and safety considerations, the solar energy industry can promote the responsible adoption of smart materials, enhancing both sustainability and public confidence in solar technologies.

7.2.4 Regulatory and Market Adoption

The adoption of smart materials in solar cell technology is significantly influenced by regulatory hurdles that must be navigated to ensure compliance with existing industry standards. As new materials and technologies are introduced, they often face rigorous scrutiny from regulatory bodies aimed at protecting public health and the environment. It is essential for manufacturers to demonstrate that these smart materials meet established safety and performance regulations, which may involve extensive testing and validation processes. Overcoming these regulatory challenges is crucial for the widespread acceptance of smart materials, as any delays or failures to comply can impede innovation and slow down the deployment of advanced solar technologies.

Market acceptance is another critical factor in the successful integration of smart materials into solar cells. Educating industry stakeholders, including manufacturers, investors, and consumers, about the advantages offered by smart materials—such as enhanced efficiency, improved performance, and long-term cost savings—is vital for fostering widespread adoption. Clear communication regarding the benefits and potential applications of these materials can help mitigate skepticism and drive demand. Furthermore, showcasing successful case studies and pilot projects that illustrate the effective use of smart materials in real-world settings can bolster confidence among consumers and industry professionals alike. By addressing both regulatory challenges and market acceptance, the path to integrating smart materials into solar cell technology can be significantly accelerated, ultimately contributing to a more sustainable energy future.

7.3 *Conclusion*

The future of solar energy is intricately linked to the advancements in smart materials, particularly their integration as nanosensors within solar cells. Despite the significant challenges posed by material stability, integration complexities, economic feasibility, and environmental impacts, the potential advantages of these technologies make this an exhilarating domain for research and development. Smart materials can enhance the performance of solar cells by improving energy conversion efficiencies, enabling real-time monitoring, and adapting to changing environmental conditions. As these materials evolve, they promise to revolutionize the solar industry by facilitating the creation of more efficient and responsive solar technologies.

Continued innovation in the integration of smart materials into solar cells holds the potential to transform solar energy into a more efficient, adaptable, and accessible resource for a sustainable future. By addressing the current challenges through interdisciplinary research and collaboration, the solar industry can leverage the capabilities of smart materials to optimize energy generation and storage. As this technology matures, it will pave the way for broader adoption and deployment, further contributing to the global transition toward renewable energy sources. Ultimately, the pursuit of smart materials in solar applications not only underscores the importance of sustainable practices but also highlights the critical role of innovation in addressing the pressing energy needs of our world.

References

1. Green MA, Dunlop ED, Hohl-Ebinger J, Yoshita M, Kopidakis N, Ho-Baillie AWY (2022) Solar cell efficiency tables (Version 60). Prog Photovolt Res Appl 30(1):687–701
2. Zhao Y, Zhu K (2022) Organic–inorganic hybrid lead halide perovskites for optoelectronic and electronic applications. Chem Soc Rev 51(10):655–689
3. Wei XK, Domingo N, Sun Y, Balke N, Dunin-Borkowski RE, Mayer J (2022) Progress on emerging ferroelectric materials for energy harvesting, storage and conversion (Adv. Energy Mater. 24/2022). Adv Energy Mater 12(24):2270102
4. Li Z, Li B, Wu X, Sheppard SA., Zhang S, Gao D, Long NJ, Zhu Z (2022) Organometallic-functionalized interfaces for highly efficient inverted perovskite solar cells. Science 376(6591):416–420
5. Su H, Zhang J, Hu Y, Yao Y, Zheng X, She Y, Jia B, Gao L, Liu S (2023) Modulation on electrostatic potential of passivator for highly efficient and stable perovskite solar cells. Adv Funct Mater 33(34):2213123
6. Polman A, Knight M, Garnett EC, Ehrler B, Sinke WC (2016) Photovoltaic materials: present efficiencies and future challenges. Science 352(6283):aad4424
7. Lin H, Yang M, Ru X, Wang G, Yin S, Peng F, Hong C, Qu M, Lu J, Fang L, Han C, Procel P, Isabella O, Gao P, Li Z, Xu X (2023) Silicon heterojunction solar cells with up to 26.81% efficiency achieved by electrically optimized nanocrystalline-silicon hole contact layers. Nat Energy 8(8):789–799
8. Liu D, Liu X, Gan Y, Liu Z, Sun G, Shen C, Peng X, Qiu W, Li D, Zhou Z, Li Z, Yip HL, Su SJ (2022) Perovskite/organic hybrid white electroluminescent devices with stable spectrum and extended operating lifetime. ACS Energy Lett 7(1):523–532

9. Jeong SY, Kim HS, Park NG (2022) Challenges for thermally stable spiro-MeOTAD toward the market entry of highly efficient perovskite solar cells. ACS Appl Mater Interfaces 14(30):34220–34227
10. Li C, Xu Y, Tu W, Chen G, Xu R (2017) Metal-free photocatalysts for various applications in energy conversion and environmental purification. Green Chem 19(4):882–899
11. Ilic S, Zoric MR, Kadel UP, Huang Y, Glusac KD (2017) Metal-free motifs for solar fuel applications. Annu Rev Phys Chem 68(1):305–331
12. Zhao JN, Liu C, Yang K, Tang R, Fu Q, Ke K, Yin B, Yang MB, Yang W (2024) Solvent-free and metal-free fabrication of multifunctional cellulose-based fabrics for wearable energy harvesting, body motion detection, and thermotherapy. ACS Sustain Chem Eng 12(50):18136–18148
13. Malhotra SS, Ahmed M, Gupta MK, Ansari A (2024) Metal-free and natural dye sensitized solar cells: recent advancement and future perspectives. Sustain Energy Fuels 8:4127–4163
14. Xia K, Yu M (2025) Highly robust and efficient metal-free water cup solid–liquid triboelectric nanogenerator for water wave energy harvesting and ethanol detection. Chem Eng J 503:157938
15. Taheri-Ledari R, Ganjali F, Zarei-Shokat S, Saeidirad M, Ansari F, Forouzandeh-Malati M, Hassanzadeh-Afruzi F, Hashemi SM, Maleki A (2022) A review of metal-free organic halide perovskite: future directions for the next generation of solar cells. Energy Fuels 36(18):10702–10720
16. Kundu S, Bramhaiah K, Bhattacharyya S (2020) Carbon-based nanomaterials: in the quest of alternative metal-free photocatalysts for solar water splitting. Nanoscale Adv 2(11):5130–5151
17. Materon EM, Gómez FR, Joshi N, Dalmaschio CJ, Carrilho E, Oliveira Jr ON (2021) Smart materials for electrochemical flexible nanosensors: advances and applications. Nanosensors Smart Manufact 347–371
18. Peixoto AC, Silva AF (2017) Smart devices: micro- and nanosensors. In: Bioinspired materials for medical applications. Woodhead Publishing, pp 297–329
19. Nazari A (2020) Nanosensors for smart cities: an introduction. In: Nanosensors for smart cities. Elsevier, pp 3–8
20. Han B, Tomer V, Nguyen TA, Farmani A, Singh PK (eds) (2020) Nanosensors for smart cities. Elsevier
21. Denizli A, Nguyen TA, Rajendran S, Yasin G, Nadda AK (eds) (2021) Nanosensors for smart agriculture. Elsevier
22. Saini RK, Bagri LP, Bajpai AK (2017) Smart nanosensors for pesticide detection. In: New pesticides and soil sensors. Academic Press, pp 519–559
23. Khan T, Civas M, Cetinkaya O, Abbasi NA, Akan OB (2020) Nanosensor networks for smart health care. In: Nanosensors for smart cities. Elsevier, pp 387–403
24. Borisov SM, Klimant I (2008) Optical nanosensors—smart tools in bioanalytics. Analyst 133(10):1302–1307
25. Atta NF (2014) Nanosensors: materials and technologies. Lulu.com
26. Michaels H, Rinderle M, Freitag R, Benesperi I, Edvinsson T, Socher R, Gagliardi A, Freitag M (2020) Dye-sensitized solar cells under ambient light powering machine learning: towards autonomous smart sensors for the internet of things. Chem Sci 11(11):2895–2906
27. Aslam A, Mehmood U, Arshad MH, Ishfaq A, Zaheer J, Khan AUH, Sufyan M (2020) Dye-sensitized solar cells (DSSCs) as a potential photovoltaic technology for the self-powered internet of things (IoTs) applications. Solar Energy 207:874–892
28. Correia SF, Bastos AR, Martins M, Macário IP, Veloso T, Pereira JL, Coutinho JA, Ventura SP, André PS, Ferreira RA (2022) Bio-based solar energy harvesting for onsite mobile optical temperature sensing in smart cities. Adv Sci 9(17):2104801
29. Liu CY, Huang CK, Huang YY, Chang KC, Yu KL, Chiang CH, Wu CG, Lee SC, Yen WY, Sheu JK, Shi JW (2022) Flexible multijunction solar cells embedded inside smart dust modules for outdoor applications to smart grids. Appl Energy 306:117970

30. Okasha AM, Ibrahim HG, Elmetwalli AH, Khedher KM, Yaseen ZM, Elsayed S (2021) Designing low-cost capacitive-based soil moisture sensor and smart monitoring unit operated by solar cells for greenhouse irrigation management. Sensors 21(16):5387
31. García E, Ponluisa N, Quiles E, Zotovic-Stanisic R, Gutiérrez SC (2022) Solar panels string predictive and parametric fault diagnosis using low-cost sensors. Sensors 22(1):332
32. Kantareddy SN, Mathews I, Sun S, Layurova M, Thapa J, Correa-Baena JP, Bhattacharyya R, Buonassisi T, Sarma SE, Peters IM (2019) Perovskite PV-powered RFID: enabling low-cost self-powered IoT sensors. IEEE Sens J 20(1):471–478
33. Mandourarakis I, Gogolou V, Agorastou Z, Voutsinas S, Rigogiannis N, Koutroulis E, Siskos S, Samarakou M, Papanikolaou N, Karystinos G (2023) Development of a smart photovoltaic cells system. Energy Convers Manage 293:117478
34. Green MA (2003) Crystalline and thin-film silicon solar cells: state of the art and future potential. Solar Energy 74(3):181–192
35. Shockley W, Queisser HJ (1961) Detailed balance limit of efficiency of p-n junction solar cells. J Appl Phys 32(3):510–519
36. Green MA (1995) Silicon solar cells advanced principles & practice. University of New South Wales Press
37. Swanson RM (2000) The promise of concentrators. Prog Photovoltaics Res Appl 8(1):93–111
38. Goetzberger A, Knobloch J, Voss B (1998) Crystalline silicon solar cells, vol 1. Wiley, Chichester, pp 114–118
39. Hegedus S, Luque A (2011) Achievements and challenges of solar electricity from photovoltaics. In: Handbook of photovoltaic science and engineering. Wiley, pp 1–38
40. Green MA (2005) Silicon photovoltaic modules: a brief history of the first 50 years. Prog Photovoltaics Res Appl 13(5):447–455
41. Rizwan M, Khan WS, Aleena S (2021) Monocrystalline silicon solar cells. Mater Sol Cell Technol I 88:148
42. Zhao J, Wang A, Green MA (1999) 24.5% efficiency silicon PERT cells on MCZ substrates and 24.7% efficiency PERL cells on FZ substrates. Prog Photovoltaics Res Appl 7(6):471–474
43. Green MA (2009) The path to 25% silicon solar cell efficiency: history of silicon cell evolution. Prog Photovoltaics Res Appl 17(3):183–189
44. Rath JK (2003) Low temperature polycrystalline silicon: a review on deposition, physical properties and solar cell applications. Sol Energy Mater Sol Cells 76(4):431–487
45. Fébba DM, Rubinger RM, Oliveira AF, Bortoni EC (2003) Impacts of temperature and irradiance on polycrystalline silicon solar cells parameters. Sol Energy 174:628–639
46. Carlson DE, Wronski CR (2005) Amorphous silicon solar cells. In: Amorphous semiconductors. Springer, pp 287–329
47. Carlson DE, Wronski CR (1976) Amorphous silicon solar cell. Appl Phys Lett 28(11):671–671
48. Street RA (1991) Hydrogenated amorphous silicon. Cambridge University Press
49. Soler Vilamitjana D (2004) Amorphous silicon solar cells obtained by hot-wire chemical vapour deposition. Universitat de Barcelona
50. Chapin DM, Fuller CS, Pearson GL (1954) A new silicon p-n junction photocell for converting solar radiation into electrical power. J Appl Phys 25(5):676
51. Green MA (2007) Thin-film solar cells: review of materials, technologies and commercial status. J Mater Sci Mater Electron 18:15–19
52. Chopra KL, Paulson PD, Dutta V (2004) Thin-film solar cells: an overview. Prog Photovoltaics Res Appl 12(2–3):69–92
53. Jackson P, Hariskos D, Lotter E, Paetel S, Wuerz R, Menner R, Wischmann W, Powalla M (2011) New world record efficiency for Cu (In, Ga) Se_2 thin-film solar cells beyond 20%. Prog Photovoltaics Res Appl 19(7):894–897
54. Razykov TM, Ferekides CS, Morel D, Stefanakos E, Ullal HS, Upadhyaya HM (2011) Solar photovoltaic electricity: current status and future prospects. Solar Energy 85(8):1580–1608
55. Staebler DL, Wronski CR (1977) Reversible conductivity changes in discharge-produced amorphous Si. Appl Phys Lett 31(4):292–294

56. Shah A (2020) Amorphous silicon solar cells. In: Solar cells and modules. Springer, pp 139–161
57. Shafarman WN, Basol BM, Britt JS, Hall RB, Rocheleau RE (1997) Semiconductor processing and manufacturing. Prog Photovoltaics Res Appl 5(5):359–364
58. Wu X (2004) High-efficiency polycrystalline CdTe thin-film solar cells. Solar Energy 77(6):803–814
59. Bonnet D, Meyers P (1998) Cadmium-telluride—material for thin film solar cells. J Mater Res 13(10):2740–2753
60. Green MA, Emery K, King DL, Hisikawa Y, Warta W (2006) Solar cell efficiency tables (version 27). Prog Photovoltaics 14(1):45–52
61. Ramanujam J, Singh UP (2017) Copper indium gallium selenide based solar cells—a review. Energy Environ Sci 10(6):1306–1319
62. Singh M, Jiu J, Sugahara T, Suganuma K (2014) Thin-film copper indium gallium selenide solar cell based on low-temperature all-printing process. ACS Appl Mater Interfaces 6(18):16297–16303
63. Keller J, Kiselman K, Donzel-Gargand O, Martin NM, Babucci M, Lundberg O, Wallin E, Stolt L, Edoff M (2024) High-concentration silver alloying and steep back-contact gallium grading enabling copper indium gallium selenide solar cell with 23.6% efficiency. Nat Energy 9(4):467–478
64. Chow TT, Li C, Lin Z (2010) Innovative solar windows for cooling-demand climate. Sol Energy Mater Sol Cells 94(2):212–220
65. Hussain AM, Hussain MM (2016) CMOS-technology-enabled flexible and stretchable electronics for internet of everything applications. Adv Mater 28(22):4219–4249
66. Tran TV, Chung WY (2016) High-efficient energy harvester with flexible solar panel for a wearable sensor device. IEEE Sens J 16(24):9021–9028
67. Green M, Dunlop E, Hohl-Ebinger J, Yoshita M, Kopidakis N, Hao X (2021) Solar cell efficiency tables (version 57). Prog Photovoltaics Res Appl 29(1):3–15
68. Green MA, Dunlop ED, Hohl-Ebinger J, Yoshita M, Kopidakis N, Hao X (2021) Solar cell efficiency tables (version 58). Prog Photovoltaics 29(7)
69. Ferry VE, Verschuuren MA, Li HB, Verhagen E, Walters RJ, Schropp RE, Atwater HA, Polman A (2010) Light trapping in ultrathin plasmonic solar cells. Opt Express 18(S2):A237–A245
70. Green MA, Emery K, Hishikawa Y, Warta W (2010) Solar cell efficiency tables (version 35). Prog Photovoltaics 18(2):144–150
71. Snaith HJ (2013) Perovskites: the emergence of a new era for low-cost, high-efficiency solar cells. J Phys Chem Lett 4(21):3623–3630
72. Shah N, Shah AA, Leung PK, Khan S, Sun K, Zhu X, Liao Q (2023) A review of third generation solar cells. Processes 11(6):1852
73. Green MA (2006) Third generation photovoltaics. Springer
74. Gevorgyan SA, Madsen MV, Roth B, Corazza M, Hösel M, Søndergaard RR, Jørgensen M, Krebs FC (2016) Lifetime of organic photovoltaics: status and predictions. Adv Energy Mater 6(2):1501208
75. Polman A, Knight M, Garnett EC, Ehrler B, Sinke WC (2016) Photovoltaic materials: present efficiencies and future challenges. Science 352(6283):aad4424
76. Green MA (2002) Third generation photovoltaics: solar cells for 2020 and beyond. Physica E 14(1–2):65–70
77. Bag M, Renna LA, Adhikari RY, Karak S, Liu F, Lahti PM, Russell TP, Tuominen MT, Venkataraman D (2015) Kinetics of ion transport in perovskite active layers and its implications for active layer stability. J Am Chem Soc 137(40):13130–13137
78. Yang WS, Noh JH, Jeon NJ, Kim YC, Ryu S, Seo J, Seok SI (2015) High-performance photovoltaic perovskite layers fabricated through intramolecular exchange. Science 348(6240):1234–1237
79. Green MA, Ho-Baillie A, Snaith HJ (2014) The emergence of perovskite solar cells. Nat Photonics 8(7):506–514

80. Jeon NJ, Noh JH, Kim YC, Yang WS, Ryu S, Seok SI (2014) Solvent engineering for high-performance inorganic–organic hybrid perovskite solar cells. Nat Mater 13(9):897–903
81. Heeger AJ (2014) 25th anniversary article: bulk heterojunction solar cells: understanding the mechanism of operation. Adv Mater 26(1):10–28
82. Li G, Zhu R, Yang Y (2012) Polymer solar cells. Nat Photon 6(3):153–161
83. Krebs FC (2009) Fabrication and processing of polymer solar cells: a review of printing and coating techniques. Sol Energy Mater Sol Cells 93(4):394–412
84. Zhang F, Johansson M, Andersson MR, Hummelen JC, Inganäs O (2002) Polymer photovoltaic cells with conducting polymer anodes. Adv Mater 14(9):662–665
85. Brabec CJ, Gowrisanker S, Halls JJ, Laird D, Jia S, Williams SP (2010) Polymer-fullerene bulk-heterojunction solar cells. Adv Mater (34):3839–3856
86. Klimov VI (2007) Spectral and dynamical properties of multiexcitons in semiconductor nanocrystals. Annu Rev Phys Chem 58(1):635–673
87. Bera D, Qian L, Tseng TK, Holloway PH (2010) Quantum dots and their multimodal applications: a review. Materials 3(4):2260–2345
88. Kamat PV (2008) Quantum dot solar cells. Semiconductor nanocrystals as light harvesters. J Phys Chem C 112(48):18737–18753
89. Semonin OE, Luther JM, Choi S, Chen HY, Gao J, Nozik AJ, Beard MC (2011) Peak external photocurrent quantum efficiency exceeding 100% via MEG in a quantum dot solar cell. Science 334(6062):1530–1533
90. Toyoda T, Shen Q (2012) Quantum-dot-sensitized solar cells: effect of nanostructured TiO_2 morphologies on photovoltaic properties. J Phys Chem Lett 3(14):1885–1893
91. Zhang X, Huang H, Zhao C, Jin L, Lee C, Li Y, Ko DH, Ma W, Wu T, Yuan J (2024) Conductive colloidal perovskite quantum dot inks towards fast printing of solar cells. Nat Energy 9:1378–1387
92. Essig S, Allebé C, Remo T, Geisz JF, Steiner MA, Horowitz K, Barraud L, Ward JS, Schnabel M, Descoeudres A, Young DL, Woodhouse M, Despeisse M, Ballif C, Tamboli A (2017) Raising the one-sun conversion efficiency of III–V/Si solar cells to 32.8% for two junctions and 35.9% for three junctions. Nat Energy 2(9):17144
93. Al-Ashouri A, Köhnen E, Li B, Magomedov A, Hempel H, Caprioglio P, Márquez JA, Morales Vilches AB, Kasparavicius E, Smith JA, Phung N (2020) Monolithic perovskite/silicon tandem solar cell with >29% efficiency by enhanced hole extraction. Science 370(6522):1300–1309
94. Basheer AA (2020) Advances in the smart materials applications in the aerospace industries. Aircr Eng Aerosp Tech 92(7):1027–1035
95. Wang W, Xiang Y, Yu J, Yang L (2023) Development and prospect of smart materials and structures for aerospace sensing systems and applications. Sensors 23(3):1545
96. del Bosque A, Vergara D, Fernández-Arias P (2025) An overview of smart composites for the aerospace sector. Appl Sci 15(6):2986
97. Hoffman AS (2012) Hydrogels for biomedical applications. Adv Drug Deliv Rev 64:18–23
98. Liu Z, Jiao Y, Wang Y, Zhou C, Zhang Z (2008) Polysaccharides-based nanoparticles as drug delivery systems. Adv Drug Deliv Rev 60(15):1650–1662
99. Aflori M (2021) Smart nanomaterials for biomedical applications—a review. Nanomaterials 11(2):396
100. Shi J, Votruba AR, Farokhzad OC, Langer R (2010) Nanotechnology in drug delivery and tissue engineering: from discovery to applications. Nano Lett 10(9):3223–3230
101. Mura S, Nicolas J, Couvreur P (2013) Stimuli-responsive nanocarriers for drug delivery. Nat Mater 12(11):991–1003
102. Zhang X, Zhang X, Tao L, Chi Z, Xu J, Wei Y (2014) Aggregation induced emission-based fluorescent nanoparticles: fabrication methodologies and biomedical applications. J Mater Chem B 2(28):4398–4414
103. Naebe M, Shirvanimoghaddam K (2016) Functionally graded materials: a review of fabrication and properties. Appl Mater Today 5:223–245
104. Askari H, Khajepour A, Khamesee MB, Wang ZL (2019) Embedded self-powered sensing systems for smart vehicles and intelligent transportation. Nano Energy 66:104103

105. Ravi S, Ravi M, Selvi JA (2024) Nano-hybrid smart coatings for automotive applications. In: Nano-hybrid smart coatings: advancements in industrial efficiency and corrosion resistance. American Chemical Society, pp 227–243
106. Behl M, Kratz K, Zotzmann J, Nöchel U, Lendlein A (2013) Reversible bidirectional shape-memory polymers. Adv Mater (Deerfield Beach, Fla.) 25(32):4466–4469
107. Sadik OA, Aluoch AO, Zhou A (2009) Status of biomolecular recognition using electrochemical techniques. Biosens Bioelectron 24(9):2749–2765
108. Gramling HM, Kiziroglou ME, Yeatman EM (2017) Nanotechnology for consumer electronics. In: Nanoelectronics: materials, devices, applications. Wiley, pp 501–526
109. Bauer S (2013) Sophisticated skin. Nat Mater 12(10):871–872
110. Otsuka K, Wayman CM (eds) (1998) Shape memory materials. Cambridge University Press
111. Bhattacharya K (eds) (2003) Microstructure of martensite: why it forms and how it gives rise to the shape-memory effect. Oxford University Press
112. Tobushi H, Hara H, Yamada E, Hayashi S (1996) Thermomechanical properties in a thin film of shape memory polymer of polyurethane series. Smart Mater Struct 5(4):483–491
113. Otsuka K, Ren X (1999) Recent developments in the research of shape memory alloys. Intermetallics 7(5):511–528
114. Salh H, Saleh DS, Smail JM, Manguri HSA, BALO ŞN, Ahmad ST (2025) Experimental and theoretical evaluation of Cu–Al–Ni shape memory alloys shielding properties at the 511–1333 keV gamma energy range. J Radioanal Nucl Chem 334(1):851–862
115. Sampath V (2007) Effect of thermal processing on microstructure and shape-memory characteristics of a copper–zinc–aluminum shape-memory alloy. Mater Manuf Processes 22(1):9–14
116. Sekhar MC, Veena E, Kumar NS, Naidu KCB, Mallikarjuna A, Basha DB (2023) A review on piezoelectric materials and their applications. Cryst Res Technol 58(2):2200130
117. Anton SR, Sodano HA (2007) A review of power harvesting using piezoelectric materials (2003–2006). Smart Mater Struct 16(3):R1
118. Sun C, Wang H, Shang G, Du J (2025) Parameters optimization for piezoelectric harvesting energy from pavement based on Taguchi's orthogonal experiment design. World J Eng Technol 3(4):149–157
119. Curie J, Curie P (1880) Development by compression of polar electricity in hemihedral crystals with inclined faces. Bulletin de minérale 3(4):90–93
120. Aliqué M, Simão CD, Murillo G, Moya A (2021) Fully-printed piezoelectric devices for flexible electronics applications. Adv Mater Technol 6(3):2001020
121. Kulkarni H, Zohaib K, Khusru A, Aiyappa KS (2018) Application of piezoelectric technology in automotive systems. Mater Today: Proc 5(10):21299–21304
122. Elahi H, Munir K, Eugeni M, Abrar M, Khan A, Arshad A, Gaudenzi P (2020) A review on applications of piezoelectric materials in aerospace industry. Integr Ferroelectr 211(1):25–44
123. Ali F, Koc M (2023) 3D printed polymer piezoelectric materials: transforming healthcare through biomedical applications. Polymers 15(23):4470
124. Safaei M, Sodano HA, Anton SR (2019) A review of energy harvesting using piezoelectric materials: state-of-the-art a decade later (2008–2018). Smart Mater Struct 28(11):113001
125. Mahapatra SD, Mohapatra PC, Aria AI, Christie G, Mishra YK, Hofmann S, Thakur VK (2021) Piezoelectric materials for energy harvesting and sensing applications: roadmap for future smart materials. Adv Sci 8(17):2100864
126. Hakami A, Srinivasan SS, Biswas PK, Krishnegowda A, Wallen SL, Stefanakos EK (2022) Review on thermochromic materials: development, characterization, and applications. J Coat Technol Res 19(2):377–402
127. Crosby PH, Netravali AN (2022) Green thermochromic materials: a brief review. Adv Sustain Syst 6(9):2200208
128. Wang X, Narayan S (2021) Thermochromic materials for smart windows: a state-of-art review. Front Energy Res 9:800382
129. Zhang J, Zou Q, Tian H (2013) Photochromic materials: more than meets the eye. Adv Mater 25(3):378–399

130. Wang S, Fan W, Liu Z, Yu A, Jiang X (2018) Advances on tungsten oxide based photochromic materials: strategies to improve their photochromic properties. J Mater Chem C 6(2):191–212
131. Ru Y, Shi Z, Zhang J, Wang J, Chen B, Huang R, Liu G, Yu T (2021) Recent progress of photochromic materials towards photocontrollable devices. Mater Chem Front 5(21):7737–7758
132. Wang MS, Xu G, Zhang ZJ, Guo GC (2010) Inorganic–organic hybrid photochromic materials. Chem Commun 46(3):361–376
133. Kobayashi Y, Abe J (2022) Recent advances in low-power-threshold nonlinear photochromic materials. Chem Soc Rev 51(7):2397–2415
134. Mortimer RJ (2011) Electrochromic materials. Annual Rev Mater Res 41(1):241–268
135. Mortimer RJ (1997) Electrochromic materials. Chem Soc Rev 26(3):147–156
136. Somani PR, Radhakrishnan SJMC (2003) Electrochromic materials and devices: present and future. Mater Chem Phys 77(1):117–133
137. Gu C, Jia AB, Zhang YM, Zhang SXA (2022) Emerging electrochromic materials and devices for future displays. Chem Rev 122(18):14679–14721
138. Wang J, Guo X, Bian C, Zhong Y, Tu J, Lee PS, Cai G (2025) Roadmap for electrochromic smart devices: from materials engineering and architectures design to multifunctional application. Prog Mater Sci 101461
139. García-Arribas A (2021) Magnetostrictive materials. In: Magnetic measurement techniques for materials characterization. Springer International Publishing, Cham, pp 727–750
140. Chen C, Yi Z (2021) Photostrictive effect: characterization techniques, materials, and applications. Adv Funct Mater 31(22):2010706
141. Zenkour AM, El-Shahrany HD (2021) Hygrothermal forced vibration of a viscoelastic laminated plate with magnetostrictive actuators resting on viscoelastic foundations. Int J Mech Mater Des 17(2):301–320
142. Dolui C, Dasadhikari S, Bose D, Roy D, Chanda CK (2024) Different structural configurations of magnetostrictive energy harvester: a comprehensive review for sensor applications. In: International symposium on sustainable energy and technological advancements. Springer Nature, Singapore, pp 381–398
143. Feng K, Wang G, Lian Q, Gámez-Valenzuela S, Li B, Ding R, Yang W, Wang K, Zeng J, Zhang Y, Jeong SY, Guo X (2025) Non-fullerene electron-transporting materials for high-performance and stable perovskite solar cells. Nat Mater 1–8
144. Mutar ZS, Mutlak FAH (2025) Fabrication and characterization of zinc oxide nanoparticles deposited on pours silicon as a photodetector and solar cell. J Opt 1–13
145. Ilager D, Kurangi B, Jalalpure SS, Pandiaraj S, Alodhayb A, Mondal K, Alshehri MA, Shetti NP (2025) Development of carbon nanosensor for vinpocetine at zinc oxide modified carbon matrix with anionic surfactant for clinical application. J Mol Struct 141967
146. Dhahi TS, Dafhalla AK, Tayfour OE, Mubarakali A, Alqahtani AS, Ahmed AE, Elobaid ME, Adam T, Gopinath SC (2024) Advances in nano sensors for monitoring and optimal performance enhancement in photovoltaic cells. Iscience
147. Malaver Rojas A, Motta N, Corke P (2011) Remote monitoring of outdoor performance of low scale dye sensitized solar cells for nanosensors nodes. In: Proceedings of international conference on remote sensing and data (ICRSD 2011). Thomson ISI Proceedings
148. Malaver A, Matteocci F, Carlo AD, Corke, P, Motta N (2011) Remote monitoring of outdoor performance of low scale dye sensitized solar cells for nanosensors nodes. In: Proceedings of international proceedings of chemical, biological & environmental engineering, Harbour Plaza Resort City, Hong Kong, p 12
149. Lay-Ekuakille A, Massaro A, Singh SP, Jabłoński I, Rahman MZU, Spano F (2021) Optoelectronic and nanosensors detection systems: a review. IEEE Sens J 21(11):12645–12653
150. Turkevych I, Kazaoui S, Shirakawa N, Fukuda N (2021) Potential of AgBiI4 rudorffites for indoor photovoltaic energy harvesters in autonomous environmental nanosensors. Jpn J Appl Phys 60(SC):SCCE06
151. Yang KJ, Kim S, Sim JH, Son DH, Kim DH, Kim J, Jo W, Yoo H, Kim J, Kang JK (2018) The alterations of carrier separation in kesterite solar cells. Nano Energy 52:38–53

152. Jiang CS, Yang M, Zhou Y, To B, Nanayakkara SU, Luther JM, Zhou W, Berry JJ, Van De Lagemaat J, Padture NP, Zhu K, Al-Jassim MM (2015) Carrier separation and transport in perovskite solar cells studied by nanometre-scale profiling of electrical potential. Nat Commun 6(1):8397
153. Jeong WL, Kim KP, Kim J, Park HK, Min JH, Lee JS, Mun SH, Kim ST, Jang JH, Jo W, Lee DS (2020) Impact of Na doping on the carrier transport path in polycrystalline flexible $Cu_2ZnSn\ (S, Se)_4$ solar cells. Adv Sci 7(21):1903085
154. Wang J, Zeng L, Zhang D, Maxwell A, Chen H, Datta K, Caiazzo A, Remmerswaal WH, Schipper NR, Chen Z, Ho K et al (2024) Halide homogenization for low energy loss in 2-eV-bandgap perovskites and increased efficiency in all-perovskite triple-junction solar cells. Nat Energy 9(1):70–80
155. Meng H, Mao K, Cai F, Zhang K, Yuan S, Li T, Cao F, Su Z, Zhu Z, Feng X, Peng W (2024) Inhibition of halide oxidation and deprotonation of organic cations with dimethylammonium formate for air-processed p–i–n perovskite solar cells. Nat Energy 9(5):536–547
156. Su LY, Huang HH, Tsai CE, Hou CH, Shyue JJ, Lu CH, Pao CW, Yu MH, Wang L, Chueh CC (2022) Improving thermal and photostability of polymer solar cells by robust interface engineering. Small 18(23):2107834
157. Jabeen M, Haxha S (2018) Increased optical absorption and light-matter interaction in Silicon thin-film solar cell nanostructure using graphene and 2-D Au/Ag nanograting. IEEE J Quantum Electron 54(6):1–12
158. Li J, Zaheer SM, Lao C, Yin Z, Gong C, Guo Y, Li A, Guo C, Bian F, Xu Z (2024) Broadband enhancement of SiO_2@TiO_2 light-trapping nanostructure for improving the efficiency of perovskite solar cells. Opt Commun 563:130542
159. Guillot SL, Mistry KS, Avery AD, Richard J, Dowgiallo AM, Ndione PF, van de Lagemaat J, Reese MO, Blackburn JL (2015) Precision printing and optical modeling of ultrathin SWCNT/C 60 heterojunction solar cells. Nanoscale 7(15):6556–6566
160. Munn C, Haran S, Seok I (2013) Fabrication of CZTS-based thin film solar cells using all-solution processing and pulsed light crystallization. In: Nanosensors, biosensors, and info-tech sensors and systems 2013, vol 8691. SPIE, pp 174–179
161. Nguyen TT, Kim J, Yi J, Wong CP (2022) High-performing UV photodetectors by thermal-coupling transparent photovoltaics. Nano Energy 100:107504
162. Jung H, Park J, Yoo ES, Han GS, Jung HS, Ko MJ, Park S, Choe W (2013) Functionalization of nanomaterials by non-thermal large area atmospheric pressure plasmas: application to flexible dye-sensitized solar cells. Nanoscale 5(17):7825–7830
163. Chatzimanolis K, Rogdakis K, Tsikritzis D, Tzoganakis N, Tountas M, Krassas M, Bellani S, Najafi L, Martín-García B, Oropesa-Nuñez R, Prato M, Kymakis E (2021) Inverted perovskite solar cells with enhanced lifetime and thermal stability enabled by a metallic tantalum disulfide buffer layer. Nanoscale Adv 3(11):3124–3135
164. Su LY, Huang HH, Lin YC, Chen GL, Chen WC, Chen W, Wang L, Chueh CC (2021) Enhancing long-term thermal stability of non-fullerene organic solar cells using self-assembly amphiphilic dendritic block copolymer interlayers. Adv Funct Mater 31(4):2005753
165. Anandan S, Wen X, Yang S (2005) Room temperature growth of CuO nanorod arrays on copper and their application as a cathode in dye-sensitized solar cells. Mater Chem Phys 93(1):35–40
166. Zhang C, Liu S, Liu X, Deng F, Xiong Y, Tsai FC (2018) Incorporation of Mn^{2+} into CdSe quantum dots by chemical bath co-deposition method for photovoltaic enhancement of quantum dot-sensitized solar cells. Roy Soc Open Sci 5(3):171712
167. Wahab R, Khan F, Alam M, Mishra YK (2023) Silver nanoparticle based selective, sensitive and instantaneous electrochemical nanosensors for the analysis of riboflavin. Mater Sci Semicond Process 153:107166
168. Sokeng Djoumessi A, Sichwardt A, Miliaieva D, Čermák J, Schaal M, Otto F, Stehlík Š, Kuliček J, Nádaždy V, Fritz T, Kromka A, Hoppe H (2023) Nanodiamonds as charge extraction layer in organic solar cells: the impact of the nanodiamond surface chemistry. Sol RRL 7(12):2201061

169. Abbas MA, Thota R, Pyo K, Lee D, Bang JH (2020) Alkali metal ions: a secret ingredient for metal nanocluster-sensitized solar cells. ACS Energy Lett 5(5):1404–1406
170. Munawar A, Ong Y, Schirhagl R, Tahir MA, Khan, WS, Bajwa SZ (2019) Nanosensors for diagnosis with optical, electric and mechanical transducers. RSC Adv 9(12):6793–6803
171. Hang X, He S, Dong Z, Minnick G, Rosenbohm J, Chen Z, Yang R, Chang L (2021) Nanosensors for single cell mechanical interrogation. Biosens Bioelectron 179:113086
172. Bruns N, Pustelny K, Bergeron LM, Whitehead TA, Clark DS (2009) Mechanical nanosensor based on FRET within a thermosome: damage-reporting polymeric materials. Angew Chem 121(31):5776–5779
173. Binh VT, Garcia N, Levanuyk AL (1994) A mechanical nanosensor in the gigahertz range: where mechanics meets electronics. Surface Sci 301(1-3):L224–L228
174. Nath B, Behera SK, Kumar J, Hemmerle A, Fontaine P, Ramamurthy PC, Mahapatra DR, Hegde G (2024) Understanding the heterointerfaces in perovskite solar cells via hole selective layer surface functionalization. Adv Mater 36(5):2307547
175. Qiu J, Zhou Q, Jia D, Wang Y, Li S, Zhang X (2022) Robust molecular-dipole-induced surface functionalization of inorganic perovskites for efficient solar cells. J Mater Chem A 10(4):1821–1830
176. Gnichwitz JF, Marczak R, Werner F, Lang N, Jux N, Guldi DM, Peukert W, Hirsch A (2010) Efficient synthetic access to cationic dendrons and their application for ZnO nanoparticles surface functionalization: new building blocks for dye-sensitized solar cells. J Am Chem Soc 132(50):17910–17920
177. Vidyasagar D, Yun Y, Cho JY, Lee H, Kim KW, Kim YT, Yang SW, Jung J, Choi WC, Kim S, Gunasekaran RK, Lee S (2024) Surface-functionalized hole-selective monolayer for high efficiency single-junction wide-bandgap and monolithic tandem perovskite solar cells. J Energy Chem 88:317–326
178. Xu X, Wang H, Wang J, Muhammad M, Wang Z, Chen P, Zhao W, Kang B, Zhang J, Li C, Duan Y (2020) Surface functionalization of a graphene cathode to facilitate ALD growth of an electron transport layer and realize high-performance flexible perovskite solar cells. ACS Appl Energy Mater 3(5):4208–4216
179. Wang L, Sun X, Wang D, Cui P, Wang J, Li Q (2024) An integrated, multi-functional intrinsic responsive foldable protective layer for space station solar panels. Energy 303:131825
180. Yu B, Wang K, Sun Y, Yu H (2024) Multi-functional interface modulation through thiol functionalized covalent organic frameworks for efficient and durable perovskite solar cells. J Mater Chem A 12(40):27577–27587
181. Yu R, Wu G, Cui Y, Wei X, Hong L, Zhang T, Zou C, Hu S, Hou J, Tan ZA (2021) Multi-functional solid additive induced favorable vertical phase separation and ordered molecular packing for highly efficient layer-by-layer organic solar cells. Small 17(44):2103497
182. Socol M, Preda N (2021) Hybrid nanocomposite thin films for photovoltaic applications: a review. Nanomaterials 11(5):1117
183. Adimule V, Suryavanshi A, BC Y, Nandi SS (2020) A facile synthesis of poly (3-octyl thiophene): $Ni_{0.4}Sr_{0.6}TiO_3$ hybrid nanocomposites for solar cell applications. Macromol Symp 392(1):2000001
184. Kirchartz T, Bisquert J, Mora-Sero I, Garcia-Belmonte G (2015) Classification of solar cells according to mechanisms of charge separation and charge collection. Phys Chem Chem Phys 17(6):4007–4014
185. Würfel U, Cuevas A, Würfel P (2014) Charge carrier separation in solar cells. IEEE J Photovoltaics 5(1):461–469
186. Choudhury KR, Subbiah J, Chen S, Beaujuge PM, Amb CM, Reynolds JR, So F (2011) Understanding the performance and loss-mechanisms in donor—acceptor polymer based solar cells: Photocurrent generation, charge separation and carrier transport. Sol Energy Mater Sol Cells 95(8):2502–2510
187. Chang CH, Huang TK, Lin YT, Lin YY, Chen CW, Chu TH, Su WF (2008) Improved charge separation and transport efficiency in poly (3-hexylthiophene)—TiO_2 nanorod bulk heterojunction solar cells. J Mater Chem 18(19):2201–2207

188. Yu Y, Zhang F, Yu H (2020) Self-healing perovskite solar cells. Sol Energy 209:408–414
189. Lang F, Nickel NH, Bundesmann J, Seidel S, Denker A, Albrecht S, Brus VV, Rappich J, Rech B, Landi G, Neitzert HC (2016) Radiation hardness and self-healing of perovskite solar cells. Adv Mater 28(39):8726–8731
190. Pallotta R, Cavalli S, Degani M, Grancini G (2024) Smart materials to empowering perovskite solar cells with self-healing capability. Small Struct 5(5):2300448
191. Finkenauer BP, Akriti, Ma K, Dou L (2022) Degradation and self-healing in perovskite solar cells. ACS Appl Mater Interfaces 14(21):24073–24088
192. Nam JS, Choi JM, Lee JW, Han J, Jeon I, Kim HD (2024) Decoding polymeric additive-driven self-healing processes in perovskite solar cells from chemical and physical bonding perspectives. Adv Energy Mater 14(12):2304062
193. Chu K, Song BG, Yang HI, Kim DM, Lee CS, Park M, Chung CM (2018) Smart passivation materials with a liquid metal microcapsule as self-healing conductors for sustainable and flexible perovskite solar cells. Adv Funct Mater 28(22):1800110
194. Xue T, Fan B, Jiang KJ, Guo Q, Hu X, Su M, Zhou E, Song Y (2024) Self-healing ion-conducting elastomer towards record efficient flexible perovskite solar cells with excellent recoverable mechanical stability. Energy Environ Sci 17(7):2621–2630
195. Xu P, Liu J, Wang S, Chen J, Han B, Meng Y, Yang S, Xie L, Yang M, Jia R, Ge Z (2023) Dynamic covalent polymer engineering for stable and self-healing perovskite solar cells. Mater Horiz 10(11):5223–5234
196. Liang FC, Akman E, Aftab S, Mohammed MK, Hegazy HH, Zhang X, Zhang F (2025) Self-healing polymers in rigid and flexible perovskite photovoltaics. InfoMat 7(1):e12628
197. Yang D, Ni Y, Su H, Shi Y, Liu Q, Chen X, He D (2021) Hybrid energy system based on solar cell and self-healing/self-cleaning triboelectric nanogenerator. Nano Energy 79:105394
198. Shinkar AR, Joshi D, Praveen RV, Rajesh Y, Singh D (2024) Intelligent solar energy harvesting and management in IoT nodes using deep self-organizing maps. In: 2024 International conference on emerging research in computational science (ICERCS). IEEE, pp 1–6
199. Li F, Sui Y, Lin H, Sui Z, Lee K, Xie S, Zeng W, Ding Z, Yip HL, Wu W (2025) Self-adaptive interfacial evaporation for high-efficiency photovoltaic panel cooling. Device 3(2):100569
200. Boutahir MK, Farhaoui Y, Salma B, Azrour M (2023) Enhancing solar power generation through threshold-based anomaly detection in Errachidia, Morocco. In: The international conference on artificial intelligence and smart environment. Springer, Cham, pp 522–530

Smart Decision for Nanosensor Industry Using Control Chart Approach

Zahid Khan

Abstract Over recent decades, the global market for nanosensors has experienced substantial growth. These sensors are employed to measure and detect chemical, physical, or biological properties at the nano level of different products. Nanosensors have a wide range of applications in fields such as medicine, manufacturing processes, transportation systems, environmental monitoring, electronics, agriculture, and the food industry. To produce technologically advanced products, it is essential to prioritize quality. In this chapter we focus on the usage of control chart approach as a diagnostic tool for monitoring quality parameters. The goal is to ensure high stability in products, processes, and systems quality while minimizing variations. This chapter presents a thorough review and detailed analysis of the control chart approach to track and illustrate changes in processes over a period of time. Various types of control charts and their applications in conjunction with nanosensor data for monitoring variations in manufactured products are described. This work has the potential to serve as a foundation for future research, which aims to develop a comprehensive report on the effectiveness of different control charts in assessing the performance of nanosensors and their production at an industrial level with high quality standards.

Keywords Nanotechnology · Nanosensors · Quality · Attribute charts · Variable charts · Memory charts

1 Introduction

Nanotechnology involves a wide range of technologies that are carried out on a very small scale, specifically at the nanometer level [1]. These technologies have numerous applications in various industries [2]. The process involves manipulating and utilizing systems at sizes as small as individual atoms or molecules, up to about 100 nm

Z. Khan (✉)
Department of Quantitative Methods, University of Pannonia, H-8200, Egyetem str. 10, Veszprem, Hungary
e-mail: zahidkhan@hu.edu.pk

I. Uddin (ed.), *Smart Nanosensors*, Smart Nanomaterials Technology,
https://doi.org/10.1007/978-981-96-3878-9_14

[3]. Moreover, it also involves integrating these nanostructures into larger systems. Nanotechnologies are seen as the big driving force for a new industrial revolution [4]. Both governments and local businesses are spending a lot of money on this technology. The nanotechnology industry plays a significant role in economic growth due to its strong connections with various industries [5–7]. Scientists, researchers, investors, and policymakers worldwide acknowledge the immense potential of this industry, as it offers diverse applications across all sectors [8–10].

The potential of nanotechnologies lies in their ability to revolutionize various industries, from healthcare and electronics to energy and manufacturing [11–13]. Governments recognize the immense benefits that nanotechnology can bring, such as improved medical treatments, enhanced materials with unprecedented properties, and more efficient energy systems. Similarly, businesses are investing heavily in nanotechnology research and development to gain a competitive edge, create innovative products, and tap into new markets [14, 15]. Nanotechnology has a significant and diverse impact in different fields. In the field of medicine, it brings about a revolution by improving treatment methods through targeted drug delivery, advanced diagnostics, and innovative therapies for diseases like cancer. The possibility of personalized medicine could be realized through the use of nanotechnology [16]. By utilizing nanotechnologies, it becomes feasible to administer medicine in precise dosages and timings tailored to individual patients [17–20]. Tiny structures created by nanotechnology can serve as carriers, transporting medication directly into cells. These minuscule particles act as vehicles, delivering medication to the appropriate location within the body and releasing it in optimal quantities when required [21]. As a result, treatments could become more efficient and customized for every individual. Additionally, it contributes to the development of nanoscale materials that assist in tissue regeneration and wound healing [22, 23]. Furthermore, nanomaterials play a crucial role in promoting environmental well-being by aiding water purification, air filtration, and pollution control [24–27]. Nanotechnology plays a crucial role in revolutionizing energy production by enabling the development of efficient solar panels and catalysts for cleaner energy. Additionally, it enhances the quality of everyday consumer products, such as stain-resistant fabrics and stronger yet lighter materials for electronics and automobiles. Moreover, nanotech advancements in information technology leads to smaller and more efficient electronic devices, resulting in faster computers, expanded data storage capabilities, and impressive advancements in sensors and displays. The widespread influence of nanotechnology highlights its immense importance in shaping diverse sectors of our lives and industries. The nanotechnology industry is highly valued for its extensive applications in various sectors and its contribution to economic growth. Professionals and policymakers worldwide recognize its immense potential, as it offers diverse applications across all industries. The industrial sector plays a vital role in driving economic growth by creating jobs, fostering technological advancements, increasing productivity, and improving living standards. To achieve higher living standards and increase output rates, it is crucial to utilize nanometer-scale material properties, particularly in the manufacturing industry. By incorporating nanotechnology in manufacturing processes, it becomes possible to control the placement of atoms during reactions, resulting in more precise and refined materials compared to traditional production methods [28]. This improved precision enhances product quality while also reducing energy consumption and ultimately lowering production costs. Nanotechnology plays a crucial role in the field of economics, as it enables firms to manufacture goods using lighter and thinner materials. This breakthrough technology is widely recognized as a fundamental pillar for economic growth across various industries in developed nations [29]. The advanced nations are engaged in a parallel and harmonious competition in the field of nanotechnology, reflecting the realities of the modern global economy. Nanotechnology is set to revolutionize various industries, including computer technology, pharmaceuticals, nerve and tissue repair, surface coatings, catalysts, sensors, materials, telecommunications, and pollution control. Furthermore, economists view nanotechnology as a crucial driver of green innovation that can help lift the global economy out of recession. Based on the data available from the source [30], the investment of some developed countries during the time span 2011 to 2021 can be seen in the following bar chart.

By observing Fig. 1, it becomes evident that developed nations have made significant advancements in both technical and economic development, as well as in the management of global markets. These achievements can be primarily attributed to their successful investments in the nanotechnology industry and their ability to leverage scientific research to drive economic and industrial growth. This has been made possible through the implementation of effective scientific policies that foster innovation and development.

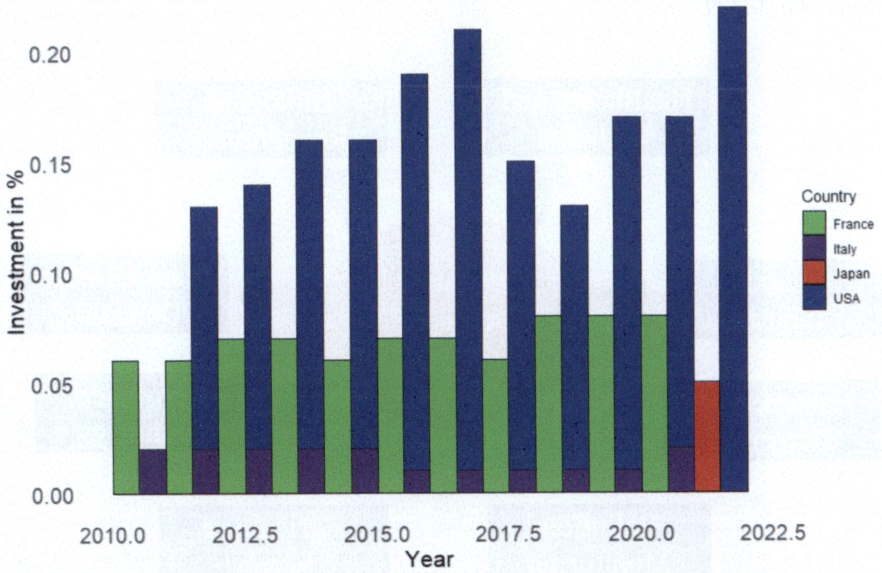

Fig. 1 Government funding for nanotechnology in time span 2011–2021

2 Role of Nanosensors in Nanotechnology

An important part of nanotechnology is nanosensors that have become a valuable tool for detecting and tracking various physical attributes at the nanoscale level, including temperature, chemical species, and nanoparticles. A nanosensor is a type of sensing device that is specifically designed to operate on the nanoscale [31]. It has a minimum dimension of 100 nm and its main purpose is to collect data from this minuscule scale. This information is then converted into data that can be further analyzed for various purposes. Researchers from diverse fields such as chemistry, medicine, pharmacy, biology, and material science and engineering have been working diligently to develop nanomaterials-based products [32]. These innovative products are anticipated to be introduced into the market in the coming years. With the rapid advancement of nanotechnology, there have been exciting developments in the design and production of Nanosensors. These sensors possess exceptional qualities such as optical, electrical, magnetic, mechanical, and thermal properties, which make them highly suitable for various scientific and technological applications. One significant advantage of Nanosensors is their ability to provide a suitable surface for target molecules in samples, resulting in enhanced sensitivity and lower detection limits [33]. Nanosensors play a crucial role in advancing nanotechnology research by enabling the detection and measurement of molecular-level phenomena. Their importance has been growing steadily as scientists recognize the need for precise and accurate analysis in this field. In recent years, the global nanosensor market has experienced significant growth. Some potential areas of nanosensor applications are depicted in Fig. 2.

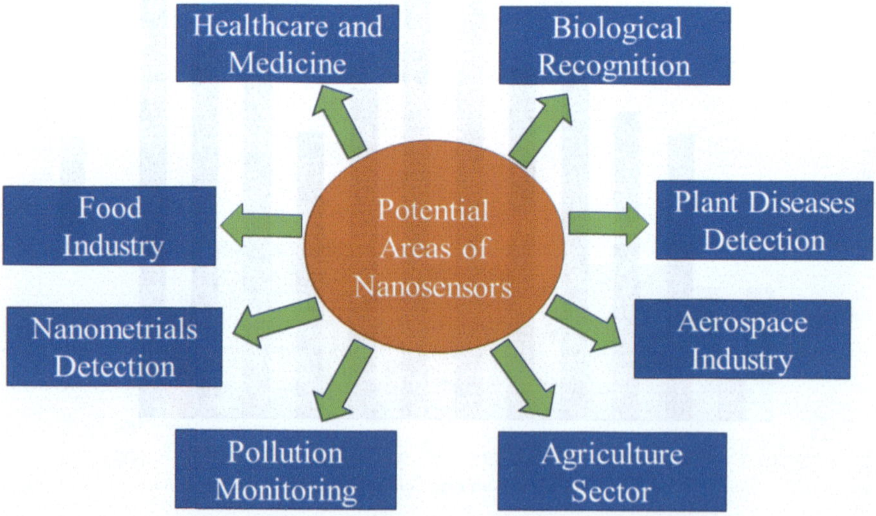

Fig. 2 Some possible applications of nanosensors

Nanosensors have found widespread application in medical diagnostics and various devices, driving innovation in the industry [34]. Additionally, nanosensors are being utilized in the design of structures and can be implemented on a large scale. This enables environmental chemists to monitor pollutants more effectively, efficiently, and selectively using cost-effective nanomaterials. In addition, the food industry also utilizes nanosensors to ensure the freshness and purity of food items. Detecting adulterants and contaminants in food using conventional detection systems is difficult, especially at low levels. To overcome this challenge, nanoparticles have been investigated as a mean to detect toxic chemicals and microorganisms with greater sensitivity. Nanosensors possess the ability to identify harmful bacteria, toxins that contaminate food, impurities, vitamins, dyes, fertilizers, pesticides, taste, and smell [35]. The freshness of food can be monitored by employing time–temperature and oxygen nanosensor indicators. Furthermore, invisible nanobarcodes can be utilized to evaluate product authenticity and ensure brand protection. In military and national security domains, nanosensors play a crucial role in detecting biotoxins like smallpox and anthrax [36]. Recently, there has been an increasing utilization of minuscule nanosensors composed of nanomaterials to facilitate the removal of toxic heavy metals from the environment. In the past two decades, there has been a notable increase in both the motivation and progress in the field of biosensor technology. Biosensors have been constructed using electrical and electrochemical technologies, which can significantly improve their functioning and overall effectiveness [37]. In addition, the early identification of plant diseases can be accomplished by employing nanosensors. Human beings face a substantial danger from environmental pollutants, as these substances can lead to numerous health hazards and cause severe harm to the ecosystem. A new method of air sampling is being implemented by researchers, utilizing nanosensors to evaluate air quality and specifically detect pollutants [38]. In short, nanosensors are widely used in medical diagnostics, environmental monitoring, and various other applications. It is worth noting that there are some promising prospects for nanosensors future potential across a wide range of applications. Designing and fabricating nanosensors present numerous challenges that need to be overcome. At present, the sensor industry lacks comprehensive regulations for the testing and utilization of nanosensors. The manufacturing process of nanosensors demands extreme precision due to their tiny size and susceptibility to various synthesis methods. Nanosensors possess certain benefits in comparison to conventional sensors, although they also have certain limitations. For example, the process of generating and utilizing them can incur significant costs and require a substantial amount of time. Moreover, certain nanomaterials employed in these sensors may pose a risk to human tissues and cell cultures, contingent upon their composition and quantity [39]. One more disadvantage of nanosensor industry is the lack of availability of real-time performance data that can be used to assess the reliability of nanosensors by utilizing some statistical method.

In the following section we propose the use of control chart approach to monitor the critical parameters of nanosensors during the fabrication process and to assess their performance.

3 Control Chart Approach

A control chart is a useful tool for visually representing process data over time [40]. It helps differentiate between common causes of variation and special causes of variation by using interpretation rules based on probability [41]. Additionally, the utilization of this method can be advantageous in evaluating the influence and effectiveness of implemented modifications in a procedure. A control chart is a vital part of statistical process control (SPC). The main objective of implementing SPC methods is to improve the understanding of a process and drive it towards desired outcomes. These methods help in reducing variations in the final-product parameters and ultimately enhance the overall performance of the process. The scientific community has recognized SPC as a valuable tool for improving manufacturing and administrative processes. By using SPC, organizations can proactively identify and tackle potential issues before they have any adverse effects on productivity and quality [42]. Process data charts are widely used in different industries as they help in understanding and improving processes. These charts, based on statistical theory, are user-friendly and can be easily interpreted by practitioners [43]. They not only provide valuable insights but also support informed decision-making and drive continuous improvements. When making decisions about which quality attributes to prioritize, there is a choice between analyzing a characteristic that currently has a high number of nonconformities or controlling a feature that is essential for maintaining quality. A critical quality attribute is a property related to the physical, chemical, biological, or microbiological aspects of a product that must meet specific limits, ranges, or distributions in order to ensure the desired level of quality [44]. To create a chart showing the measurements of a quality characteristic, you need to take samples from the process during a specific period [45]. There are different types of control charts available, but the most common one is the Shewhart control chart, which is depicted in Fig. 3. This type of chart provides a standard format and interpretation for analyzing the data.

Figure 3 depicts the measurements for a quality characteristic are plotted based on a specific number of samples taken within a defined period of time (such as hour, day or week etc., ...) in the process. The chart includes a central line (CL) as well as upper and lower control limits (UCL and LCL). The UCL and LCL are at equal distance from the CL which may be $\mp 3\sigma$ or $\mp 2\sigma$ or are set up based on probability limit (prob limit) [45]. The probability limits are usually preferable in case of asymmetric distribution such as gamma distribution beta distribution or exponential distribution etc. [46, 47]. This chart is utilized to monitor processes where data is gathered at specific intervals across time, organized into logical and sequential subgroups. The control chart's CL represents the anticipated position of a process characteristic when no abnormal sources of variability are present. Occasionally, this CL aligns with the desired value for specific quality features. Control limits (UCL and LCL) are computed to include nearly all observations within the interval (UCL–LCL) when historical data is available. These control limitations help to spot variations that could point to a problem with the process [48].

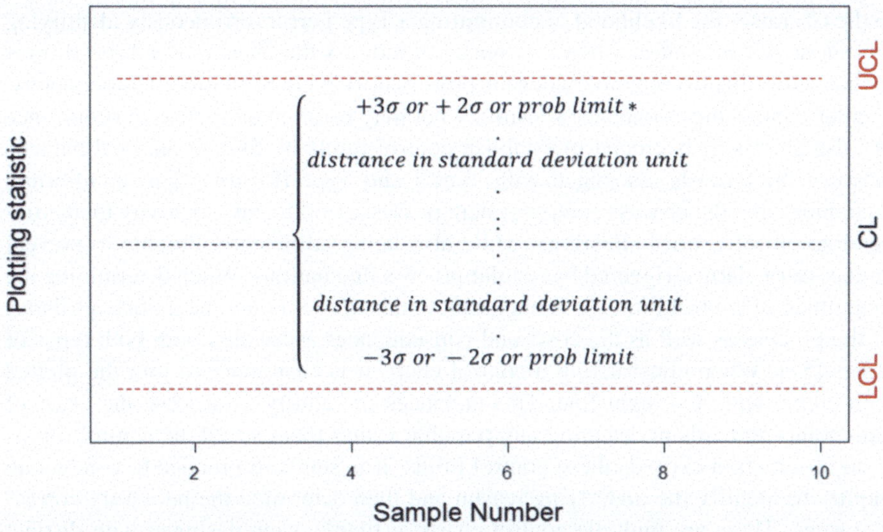

Fig. 3 A typical format of the control chart for monitoring inspection process

Furthermore, predetermined specification limits, such as the upper specification limit (USL) and lower specification limit (LSL), are determined according to the specifications of the product or process. Customers, regulatory agencies, or the manufacturer itself establish these values to guarantee that the product or process adheres to specific standards or requirements. The specification limits serve as guidelines for the allowable range in which the product or process must operate in order to achieve the established standards. A process is considered to be in statistical control when all the subgroup values plotted on the control chart consistently remain within the control boundaries for an adequate length of time [49]. Furthermore, it is imperative that there are no signs of unusual patterns or behaviours detected within these established control limits. Instances of such abnormal behaviours contain patterns, recurring patterns, noticeable changes, and other non-random variation of non-random fluctuation. The particular non-random variation is frequently identified and described by supplemental regulations as specified in the recommendations cited in [50]. In order to guarantee the process's stability and predictability, it is essential to reach and maintain this level of statistical control, which shows that any changes are within predicted bounds and do not originate from problematic sources. Accurately defining control limits is essential for minimizing both types of errors in control chart methodology. By increasing the distance between the control limits and from the CL, the probability of a Type I error (false alarm) diminishes. This means that the occurrence of an out-of-control state, when there is no genuine assignable cause, is less likely. Expanding the control boundaries, however, raises the probability of committing a Type II error, which involves failing to detect a genuine indication of an out-of-control state. This occurs because a data point may lie inside the broader limits, even if the process is truly out of control. In contrast, setting control limits in close proximity to the CL raises the likelihood of committing a Type I error (erroneously identifying a point as out of control when it is not), but reduces the chance of a Type II error (enhanced ability to recognize when the process is truly out of control). Consequently, smaller control limits enhance sensitivity but may result in an increased occurrence of false alarms. It is crucial to establish control limits by finding an ideal balance between the hazards associated with Type I and Type II errors. This equilibrium guarantees that the process control system possesses sufficient sensitivity to identify genuine out-of-control situations, while also being sufficiently durable to prevent unnecessary alarms triggered by regular process fluctuations. When determining the magnitude of control limits, it is important to take into account the unique attributes of the process as well as the costs and consequences associated with both types of errors [51]. When constructing a control chart, it is customary to join the plotted data points with a straight line. This enhances the ability to observe the order of data points and aids in detecting any possible values that exceed the control limits. If an observation exceeds these control limits, it is standard practice to conduct an inquiry to identify the underlying reason and then determine the necessary corrective steps. There are multiple control charts available, each designed with distinct formulas that are appropriate for various types of data. The selection of the suitable control chart is dependent upon the characteristics of the data, such as its continuity or discreteness, as well as the distributional assumptions of the underlying process. The best control chart is determined by a variety of statistical distributions, including the normal, binomial, Poisson, and geometric distributions [52]. Selecting the appropriate control chart requires knowing the data's underlying distribution. For example, continuous data sets are typically associated with the normal distribution, whereas discrete data sets are linked to the binomial and Poisson distributions. To monitor and manage the process or product quality that is being evaluated, the best control chart may be chosen by knowing the characteristics of the data and how they are distributed statistically. In many situations, the facts at hand may not provide sufficient evidence to support certain hypothetical distribution. Consequently, there is an increasing demand for control charting methods with no distributional assumptions (Fig. 4) [53]. Nonparametric or distribution-free control charts have been introduced in recent years in response to this demand. Numerous other types of control charts exist, Fig. 5 can be useful a reasonably suitable choice for describing and monitoring many types of processes.

Figure 5 illustrates the categorization of all established control charts according to three prominent structures: the Shewhart model, the exponentially weighted moving average (EWMA) model, and the cumulative sum (CUSUM) model. These three structures function as universal frameworks that classify and facilitate comprehension of different types of control charts, whether they pertain to univariate or multivariate quality characteristics. Specific formulae corresponding to each type of chart can usually be found in respected quality control literature [54–56].

Products currently on the market have already benefited from the implementation of nanotechnology advancements. Nanotechnology advancements have already proven beneficial for products available in the market. The focus lies on addressing the challenge of creating nanomaterials with unique physical properties and finding

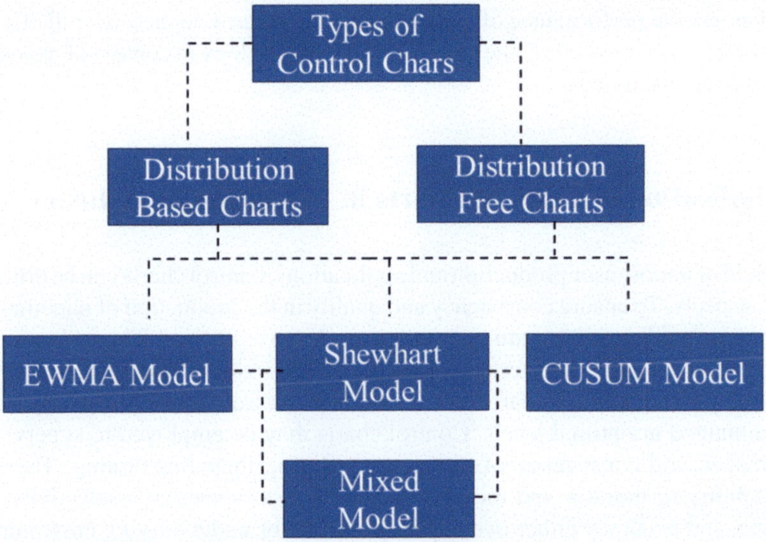

Fig. 4 Classification of control charts based on designed model

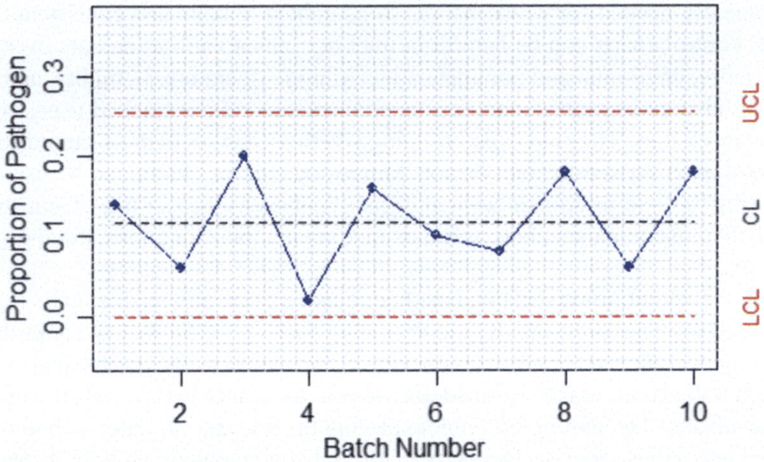

Fig. 5 Performance of food nanosensors using P chart

ways to manufacture them effectively. The importance of reliability, a crucial aspect of technology, is often overlooked in nanosensor manufacturing. In order to ensure progress and the effective use of nanosensor technology, it is essential to prioritize high levels of reliability. To evaluate the reliability of nanosensors at the manufacturing level, this study utilizes a control chart approach. By utilizing this approach, we can assess the performance of these sensors and determine their overall effectiveness. In the following section, the control chart methodology in context of nanosensor data has been discussed.

4 Application of Control Charts in Nannosensor Industry

In the field of nanosensor production and applications, control charts can be utilized in various aspects. To ensure consistency and quality in the production of nanomaterials used for sensor fabrication, control chart approach can be utilized. These charts enable monitoring of various characteristics such as particle size, composition, surface area, etc. By employing control charts, the synthesis processes can be effectively tracked and maintained at optimal levels. Control charts may be employed to supervise the performance and consistency of nanosensors during their functioning. They have the capability to monitor and measure factors such as sensitivity, selectivity, reaction time, and accuracy either over a period of time or under varying environmental conditions. Control charts can be used for mass production of nanosensors to identify differences between batches. They assist in locating any irregularities or discrepancies in the manufacturing procedure. In addition, calibration and routine maintenance are frequently needed for nanosensors. To guarantee reliable and consistent observations, control charts can be helpful in tracking sensor calibration data over time. Furthermore, environmental parameters including temperature, humidity, and pressure that have an impact on nanosensor performance can be tracked using control chart methodology. They aid in determining whether changes to these circumstances have an impact on sensor data or not. Moreover, control charts can be useful for monitoring the deterioration or wear and tear of nanosensors over time. Contributing to predictive maintenance plans, this aids in knowing their long-term durability and reliability.

In our viewpoint, some statistical methods need to be considered for analyzing data transmitted from nanosensors. By incorporating such methods, we can gain valuable insights and ensure accurate assessments of the nanosensors' performance. This approach will help us make informed decisions and enhance the overall effectiveness of these advanced technologies. After exploring the relevant literature and to best of authors knowledge, there is a lack of available data for conducting reliability analysis of nanosensors. As a result, we must depend on simulated computer-generated data to explore the suitability of control chart methodology for consistent performance of nanosensors and assess the quality of the nanomaterial that are utilized in the production of nanosensors. How this methodology can be useful, some examples are presented below.

In the food industry, one of the major challenges is the presence of microorganisms that can spoil food, change its taste, and even lead to infections in humans. These health risks can be severe and sometimes even fatal. Another concern within the industry is food adulteration and maintaining brand integrity, which are important aspects to address in order to ensure consumer safety and trust. Nanosensors have the ability to identify a wide range of food contaminants, including toxins, adulterants, pathogenic bacteria, vitamins, dyes, fertilizers, and pesticides. Additionally, they can determine the freshness of food by detecting its aroma and flavor properties. This ensures that the food remains fresh and maintains its quality.

Let us consider a situation where food samples are tested for the existence of a specific pathogen using a nanosensor. The binary output would be resulted from the sensor, which indicates the presence (1) or absence (0) of the pathogen in each sample. We will be utilizing this binary information to develop a P-chart, which is specifically designed to monitor the proportion of occurrences of this pathogen in food samples. Suppose that a total of 10 sample batches were collected for testing using the food nanosensor. Each sample batch consisted of 50 food items randomly selected. The resulting data on nonconforming items (i.e., affected samples), are presented in Table 1 along with the corresponding proportions of nonconforming items for each sample batch.

The sample proportion from each sample is calculated as:

$$P_i = \frac{X_i}{n};\ i = 1, 2, \ldots 10 \tag{1}$$

where X_i is a binary vector of zero and one i.e., $X_i = [0, 1, 0, \ldots .1]$ and n is sample batch size which is 50 in this case.

The P-chart based on sample proportions given Table 1 can be calculated as follows:

$$\left.\begin{aligned} CL &= \overline{p} = 0.122 \\ UCL &= 0.12 + 3 * \sqrt{\tfrac{0.12*(1-0.12)}{50}} = 0.261 \\ LCL &= 0.12 - 3 * \sqrt{\tfrac{0.12*(1-0.12)}{50}} = -0.017 \end{aligned}\right\} \tag{2}$$

The lower control limit can be set to zero since the sample proportions cannot be negative. Figure 1 displays the P chart that is derived from the computation provided above.

The stability and consistency of the nanosensor's performance are reflected in the P-chart's representation of the proportions of nonconforming samples across batches. Consistent outcomes between batches show stable and reliable sensor performance in the pathogen detection. Deviations from the observed proportions can be indicated by any data points on the control chart that fall outside the control limits. Such variations may be an indication of inconsistence in the sensor readings between

Table 1 Nanosener data for testing the presence of pathogen

Sample batch	1	2	3	4	5	6	7	8	9	10
Poisoned samples (X_i)	8	4	11	2	9	6	5	10	4	2
Proportion (P_i)	0.16	0.08	0.22	0.04	0.18	0.12	0.10	0.20	0.08	0.04

batches. This may suggest that there are problems with the precision or consistency of the nanosensor detection ability. Thus, on one hand, the P chart can be applied to assess the consistent performance of nanosensors in terms of stability and consistency of nanosensors. On the other hand, the P chart can be utilized by the food industry to effectively monitor and compare the proportions of nonconforming samples in different batches based on nanosensor data. This approach allows for the timely identification of batch-to-batch variations, enabling necessary interventions or improvement in food safety practices when deviations are observed.

The P chart is used as an example to demonstrate the application of a control chart. However, other types of control charts can be used depending on the type of variable and assumptions. They provide a visual representation of the data, allowing for the identification of any patterns or trends that may indicate a process is out of control. To further demonstrate the procedure, we add another simple example of an $\overline{X}$ chart for nanosensor data. This will help to enhance understanding of the utility part of the proposed control chart methodology in nanosensor industry. Nanostructured materials such as nanoscale wires, carbon nanotubes, thin films, metal and metal oxide nanoparticles, polymer, and biomaterials are commonly used in the production of nanosensors [34]. The quality of nanosensors is thus greatly influenced by the choice of materials used during manufacturing. To illustrate the utilization of a control chart, consider the scenario of monitoring particle size in the production of nanoparticles for nanosensors. We wish to set up the statistical control of the nanoparticle size in this manufacturing process using $\overline{X}$ and R charts. To ensure consistency in the production of nanoparticles with an average diameter of 50 nm, the quality team collects data from 15 different batches each with 3 samples used in the nanosensor manufacturing process in order to control mean and variability. The particle sizes of the nanoparticles produced in these batches are presented in Table 2 for analysis.

Mean size and range of each sample batch are calculated and given in the last column of Table 2. The trial control limits for the given data based on $\overline{X}$ and R monitoring schemes can be calculated as:

<u>For $\overline{X}$ chart</u>

$A_2 = 1.023$ (From the Table VI [42])

$$\left.\begin{aligned} CL &= \overline{\overline{X}} = 50.65 \\ UCL &= \overline{\overline{X}} + 3 * A_2 = 53.72 \\ LCL &= \overline{\overline{X}} - 3 * A_2 = 47.58 \end{aligned}\right\} \quad (3)$$

<u>For R chart</u>

To set up the R chart we typical need constants D_3 and D_4 whose values based on sample size. For $n = 3$, $D_3 = 0$ and $D_4 = 2.574$ (From the Table VI [42]).

Thus,

Table 2 Data on size of nanoparticles used in the production of nanosensors

Batch number	Nanoparticle sizes				
	I	II	III	Mean	Range
1	51.22	49.21	50.17	50.20	2.01
2	50.95	50.28	50.67	50.63	0.67
3	50.67	51.67	49.29	50.54	2.38
4	48.15	46.98	50.05	48.39	3.07
5	49.97	52.20	52.86	51.67	2.89
6	51.43	51.08	52.25	51.58	1.17
7	49.51	51.18	50.00	50.23	1.67
8	50.65	52.73	51.65	51.67	2.08
9	51.70	50.93	49.80	50.81	1.90
10	52.77	52.70	48.01	51.16	4.76
11	48.26	52.59	49.77	50.20	4.33
12	49.85	51.07	48.36	49.76	2.71
13	52.27	48.51	48.94	49.90	3.76
14	52.34	51.01	51.45	51.60	1.33
15	52.22	48.85	52.81	51.29	3.96

$$\left.\begin{aligned} CL &= \overline{R} = 2.48 \\ UCL &= \overline{R}D_4 = 6.38 \\ LCL &= \overline{R}D_3 = 0 \end{aligned}\right\} \tag{4}$$

The $\overline{X}$ and R are shown in Figs. 6 and 7 respectively.

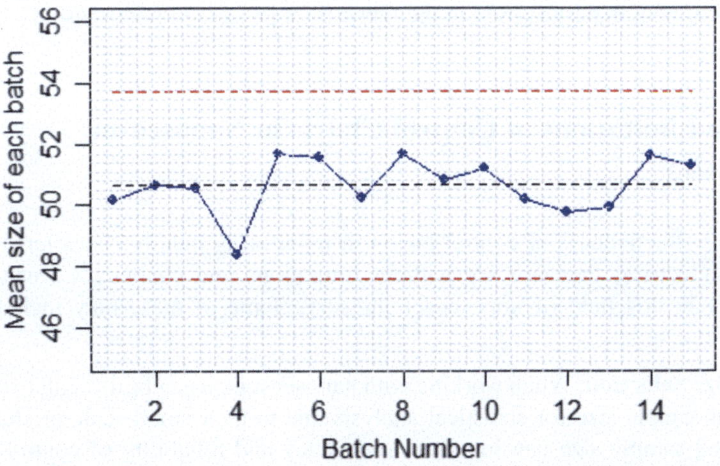

Fig. 6 $\overline{X}$ chart for nanoparticles used in the production process of nanosensors

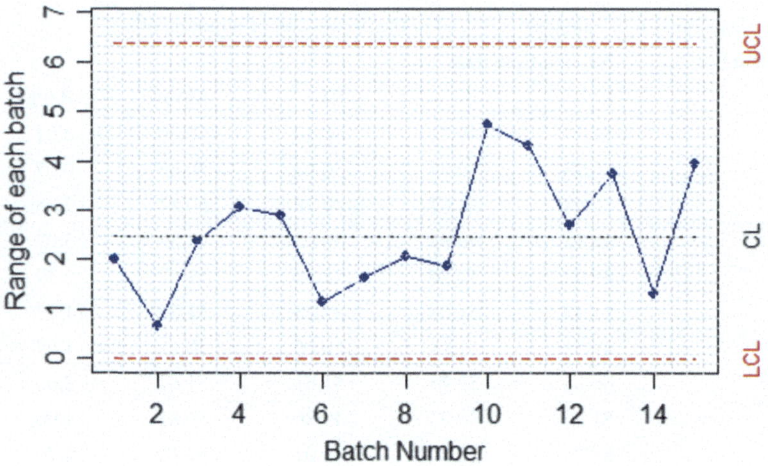

Fig. 7 R chart for nanoparticles used in the production process of nanosensors

After plotting the preliminary sample averages on the $\overline{X}$ chart and ranges on R chart, it is clear that there is no sign of an out-of-control condition. Therefore, we can conclude that both the $\overline{X}$ and R charts exhibit in-control state, indicating that the nanosesnor manufacturing process is in control at the specified levels. As a result, we will use the trial control limits for phase II, where we can monitor future production of nanosensors. By visually representing the batch numbers against their mean particle sizes, the control chart effectively displays any potential variations in the manufacturing process. If any point on the control chart falls outside the control limits, it indicates that there may be deviations in the manufacturing process. In short control charts are effective tools for assessing characteristics of nanosensors over time or under different environmental conditions. They can be used to assess the quality of nanomaterials used in the fabrication process of nanosensors.

5 Some Limitations of Control Charts in Nanosensor Industry

Despite the effectiveness of control charts in monitoring quality characteristics in different industries, they are not without limitations and challenges when used specifically in this field of nonosensor industry. Some of these may include the following.

Sample Size Selection: When working with nanosensors, it can be difficult to obtain a sufficient sample size for statistical analysis due to their small-scale production. This limited sample size can impact the accuracy and reliability of control chart analysis, making it more challenging to identify small variations or trends.

Measurement Accuracy and Precision: In any quality system, measurements play a crucial role. The impact of an inefficient measurement system on business performance is substantial, as it often results in uninformed and poor decision-making. Thus, ensuring an effective measurement system is essential for optimal business outcomes. The effectiveness of control charts can be affected by the accuracy and precision of measurement instruments used in nanosensor production. If there are measurement errors or limitations in detecting extremely small variations, it could impact the reliability of the control chart results. To assess the measurement system's capability in the context of nanosensor data, it is crucial to consider both repeatability and reproducibility. Repeatability refers to obtaining consistent observed values when measuring the same unit multiple times under identical conditions. On the other hand, reproducibility takes into account the variation in observed values when units are measured under different conditions, such as different operators or time periods. These two factors play a significant role in evaluating the overall performance of the measurement system. Measurement accuracy and precision play crucial roles in nanosensor data collection as even slight variations can significantly influence decision-making. Therefore, it is imperative to ensure the utmost accuracy and precision in measuring these highly sensitive data. When conducting measurements, it is important to consider that some of the variability observed is due to the inherent nature of the units or items being measured, while some variability is a result of the measurement system used. In the case of nanosensors, performing a measurement system capability study can be a challenging task.

External variability: Nanosensor production can be greatly affected by external factors such as environmental conditions, variations in raw materials, and equipment sensitivity. The use of control charts may face challenges in differentiating between natural process variability and deviations caused by these external factors.

Real-time monitoring challenge: Real-time monitoring is essential for nanosensors as certain characteristics may require immediate attention. Any delays in measurement or analysis can significantly affect the effectiveness of quality control using control charts.

Failure cost: Internal failure costs arise when there are discrepancies in the quality of products, components, materials, or services, which are identified before the customer receives the final product. These costs would be eliminated if there were no defects in the product. However, in the manufacturing of nanosensors, implementing a control chart approach may lead to an increase in internal failure costs.

Economic design: Control chart approach discussed above is primarily based on statistical criteria. Typically, this involved determining the sample size and control limits to ensure that the chart has a specific average run length for detecting quality characteristic shifts and maintaining a specified average run length when the process is under control. However, considering the economic implications, it is important to design a control chart carefully in nanosensor industry as it directly impacts various costs. These costs include expenses related to sampling, testing, investigating out-of-control signals, correcting assignable causes, and allowing nonconforming units to reach consumers. Thus, it is also advisable to approach the design of a control chart from an economic perspective.

6 Concluding Remarks

This research has studied the role of nanosensor market and its significance across different fields. The substantial expansion witnessed in the nanosensor market highlights its crucial role in monitoring and detecting characteristics at the nano size. In order to guarantee the production of outstanding products of high quality, it is crucial to implement a rigorous approach to statistical process control. This chapter has assessed the effectiveness of control chart methodology in monitoring and maintaining the quality standards in products, processes and systems that depend on data from nanosensors. To ensure stability and uniformity in data provided by nanosensors or materials used in their fabrication, it is recommended to use the visualization method of control chart to observe and depict any alterations in processes. Our research investigated several control charts and their applications by utilizing some examples data to demonstrate their efficacy in monitoring deviations. We have illustrated the implementation of control chart methodology to analyze the performance of nanosensors in different scenarios using simulated data. By evaluating the effectiveness of these nanosensors, we can determine their reliability and suitability for various applications.

In this way, this chapter has established a basis for future research that seeks to thoroughly examination of nanosensors reliability and their performance in real-time.

References

1. Bastos NA, Bitencourt SB, Martins EA, De Souza GM (2021) Review of nano-technology applications in resin-based restorative materials. J Esthet Restor Dent 33(4):195–197. https://doi.org/10.1111/jerd.12699
2. Chakraborty K, Shivakumar A, Ramachandran S (2016) Nano-technology in herbal medicines: a review. Int J Herb Med 4(3):195–197. https://doi.org/10.22271/flora.2016.v4.i3.05
3. Amudhavalli NK, Ravi C A review on nano technology in concrete. Int J Civ Eng Technol 10(1):195–197
4. Ghosal M, Chakraborty A (2021) The growing use of nanotechnology in the built environment: a review. IOP Conf Ser Mater Sci Eng 1170(1):195–197. https://doi.org/10.1088/1757-899x/1170/1/012007
5. Haroun A et al (2021) Progress in micro/nano sensors and nanoenergy for future AIoT-based smart home applications. Nano Express 2(2):195–197. https://doi.org/10.1088/2632-959X/abf3d4
6. Ye Z et al (2021) Laser nano-technology of light materials: precision and opportunity. Opt Laser Technol 139:195–197. https://doi.org/10.1016/j.optlastec.2021.106988
7. Xu H, Gao M, Tang X, Zhang W, Luo D, Chen M (2020) Micro/nano technology for next-generation diagnostics. Small Methods 4(4):195–197. https://doi.org/10.1002/smtd.201900506

8. Singh R, Kumar S (2023) Intelligent nano-based sensor for quality detection of food products. In: Nanotechnology advancement in agro-food ındustry.https://doi.org/10.1007/978-981-99-5045-4_4
9. Rahmat M, Naz S, Kiran S (2023) Nano-trackers (nano-sensors) for forensics investigation. In: Modeling and simulation of functional nanomaterials for forensic investigation. https://doi.org/10.4018/978-1-6684-8325-1.ch005
10. Lagopati N et al (2023) The role of nano-sensors in breath analysis for early and non-ınvasive disease diagnosis. Chemosensors 11(6):195–197. https://doi.org/10.3390/chemosensors11060317
11. Liu Y, Li X, Chen J, Yuan C (2020) Micro/Nano electrode array sensors: advances in fabrication and emerging applications in bioanalysis. Front Chem 8:195–197. https://doi.org/10.3389/fchem.2020.573865
12. Hosseini-Ara R, Mokhtarian A, Karamrezaei AH, Toghraie D (2022) Computational analysis of high precision nano-sensors for diagnosis of viruses: effects of partial antibody layer. Math Comput Simul 192:195–197. https://doi.org/10.1016/j.matcom.2021.09.009
13. Hingangavkar GM, Kadam SA, Ma YR, Mulik RN, Patil VB (2023) Highly selective and sensitive MoS_2 nano-sensor for H_2S detection. J Alloys Compd 941:195–197. https://doi.org/10.1016/j.jallcom.2023.168849
14. Rawat B, Bist AS, Supriyanti D, Elmanda V, Sari SN (2022) AI and nanotechnology for healthcare: a survey. APTISI Trans Manag 7(1):195–197. https://doi.org/10.33050/atm.v7i1.1819
15. Sim S, Wong NK (2021) Nanotechnology and its use in imaging and drug delivery (review). Biomed Rep 14(5):195–197. https://doi.org/10.3892/br.2021.1418
16. Sarkar MKI et al (2018) Effect of nanotechnology on cancer disease. J Bionanosci 12(3):195–197. https://doi.org/10.1166/jbns.2018.1532
17. Malik S, Muhammad K, Waheed Y (2023) Emerging applications of nanotechnology in healthcare and medicine. Molecules 28(18):195–197. https://doi.org/10.3390/molecules28186624
18. Ibrahim H (2020) Nanotechnology and ıts applications to medicine: an overview. QJM Int J Med 113(Supplement_1):195–197. https://doi.org/10.1093/qjmed/hcaa060.008
19. Shandilya R, Mishra PK, Pathak N, Lohiya NK, Sharma RS (2020) Nanotechnology in reproductive medicine: opportunities for clinical translation. Clin Exp Reprod Med 47(4):195–197. https://doi.org/10.5653/cerm.2020.03650
20. Boisseau P, Loubaton B (2011) Nanomedicine, nanotechnology in medicine. Comptes Rendus Phys 12(7):195–197. https://doi.org/10.1016/j.crhy.2011.06.001
21. Alba-Patiño A, Vaquer A, Barón E, Russell SM, Borges M, de la Rica R (2022) Micro- and nanosensors for detecting blood pathogens and biomarkers at different points of sepsis care. Microchim Acta 189(2):195–197. https://doi.org/10.1007/s00604-022-05171-2
22. Rabbani A et al (2021) Development of bactericidal spinel ferrite nanoparticles with effective biocompatibility for potential wound healing applications. RSC Adv 11(3):195–197. https://doi.org/10.1039/d0ra08417d
23. Archana D, Dutta J, Dutta PK (2013) Evaluation of chitosan nano dressing for wound healing: characterization, in vitro and in vivo studies. Int J Biol Macromol 57:195–197. https://doi.org/10.1016/j.ijbiomac.2013.03.002
24. Mehdinia A, Rostami S (2020) Green synthesis of plasmonic metal nanoparticles and their application as environmental sensors. In: Nanotechnology in the life sciences, pp 219–257. https://doi.org/10.1007/978-3-030-45116-5_8
25. Mohamed EF, Awad G (2022) Development of nano-sensor and biosensor as an air pollution detection technique for the foreseeable future. Compr Anal Chem 99:163–188. https://doi.org/10.1016/bs.coac.2021.11.003
26. García-Guzmán JJ, López-Iglesias D, Bellido-Milla D, Palacios-Santander JM, Cubillana-Aguilera L (2020) Green synthesis of nanomaterials for biosensing. In: Nanotechnology in the life sciences, pp 135–217. https://doi.org/10.1007/978-3-030-45116-5_7

27. Kumar V, Raghuwanshi SK, Kumar S (2022) Advances in nanocomposite thin-film-based optical fiber sensors for environmental health monitoring—a review. IEEE Sens J 22(15):195–197. https://doi.org/10.1109/JSEN.2022.3185004
28. Ariga K, Ji Q, Hill JP, Bando Y, Aono M (2012) Forming nanomaterials as layered functional structures toward materials nanoarchitectonics. NPG Asia Mater 4(5):195–197. https://doi.org/10.1038/am.2012.30
29. Olawoyin R (2018) Nanotechnology: the future of fire safety. Saf Sci 110:195–197. https://doi.org/10.1016/j.ssci.2018.08.016
30. OECD, Key Nanotechnology Indicators, oe.cd/kni; and OECD, Main Science and Technology Indicators Database. www.oecd.org/sti/msti.htm
31. Javaid M, Haleem A, Singh RP, Rab S, Suman R (2021) Exploring the potential of nanosensors: a brief overview. Sens Int 2:100130. https://doi.org/10.1016/j.sintl.2021.100130
32. Varadan VK, Chen L, Xie J (2008) Nanomedicine: design and applications of magnetic nanomaterials, nanosensors and nanosystems. Wiley
33. Kumar M, Agrawal AV, Moradi M, Yousefi R (2020) Nanosensors for gas sensing applications. In: Nanomaterials for air remediation.https://doi.org/10.1016/B978-0-12-818821-7.00006-3
34. Kumar A (2018) Nanosensors: applications and challenges. Int J Sci Res 1472–1474
35. Patel G, Pillai V, Bhatt P, Mohammad S (2020) Application of nanosensors in the food industry. In: Nanosensors for smart cities.https://doi.org/10.1016/B978-0-12-819870-4.00020-7
36. Lichtenstein A et al (2014) Supersensitive fingerprinting of explosives by chemically modified nanosensors arrays. Nat Commun 5:4195. https://doi.org/10.1038/ncomms5195
37. Avelino KYPS, Silva RR, da Silva Junior AG, Oliveira MDL, Andrade CAS (2017) Smart applications of bionanosensors for BCR/ABL fusion gene detection in leukemia. J King Saud Univ Sci 29(4). https://doi.org/10.1016/j.jksus.2017.08.002
38. Santos JP, Sayago I, Aleixandre M (2020) Air quality monitoring using nanosensors. In: Nanomaterials for air remediation. https://doi.org/10.1016/B978-0-12-818821-7.00002-6
39. Kaur R, Tripathy SK, Sharma SK (2019) Advantages and limitations of environmental nanosensors. In: Advances in nanosensors for biological and environmental analysis.https://doi.org/10.1016/B978-0-12-817456-2.00007-3
40. Suman G, Prajapati DR (2018) Control chart applications in healthcare: a literature review. Int J Metrol Qual Eng 9:5. https://doi.org/10.1051/IJMQE/2018003
41. Smajdorová T, Noskievičová D (2022) Analysis and application of selected control charts suitable for smart manufacturing processes. Appl Sci 12(11). https://doi.org/10.3390/app12115410
42. Montgomery DC (2009) Introduction to statistical quality control, 6th edn. Wiley, New York
43. Bisiotis K, Psarakis S, Yannacopoulos AN (2022) Control charts in financial applications: an overview. Qual Reliab Eng Int 38(3):1441–1462. https://doi.org/10.1002/QRE.2962
44. Williams E (2018) Understanding variation-part 2: the control chart. Curr Probl Pediatr Adolesc Health Care 48(8):202–205. https://doi.org/10.1016/j.cppeds.2018.08.009
45. Mehmood R, Riaz M, Does RJMM (2013) Control charts for location based on different sampling schemes. J Appl Stat 40(3):483–494. https://doi.org/10.1080/02664763.2012.740624
46. Park C, Ouyang L, Wang M (2023) A study on the performance of the probability-limit control charts based on the geometric and negative binomial distributions. Comput Ind Eng 180. https://doi.org/10.1016/j.cie.2023.109275
47. Shah F, Khan Z, Aslam M, Kadry S (2021) Statistical development of the VSQ -control chart for extreme data with an application to the carbon fiber ındustry. Math Probl Eng 2021. https://doi.org/10.1155/2021/9766986
48. Derya K, Glu Canan H (2012) Control charts for skewed distributions. Adv Meth Stat 9(2):95–106. https://doi.org/10.51936/GHAA8860
49. Kosztyán ZT, Katona AI (2018) Risk-based X-bar chart with variable sample size and sampling interval. Comput Ind Eng 120. https://doi.org/10.1016/j.cie.2018.04.052
50. MacCarthy BL, Wasusri T (2002) A review of non-standard applications of statistical process control (SPC) charts. J Qual Reliab Manag 19(3). https://doi.org/10.1108/02656710210415695

51. Katona AI, Saghir A, Hegedűs C, Kosztyán ZT (2023) Design of risk-based univariate control charts with measurement uncertainty. IEEE Access 11. https://doi.org/10.1109/access.2023.3312226
52. Lee Ho L, Fernandes FH, Bourguignon M (2019) Control charts to monitor rates and proportions. Qual Reliab Eng Int 35(1):74–83. https://doi.org/10.1002/QRE.2381
53. Chakraborti S, Graham MA (2019) Nonparametric (distribution-free) control charts: an updated overview and some results. Qual Eng 31(4):523–544. https://doi.org/10.1080/08982112.2018.1549330
54. Ryan T (2011) Statistical methods for quality improvement. Wiley
55. Chakraborti S, Graham M (2019) Nonparametric statistical process control. Wiley
56. Qiu P (2013) Introduction to statistical process control. CRC Press

Rare-Earth Based Nanomaterials for Sensors: Experimental and Application Perspectives

Naveen Tiwari, Edgar R. Lopez-Mena, Diego E. Navarro-López, and Massimo Lazzari

Abstract Nanoparticles made from rare earth elements (REEs) have attracted significant attention due to their distinct quantum properties, surpassing the capabilities of larger rare earth materials. These exceptional characteristics make them ideal for a wide array of applications, particularly in sensor and biomedical applications where their remarkable nanomaterials attributes shine. Consequently, conducting thorough research, development, synthesis, and analysis of these rare earth nanoparticles is crucial. This chapter offers a comprehensive exploration into the nanomaterial advancements involving rare earth nanoparticles, encompassing their production techniques, characterization, electrochemical traits, and their application within various sensors to identify various biomolecules and hazardous substances. The examination also delves into how the size of the nanoparticles and modifications to the electrode surface impact their electrochemical performance. Moreover, it highlights the sensing mechanisms, selectivity, sensitivity, linear range, and detection limits of these sensors. Herein, the chapter underscores the highly sensitive nature of the explored methods, emphasizing their potential for precise and accurate identification of target substances. The broad spectrum of potential applications for detecting diverse biomolecules and hazardous materials is underscored, demonstrating the adaptability and significant potential of nanoparticles derived from rare earth elements across multiple domains. The insights provided in this section furnish a valuable source for future research endeavors, offering a glimpse into the present research status and the potential directions within the realm of rare earth element nanoparticles and electrochemical nanosensors.

Keywords Rare-earth materials · Sensor · Nanomaterials · Biomedical · Electrochemical

N. Tiwari (✉) · M. Lazzari
Center for Research in Biological Chemistry and Molecular Materials (CiQUS), University of Santiago de Compostela, Santiago de Compostela, Spain
e-mail: naveen.tiwari@usc.es; naveen.tiwari@tuni.fi

E. R. Lopez-Mena · D. E. Navarro-López
Tecnologico de Monterrey, Escuela de Ingeniería y Ciencias, Colonia Nuevo México, Zapopan, Jalisco, Mexico

I. Uddin (ed.), *Smart Nanosensors*, Smart Nanomaterials Technology,
https://doi.org/10.1007/978-981-96-3878-9_15

1 Introduction

The lanthanide (Ln) or lanthanoid series of chemical elements, also called rare earth elements (RE), comprises 15 metallic chemical elements with atomic numbers 57–70, from lanthanum through ytterbium. In the periodic table, they fill the 4f orbitals [1]. Due to their f-orbitals, these elements possess unique electronic and stereochemical properties and have great potential as reagents and catalysts in organic synthesis [2]. In the pursuit of pushing the boundaries of sensor technology, researchers have turned their attention to an extraordinary class of materials—rare-earth-based nanomaterials [3]. The allure of these nanomaterials lies in their unique and multifaceted properties derived from the elusive rare-earth elements, captivating the scientific community with the promise of revolutionizing sensor applications. As we stand at the intersection of nanoscience and sensing technology, the potential of rare-earth based nanomaterials to redefine the capabilities of sensors is nothing short of groundbreaking [4].

Sensors consist of a device capable of transforming a change of physical parameter into a measurable magnitude [5]. Sensors can perform this task using a transducer, a component capable of transforming one type of energy into another. A wide variety of sensors can be classified for many different purposes, including the type of measurement, signal conversion, output, material, etc. [6]. New technologies aim for the enhancement of some fundamental performance properties: sensitivity (the ability to detect the desired physical parameter in small quantities), selectivity (the ability to recognize the desired physical parameter), and stability (to detect the physical parameter continuously with the desired output) [2]. From enhancing sensitivity to elevating selectivity, rare-earth-based nanomaterials are crucial to addressing the limitations that have long constrained traditional sensors. A biosensor is a device that combines a biological element, like an enzyme or biomaterial, with an electronic component to produce a quantifiable signal. Biosensors consist of four main components: an analyte (the desired agent to detect), the bioreceptor (responsible for the necessary specificity for the analytical assessment within the biosensor method), the transducer (component capable of converting alterations caused by the interaction between the bioreceptor and from the analyte) and the electronics (responsible for detecting and manipulating the electric signal generated by the transducer for it to be processed) (Fig. 1) [7, 8].

Electrochemical biosensors have become very interesting due to their stability and reproducibility measurements [9]. To date, four generations of the electrochemical (glucose) sensor have been described: (1) glucose oxidase (GOx) based sensor. (2) A new synthetic mediator was introduced in the device. (3) The generation uses immobilized mediators to facilitate effective electron transport. (4) Finally, the actual generation uses several materials and complex architecture to improve efficiency (Fig. 2) [10].

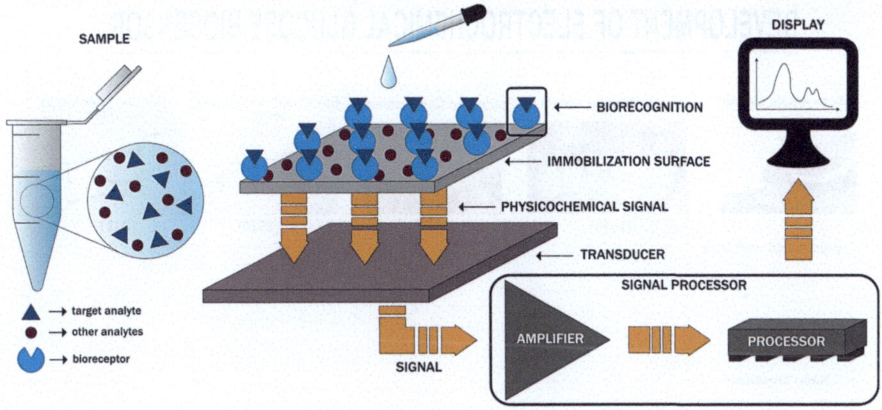

Fig. 1 Illustration of the main components of biosensors and the basic working principle [8]

2 Synthesis of Rare-Earth-Based Nanomaterials

2.1 Sol–Gel

The sol–gel, solution-based method is a prevalent technique for preparing metal-oxide nanoparticles. The method involves a series of chemical reactions, which include hydrolysis, gelation, drying, and annealing. In the initial steps, the sol–gel process involves the formation of sol through hydrolysis and polymerization reactions [11]. This allows the formation of a rigid porous gel. After that, the gel is dried to remove the solvent and residuals, usually at a low temperature. The resulting powders are calcined to obtain the desired phase. This method makes it possible to obtain high-quality nanoparticles on an industrial scale [12]. In recent work, the effect of the Er^{3+}, Yb^{3+}-doping, and heat treatment on suspension stability of Y_2O_3 nanoparticles elaborated by the sol–gel method was analyzed [13]. In this case, methanol was used as a solvent, and acetic acid as a catalyst. The results indicate that the incorporation of Er and Yb ions reduces the grain size and the z-potential. The resulting microstructure consists of uniform spheres. Similarly, Nd^{3+}, Gd^{3+}, and Dy^{3+} were incorporated into $BiFe_{0.95}Mn_{0.05}O_3$ (BFM) nanoparticles for this method, where tartaric acid was used to improve the el formation [14]. Incorporating the RE increases the crystallite size (Nd > Dy > Gd), and nanoparticles were obtained. The authors conclude that RE-doped BFM are suitable for biomedical applications. In work related to water pollution, Sm^{3+}, Nd^{3+}, and Dy^{3+} were used to improve the photocatalytic activity of ZnO nanoparticles against tetracycline, malachite green, and crystal violet dyes [15]. This preparation method allowed the obtaining of dense-agglomerated microstructure. The incorporation of RE promotes the creation of structural defects and, consequently, the improvement of the activity. Finally, in another work, nanocubes were obtained by adding ethanol and glacial acetic acid to a titanium isopropoxide solution

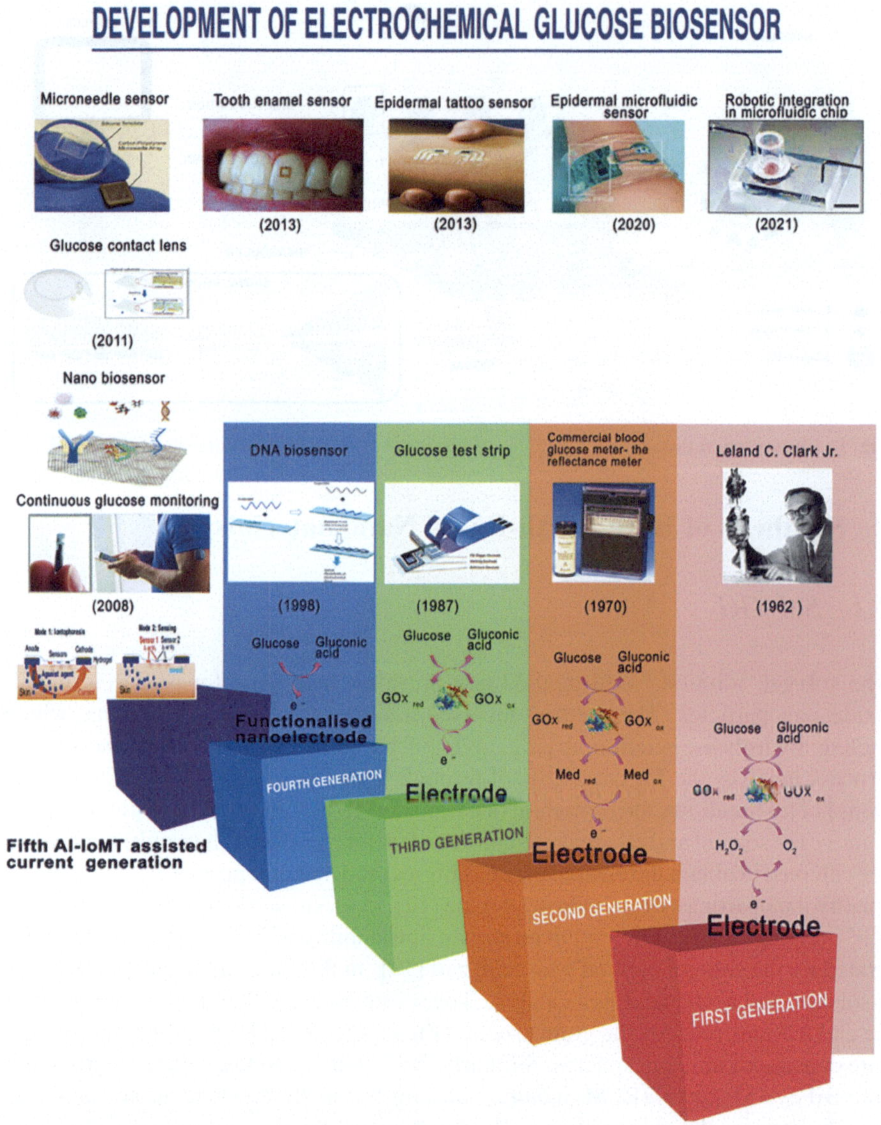

Fig. 2 Evolution of electrochemical biosensor for glucose monitoring [10]

[16]. The Er^{3+}-TiO_2 nanocubes showed more significant photocatalytic activity than pristine TiO_2 for methylene blue degradation (Fig. 3).

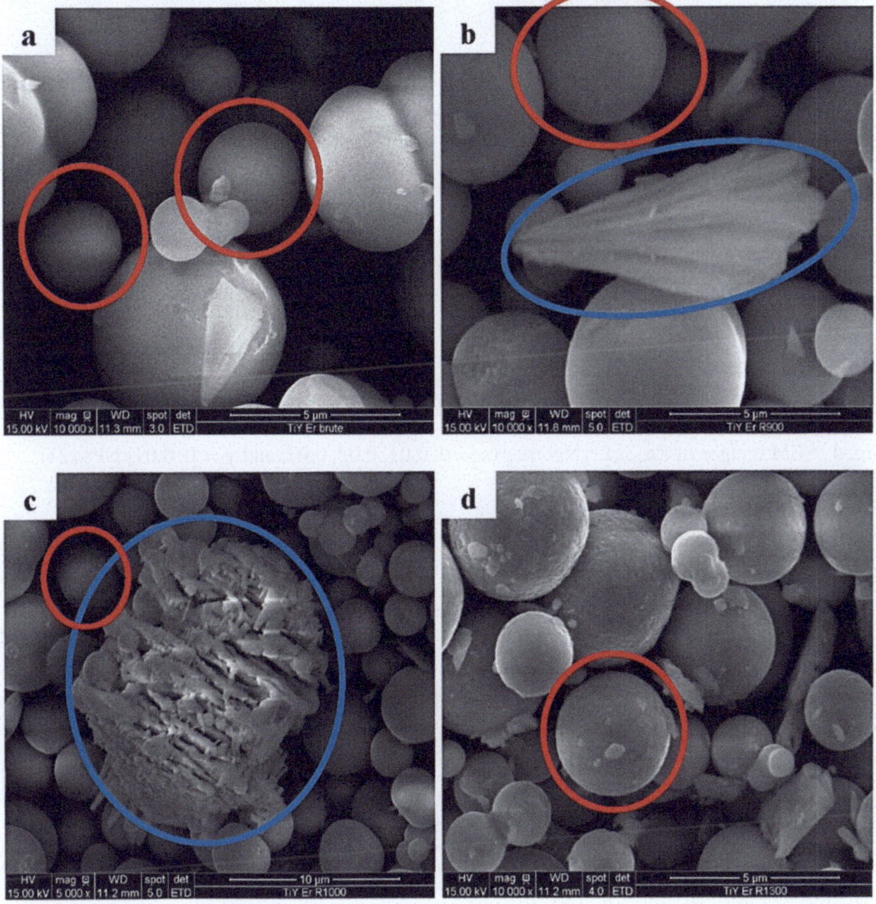

Fig. 3 Microstructure (SEM) images of the samples 5%Er:YTO at various temperature and time [13]

2.2 *Coprecipitation*

The experimental procedure of the coprecipitation method consists of aqueous metal salts mixed with a base, acting as a precipitating agent. After that, the solvent is evaporated, and the precipitated powder is calcined. In this method, the chemical reactions include the simultaneous occurrence of nucleation, growth, coarsening, and aggregation processes [17]. Similarly, Yb^{3+} was used as a doping agent of $MnFe_2O_4$ [18]. Incorporating Yb^{3+} changes the morphology from faceted nanoparticles for undoped samples and spherical shapes for doped materials. This study reveals that the $MnFe_2O_4$ can be potentially used in permanent magnets at low temperatures after doping. NaOH was used to precipitate the starting solution.

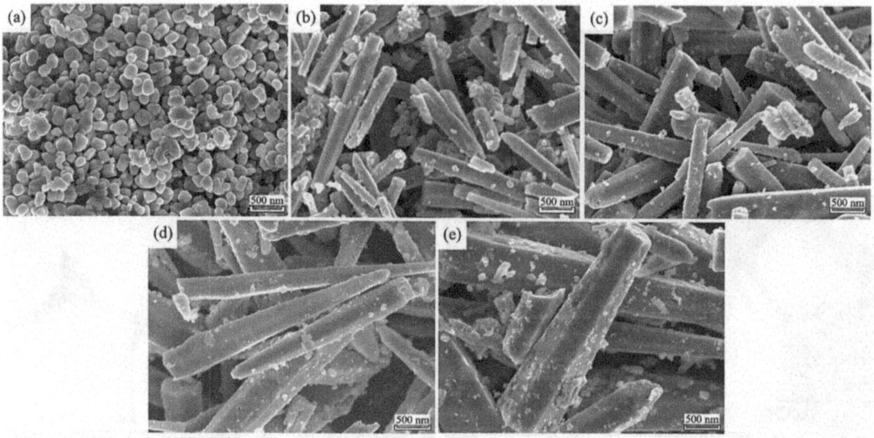

Fig. 4 SEM images of $Zn_{1-x}Er_xNyO_{1-y}$ (x = 0, 0.01, 0.02, 0.03, and y = 0, 0.01) NPs [21]

Finally, Tb-doped ZnO was produced using the same method [19]. Dense nanoparticles were obtained, and no variation in the microstructure or size was reported. The antimicrobial activity was evaluated against *Staphylococcus aureus* and *Klebsiella pneumoniae*. Tb-ZnO nanoparticles were found to have significantly enhanced antibacterial activity compared to undoped ZnO nanoparticles. NaOH was used as a precipitating agent. Finally, the effect of coprecipitation synthesis parameters on gadolinium aluminate nanoparticles [20]. In this work, NaOH was used as a precipitating agent, and sodium o-dodecyl benzenesulfonate was used as a dispersant. The author reported that this preparation route allowed for obtaining high-purity materials and was completely nanostructured (Fig. 4).

2.3 *Microemulsion*

As per the widely accepted definition, emulsions are dispersed systems where one liquid phase is dispersed within another. Typically, emulsions consist of mutually insoluble liquids, often a polar solvent like water and a hydrophobic solvent known as oil [22]. Emulsions are classified based on the continuous phase: direct emulsions (o/w), where the hydrophobic solvent disperses in water, and reverse emulsions (w/o), where water particles disperse in oil. Different variations of this method have been used to prepare rare-earth-based materials. A water-in-oil microemulsion prepared RE-doped $MSeO_4$ (M = Ca, Sr, Ba) [23]. For this purpose, ODC was mixed with CTAB (surfactant) and 1-hexanol. After the addition of Na_2SeO_4 in an aqueous phase, a transparent and stable microemulsion was obtained. The alkaline earth chloride and europium chloride were dissolved in HCl and added dropwise at 60 °C. Regarding microstructure, images of selenate NPs depict rod-like shapes for the Ca and fiber shapes for the Sr compound, while spherical particles for $BaSeO_4$

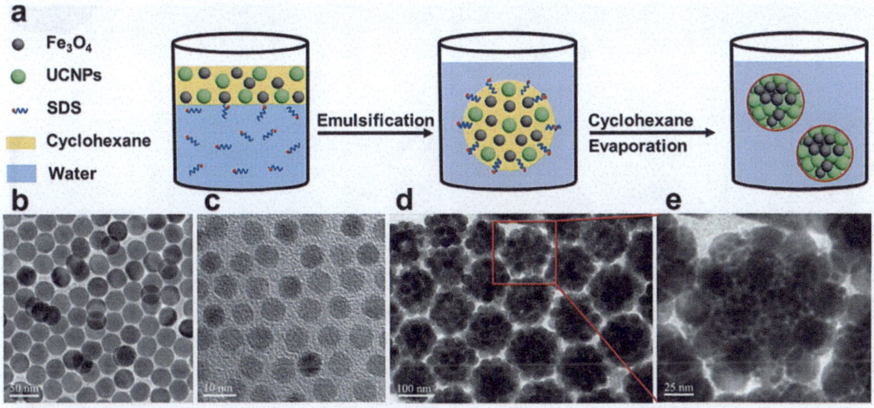

Fig. 5 **a** Schematic showing the micro emulsion-based synthesis method for the coassembly of upconversion nanoparticles (UCNPs) and Fe_3O_4 nanoparticles, **b** TEM images of UCNPs, **c** Fe_3O_4 nanoparticles, **d** MFSPs, **e** and magnified figure of MFSPs [25]

were observed. In other work, a (Tb-Eu)-terephthalate complex was incorporated into iron oxide nanocubes [24]. For this, Igepal-CO-50 was dissolved in cyclohexane in an ultrasonic bath. Then Fe_3O_4-OA and Tb-TA/Eu-TA were dispersed ultrasonically in cyclohexane. Then, NH_4OH (25%) and TEOS were added. Finally, the mixture was agitated using a rotational at room temperature to coat the SiO_2 onto the magnetite (Fe_3O_4) particles. The final nanocomposites were washed with ethanol using magnetic decantation to remove the surfactants and any non-magnetic silica or Tb-TA complex (Fig. 5).

2.4 *Other Methods*

Several other synthesis methods have been developed to obtain different microstructures. To mention some examples, the solution-polymerization methods were used to fabricate RE-doped ZnO NPs ad biotechnological agents. In this method, polyvinyl alcohol was used as binder agent, and sucrose as chelating agent. In these works, the role of the incorporation of RE (La, Ce, Er, Yb) elements was systematically analyzed. The incorporation of RE onto ZnO does not improve the antibacterial activity. However, the cytotoxicity and antioxidant activity were increased (Fig. 6) [26–31].

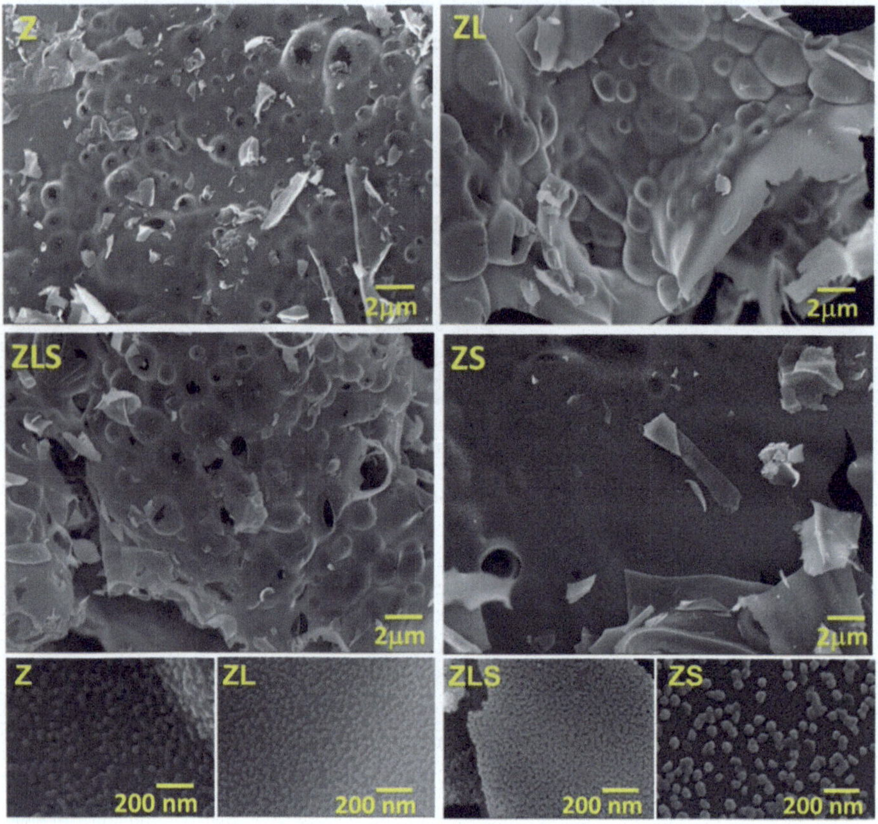

Fig. 6 SEM observation of Z, ZL, ZLS, and ZS and the effect of the ultrasound-assisted synthesis route in their microstructure at low and high magnification [30]

3 Sensor Characterization

Sensor characterization consists of determining the biosensor's performance to determine the sensor's detection parameters. Sensor characterization is vital to obtaining critical parameters of the sensor itself [32]. These properties are sensitivity, selectivity, reproducibility, and response time precision, among other interest parameters. Characterization consists of performing continuous tests under the same conditions to normalize each parameter's values. An example of the characterization of a sensor is the electric characterization of an electrochemical biosensor. Electrochemical biosensors are classified as amperometric, voltammetric, electrochemical impedance spectroscopy, and field effect transistors [33, 34]. This type of characterization consists of applying a voltage through the sensor before and after placing the analyte in question. This way, we can compare the difference between each measurement and obtain parameters such as sensitivity, selectivity, and accuracy. Something to keep in mind

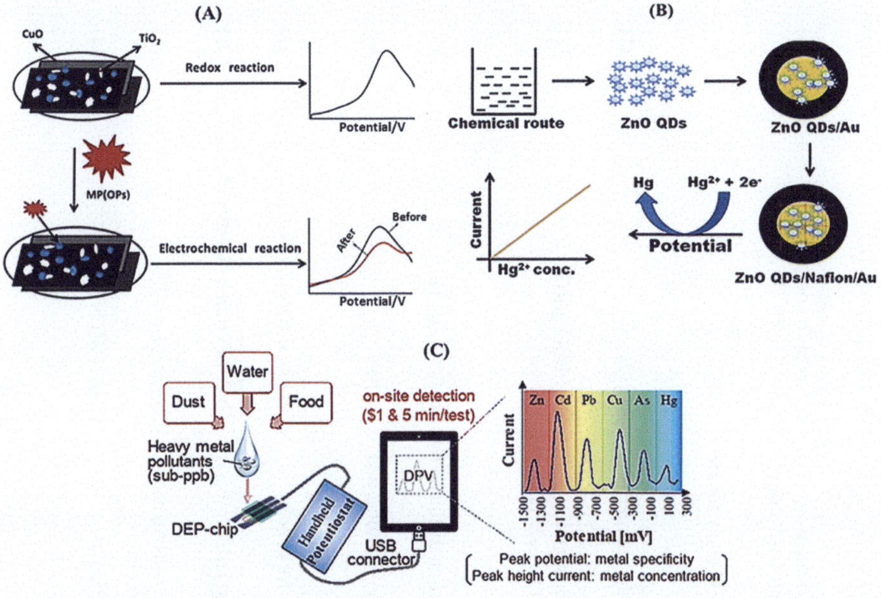

Fig. 7 Various schematics representation of the sensing platform showing the electrochemical sensing mechanism of different samples [33]

while performing a sensor characterization is to maintain the initial parameters with each measurement, as a change in any condition could lead to a change in the results obtained. Amongst parameters to keep in mind in characterization are temperature, light, analyte quantity and purity, uniformity between sensors employed, and initial test values, among others (Fig. 7).

4 Rare Earth-Based Materials for Sensor Applications

The new table summarizes the application of rare-earth-based materials for sensor applications.

Atomic number	Element	Application	RE-based material	Refs.
57	La (Lanthanum)	Highly specificity as chemosensor for determination of macrofurans in pharmaceutical formulation and for different biological samples as whole blood-serum-plasma	Lanthanum-complex	[35]
58	Ce (Cerium)	Detection of pathogenic bacteria (*Escherichia coli*) in milk and whey samples through the use of FRET pair azithromycin-functionalized CeO_2 nanoparticles and aptamer AuNPPS	CeO_2	[36]
59	Pr (Praseodymium)	Dopamine electrochemical detection with nano-sized praseodymiumferrite material-based sensor	$PrFeO_3$	[37]
60	Nd (Neodymium)	Detection of carcinogenic organic compound (Metol) in water samples throughout electrochemical sensor based on neodymium molibdate/reduced graphene oxide	$Nd_2Mo_3O_{12}$/RGO	[38]
61	Pm (Promethium)	NI	NI	
62	Sm (Samarium)	Disease biomarker organic compound (quinolinic acid) detection mediated by samarium metal–organic framework	Sm-MOF (meta-organic framework)	[39]
63	Eu (Europium)	Drug detection in human hair samples of Morphine, Methamphetamine, and Ketamine using europium nanoparticles-based fluorescence immunochromatographic assay (EuNPs-FIA)	Eu-NPs	[40]
64	Gd (Gadolinium)	Nanosystems complex based on Gd/photosensitizers hybrids for cancer imaging	Gd-based NPs	[41]
65	Tb (Terbium)	Detection of the biological organic compound bilirubin using Terbium-based bright green, fluorescent lanthanide complex (Tb-DPA)	Tb-2,6-Pyridine dicarboxylic acid	[42]
66	Dy (Dysprosium)	Carbofuran plant pesticide detection through electrochemical sensing using graphene oxide incorporated with cysprosium stannate nanoplatelets	$Dy_2Sb_2O_7$/GO	[43]
67	Ho (Holmium)	Biocompatiple HoF_3 nanoparticles with modifications with PEG as dual-modal contrast agent for cancer detection	PEG-HoF_3 NPs	[44]

(continued)

(continued)

Atomic number	Element	Application	RE-based material	Refs.
68	Er (Erbium)	Nanosensing probe for Fe^{3+} detection in aqueous and human serum samples using erbium-doped graphene quantum dots	Er-doped graphene quantum dots (Er-GQDs)	[45]
69	Tm (Thulium)	Detection of roxarsone (ROS) as a arsenic toxicity marker using Thulium(III)-benzene 1,3,5-tricarboxylic acid-based metal–organic framework	Thulium(III)-benzene 1,3,5-tricarboxylic acid-based metal–organic framework (Tm-BTC MOF)	[46]
70	Yb (Ytterbium)	Dopamine detection through the use of (3-mercaptopropyl)trimethoxysilane-functionalized Ytterbium oxide nanoparticles biosensor	Yb_2O_3	[47]

References

1. Patyal M, Kaur K, Bala N et al (2023) Innovative lanthanide complexes: shaping the future of cancer/tumor chemotherapy. J Trace Elements Med Biol 80
2. Lee YY, Sriram B, Wang SF et al (2023) A comprehensive review on emerging role of rare earth oxides in electrochemical biosensors. Microchem J 193
3. Hossain MK, Ahmed MH, Khan MI et al (2021) Recent progress of rare earth oxides for sensor, detector, and electronic device applications: a review. ACS Appl Electron Mater 3:4255–4283
4. Shin WR, Ahn G, Lee JP et al (2023) Recent advances in engineering aptamer-based sensing and recovery of heavy metals and rare earth elements for environmental sustainability. Chem Eng J 472
5. Naresh V, Lee N (2021) A review on biosensors and recent development of nanostructured materials-enabled biosensors. Sensors (Switzerland) 21:1–35
6. Malik S, Singh J, Goyat R et al (2023) Nanomaterials-based biosensor and their applications: a review. Heliyon 9
7. Solanki PR, Kaushik A, Agrawal VV, Malhotra BD (2011) Nanostructured metal oxide-based biosensors. NPG Asia Mater 3:17–24
8. Chircov C, Bîrcă AC, Grumezescu AM, Andronescu E (2020) Biosensors-on-chip: an up-to-date review. Molecules 25(24):6013
9. Wu J, Liu H, Chen W et al (2023) Device integration of electrochemical biosensors. Nat Rev Bioeng 1:346–360. https://doi.org/10.1038/s44222-023-00032-w
10. Sumitha MS, Xavier TS (2023) Recent advances in electrochemical biosensors—a brief review. Hybrid Adv 2:100023. https://doi.org/10.1016/j.hybadv.2023.100023
11. Bokov D, Turki Jalil A, Chupradit S et al (2021) Nanomaterial by sol-gel method: synthesis and application. Adv Mater Sci Eng
12. Ugemuge N, Parauha YR, Dhoble SJ (2021) Chapter 15—Synthesis and luminescence study of silicate-based phosphors for energy-saving light-emitting diodes. In: Dhoble SJ, Kalyani NT, Vengadaesvaran B, Kariem Arof A (eds) Energy materials. Elsevier, pp 445–480
13. Benammar I, Salhi R, Deschanvres JL, Maalej R (2023) The effect of rare earth (Er, Yb) element doping on the crystallization of $Y_2Ti_2O_7$ pyrochlore nanoparticles developed by hydrothermal-assisted-sol-gel method. J Solid State Chem 320:123856. https://doi.org/10.1016/j.jssc.2023.123856
14. Dubey A, Salamon S, Attanayake SB et al (2022) Rare-earth doped $BiFe_{0.95}Mn_{0.05}O_3$ nanoparticles for potential hyperthermia applications. Front Bioeng Biotechnol 10. https://doi.org/10.3389/fbioe.2022.965146
15. Dhiman P, Rana G, Kumar A et al (2023) Rare earth doped ZnO nanoparticles as spintronics and photo catalyst for degradation of pollutants. Molecules 28.https://doi.org/10.3390/molecules28062838
16. El-Kholy RA, Isawi H, Zaghlool E et al (2023) Preparation and characterization of rare earth element nanoparticles for enhanced photocatalytic degradation. Environ Sci Pollut Res 30:69514–69532. https://doi.org/10.1007/s11356-023-27090-2
17. Tolosana-Moranchel Á, Faraldos M, Bahamonde A (2021) Chapter 5—Methodologies of synthesis of titania and titania-graphene photocatalysts. In: García-López EI, Palmisano L (eds) Materials science in photocatalysis. Elsevier, pp 83–94
18. Gulati S, Gokhale S, Luthra V (2023) Effects of Yb^{3+} doping on structural, morphological, and temperature dependent magnetic properties of $MnFe_2O_4$ nanoparticles. J Supercond Nov Magn 36:1033–1041. https://doi.org/10.1007/s10948-023-06534-3
19. Shivaraj B, Prabhakara MC, Bhojya Naik HS et al (2023) Fabrication of Tb doped ZnO nanoparticle via co-precipitation technique for multifunctional applications. Chem Phys Lett 818. https://doi.org/10.1016/j.cplett.2023.140421
20. Qi P, Cong Y, Yu L et al (2023) Effect of co-precipitation synthesis parameters on gadolinium aluminate nanoparticles. Mater Lett 341. https://doi.org/10.1016/j.matlet.2023.134163

21. Mondal S, Jamal M, Ayon SA, Anik MJF, Billah MM (2023) Synergistic enhancement of photocatalytic and antimicrobial efficacy of nitrogen and erbium co-doped ZnO nanoparticles. J Rare Earths
22. Nikolaev B, Yakovleva L, Fedorov V et al (2023) Nano- and microemulsions in biomedicine: from theory to practice. Pharmaceutics 15
23. Kuhlmann N, Wickleder C (2023) Glowing selenates: novel alkaline earth nanoparticles. RSC Adv 13:21225–21230. https://doi.org/10.1039/d3ra01669b
24. Rahman MM, Nguyen C, Yan Z, et al (2024) Synthesis of iron oxide nanocubes and their incorporation with luminescent lanthanide (Ln = Tb/ Eu)-Terephthalate complex in silica nanospheres using a reverse microemulsion method. J Solid State Chem 329.https://doi.org/10.1016/j.jssc.2023.124406
25. Liu Y, Liang Y, Lei P, Zhang Z, Chen Y (2023) Multifunctional superparticles for magnetically targeted NIR-II imaging and photodynamic therapy. Adv Sci 10(2):2203669
26. Navarro-López DE, Garcia-Varela R, Ceballos-Sanchez O et al (2021) Effective antimicrobial activity of ZnO and Yb-doped ZnO nanoparticles against *Staphylococcus aureus* and *Escherichia coli*. Mater Sci Eng C 123:112004–112015. https://doi.org/10.1016/j.msec.2021.112004
27. Navarro-López DE, Sánchez-Huerta TM, Flores-Jimenez MS et al (2022) Nanocomposites based on doped ZnO nanoparticles for antibacterial applications. Colloids Surf A Physicochem Eng Asp 652.https://doi.org/10.1016/j.colsurfa.2022.129871
28. Perfecto-Avalos Y, Navarro-López DE, Martínez-Beltrán S et al (2023) Data-driven machine learning to predict antibacterial activity of cerium-doped nanoparticles. ACS Appl Nano Mater. https://doi.org/10.1021/acsanm.3c03651
29. Sánchez-López AL, Perfecto-Avalos Y, Sanchez-Martinez A et al (2022) Influence of erbium doping on zinc oxide nanoparticles: structural, optical and antimicrobial activity. Appl Surf Sci 575:151764. https://doi.org/10.1016/j.apsusc.2021.151764
30. Mejía-Méndez JL, Navarro-López DE, Sanchez-Martinez A et al (2024) Lanthanide-doped ZnO nanoparticles: unraveling their role in cytotoxicity, antioxidant capacity, and nanotoxicology. Antioxidants 13:213. https://doi.org/10.3390/antiox13020213
31. Navarro-López DE, Perfecto-Avalos Y, Zavala A et al (2024) Unraveling the complex interactions: machine learning approaches to predict bacterial survival against ZnO and lanthanum-doped ZnO nanoparticles. Antibiotics 13:220. https://doi.org/10.3390/antibiotics13030220
32. Afsarimanesh N, Nag A, Sarkar S et al (2020) A review on fabrication, characterization and implementation of wearable strain sensors. Sens Actuators A Phys 315
33. Khanmohammadi A, Jalili Ghazizadeh A, Hashemi P, Afkhami A, Arduini F, Bagheri H (2020) An overview to electrochemical biosensors and sensors for the detection of environmental contaminants. J Iran Chem Soc 17(10):2429–2447
34. Choi HK, Choi JH, Yoon J (2023) An updated review on electrochemical nanobiosensors for neurotransmitter detection. Biosensors (Basel) 13
35. Sheta SM, Abd-Elzaher MM, El-Sheikh SM (2021) A novel nano-lanthanum complex: synthesis, characterization and application as a macrofuran chemosensor in pharmaceutical, biological and environmental samples. RSC Adv 11:9675–9681. https://doi.org/10.1039/d0ra10116h
36. Ghayyem S, Faridbod F (2022) Detection of pathogenic bacteria in milk and whey samples using a fluorescence resonance energy transfer aptasensor based on cerium oxide nanoparticles. Anal Methods 14:813–819. https://doi.org/10.1039/d1ay02023d
37. Kumar Y, Vashishtha VK, Singh PP et al (2020) Praseodymium ferrite nano-particles based modified electrode and its application in determination of dopamine. Biointerface Res Appl Chem 10:5855–5859. https://doi.org/10.33263/BRIAC104.855859
38. Rex Shanlee SS, Sundaresan R, Chen SM et al (2023) High-performance electrochemical sensor based on neodymium molybdate/reduced graphene oxide ($Nd_2Mo_3O_{12}$/RGO) for rapid detection of carcinogenic organic pollutants in water samples. Surf Interfaces 40.https://doi.org/10.1016/j.surfin.2023.103020

39. Zhang BD, Liu B, Yin YX, Xu L (2023) Samarium-based turn-off fluorescence sensor for sensitive and selective detection of quinolinic acid in human urine and serum. Inorg Chem 62:1007–1017. https://doi.org/10.1021/acs.inorgchem.2c03926
40. Xu S, Ma B, Li J et al (2023) Europium nanoparticles-based fluorescence immunochromatographic detection of three abused drugs in hair. Toxics 11. https://doi.org/10.3390/toxics11050417
41. Zeng Y, Li H, Li Z et al (2020) Engineered gadolinium-based nanomaterials as cancer imaging agents. Appl Mater Today 20
42. Abraham MK, Gouri Krishna K, Madanan AS et al (2024) Development of Terbium based luminescent platform as a "Turn-off" sensor for the detection of bilirubin. J Photochem Photobiol A Chem 449.https://doi.org/10.1016/j.jphotochem.2023.115424
43. Thirumalraj B, Krishnapandi A, Chen SM et al (2020) Rational design and interlayer effect of dysprosium-stannate nanoplatelets incorporated graphene oxide: a versatile and competent electrocatalyst for toxic carbamate pesticide detection in vegetables. ACS Sustain Chem Eng 8:17882–17892. https://doi.org/10.1021/acssuschemeng.0c07973
44. Zhang T, Deng M, Zhang L et al (2021) Facile synthesis of holmium-based nanoparticles as a CT and MRI dual-modal imaging for cancer diagnosis. Front Oncol 11.https://doi.org/10.3389/fonc.2021.741383
45. Huynh TV, Anh NTN, Darmanto W, Doong RA (2021) Erbium-doped graphene quantum dots with up- and down-conversion luminescence for effective detection of ferric ions in water and human serum. Sens Actuators B Chem 328.https://doi.org/10.1016/j.snb.2020.129056
46. Chinnapaiyan S, Rajaji U, Chen SM et al (2022) Fabrication of thulium metal–organic frameworks-based smartphone sensor towards arsenical feed additive drug detection: applicable in food safety analysis. Electrochim Acta 401.https://doi.org/10.1016/j.electacta.2021.139487
47. Kumar S, Ram Chaudhary G, Chaudhary S (2022) Designing of surface engineered Ytterbium oxide nanoparticles as effective electrochemical sensing platform for dopamine. J Mol Liq 355.https://doi.org/10.1016/j.molliq.2022.118929

Fabrication and Functionalization of Bismuth Chalcogenide Nanostructures for Biomedical and Toxic Gas Sensing Applications

Shaik M. Abzal, Kurapati Kalyan, Sai Lakshmi Janga, A. Arshad Ahamed, and Jatis Kumar Dash

Abstract Living organisms increasingly grapple with the impact of rapid environmental changes, including climate shifts, pollution, and resource scarcity. To effectively address these challenges, advanced monitoring tools are essential. Traditional methods like gas chromatography, high-performance liquid chromatography, and mass spectrometry, while effective, suffer from high costs, time-intensive procedures, and reliance on specialized equipment and expertise. In this context, nanoparticle synthesis technology emerges as a promising solution. By precisely controlling size, shape, and properties at the nanoscale, it enables the development of advanced sensors. These nanosensors offer rapid response times, heightened sensitivity, and selectivity, making them ideal for accurate environmental and health monitoring. Nanomaterials (NMs), with their unique attributes, such as a high surface area-to-volume ratio and exceptional electronic, optical, and mechanical properties—outperform traditional methods in detecting minute quantities of contaminants and environmental fluctuations. This chapter explores the use of bismuth chalcogenides nanostructures in sensing applications. It covers synthesis methods, sensor fabrication, and characterization techniques, highlighting bismuth chalcogenides such as Bi_2S_3, Bi_2Se_3, and Bi_2Te_3 for their high carrier mobility, extensive surface area, and tunable properties, making them promising candidates for cost-effective, sensitive sensors.

Keywords Nanomaterials · Sensors · Bio-sensors · Electrochemical sensor · Bismuth chalcogenides

S. M. Abzal · K. Kalyan · S. L. Janga · A. Arshad Ahamed · J. K. Dash (✉)
Department of Physics, SRM University-AP, Amaravati, India
e-mail: jatis.d@srmap.edu.in; jatiskumar@gmail.com

I. Uddin (ed.), *Smart Nanosensors*, Smart Nanomaterials Technology,
https://doi.org/10.1007/978-981-96-3878-9_16

1 Introduction

Over the past several decades, living beings have been facing numerous challenges due to climate change, pollution, and resource depletion, which have significantly impacted ecosystems, human health, and global economies [1, 2]. To address these issues, there is a critical need for advanced monitoring and diagnostic tools that can provide timely information and early warnings about environmental changes and potential health risks [3, 4]. Currently, various methods are available for monitoring environmental changes and potential health risks, including gas chromatography, high-performance liquid chromatography, and mass spectrometry [5]. While these techniques can effectively detect and quantify contaminants, they are often limited by time consumption, high costs, and the need for sophisticated instruments and skilled personnel [6]. Consequently, there is a growing demand for simpler, faster, and more affordable monitoring solutions [7, 8]. Modern nano synthesis technology has emerged as a promising solution to address the limitations of traditional monitoring methods [9].

Nanotechnology has led to significant advancements across various sectors by enabling the development of materials with precise control over their size, shape, and tunable chemical, physical, and mechanical properties at the nanoscale level, typically ranging from 1 to 100 nm [10]. This has led to the development of a new generation of sensors using nanomaterials. These nanosensors offer superior performance with rapid response times, high sensitivity, and selectivity, making them ideal tools for quickly and accurately monitoring environmental conditions and potential health risks [11].

Sensors made from nanomaterials (NMs) possess unique properties, such as a high surface area-to-volume ratio, enhanced reactivity, and exceptional electronic, optical, and mechanical characteristics [12]. These properties differ significantly from their bulk (micron-sized) counterparts due to the small size (1–100 nm) [13]. As a result, they exhibit enhanced surface reactivity, quantum confinement effects, improved electrical conductivity, and increased magnetic properties, among other unique characteristics [14]. These attributes allow sensors to detect minute quantities of contaminants and changes in environmental conditions with a level of accuracy and speed that conventional methods cannot achieve [15, 16]. The distinctive physical and chemical properties of NMs make them suitable for various applications, including catalysis, sensing, biomedical applications, water treatment, agriculture, energy storage and conversion, and many more [17–19]. The general properties of nanomaterials and bulk materials differ markedly, highlighting the advantages of using NMs in advanced sensor technology as shown in Fig. 1.

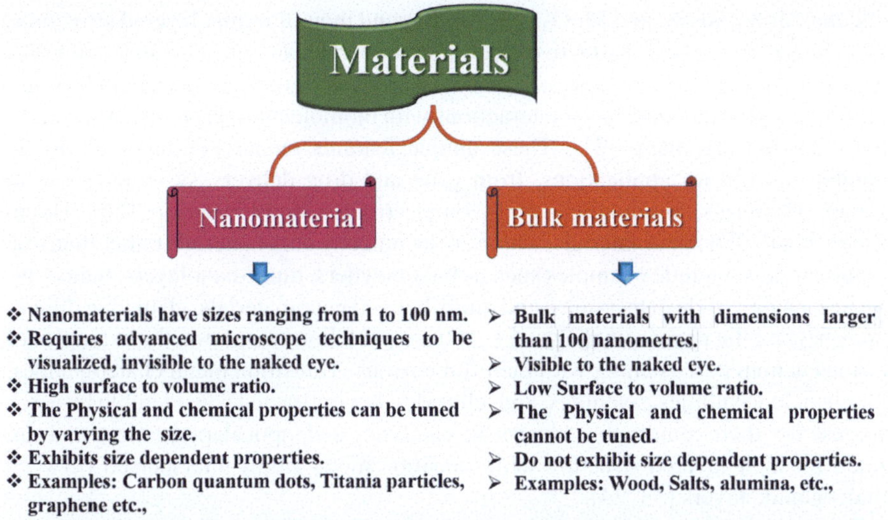

Fig. 1 Nanomaterials versus bulk comparison based on their properties

1.1 Materials Classification

Nanomaterials are fundamentally categorized into four types based on their dimensions such as, zero-dimensional, one-dimensional, two-dimensional, and three-dimensional [20], as shown in the Fig. 2. Zero-dimensional nanomaterials (0D nanomaterials) have all three dimensions-length, width, and height-within the nanoscale, typically ranging from 1 to 100 nm. Examples include molecules, quantum dots, fullerenes (distinct buckyball-shaped structures), and clusters (small collections of atoms) [21]. Graphene and carbon quantum dots are also examples of 0D nanomaterials [22]. These materials can be either crystalline or amorphous and may vary in shape and size, exhibiting metallic or ceramic properties [23]. This versatility makes them suitable for a wide range of applications. One-dimensional nanomaterials (1D nanomaterials) have two dimensions within the nanoscale range of 1–100 nm, with an elongated structure [24]. Examples of one-dimensional (1D) nanomaterials include nanotubes, nanorods, nanowires, and nanofibers, such as multi-walled (MWNT) and single-walled (SWNT) carbon nanotubes, as well as gold and ZnO nanorods [25]. Two-dimensional (2D) nanomaterials, with their sheet-like structures, exhibit a wide array of properties that make them ideal for diverse applications [26]. These atomically thin materials boast large surface areas, exceptional conductivity, and tunable characteristics, offering significant potential for advancements in sensing, energy storage and antibacterial applications [27]. Within these 2D materials, in-plane atoms are bonded covalently, while out-of-plane atoms are held together by weak van der Waals forces [28]. The spectrum of 2D materials encompasses metals, insulators, conductors, and semiconductors, including elemental forms, composites, transition metal dichalcogenides, and transition metal oxides, such as graphene, borophene, silicene, MoS_2, Bi_2S_3, and MoO_3 [29, 30]. Beyond their ultrathin, layered structures, these materials possess distinctive electronic properties and physicochemical traits, such as high specific surface areas, flexible layered configurations, excellent biocompatibility, and strong interfacial interactions with biomolecules, proteins, DNA, cells, and other bio-organisms [31]. These unique features render 2D materials highly suitable for various applications, from gene and drug delivery systems to cancer therapy platforms, bioimaging agents, bioelectronics, and biosensors [26]. Three-dimensional (3D) nanomaterials, which are technically bulk materials rather than true nanomaterials, include examples such as bulk powders, multi-nanolayers, nanocomposites, nanoparticle dispersions, and multi-layered nanomaterials, all having dimensions beyond the nanoscale range of 1–100 nm [32]. Figure 3 shows the examples of various nanomaterials along with their dimensions. Transition metal chalcogenides, which include sulfides, selenides, and tellurides, have attracted considerable research interest for their remarkable electrical, catalytic, and optical properties, and are noted for their layered structure with transition metal atoms interspersed between chalcogenide layers [33, 34].

Bismuth chalcogenides (Bi_2X_3, where X = S, Se, or Te) are distinguished among 2D materials for their eco-friendly nature and versatile properties, such as thermoelectric behavior, optical characteristics, and low thermal conductivity, which, combined with their layered structure of strong in-plane covalent bonds and weak van der Waals forces between layers, make them ideal for applications in optoelectronics, thermoelectric devices, photocatalysis, energy storage, and sensing [35, 36].

This chapter focuses into the application of bismuth chalcogenides in sensing. It explores synthesis methods, including solvothermal and solution-based approaches,

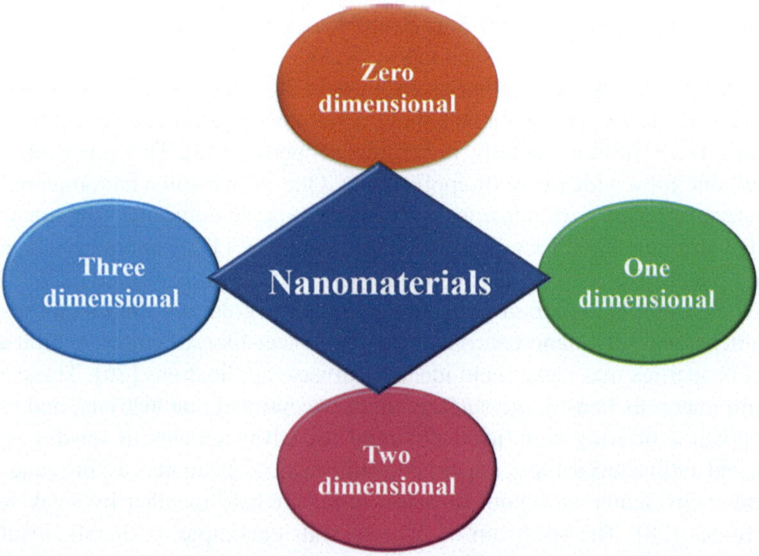

Fig. 2 Nanomaterials can be categorized based on their dimensions

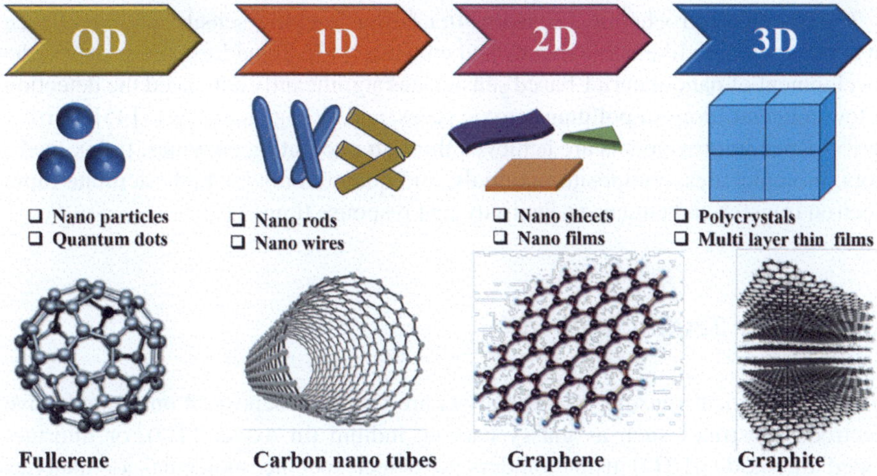

Fig. 3 Examples of nanomaterials based on their dimensions

as well as thin film deposition techniques. Additionally, the chapter covers sensor fabrication, characterization methods to evaluate material properties, and methodologies for assessing sensor performance.

1.2 Sensors

Sensors are devices that convert physical or chemical inputs such as motion, pressure, or chemical substances into measurable signals, with physical sensors measuring quantities like distance, pressure, temperature, and mass, and chemical sensors detecting substances through changes in chemical parameters, such as toxic gases [37].

In the past twenty years, sensor technology has attracted significant research interest, especially in areas like environmental monitoring, biomolecule detection, food safety, and medical diagnostics [38, 39]. Electrochemical sensors detect specific analytes or biomolecules and convert this interaction into measurable electronic signals using a signal transducer, offering high accuracy, selectivity, and rapid response, and are widely used in environmental and health monitoring, electrochemical biosensors, in particular, are noted for their capability to directly convert biological processes into electrical signals such as current or voltage [40, 41]. In 1916, biosensors were first developed by Clark for detecting glucose using amperometry, achieved through protein immobilization on activated charcoal [42]. Subsequently, enzymatic and non-enzymatic field-effect transistors (FETs) were developed to detect various biological aspects [43].

Traditional electrochemical sensors often suffer from low sensitivity and reliance on costly materials like noble metals and enzymes [44]. To address these issues, the development of nanomaterial-based sensors has significantly enhanced the detection of low concentrations of pollutants, toxic gases, and biological markers [45]. Further performance improvements are achieved through integrating enzymes, redox mediators, biomolecules, composite materials, and quantum dots, which facilitate rapid electron transfer and enhance sensitivity and response times [46].

1.3 Sensor Testing Methods

In electrochemical sensors, nanomaterials are typically deposited onto conductive electrode substrates such as glassy carbon, indium tin oxide (ITO), or fluorine-doped tin oxide (FTO) using binders to detect specific molecules or analytes [47]. Electrochemical sensor performance is typically assessed using either a two-electrode system, consisting of a working electrode and a reference electrode, or a three-electrode system that includes a counter electrode [48]. In electrochemical sensing applications, frequently used solvents and electrolytes include ferro/ferric cyanide, phosphate buffer solution (PBS), sodium sulfate (Na_2SO_4), sulfuric acid (H_2SO_4), potassium hydroxide (KOH), and sodium hydroxide (NaOH), often used in conjunction with the analyte.

Electrochemical sensors are classified according to the electrical signal they detect potentiometric sensors gauge the potential difference between a working electrode and a reference electrode, voltammetric sensors apply a series of potentials to measure the resulting currents, amperometric sensors use a constant potential to monitor current responses and conductometric sensors assess the conductance of the cell, which is useful for determining ionic analyte concentrations [49, 50]. Electrochemical sensors are utilized for detecting a range of substances including nitrogen dioxide, hydrogen peroxide, heavy metals, dyes, glucose, ascorbic acid, and dopamine [51]. These sensors utilize two main techniques to evaluate their response to specific analytes cyclic voltammetry and electrochemical impedance spectroscopy (EIS). Cyclic voltammetry involves applying a range of voltages to the sensor while measuring the resulting current, providing insights into the analyte's interaction with the sensor surface, including its redox behavior and electron transfer capabilities [52]. Electrochemical impedance spectroscopy (EIS) assesses the sensor's response to a small applied electrical signal by measuring resistance and capacitance at different frequencies, offering an understanding of charge transfer kinetics and the rate of electron exchange between the analyte and the electrode [53].

Three key parameters are crucial for the accuracy and reliability of electrochemical sensors sensitivity, selectivity, and limit of detection (LOD) [54]. Sensitivity measures the sensor's ability to detect small changes in the concentration of the target molecule, producing a substantial signal in response to minute quantities of the analyte [55]. Selectivity indicates the sensor's ability to distinguish the target molecule from other substances in the sample, reducing responses to interfering

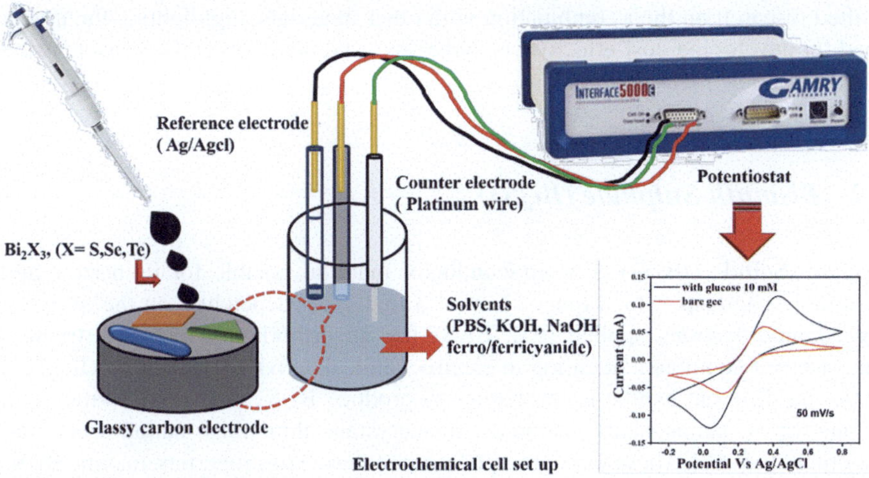

Fig. 4 Schematic of electrochemical sensing set up

components [56]. The limit of detection (LOD) defines the smallest concentration of the target molecule that the sensor can accurately detect, with a lower LOD demonstrating the sensor's capacity to identify even trace amounts of the analyte with high precision [57]. The schematic of electrochemical set up is shown in Fig. 4.

2 Bismuth Chalcogenides

Bismuth chalcogenides, including Bi_2S_3, Bi_2Se_3, and Bi_2Te_3, are emerging as highly promising materials for sensing applications due to their unique characteristics [58]. Bismuth chalcogenides exhibit notable topological insulating properties on their surfaces, characterized by high carrier mobility, topological protection of surface states from chemical contamination, and delocalized surface states that enhance the surface diffusion of adsorbed molecules [59]. The layered structure of bismuth chalcogenides facilitates ion intercalation and adsorption, boosting sensitivity and selectivity, while their extensive surface area offers numerous active sites for analyte interactions [60]. The electronic properties of bismuth chalcogenides can be tuned through surface modification, doping, and structural engineering to enhance sensor performance, making them ideal for electrochemical sensing due to their high electron mobility and delocalized surface states that improve interfacial charge transfer [61].

Currently, enzyme-based electrochemical sensors rely on expensive precious metals and enzymes, which suffer from high costs and limited thermal and chemical stability, thus constraining their diagnostic applications [62]. Despite their promise, bismuth chalcogenides have been rarely explored in electrochemical sensors, with limited research on their combination with other materials, highlighting the urgent need for developing cost-effective, enzyme-free, and highly sensitive sensors using these non-noble metals.

2.1 Bismuth Sulphide (Bi_2S_3)

Bismuth Sulfide (Bi_2S_3) is a semiconductor material notable for its narrow and adjustable bandgap, which ranges from 1.3 to 1.7 eV depending on the morphological modifications applied [63]. Bi_2S_3 has an orthorhombic crystal structure has garnered significant attention in electrochemical activity [64]. Researchers are exploring different synthesis techniques to produce Bi_2S_3 in various forms, such as nanowires, nanospheres, nanorods, nanocrystals, thin films, nanoflowers, and nanotubes [65–67]. These varied morphologies enhance its properties, making Bi_2S_3 ideal for numerous applications.

Its high surface area and electronic activity, combined with its layered structure that enables efficient charge transport, make Bi_2S_3 an ideal candidate for advanced sensor development [68]. Bi_2S_3's rapid electron–hole pair recombination enhances its electrochemical activity and responsiveness, making it a popular choice for electrochemical sensors owing to its high surface area and numerous active sites [69]. Bi_2S_3 sensors are capable of detecting glucose, dopamine, uric acid, and ascorbic acid in biological fluids [70]. The extensive surface area of Bi_2S_3 enhances sensitivity, allowing for the detection of trace amounts of target analytes, making these sensors effective in identifying a variety of biomolecules and environmental pollutants. Table 1 illustrates the electrochemical detection of various biological substances using Bi_2S_3, highlighting its suitability for sensing applications due to its high surface area and reactive properties.

Table 2 demonstrates the electrochemical detection of gases like ammonia (NH_3) and nitrogen dioxide (NO_2) using Bi_2S_3 sensors [79, 80]. These sensors can operate at relatively low temperatures, which is advantageous for practical applications in various environmental conditions. Bi_2S_3 nanomaterials can be synthesized in various morphologies and sizes using techniques such as hydrothermal synthesis, chemical vapor deposition (CVD), and electrochemical deposition [81]. Surface modification strategies, such as doping with other elements or combining Bi_2S_3 with other nanomaterials, can further enhance the material's sensing performance by improving its selectivity and sensitivity.

2.2 Bismuth Selenide (Bi_2Se_3)

Bismuth Selenide (Bi_2Se_3) is a notable n-type topological insulator with a 0.35 eV band gap and a rhombohedral layered crystal structure, recognized for its distinctive electronic and optical properties [86]. Bi_2Se_3 is distinguished by its robust surface

Table 1 Demonstrates the electrochemical detection of various biological substances utilizing Bi_2S_3

Material	Detected molecule	Solvent	Detection range (μM)	Sensitivity (μA mM^{-1} cm^{-2})	Limit of detection (LOD) (μM)	References
Bi_2S_3/rGO	Ascorbic acid	0.1 M PBS (pH = 7.0)	5.0–1200	268.8	2.9	[71]
ZnO/Bi_2O_3/Bi_2S_3	Thiamethoxam	0.2 M Na_2SO_4 and 0.1 M PBS (pH = 7.4)	–	–	3.32	[72]
Bi_2S_3/$ZnIn_2S_4$	Rutin	(0.10 M pH 7.0 PBS + 0.10 M AA	0.01–100.0	–	0.003	[73]
Bi_2S_3	H_2O_2	PBS (0.1 M, pH 7.4)	10–5000	2.95	2.5	[74]
1D/2D Bi_2S_3/g-C_3N_4	H_2O_2	0.1 M NaOH	0.5–950	1011	78	[75]
BiOBr/Bi_2S_3	Malathion	0.1 M PBS	0.001–1000	–	0.12	[76]
rGO/Bi_2S_3	Dopamine	0.1 M PBS	0.01– 40	–	12.3	[77]
Bi_2S_3	Promethazine hydrochloride (PMZH)	PBS (0.1 M, pH 7.0)	0.049–79.9	3.69	5.9	[78]

Table 2 Shows the detection of different gases by using Bi_2S_3

Material	Morphology	Detected gas	Detection range (ppm)	Response time (s)	Lower detection limit (ppb)	References
Bi_2S_3	Nanowires	Ammonia (NH_3)	30–350	95	30	[82]
Bi_2S_3	Nanowires	H_2	–	–	10	[83]
rGO/Bi_2S_3	Flower-like nanostructures	NO_2	25–1	22	1	[84]
Bi_2S_3	Nanorods	NO_2	2–20	12.2	10	[85]

states, which arise from spin–orbit coupling [87, 88]. Its unique features, such as a layered structure, large specific surface area, numerous adsorption sites, high carrier mobility, and remarkable mechanical flexibility, make it particularly suitable for gas sensing applications. These attributes enable Bi_2Se_3 to operate effectively at low temperatures while maintaining high sensitivity, excellent compatibility, and efficiency [89–91].

Table 3 shows the electrochemical detection of various biological molecules using Bi_2Se_3.

Furthermore, its inherent hydrophobicity ensures impressive resistance to humidity in various environmental conditions [96]. Collectively, these characteristics make's Bi_2Se_3 as potential material for a wide range of sensor applications. Bi_2Se_3 has a layered crystal structure composed of quintuple layers arranged in the sequence Se_1-Bi-Se_2-Bi-Se_1, within these layers, atoms are covalently bonded, while the layers themselves are weakly held together by van der Waals forces [97]. This layered arrangement contributes to its high electrical conductivity and makes it an effective material for sensing applications.

Bi_2Se_3 sensors are effective for detecting various gases due to their high surface-to-volume ratio and electrical conductivity [98, 99]. They have been used to sense gases such as nitrogen dioxide (NO_2) and ammonia (NH_3) with high sensitivity and selectivity. Table 4 present the electrochemical detection of various toxic gases using Bi_2Se_3. The material's electronic structure allows for precise detection of these gases even at low concentrations. Bi_2Se_3 is utilized in electrochemical sensors for detecting ions and biomolecules. Its high electron mobility enhances the performance of sensors designed to detect substances like glucose, dopamine, and hydrogen peroxide. The material's layered structure enables efficient electron transfer, which is crucial for accurate electrochemical measurements. Bi_2Se_3 sensors are typically synthesized using methods such as solvothermal, chemical vapor deposition (CVD), thermal evaporation, and electrochemical deposition [100–103]. These techniques allow for precise control over the material's morphology and properties. Surface modification strategies, such as doping with other elements or coating with nanomaterials, can further enhance the sensor's performance by improving sensitivity and selectivity.

Table 3 Presents the electrochemical sensing of toxic gases by Bi_2Se_3

Material	Morphology	Detected gas	Detection range (ppm)	Response time (s)	Lower detection limit (ppb)	References
Bi_2Se_3/MoO_3	Nanosheet/nanobelt	Triethylamine	151–10	43	5.5	[92]
Bi_2Se_3 film	Nanoplates	NO_2	100–12.5	81	100	[93]
Bi_2Se_3/Bi_2O_3	Nanosheets	NH_3	5–180	8.5	5	[94]
Bi_2Se_3	Nanosheets	NO_2	0.5–50	12.2	0.06	[95]

Table 4 Shows the sensing of different biological molecules by using the Bi_2Se_3

Material	Detected molecule	Solvent	Detection range (µM)	Sensitivity (µA mM^{-1} cm^{-2})	Limit of detection (LOD) (µM)	References
Bi_2Se_3-FL	Glucose	0.1 M PBS (pH = 7.0)	10–100	0.112	6.1	[104]
Au/Bi_2Se_3	HIV-1 DNA	–	0.1–1	–	50	[105]
Bi_2Se_3 NPs/rGO	Acetaminophen (ACT) and L-DOPA (LD)	0.1 M pH 7.0 PBS	0.006–0.25) and (0.0045–0.14	–	0.23 and 0.17	[106]
Bi_2Se_3 nanostructures	H_2O_2	0.1 M PBS (pH 7.0)	2.0–1.0	0.112	6.3	[107]
Bi_2Se_3@Au	H_2O_2	0.1 M PBS (pH 7.0)	0.10–27.30	–	10	[108]
Bi_2Se_3–PVP	Glucose	0.1 M PBS(pH 7.4)	1–10	–	1.05	[109]

2.3 *Bismuth Telluride (Bi_2Te_3)*

Bismuth Telluride (Bi_2Te_3) is a well-known thermoelectric material with a narrow band gap and a rhombohedral crystal structure, belonging to the $R\bar{3}m$ space group, and exhibiting a hexagonal unit cell [110]. Various surfactants are employed in its synthesis to achieve diverse morphologies. Bi_2Te_3 is an excellent candidate for electrochemical sensors, owing to its high electrical conductivity, significant theoretical capacity, and remarkable electron mobility of 22,000 $cm^2\ V^{-1}\ s^{-1}$ [111]. Its distinctive properties have led to extensive research, especially in the area of sensor technology. Table 5 presents the electrochemical detection of various biological substances using Bi_2Te_3.

The crystal structure of Bi_2Te_3 consists of quintuple layers (QLs) with an arrangement of Te(1)-Bi-Te(2)-Bi-Te(1), where atoms within each QL are covalently bonded, while adjacent QLs are held together by weak van der Waals forces [117]. This structure contributes to its unique electronic properties and potential for various applications. Various methods for synthesizing Bi_2Te_3 have been explored, including solution-based techniques, flash sintering, thermal evaporation, electrochemical deposition, chemical solution processes, and liquid exfoliation [118, 119]. Additionally, Bi_2Te_3 is recognized as a leading topological insulator with significant surface electrical conductivity, making it ideal for various sensor applications, enabling the detection of a wide range of chemical and biological analytes with high sensitivity and selectivity [120].

3 Synthesis Methods

There are numerous methods available for synthesizing bismuth chalcogenides. The detailed procedures for these synthesis methods are outlined below. Figure 5 illustrates the various types of synthesis methods for bismuth chalcogenides.

3.1 *Solvothermal Method*

The solvothermal method is highly valued in material synthesis for its versatility and effectiveness in creating materials with unique properties [121]. The solvothermal method allows precise control over material characteristics such as high crystallinity, particle size, and morphology, tailoring the structural and functional attributes of the resulting materials through specific solvents and reaction conditions, making it essential for synthesizing advanced materials with desired specifications [122]. The schematic for the synthesis of solvothermal method is shown in Fig. 6.

Table 5 Shows the detection of biomolecules by Bi_2Te_3

Material	Detected molecule	Solvent	Detection range (μM)	Sensitivity (μA mM^{-1} cm^{-2})	Limit of detection (LOD) (μM)	References
Bi_2Te_3 flakes	H_2O_2	0.1 M K_2SO_4	10–100	4900	0.016	[112]
Bi_2Te_3/rGO	Dopamine	0.1 M PBS pH 7.03	10–1000	222.93	0.06	[113]
Bi_2Te_3 nanosheets	miRNA-21	1.0 M KCl with PBS solution	10–100	–	3.3	[114]
BP/RP heterojunction@Bi_2Te_3	piRNA-823	0.1 M PBS (pH 7.0)	0.1	–	0.016	[115]
Bi_2Te_3/g-C_3N_4	Ractopamine (RAC)	0.05 M PB (pH 7.0)	0.015–456.4	–	1.07	[116]

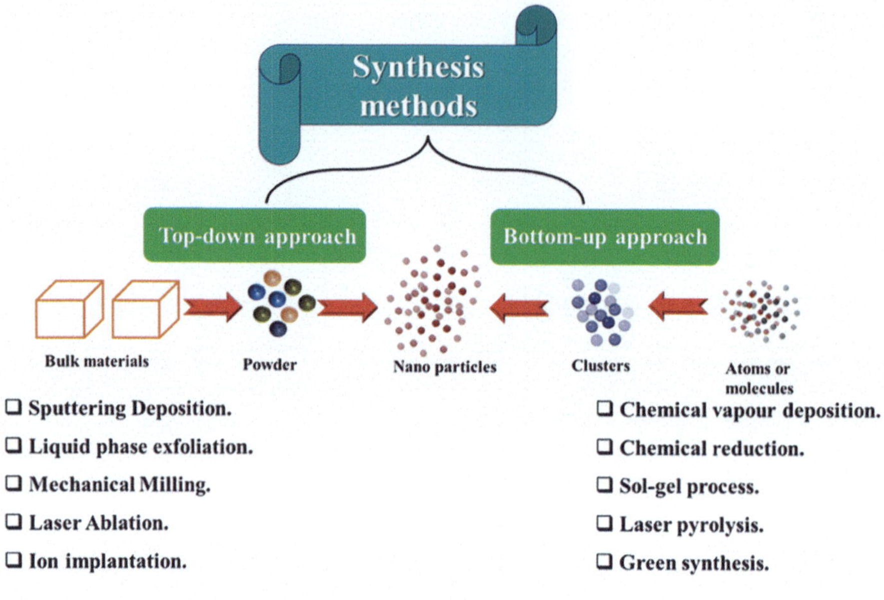

Fig. 5 Shows the different types of synthesis methods

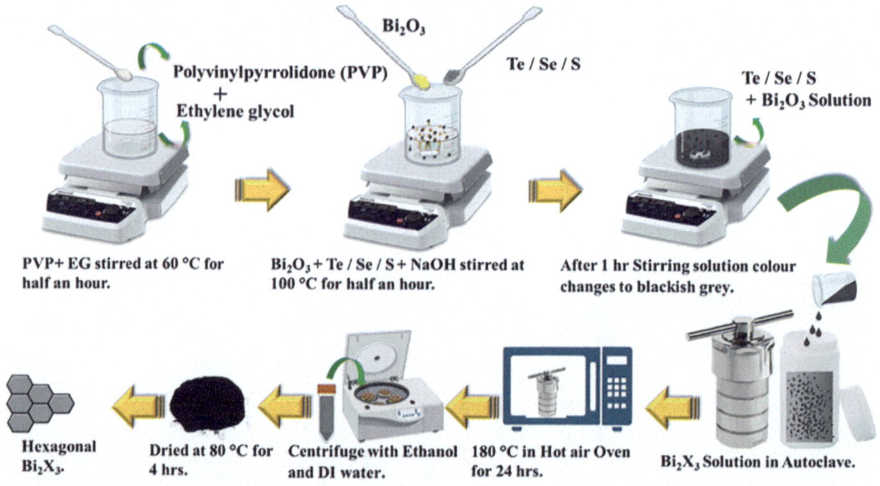

Fig. 6 Schematic of synthesis of Bi_2X_3

3.2 *Solution Route*

The solution-based synthesis method is a prominent and versatile technique for synthesizing nanoparticles, offering increased capability and flexibility compared to dry synthetic routes, as it is easy, efficient, requires no sophisticated instruments, and can be conducted at low temperatures [123, 124]. In this method, reactants are dissolved in a solvent to form a homogeneous mixture, which promotes efficient and uniform chemical reactions.

3.3 Thin Film Deposition

Thin film deposition techniques are crucial across fields such as electronics, optics, materials science, and nanotechnology, as they allow for the deposition of ultra-thin films, from a few nanometers to several micrometers, onto substrates [125]. A variety of thin-film deposition techniques exist, such as Physical vapor deposition (PVD) and chemical vapor deposition (CVD) are two techniques that can be used depositing thin films [126]. The vapor then condenses on the substrate, forming a thin film. This method is simple and cost-effective but may not be suitable for materials with high melting points. Where as in Chemical Vapor Deposition (CVD), process, gaseous precursors react chemically on the substrate surface, resulting in the deposition of a thin film [127]. High-quality films, excellent conformal coverage, and ability to deposit a wide variety of materials. In addition to these synthesis techniques, other methods such as flash sintering, liquid exfoliation, electrochemical deposition, and chemical solution processes are also used for materials synthesis [128].

4 Characterization Techniques

After synthesis, materials are characterized using various techniques: Powder X-ray Diffraction (PXRD) is employed for structural analysis, Scanning Electron Microscopy (SEM) provides insights into surface morphology, and High-Resolution Transmission Electron Microscopy (HRTEM) examines the crystalline structure. Raman Spectroscopy and Fourier Transform Infrared Spectroscopy (FT-IR) are utilized to identify functional groups and molecular vibrations, while X-ray Photoelectron Spectroscopy (XPS) is used to determine oxidation states and chemical composition [129, 130]. The different types of characterization techniques were showed in Fig. 7.

5 Challenges and Future Directions

Bismuth chalcogenides exhibit excellent electrochemical properties, but their low intrinsic conductivity can restrict sensor sensitivity and performance. Enhancing conductivity without affecting other properties poses a significant challenge. Current synthesis methods are often complex, time-consuming, and expensive, hindering large-scale production and commercial feasibility. Scaling up bismuth chalcogenide

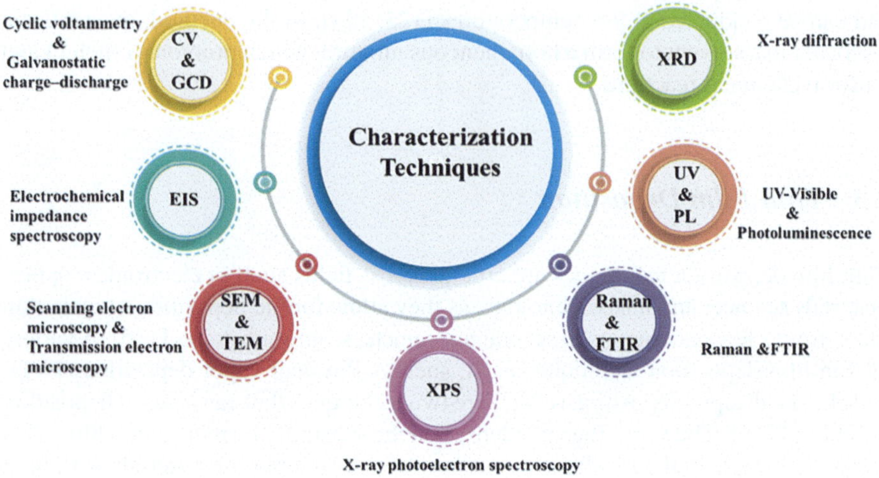

Fig. 7 Various analytical characterization techniques

production while maintaining consistent quality and desirable nanostructures is a major obstacle. Additionally, ensuring the long-term stability and durability of these materials in various environmental conditions is critical, as factors like oxidation, thermal degradation, and mechanical wear can impact sensor performance over time. Effective surface functionalization to boost selectivity and sensitivity is challenging, requiring uniform and reproducible modification of the sensor surface.

Moreover, integrating bismuth chalcogenide-based sensors with existing electronic and sensing technologies necessitates overcoming compatibility issues, ensuring seamless operation with current sensor platforms and signal processing systems.

Developing scalable and cost-effective synthesis methods for high-quality bismuth chalcogenide nanomaterials is essential. Techniques like chemical vapor deposition (CVD), hydrothermal synthesis, and electrochemical deposition must be refined to achieve better control over material properties. Enhancing the conductivity of these materials is crucial, which could involve doping, creating composites, or utilizing innovative nano structuring methods to improve charge transport. Improving selectivity towards target analytes through advanced surface modification is also necessary; this can be achieved by functionalizing surfaces with specific chemical groups or biomolecules. Combining bismuth chalcogenides with other conductive or catalytic materials can further boost sensor performance, while integrating with graphene or carbon nanotubes may enhance conductivity and stability. Ensuring environmental stability through protective coatings or encapsulation is vital for long-term sensor applications. Exploring the use of bismuth chalcogenide sensors in flexible and wearable electronics offers new possibilities for real-time monitoring, requiring these sensors to be sensitive, selective, and durable. Tailoring these sensors for specific applications, such as medical diagnostics, environmental monitoring, and industrial process control, involves targeted research to optimize their performance. Employing machine learning and advanced data analysis techniques can improve sensor functionality and decision-making accuracy. Despite their potential, challenges remain in adopting bismuth chalcogenides widely, including the need for effective microfabrication and device architecture development. Addressing these challenges through ongoing research could establish bismuth chalcogenides as a foundation for a new generation of highly sensitive, selective, and reliable sensors across various fields.

6 Conclusions

In summary, the integration of nanotechnology into sensor development, particularly through the use of bismuth chalcogenides, represents a promising advancement for addressing contemporary environmental and health monitoring challenges. Bismuth chalcogenides, including Bi_2S_3, Bi_2Se_3, and Bi_2Te_3, offer exceptional properties such as high surface area, unique electronic characteristics, and tunable reactivity, making them highly suitable for sensing applications.

The distinctive properties of these materials allow for high sensitivity, rapid response times, and the ability to detect minute quantities of contaminants and biomolecules, surpassing the limitations of traditional sensor technologies. Bi_2S_3, with its adjustable bandgap and various morphologies, proves effective for detecting a range of biological and environmental substances. Bi_2Se_3, known for its topological insulating properties and high carrier mobility, excels in gas sensing applications. Bi_2Te_3, with its impressive thermoelectric and electronic properties, shows promise for diverse sensor applications. However, challenges such as enhancing intrinsic conductivity, scaling up production, and ensuring long-term stability must be addressed to fully exploit the potential of bismuth chalcogenides in sensor technology. Future research should focus on optimizing synthesis methods, improving material stability, and developing cost-effective, scalable production techniques to make these advanced sensors more accessible for widespread use.

Acknowledgements S.M.A. acknowledges the SRM University-AP for the financial support to carry out this work. JKD acknowledges the partial financial support from UGC-DAE CSR through a Collaborative Research Scheme (CRS) (project number: CRS/2022-23/1092).

References

1. Kumar Rai P, Singh JS (2020) Invasive alien plant species: their impact on environment, ecosystem services and human health. Ecol Indic 111:106020. https://doi.org/10.1016/j.ecolind.2019.106020

2. Lustig WP, Mukherjee S, Rudd ND et al (2017) Metal-organic frameworks: functional luminescent and photonic materials for sensing applications. Chem Soc Rev 46:3242–3285. https://doi.org/10.1039/c6cs00930a
3. Wang CX, Xiu LS, Hu QQ et al (2024) Advancing early warning and surveillance for zoonotic diseases under climate change: interdisciplinary systematic perspectives. Adv Clim Change Res 14:814–826. https://doi.org/10.1016/j.accre.2023.11.014
4. Fang X, Zong B, Mao S (2018) Metal-organic framework-based sensors for environmental contaminant sensing. Nanomicro Lett 10:1–19. https://doi.org/10.1007/s40820-018-0218-0
5. Popp M, Hann S, Koellensperger G (2010) Environmental application of elemental speciation analysis based on liquid or gas chromatography hyphenated to inductively coupled plasma mass spectrometry—a review. Anal Chim Acta 668:114–129. https://doi.org/10.1016/j.aca.2010.04.036
6. Umesha S, Manukumar HM (2018) Advanced molecular diagnostic techniques for detection of food-borne pathogens: current applications and future challenges. Crit Rev Food Sci Nutr 58:84–104. https://doi.org/10.1080/10408398.2015.1126701
7. Astill J, Dara RA, Fraser EDG et al (2020) Smart poultry management: smart sensors, big data, and the internet of things. Comput Electron Agric 170:105291. https://doi.org/10.1016/j.compag.2020.105291
8. Hoekstra R, Blondeau P, Andrade FJ (2018) Distributed electrochemical sensors: recent advances and barriers to market adoption. Anal Bioanal Chem 410:4077–4089. https://doi.org/10.1007/s00216-018-1104-9
9. Tovar-Lopez FJ (2023) Recent progress in micro- and nanotechnology-enabled sensors for biomedical and environmental challenges. Sensors 23:47–49. https://doi.org/10.3390/s23125406
10. Chakraborty M, Jain S, Rani V (2011) Nanotechnology: emerging tool for diagnostics and therapeutics. Appl Biochem Biotechnol 165:1178–1187. https://doi.org/10.1007/s12010-011-9336-6
11. Moldovan O, Iñiguez B, Deen MJ, Marsal LF (2015) Graphene electronic sensors—review of recent developments and future challenges. IET Circuits Devices Syst 9:446–453. https://doi.org/10.1049/iet-cds.2015.0259
12. Thakur A, Kumar A (2022) Recent advances on rapid detection and remediation of environmental pollutants utilizing nanomaterials-based (bio)sensors. Sci Total Environ 834:155219. https://doi.org/10.1016/j.scitotenv.2022.155219
13. Valtchev V, Tosheva L (2013) Porous nanosized particles: preparation, properties, and applications. Chem Rev 113:6734–6760. https://doi.org/10.1021/cr300439k
14. Prakash Sharma V, Sharma U, Chattopadhyay M, Shukla VN (2018) Advance applications of nanomaterials: a review. Mater Today Proc 5:6376–6380. https://doi.org/10.1016/j.matpr.2017.12.248
15. Chapman J, Truong VK, Elbourne A et al (2020) Combining chemometrics and sensors: toward new applications in monitoring and environmental analysis. Chem Rev 120:6048–6069. https://doi.org/10.1021/acs.chemrev.9b00616
16. Kumar V, Vaid K, Bansal SA, Kim KH (2020) Nanomaterial-based immunosensors for ultrasensitive detection of pesticides/herbicides: current status and perspectives. Biosens Bioelectron 165:112382. https://doi.org/10.1016/j.bios.2020.112382
17. Thanh TD, Balamurugan J, Lee SH et al (2016) Effective seed-assisted synthesis of gold nanoparticles anchored nitrogen-doped graphene for electrochemical detection of glucose and dopamine. Biosens Bioelectron 81:259–267. https://doi.org/10.1016/j.bios.2016.02.070
18. Abzal SM, Janga SL, Bhaskara Rao Y et al (2024) Solvothermally synthesized bismuth telluride hexagonal platelets as an efficient anode material for lithium- and sodium-ion batteries. J Mater Sci 59:6879–6893. https://doi.org/10.1007/s10853-024-09571-y
19. Majji M, Abzal SM, Jacob N et al (2024) Efficient photocatalytic green hydrogen production using crystalline elemental boron nanostructures under visible light. Int J Hydrogen Energy 56:338–347. https://doi.org/10.1016/j.ijhydene.2023.12.113

20. Naguib G, Maghrabi AA, Mira AI et al (2023) Influence of inorganic nanoparticles on dental materials' mechanical properties. A narrative review. BMC Oral Health 23:1–18. https://doi.org/10.1186/s12903-023-03652-1
21. Tiwari JN, Tiwari RN, Kim KS (2012) Zero-dimensional, one-dimensional, two-dimensional and three-dimensional nanostructured materials for advanced electrochemical energy devices. Prog Mater Sci 57:724–803. https://doi.org/10.1016/j.pmatsci.2011.08.003
22. Deka MJ (2023) Recent advances in fluorescent 0D carbon nanomaterials as artificial nanoenzymes for optical sensing applications. Int Nano Lett 13:1–14. https://doi.org/10.1007/s40089-022-00381-1
23. Mekuye B, Abera B (2023) Nanomaterials: an overview of synthesis, classification, characterization, and applications. Nano Select 4:486–501. https://doi.org/10.1002/nano.202300038
24. Xia BY, Yang P, Sun Y et al (2003) <Xia-Adv_Mat-2003.Pdf>. Adv Mater 15:353–389
25. Ziegler JM, Andoni I, Choi EJ et al (2021) Sensors based upon nanowires, nanotubes, and nanoribbons: 2016–2020. Anal Chem 93:124–166. https://doi.org/10.1021/acs.analchem.0c04476
26. Li S, Ma L, Zhou M et al (2020) New opportunities for emerging 2D materials in bioelectronics and biosensors. Curr Opin Biomed Eng 13:32–41. https://doi.org/10.1016/j.cobme.2019.08.016
27. Sinha S, Kim H, Robertson AW (2021) Preparation and application of 0D–2D nanomaterial hybrid heterostructures for energy applications. Mater Today Adv 12:100169. https://doi.org/10.1016/j.mtadv.2021.100169
28. Lin H, Zhang Z, Zhang H et al (2022) Engineering van der Waals materials for advanced metaphotonics. Chem Rev 122:15204–15355. https://doi.org/10.1021/acs.chemrev.2c00048
29. Lin YC, Torsi R, Younas R et al (2023) Recent advances in 2D material theory, synthesis, properties, and applications. ACS Nano 17:9694–9747. https://doi.org/10.1021/acsnano.2c12759
30. Kumar S, Pavelyev V, Mishra P et al (2020) A review on 2D transition metal di-chalcogenides and metal oxide nanostructures based NO_2 gas sensors. Mater Sci Semicond Process 107:104865. https://doi.org/10.1016/j.mssp.2019.104865
31. Ulhakim MT, Rezki M, Dewi KK et al (2020) Review—recent trend on two-dimensional metal-organic frameworks for electrochemical biosensor application. J Electrochem Soc 167:136509. https://doi.org/10.1149/1945-7111/abb6cc
32. Adul-Rasool AA, Athair DM, Zaidan HK et al (2024) 0, 1, 2, 3D nanostructures, types of bulk nanostructured materials, and drug nanocrystals: an overview. Cancer Treat Res Commun 40:100834. https://doi.org/10.1016/j.ctarc.2024.100834
33. Bhattacharya D, Mukherjee S, Mitra RK, Ray SK (2023) Superior piezoelectric performance of chemically synthesized transition metal dichalcogenide heterostructures for self-powered flexible piezoelectric nanogenerator. Nanotechnology 34. https://doi.org/10.1088/1361-6528/ace97d
34. Shao B, Chen Z, Su H et al (2024) The latest advances in ink-based nanogenerators: from materials to applications. Int J Mol Sci 25. https://doi.org/10.3390/ijms25116152
35. Huang J, Huang Q, Liu M et al (2022) Emerging bismuth chalcogenides based nanodrugs for cancer radiotherapy. Front Pharmacol 13:1–8. https://doi.org/10.3389/fphar.2022.844037
36. Hegde GS, Prabhu AN (2022) A review on doped/composite bismuth chalcogenide compounds for thermoelectric device applications: various synthesis techniques and challenges. J Electron Mater 51:2014–2042. https://doi.org/10.1007/s11664-022-09513-x
37. Hossain N, Rimon MIH, Mimona MA et al (2024) Prospects and challenges of sensor materials: a comprehensive review. e-Prime Adv Electr Eng Electron Energy 7:100496. https://doi.org/10.1016/j.prime.2024.100496
38. Perumal J, Wang Y, Attia ABE et al (2021) Towards a point-of-care SERS sensor for biomedical and agri-food analysis applications: a review of recent advancements. Nanoscale 13:553–580. https://doi.org/10.1039/d0nr06832b

39. Chadha U, Bhardwaj P, Agarwal R et al (2022) Recent progress and growth in biosensors technology: a critical review. J Ind Eng Chem 109:21–51. https://doi.org/10.1016/j.jiec.2022.02.010
40. Thulasiprevinnah S, Bashir S, Ramesh K, Ramesh S (2024) Recent advances in electrochemical biosensors for the determination of biomolecules on modified and unmodified electrodes. J Iran Chem Soc 21:1739–1768. https://doi.org/10.1007/s13738-024-03050-6
41. Shanbhag MM, Manasa G, Mascarenhas RJ et al (2023) Fundamentals of bio-electrochemical sensing. Chem Eng J Adv 16:100516. https://doi.org/10.1016/j.ceja.2023.100516
42. Chen T, Liu D, Lu W et al (2016) Three-dimensional Ni2P nanoarray: an efficient catalyst electrode for sensitive and selective nonenzymatic glucose sensing with high specificity. Anal Chem 88:7885–7889. https://doi.org/10.1021/acs.analchem.6b02216
43. Syedmoradi L, Ahmadi A, Norton ML, Omidfar K (2019) A review on nanomaterial-based field effect transistor technology for biomarker detection. Microchim Acta 186. https://doi.org/10.1007/s00604-019-3850-6
44. Romanholo PVV, Razzino CA, Raymundo-Pereira PA et al (2021) Biomimetic electrochemical sensors: new horizons and challenges in biosensing applications. Biosens Bioelectron 185:113242. https://doi.org/10.1016/j.bios.2021.113242
45. Saputra HA (2023) Electrochemical sensors: basic principles, engineering, and state of the art. Monatsh Chem 154:1083–1100. https://doi.org/10.1007/s00706-023-03113-z
46. Wang T-T, Huang X-F, Huang H et al (2022) Nanomaterial-based optical- and electrochemical-biosensors for urine glucose detection: a comprehensive review. Adv Sens Energy Mater 1:100016. https://doi.org/10.1016/j.asems.2022.100016
47. Ahmad R, Wolfbeis OS, Hahn YB et al (2018) Deposition of nanomaterials: a crucial step in biosensor fabrication. Mater Today Commun 17:289–321. https://doi.org/10.1016/j.mtcomm.2018.09.024
48. Gutiérrez-Capitán M, Baldi A, Merlos Á, Fernández-Sánchez C (2022) Array of individually addressable two-electrode electrochemical cells sharing a single counter/reference electrode for multiplexed enzyme activity measurements. Biosens Bioelectron 201. https://doi.org/10.1016/j.bios.2021.113952
49. Labib M, Sargent EH, Kelley SO (2016) Electrochemical methods for the analysis of clinically relevant biomolecules. Chem Rev 116:9001–9090. https://doi.org/10.1021/acs.chemrev.6b00220
50. Barhoum A, Hamimed S, Slimi H et al (2023) Modern designs of electrochemical sensor platforms for environmental analyses: principles, nanofabrication opportunities, and challenges. Trends Environ Anal Chem 38:e00199. https://doi.org/10.1016/j.teac.2023.e00199
51. Jiao L, Xu W, Wu Y et al (2023) On the road from single-atom materials to highly sensitive electrochemical sensing and biosensing. Anal Chem 95:433–443. https://doi.org/10.1021/acs.analchem.2c01722
52. Nur Ashakirin Binti Mohd Nashruddin S, Hani Mohamed Salleh F, Ayunni Mohd Raub A (2024) Early detection of kidney problems through voltammetry, potentiometry, amperometry, and impedance electrochemical techniques: a comprehensive review. Measurement (Lond) 230:114475. https://doi.org/10.1016/j.measurement.2024.114475
53. Magar HS, Hassan RYA, Mulchandani A (2021) Electrochemical impedance spectroscopy (Eis): principles, construction, and biosensing applications. Sensors 21. https://doi.org/10.3390/s21196578
54. Justino CIL, Rocha-Santos TA, Duarte AC (2010) Review of analytical figures of merit of sensors and biosensors in clinical applications. TrAC Trends Anal Chem 29:1172–1183. https://doi.org/10.1016/j.trac.2010.07.008
55. Flynn CD, Chang D, Mahmud A et al (2023) Biomolecular sensors for advanced physiological monitoring. Nat Rev Bioeng 1:560–575. https://doi.org/10.1038/s44222-023-00067-z
56. Bucur B, Purcarea C, Andreescu S, Vasilescu A (2021) Addressing the selectivity of enzyme biosensors: solutions and perspectives. Sensors 21. https://doi.org/10.3390/s21093038
57. Macchia E, Manoli K, Di Franco C et al (2020) New trends in single-molecule bioanalytical detection. Anal Bioanal Chem 412:5005–5014. https://doi.org/10.1007/s00216-020-02540-9

58. Ni J, Bi X, Jiang Y et al (2017) Bismuth chalcogenide compounds Bi_2X_3 (X = O, S, Se): applications in electrochemical energy storage. Nano Energy 34:356–366. https://doi.org/10.1016/j.nanoen.2017.02.041
59. Ikram M, Raza A, Ahmad SOA et al (2023) Solar-triggered engineered 2D-materials for environmental remediation: status and future insights. Adv Mater Interfaces 10. https://doi.org/10.1002/admi.202202172
60. Lazanas AC, Prodromidis MI (2021) Two-dimensional inorganic nanosheets: production and utility in the development of novel electrochemical (bio)sensors and gas-sensing applications. Microchimica Acta 188. https://doi.org/10.1007/s00604-020-04674-0
61. Yao J, Koski KJ, Luo W et al (2014) Optical transmission enhacement through chemically tuned two-dimensional bismuth chalcogenide nanoplates. Nat Commun 5:1–7. https://doi.org/10.1038/ncomms6670
62. Govindaraj M, Srivastava A, Muthukumaran MK et al (2023) Current advancements and prospects of enzymatic and non-enzymatic electrochemical glucose sensors. Int J Biol Macromol 253:126680. https://doi.org/10.1016/j.ijbiomac.2023.126680
63. Uddin I, Abzal SM, Kalyan K et al (2022) Starch-assisted synthesis of Bi_2S_3 nanoparticles for enhanced dielectric and antibacterial applications. ACS Omega 7:42438–42445. https://doi.org/10.1021/acsomega.2c05593
64. Jabeen N, Ul Hassan N, Bokhari A et al (2023) High performance δ-Bi_2O_3 nanosheets transformed Bi_2S_3 nanoflakes interconnected nanosheets as negative electrode for supercapacitor applications. Fuel 347:128392. https://doi.org/10.1016/j.fuel.2023.128392
65. Moyseowicz A (2019) Scalable one-pot synthesis of bismuth sulfide nanorods as an electrode active material for energy storage applications. J Solid State Electrochem 23:1191–1199. https://doi.org/10.1007/s10008-019-04215-7
66. Ajiboye TO, Onwudiwe DC (2021) Bismuth sulfide based compounds: properties, synthesis and applications. Results Chem 3:100151. https://doi.org/10.1016/j.rechem.2021.100151
67. Yu X, Cao C (2008) Photoresponse and field-emission properties of bismuth sulfide nanoflowers. Cryst Growth Des 8:3951–3955. https://doi.org/10.1021/cg701001m
68. Wang L, Gui WK, Jiang S et al (2024) Bi_2S_3 nanofiber bunch for highly efficient CO_2 electroreduction to formate at low overpotential. Rare Met 43:3391–3399. https://doi.org/10.1007/s12598-024-02665-1
69. Li M, Zhong LX, Chen W et al (2021) Regulating the electron–hole separation to promote selective oxidation of biomass using ZnS@Bi_2S_3 nanosheet catalyst. Appl Catal B 292:120180. https://doi.org/10.1016/j.apcatb.2021.120180
70. Zhou R, Zhou Q, Li X et al (2022) Applications of Bi_2S_3-based nanomaterials in photoelectric chemical sensors and biosensors. FlatChem 36:100440. https://doi.org/10.1016/j.flatc.2022.100440
71. Qu C, Li H, Zhou S et al (2021) Bi_2S_3/rGO composite based electrochemical sensor for ascorbic acid detection. Chemosensors 9:1–15. https://doi.org/10.3390/chemosensors9080190
72. Xiao W, Wang L, Wei X, Li J (2022) Chitosan-based molecularly imprinted photoelectric sensor with ZnO/Bi_2O_3/Bi_2S_3 sensing layer for thiamethoxam determination. Microchimica Acta 189
73. Xi J, Wang H, Zhang B et al (2020) Novel molecularly imprinted photoelectrochemical sensor for rutin based on Bi_2S_3/$ZnIn_2S_4$ heterojunction. Sens Actuators B Chem 320:128409. https://doi.org/10.1016/j.snb.2020.128409
74. Dutta AK, Maji SK, Mitra K et al (2014) Single source precursor approach to the synthesis of Bi_2S_3 nanoparticles: a new amperometric hydrogen peroxide biosensor. Sens Actuators B Chem 192:578–585. https://doi.org/10.1016/j.snb.2013.11.030
75. Othmani A, Derbali M, Kalfat R et al (2021) A novel 1D/2D Bi_2S_3/g-C_3N_4 core–shell nanocomposite as a highly performing H_2O_2 non-enzymatic electrochemical sensor. J Market Res 15:5762–5775. https://doi.org/10.1016/j.jmrt.2021.10.140

76. Li J, Xiong P, Tang J et al (2021) Biocatalysis-induced formation of BiOBr/Bi_2S_3 semiconductor heterostructures: a highly efficient strategy for establishing sensitive photoelectrochemical sensing system for organophosphorus pesticide detection. Sens Actuators B Chem 331:129451. https://doi.org/10.1016/j.snb.2021.129451
77. Yan X, Gu Y, Li C et al (2018) Morphology-controlled synthesis of Bi_2S_3 nanorods-reduced graphene oxide composites with high-performance for electrochemical detection of dopamine. Sens Actuators B Chem 257:936–943. https://doi.org/10.1016/j.snb.2017.11.037
78. Ganesamurthi J, Veerakumar P, Liu TY, Juang RS (2024) Electrochemical detection and photodegradation of antihistamine promethazine hydrochloride in water using nanostructured Bi2S3 catalysts. J Environ Chem Eng 12:111962. https://doi.org/10.1016/j.jece.2024.111962
79. Liu X, Xu X, Li F et al (2022) Heterostructured Bi_2S_3/MoS_2 nanoarrays for efficient electrocatalytic nitrate reduction to ammonia under ambient conditions. ACS Appl Mater Interfaces. https://doi.org/10.1021/acsami.2c10323
80. Feng Q, Xie X, Zhang M, Liao N (2023) First-principles investigation of Bi_2S_3 as sensitive and selective NO_2 sensor upon humidity exposure. J Mater Sci 58:2198–2208. https://doi.org/10.1007/s10853-023-08141-y
81. Darji V, Desai P, Deshpande MP et al (2023) Nickel doped bismuth sulfide nanomaterials: synthesis, characterization, and enhancement of anti-microbial activity. Mater Chem Phys 295:127049. https://doi.org/10.1016/j.matchemphys.2022.127049
82. Fu TX (2018) Gas sensor based on three dimensional Bi_2S_3 nanowires network for ammonia detection at room temperature. Mater Res Bull 99:460–465. https://doi.org/10.1016/j.materresbull.2017.11.038
83. Yao K, Gong WW, Hu YF et al (2008) Individual Bi_2S_3 nanowire-based room-temperature H_2 sensor. J Phys Chem C 112:8721–8724. https://doi.org/10.1021/jp8022293
84. Yang Y, Zhu M, Zhang H et al (2024) Room temperature gas sensor based on rGO/Bi_2S_3 heterostructures for ultrasensitive and rapid NO_2 detection. Chem Eng J 490:151872. https://doi.org/10.1016/j.cej.2024.151872
85. Kan H, Yang W, Guo Z, Li M (2024) Highly sensitive room-temperature NO_2 gas sensor based on Bi_2S_3 nanorods. J Mater Sci: Mater Electron 35
86. Mazumder K, Shirage PM (2021) A brief review of Bi_2Se_3 based topological insulator: from fundamentals to applications. J Alloys Compd 888:161492. https://doi.org/10.1016/j.jallcom.2021.161492
87. Cava RJ, Ji H, Fuccillo MK et al (2013) Crystal structure and chemistry of topological insulators. J Mater Chem C Mater 1:3176–3189. https://doi.org/10.1039/c3tc30186a
88. Dun C, Hewitt CA, Huang H et al (2015) Layered Bi_2Se_3 nanoplate/polyvinylidene fluoride composite based n-type thermoelectric fabrics. ACS Appl Mater Interfaces 7:7054–7059. https://doi.org/10.1021/acsami.5b00514
89. Hsieh YY, Tuan HY (2022) Architectural van der Waals Bi_2S_3/Bi_2Se_3 topological heterostructure as a superior potassium-ion storage material. Energy Storage Mater 51:789–805. https://doi.org/10.1016/j.ensm.2022.07.020
90. Paul S, Hossain MT, Mia AK, Giri PK (2021) Ultrabroadband absorption and high-performance photodetection in europium-doped 2D topological insulator Bi_2Se_3 nanosheets. ACS Appl Nano Mater 4:12527–12540. https://doi.org/10.1021/acsanm.1c03055
91. Liu JL, Chen H, Li X et al (2020) Ultra-fast and high flexibility near-infrared photodetectors based on Bi_2Se_3 nanobelts grown via catalyst-free van der Waals epitaxy. J Alloys Compd 818. https://doi.org/10.1016/j.jallcom.2019.152819
92. Tan X, Wang L, Chen X et al (2023) A heterojunction composite of Bi_2Se_3 nanosheets and MoO_3 nanobelts for a high-performance triethylamine sensor. J Mater Chem C Mater 11:16356–16367. https://doi.org/10.1039/d3tc03187j
93. Wang Y, Tang C, Su M et al (2023) Highly sensitive, humidity-tolerant and flexible NO_2 sensors based on nanoplate Bi_2Se_3 film. Chin Chem Lett 34:32–35. https://doi.org/10.1016/j.cclet.2022.107981
94. Das B, Riyajuddin S, Ghosh K, Ghosh R (2023) Room-temperature ammonia detection using layered Bi_2Se_3/Bi_2O_3: a next-generation sensor. ACS Appl Electron Mater 5:948–956. https://doi.org/10.1021/acsaelm.2c01496

95. Du B, Kang W, He Y et al (2023) Topological insulator Bi_2Se_3 for highly sensitive, selective and anti-humidity gas sensors. iScience 26:106387. https://doi.org/10.1016/j.isci.2023.106387
96. Wang Y, Zhou Y (2022) Recent progress on anti-humidity strategies of chemiresistive gas sensors. Materials 15. https://doi.org/10.3390/ma15248728
97. Samanta M, Biswas K (2020) 2D nanosheets of topological quantum materials from homologous $(Bi_2)M(Bi_2Se_3)n$ heterostructures: synthesis and ultralow thermal conductivity. Chem Mater 32:8819–8826. https://doi.org/10.1021/acs.chemmater.0c02129
98. Shin HS, Hamdou B, Reith H et al (2016) The surface-to-volume ratio: a key parameter in the thermoelectric transport of topological insulator Bi_2Se_3 nanowires. Nanoscale 8:13552–13557. https://doi.org/10.1039/c6nr01716a
99. Bin MK, Kumar P, Giri R, Singh J (2020) Controlled synthesis of two-dimensional (2-D) ultra-thin bismuth selenide (Bi_2Se_3) nanosheets by bottom-up solution-phase chemistry and its electrical transport properties for thermoelectric application. FlatChem 21:100165. https://doi.org/10.1016/j.flatc.2020.100165
100. Patel NF, Bhakhar SA, Jagani HS et al (2023) Synthesis, characterization and optoelectronic application of Bi_2Se_3 thin film prepared by thermal evaporation technique. Opt Mater (Amst) 136:113403. https://doi.org/10.1016/j.optmat.2022.113403
101. Ahmed R, Lin Q, Xu Y, Zangari G (2019) Growth, morphology and crystal structure of electrodeposited Bi_2Se_3 films: influence of the substrate. Electrochim Acta 299:654–662. https://doi.org/10.1016/j.electacta.2019.01.053
102. Kulsi C, Ghosh A, Mondal A et al (2017) Remarkable photo-catalytic degradation of malachite green by nickel doped bismuth selenide under visible light irradiation. Appl Surf Sci 392:540–548. https://doi.org/10.1016/j.apsusc.2016.09.063
103. Nandi S, Ghosh K, Meyyappan M, Giri PK (2023) 2D MXene electrode-enabled high-performance broadband photodetector based on a CVD-grown 2D Bi_2Se_3 ultrathin film on silicon. ACS Appl Electron Mater 5:6985–6995. https://doi.org/10.1021/acsaelm.3c01363
104. Savariraj AD, Vinoth V, Mangalaraja RV et al (2020) Microwave-assisted synthesis of localized surface plasmon resonance enhanced bismuth selenide (Bi_2Se_3) layers for non-enzymatic glucose sensing. J Electroanal Chem 856. https://doi.org/10.1016/j.jelechem.2019.113629
105. Xiong X, Zhu P, Li S et al (2022) Electrochemical biosensor based on topological insulator Bi_2Se_3 tape electrode for HIV-1 DNA detection. Microchim Acta 189:1–10. https://doi.org/10.1007/s00604-022-05365-8
106. Gorle G, Bathinapatla A, Kanchi S et al (2022) Low dimensional Bi_2Se_3 NPs/reduced graphene oxide nanocomposite for simultaneous detection of L-Dopa and acetaminophen in presence of ascorbic acid in biological samples and pharmaceuticals. J Nanostructure Chem 12:513–528. https://doi.org/10.1007/s40097-021-00428-3
107. Fan H, Zhang S, Ju P et al (2012) Flower-like Bi_2Se_3 nanostructures: synthesis and their application for the direct electrochemistry of hemoglobin and H_2O_2 detection. Electrochim Acta 64:171–176. https://doi.org/10.1016/j.electacta.2012.01.010
108. Mohammadniaei M, Yoon J, Lee T et al (2018) Electrochemical biosensor composed of silver ion-mediated dsDNA on Au-encapsulated Bi_2Se_3 nanoparticles for the detection of H_2O_2 released from breast cancer cells. Small 14:1–13. https://doi.org/10.1002/smll.201703970
109. Chen S, Fang YM, Li J et al (2013) Study on the electrochemical catalytic properties of the topological insulator Bi_2Se_3. Biosens Bioelectron 46:171–174. https://doi.org/10.1016/j.bios.2013.02.019
110. Liu F, Zhang M, Nan P et al (2023) Unraveling the origin of donor-like effect in bismuth–telluride-based thermoelectric materials. Small Sci 1–11. https://doi.org/10.1002/smsc.202300082
111. Liu M, Yin S, Ren T et al (2021) Two-dimensional heterojunction electrocatalyst: Au-Bi_2Te_3 nanosheets for electrochemical ammonia synthesis. ACS Appl Mater Interfaces 13:47458–47464. https://doi.org/10.1021/acsami.1c11246
112. Zhao F, Zhou S, Zhang Y (2021) Ultrasensitive detection of hydrogen peroxide using Bi_2Te_3-electrochemical sensors. ACS Appl Mater Interfaces 13:4761–4767. https://doi.org/10.1021/acsami.0c19911

113. Shen H, Jang B, Park J et al (2022) In situ synthesis of a Bi_2Te_3-nanosheet/reduced-graphene-oxide nanocomposite for non-enzymatic electrochemical dopamine sensing. Nanomaterials 12. https://doi.org/10.3390/nano12122009
114. Yuan Y, Hu T, Zhong X et al (2020) Highly sensitive photoelectrochemical biosensor based on quantum dots sensitizing Bi_2Te_3 nanosheets and DNA-amplifying strategies. ACS Appl Mater Interfaces 12:22624–22629. https://doi.org/10.1021/acsami.0c04536
115. Sun Z, Tong Y, Li J et al (2022) Ultrasensitive photoelectrochemical biosensor based on black/red phosphorus heterojunction@Bi_2Te_3 hybrid and enzymatic signal amplification for the detection of colorectal cancer-related piRNA-823. Sens Actuators B Chem 368:132244. https://doi.org/10.1016/j.snb.2022.132244
116. Rajaji U, Chen TW, Chinnapaiyan S et al (2020) Two-dimensional binary nanosheets (Bi_2Te_3@g-C_3N_4): application toward the electrochemical detection of food toxic chemical. Anal Chim Acta 1125:220–230. https://doi.org/10.1016/j.aca.2020.05.033
117. Huang X, Zhou X, Wu L et al (2024) A nanotwin-based physical model for designing robust layered bismuth telluride thermoelectric semiconductor. Cell Rep Phys Sci 5:101841. https://doi.org/10.1016/j.xcrp.2024.101841
118. Mamur H, Bhuiyan MRA, Korkmaz F, Nil M (2018) A review on bismuth telluride (Bi_2Te_3) nanostructure for thermoelectric applications. Renew Sustain Energy Rev 82:4159–4169. https://doi.org/10.1016/j.rser.2017.10.112
119. Tang X, Li Z, Liu W et al (2022) A comprehensive review on Bi_2Te_3-based thin films: thermoelectrics and beyond. Interdiscip Mater 1:88–115. https://doi.org/10.1002/idm2.12009
120. Netsou AM, Muzychenko DA, Dausy H et al (2020) Identifying native point defects in the topological insulator Bi_2Te_3. ACS Nano 14:13172–13179. https://doi.org/10.1021/acsnano.0c04861
121. Samriti TR, Ruzimuradov O, Prakash J (2023) Fabrication methods and mechanisms for designing highly-efficient photocatalysts for energy and environmental applications. Mater Chem Phys 307:128108. https://doi.org/10.1016/j.matchemphys.2023.128108
122. Gholami T, Seifi H, Dawi EA et al (2024) A review on investigating the effect of solvent on the synthesis, morphology, shape and size of nanostructures. Mater Sci Eng B 304:117370. https://doi.org/10.1016/j.mseb.2024.117370
123. Uddin I, Abzal SM, Kalyan K et al (2023) Starch-assisted stable synthesis of CdS nanoparticles for enhanced electrical and optical properties. J Electron Mater 52:1710–1716. https://doi.org/10.1007/s11664-022-10198-5
124. Harish V, Ansari MM, Tewari D et al (2023) Cutting-edge advances in tailoring size, shape, and functionality of nanoparticles and nanostructures: a review. J Taiwan Inst Chem Eng 149:105010. https://doi.org/10.1016/j.jtice.2023.105010
125. Rajan SP, Keloth Paduvilan J, Velayudhan P et al (2024) Progress in 2D/3D nanomaterials incorporated polymer thin films for refractive index engineering: a critical review. J Polym Res 31:1–16. https://doi.org/10.1007/s10965-024-03967-9
126. Mostafavi AH, Mishra AK, Gallucci F et al (2023) Advances in surface modification and functionalization for tailoring the characteristics of thin films and membranes via chemical vapor deposition techniques. J Appl Polym Sci 140:1–53. https://doi.org/10.1002/app.53720
127. Pedersen H, Elliott SD (2014) Studying chemical vapor deposition processes with theoretical chemistry. Theor Chem Acc 133:1–10. https://doi.org/10.1007/s00214-014-1476-7
128. Nandihalli N, Gregory DH, Mori T (2022) Energy-saving pathways for thermoelectric nanomaterial synthesis: hydrothermal/solvothermal, microwave-assisted, solution-based, and powder processing. Adv Sci 9:1–61. https://doi.org/10.1002/advs.202106052
129. Alqaheem Y, Alomair AA (2020) Microscopy and spectroscopy techniques for characterization of polymeric membranes
130. Salame PH, Pawade VB, Bhanvase BA (2018) Characterization tools and techniques for nanomaterials. Elsevier Inc.

Nanobiosensors for Antimicrobial Resistance Surveillance: Possibilities and Challenges

Tamilmaran Nagarajan and Santhoshkumar Jayakodi

Abstract The emergence of antimicrobial resistance (AMR) has become an alarming problem in health care systems; pressing a threat to treatment of infectious diseases. AMR has been announced as one among the top ten threats the world needs to address. The continued surge in the antibiotic usage leads to the emergence of antibiotic resistance bacteria (ARB) and antibiotic resistance genes (ARG) in the ecosystem and disseminates unprecedently. The nexus of ARB and the rate of dissemination demands innovative rapid approaches for their precise detection in their environment. Traditional laboratory-based methods for AMR detection, such as culture-based techniques and molecular assays, are accurate but often costly, labour-intensive, and time-consuming, limiting their use for high-throughput onsite monitoring in resource-limited settings. Nanobiosensors are promising technologies offer precise detection of ARB in a scalable manner. Nanobiosensors are hybrid instruments that seam the advantage of nanomaterials with biology (biological recognition elements). Nanomaterials such as gold nanoparticles, carbon nanotubes, quantum dots, and nanowires enhance the sensor's performance in terms of sensitivity and speed. Biological recognition elements in turn improves the specificity through molecular interactions and reinforce precise detection of ARB. This chapter aims to leverage the adoption of nanobiosensor technology for surveillance of AMR. Providing a comprehensive overview of the emergence and spread of antimicrobial resistance, this chapter delves into the molecular basis of resistance mechanisms and addresses the limitations of traditional detection methods. Additionally, the chapter explores the critical components of nanobiosensors and highlights the potential bacterial biomarkers for targeting. The chapter concludes by discussing the key challenges and limitations that must be addressed to advance this promising technology for AMR detection and management.

Keywords Antibiotics · Bacterial biomarkers · Nanomaterials · Biosensors · Resistance surveillance

T. Nagarajan (✉) · S. Jayakodi
Department of Biotechnology, Saveetha School of Engineering, Saveetha Institute of Medical and Technical Science (SIMATS), Chennai, TN, India
e-mail: nagarajantamilmaran.sse@saveetha.com

I. Uddin (ed.), *Smart Nanosensors*, Smart Nanomaterials Technology,
https://doi.org/10.1007/978-981-96-3878-9_17

1 Introduction

Antibiotics/antimicrobials are considered as the most successful form of chemotherapy ever since the development of modern medicine. The success of antibiotics in eliminating against pathogens gave a great advantage in treating many diseases [1, 2]. Additionally antibiotic treatment serves as an adjuvant role in other advanced medical procedures (including surgery, organ transplantation and so on) [3, 4]. The success of antibiotic is temporary due to the emergence of resistant bacteria. The emergence of antimicrobial resistance (AMR) is rapid and pose a serious threat to the global health care system. Antimicrobial resistance refers to the capacity of an organism to with stand and proliferate in lethal dosage of antimicrobials. This often renders antimicrobials ineffective resulting in treatment failure. Additionally, AMR is associated with long hospital stays and increased mortality [5, 6]. If the scenario of AMR progress unrestrained, then the decades of medical progress would be retrogressive, putting high risk on treating a minor infection.

Bacteria acquire resistance by a wide variety of mechanisms [7] and require appropriate ways to restrain it. Evidence-based approach, based on molecular understanding potentiates action plan against AMR. Once emerged, factors like over usage of antibiotics, horizontal gene transfer, environmental contamination and unhygienic practices promotes the dissemination of antibiotic resistance bacteria (ARB) [8, 9]. The global movement of people, animals, and goods also accelerates the spread of resistant pathogens across regions and borders, making AMR a truly international concern [10]. Addressing AMR requires strict stewardship which requires monitoring the dissemination of ARB in environment and community. Initially the detection was based on culture-based methodologies which required growing bacterial cultures. Limited sensitivity along with long processing time made their application very limited. Subsequently molecular assays (which are culture independent) were adopted to characterize AMR. Polymerase chain reaction (PCR), ELISA, Microarray are some of the assays which are efficient in both identifying and understanding of AMR pathways [11, 12]. These methods depend on the detection of various AMR-related proteins, toxins, nucleic acids, and antibodies for the diagnosis of ARB [13]. Though efficient, most of the molecular assays often involve complex procedures which demand specialized infrastructure and skilled personnels. These limitations hinder the implementation of large-scale surveillance systems, particularly in resource-limited settings where the need for rapid, accurate diagnostics is critical [13]. WHO defines ASSURED criteria for an efficient diagnostic kit which must have the following properties (i) Affordable, (ii) Sensitive, (iii) Specific, (iv) User-friendly, (v) Rapid and Robust, (vi) Equipment-free, and (vii) Deliverable to end users [14]. Therefore, there is a growing need for innovative and accessible technologies that can detect AMR rapidly, promptly and effectively, even in resource limiting conditions.

Nanobiosensors are emerging as a promising technology in biosensing. Combing the high sensitivity of nanomaterials with the specificity of the biological recognition elements, nanobiosensors are a formidable defence against ARB [15–17]. Biosensors designed to detect biomarkers (genes or proteins associated with antibiotic resistance) would bring a revolutionary change in AMR surveillance. They can detect ARB by sensing even the minute concentrations of biomarkers from the samples. This property is highly valuable in AMR surveillance, due to their precise detection and versatility. Despite several advantages, the adoption of nanobiosensors for AMR surveillance is still under developmental and required substantial improvements. This chapter discuss the emergence of antibiotic resistance bacteria, and the molecular mechanism behind it. We also highlighted the limitations of traditional diagnostic methods and examine the potential of nanobiosensors for AMR surveillance. Through this, we aim to provide insights into how nanobiosensor technology can contribute to the global fight against antimicrobial resistance and improve the stewardship of resistant pathogens in diverse settings.

2 Bacterial Resistance Mechanisms to Antibiotics

Bacteria resistance is multifaceted often contains sophisticated mechanisms (Fig. 1). Resistance bacteria produce several mechanisms which counteract the toxicity of bacteria by diverse mechanisms. They are broadly classified in into two types; (i) Intrinsic resistance and (ii) acquired resistance.

2.1 Intrinsic Resistance Is Inherent

Antibiotics exert their activity by specifically inhibiting the function of target molecule. The bactericidal spectrum of any given antibiotic contains a list of susceptible and non-susceptible bacteria. This difference primarily stems from the inherent structural or functional attributes present in the bacterial cell. Thus, the inherent characteristics of certain bacterial species, to resistant antibiotics without prior exposure is called Intrinsic resistance [18]. This form of resistance is commonly observed in gram-negative bacteria [19, 20]. Gram-negative members exhibit insensitivity to many of the antibiotics due to the presence of additional membrane and efflux pumps [19, 21]. Presence of AcrAB-TolC efflux pumps in *E. coli* confer growth advantage in presence of antimicrobial compound sanguinarine [22]. Modification of membrane components (i.e. lipopolysaccharides, proteins) is associated with intrinsic resistance to antibiotics like tetracycline, glycycline and polymixins [23, 24] in gram negative bacteria. *Pseudomonas aeruginosa* a well-known gram-negative bacterium well known for their intrinsic resistance to various antibiotics. A recently study reported that ABC transporters, a large family of transporter proteins are responsible for their intrinsic resistance [25, 26]. Adaptation of genome wide analysis bring the elaborate genetic network of intrinsic resistance to limelight. Col and coworkers, demonstrated the involvement of 18 genes involved in different process, conferring resistance to Boric acid [27]. Similarly, proteins involved in spatial regulation of

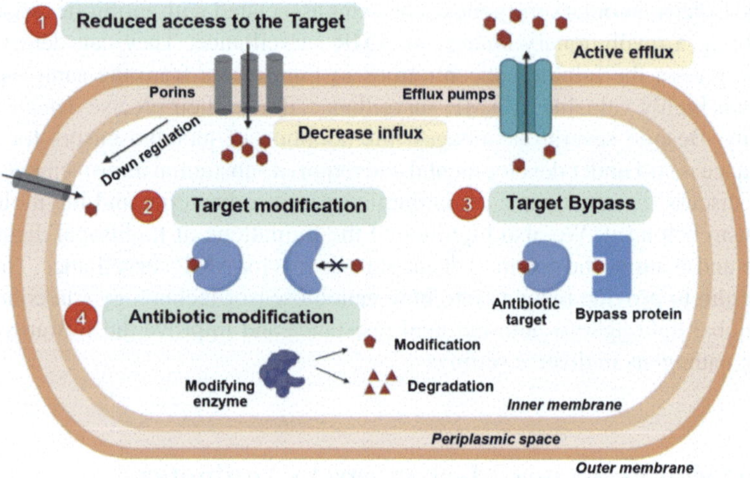

Fig. 1 Antibiotic resistance mechanisms. Resistance mechanisms are broadly classified into four major classes: (i) reducing the access to the target; (ii) modification of target; (iii) Production or bypass proteins and (iv) modification of antibiotics. Reducing the access to the target is achieved by the action of efflux pumps and modulation of the expression of porins. Target modification refers to mutational change of the target where antibiotics cannot bind. Bypass proteins/binding proteins have higher affinity to the antibiotics which reduce the availability of antibiotics to the target. Chemical modification or degradation of antibiotics is another way to alleviate the toxicity of antibiotics

cell division (MinCDE) is associated with resistant to streptomycin and gentamicin resistance [28]. A comprehensive genetic analysis of *Streptococcus pneumoniae* treated with 20 antibiotics revealed that modulation in cellular process like nucleotide metabolism, alarmone synthesis, ATP biosynthesis, cellular transport, transcription and translation had a profound effect on the antibiotic susceptibility [29]. Despite several research attempts, the diversity and molecular mechanism behind intrinsic resistance remains elusive.

2.2 Acquired Resistance Is Adaptive

Acquired resistant in contrast is adaptive and developed by the bacteria up on exposure to the antibiotics. They are dynamic response where the bacteria alter their behaviour and cellular processes to survive antibiotic exposure. Though, adaptive resistance mechanisms can be classified in detail into many grouped, they are broadly categorised into three broad classes; (i) modification of target protein, (ii) reduced access to the target and (iii) inactivation of antibiotics (see Fig. 1). Target specific binding is crucial for the bactericidal property of antibiotics. Binding of the antibiotic tor the target interfere with their function resulting in inhibition of growth or cell death. Since, the targets are essential for cellular process, bacteria cannot evade from the antibiotic activity. One of the prevalent ways of surviving antibiotic is to change the target through mutation. A mutated target may not affinity to the antibiotics and thereby exhibit limited sensitivity to it yet catalytically active [30]. Resistance to rifampicin (a commonly used antibiotic to treat tuberculosis) emerge due to mutational changes in the RNA polymerase subunits. The mutation renders RNA polymerase insensitive to antibiotic, yet can carry out transcription [31–33]. Similarly, mutations in gyrase subunits and ribosome subunits confer resistance to fluoroquinolones [34, 35] and aminoglycosides [36–38] respectively. Reducing the access of antibiotics to their targets is very effective strategy in reducing the toxicity. Bacteria achieve this primarily by modifying permeability of the cell wall. Porins are outer membrane channel-forming proteins responsible for the passive transport (import and export) of small molecules (including nutrients, antibiotics etc.) across membranes [39]. Modulation in the function of porins modify cell permeability and implicate antibiotic resistance [40].

β-lactam antibiotics enter cell through the porin OmpF. Cells defective in production of OmpF exhibit insensitivity to β-lactams [41]. In another study, *Enterobacteria aerogenes* acquire multidrug resistance due to the presence of conductance defective porin (OmpC/OmpF) variant, which retains 30% of its activity compared to wild-type allele [42]. Apart from porins, other proteins which is involved in maintaining cell permeability is also the site for development of resistance. Vibrio alginolyticus, become resistant to the antibiotic kanamycin by changing the membrane permeability using a membrane associated protein ArmPT [43]. In another study, elevated membrane potential confers resistance against tetracycline and gentamycin in *E. coli* [44]. Thus, proteins involved in membrane permeability are candidates for the occurrence of mutations and thereof generation of antibiotic resistance.

Bacteria can continuously pump the incoming antibiotic molecules thereby reducing the intracellular concentrations to nontoxic. Efflux pumps are integral membrane proteins present in most of the bacteria. These pumps actively export a wide variety of substrates (including metabolic waste, toxins and antibiotics) out of the cell and maintain cellular homeostasis [45, 46]. Resistance would arise due to the changes in their expression, formation of variants due to mutations and asymmetric accumulation [46, 47]. Antibiotic resistant pumps are broadly classified into six categories namely; (i) ABC transporters, (ii) resistance-nodulation-cell division (RND) superfamily, (iii) proteobacterial antimicrobial compound efflux (PACE) family, (iv) major facilitator superfamily (MFS), (v) small multidrug resistance (SMR) family and (vi) multidrug and toxin extrusion (MATE) family [19, 48]. The function of efflux pumps is ATP dependent and drives by the confirmational change in their structures. The structure and working mechanism of efflux pumps differs significantly and can be more complex. However generally, they work by the following method. Efflux pumps contain distinct sites for binding of substrates which capture the substrates. Substrate binding induces a confirmational change which in turn moves substrate across the membrane and release outside the cell. Efflux pumps can offer broad spectrum antibiotic resistance and considered one of the most effective forms of resistance. Weakening the activity of efflux pumps in vibrio species increases their susceptibility to antimicrobial peptides [49]. In *Campylobacter* spp. resistance-nodulation-cell division (RND) superfamily of efflux pumps confers resistance against antibiotics like erythromycin, ciprofloxacin and tetracycline [50]. The function of the any antibiotic depends on their structure which is required for target inactivation. Disruption of the antibiotic structure either by modification or by degradation make the antibiotics unable to bind with the target protein and thus rending the bacteria insensitive to the antibiotic. Bacteria harbours enzymes for both antibiotic modification and degradation which confer antibiotic resistance.

Group transfer enzymes like methyl transferases, acetyl transferases, kinases, adenylyl-transferases, ribosyl transferases and glycosyl transferases are major antibiotic modifying enzymes which confer antibiotic resistance. By transferring a chemical group, they change the structure of antibiotic and inhibit target binding [51, 52]. Multidrug resistant bacteria Acinetobacter baumannii, harbours genes encoding for chloramphenicol acetyl transferase for chloramphenicol resistance [53, 54]. Similarly, pathogenic bacteria acquire amikacin resistance through aminoglycoside 6′-N-acetyltransferase (type IB) enzyme [55–57]. Resistance to antibiotics like β-lactams, aminoglycosides, and macrolides were achieved primarily through modification/degradation of the antibiotics. β-lactamases are a large class of hydrolytic enzymes capable of degrading β-lactam antibiotics (i.e. penicillin and its derivatives) [58]. The emergence of β-lactamases severely hampered the usage of penicillin derivatives in health care settings [59, 60]. Evolution of β-lactamases for broad spectrum substrate specificity resulted in Extended spectrum beta-lactamases (ESBL) are multi drug resistant variant prevalent in pathogens like Escherichia coli and klebsiella pneumoniae responsible for resistance against a broad spectrum of β-lactam antibiotics including penicillins, cephalosporins, and aztreonam [58, 61, 62]. Several classes of β-lactamases have been emerged to counteract the toxicity of diverse β-lactam antibiotics, but with the common mechanism of antibiotic degradation has been very effective in clinical settings [59]. Finally, bacteria can critically sequester the antibiotic targets by producing additional molecules which binds with antibiotic more efficiently than the target ones [63].

3 Antibiotic Resistance Bacteria (ARB) Has Become Ubiquitous

Antimicrobial resistance is emerging rapidly due to several factors primarily being increased antibiotic consumption. Antimicrobial agents are indiscriminately used ever since its discovery [64, 65]. The global antibiotic consumption across 204 countries witnessed an alarming 46% increase in last two decades alone [66]. Specifically, developing countries like India, China, Brazil, Russia and South Africa accounts for the 76% of the global increase in antibiotic consumption [67], that impact the emergence of antibiotic resistance bacteria. The increase utilization of antibiotics helps in the emergence and enrichment of antibiotics which are disseminated though various biotic and abiotic factors (see Fig. 2). Most of the antibiotics are consumed by health care sectors, in farm practices (animal husbandry, aquaculture and agriculture) and are candidates for the origin of ARB [68]. Health care environments rich in antibiotics and serve as a potent source of ARB. Dissemination of ARB from hospitals become is inevitable through direct contact or by contaminated surfaces. Several studies reported that the work surfaces and other utensils are susceptible to ARB colonization which are the source of hospital born infection [69, 70]. A study conducted by He and coworkers [71] identified the presence of 643 antibiotic resistance gene subtypes (ARG) belongs to 22 broad ARG categories in various places of an urban hospital environment. Livestock and farming environments are another source of extensive antibiotic usage. In farming practices, antibiotics are largely used to control diseases and boost the yield. Thus, farm environments become highly contaminated with antibiotics. Furthermore, the nutrient rich environment accompanied with high bacterial count make them a source of ARB development. The aerosol, animal manure, soil and water of the farms are rich in ARB. Antibiotic resistance genes like extended spectrum beta lactamases, methicillin resistance, colistin resistance and tigecycline resistance were commonly reported among the pathogens infecting both farm animals and humans insisting their spread through exposure [72, 73]. Several studies reported that the wastewater from hospitals and farms often contains high concentrations of antibiotic residue which ended up contaminating the surrounding soil and water ecosystems. The antibiotic consumption of both human health care system and animal husbandries are surplus, and they act synergistically in the maintenance of ARBs in the environment [74].

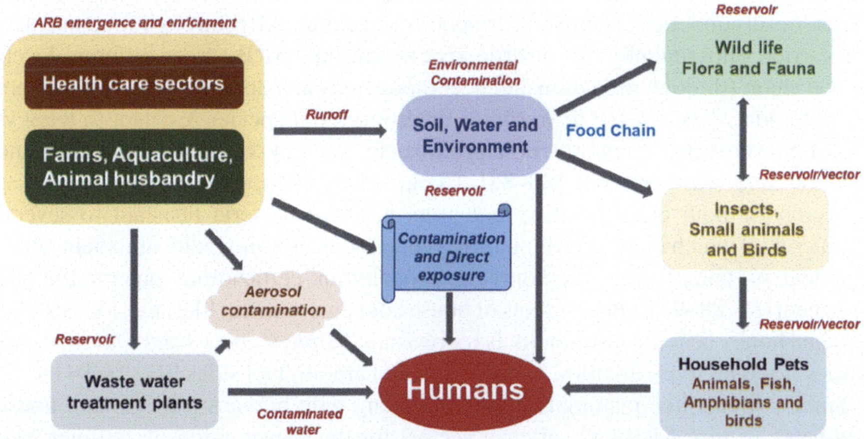

Fig. 2 The nexus of antimicrobial resistance dissemination. Hospitals, and farms are the places where ARB emerge due to the heavy antibiotic usage. ARB dissemination pathways are various from direct contact to biomagnification through food chain. All the intermediate animals serve as reservoirs/host. Even aerosols microbiome in critical regions of the community is rich in ARB and helps in dissemination

While human consumption is associated with the emergence of ARBs, consumption in livestock helps in existence of specific strains [74–76]. Even waste water treatment plants (WWPTs) could not treat eradicate the antibiotics from waste water. A conventional WWTPs can retain up to 40–90% of the pharmaceutical residues after primary treatment whereas with advanced treatment strategies, elimination cannot be achieved above 90% [77]. Thus, WWPTs became favourable environments for the selection and proliferation of ARB [78]. Additionally antimicrobial agents are also used in the manufacturing of personal hygiene other products, which further increase the exposure of antibiotics to the environment and contribute to the emergence of ARB [79]. Once when the ARB emerged, the dissemination two hierarchies. At the level of bacteria, lateral transfer of genes occurs, where through known mechanisms (like transduction, transformation and conjugation) ARG spread making the recipient resistant ones [8, 80, 81]. Secondly, the emerged ARB spread in the communities and in the environment driven by the contagion [74]. Interestingly, the spreading of the ARB and the lateral transfer of ARG happens mutually making the dissemination network intricate and difficult to track [82–84]. However, the level of environmental contamination is directly correlated with the emergence of ARB in the ecosystem [74]. Through the source of emergence of ARGs can be traced back, the dissemination of ARG and ARB in the ecosystems is complicated and beyond the limit of surveillance.

ARB once emerged, persist in the environment for a significant longer period and cause infection by surface exposure or through food chain. Food and food chain plays a crucial role in the dissemination of ARB in the ecosystem and it is a well-documented phenomenon. Various fauna (including insects, small animals, and birds) acquires ARB through direct contact or through food which subsequently become carriers/reservoirs. For example, arthropods can acquire ARB through food and acts as a carrier. Subsequently, the arthropods can transmit ARB across different levels in food chain (through predation and decomposition) in a dosage dependent manner [85]. In house flies (*Musca domestica*), Enterobacterial species resistant to tetracycline, erythromycin, streptomycin, ciprofloxacin, kanamycin, chloramphenicol and nalidixic acid was observed [86–88]. Cockroaches (*Periplaneta americana*) acts as vector for both gram positive and gram-negative bacteria resistant to several antimicrobial which includes Ampicillin, gentamycin, ciprofloxacin, ofloxacin, chloramphenicol, tetracycline, streptomycin, cephalothin, ceftazidime, piperacillin and imipenem [85, 89–92]. Other insects of house hold environment like ants, locusts, bed bugs and honey bees are also sentinels for resistant *Staphylococcus* and *Enterococcus* species resistant to methicillin, vancomycin and aminoglycosides [91, 93, 94].

Small animals like mammals that thrive along with humans (rodents and house hold pets) acquire ARB and serve as vectors for the transmission of antimicrobial resistance. Contagion and food chain are familiar ways they acquire ARB. Microbial species like *Staphylococcus aureus*, *E. coli*, *Pasteurella pneumutropica*, *Klebsiella pneumoniae* and *Aeromonas* species resistant to frontline antibiotics are observed in house hold mice [95]. Pet animals like dogs, cats, rabbits, birds and amphibians are another line or AMR reservoirs which contribute to continued sustainability of the ARB in house hold environment [96–98]. A steady increase in the occurrence of multidrug pathogens (i.e. extended-spectrum beta-lactamase) among house hold pest was observed in past decade [98–100]. Disturbingly even the air we breathe could not free from resistant pathogens. Metagenomic analysis of air bone particles confirm the presence of high level of ARBs is urban hot spots like hospitals, WWTPs and other critical locations [101]. A survey of air metagenome analysis in the environment of health care sector revealed the presence of more than 150 different antimicrobial resistant genes present in both commensal and pathogenic bacteria [101, 102]. Similarly, more than 200 types of ARGs belong to 11 antibiotic categories are reported to be present among the particulate matter of in a smog [103]. This undoubtfully suggests the complex nexus of the dissemination pathways of ARG and demands strict vigilance in the spread of AMR challenging.

4 Biosensing of Antibacterial Resistance—Genetic Versus Phenotypic Characterization

Sequence information of any protein is coded in the genetic molecule DNA and this applies for antimicrobial resistance mechanisms [104]. All resistance mechanisms arise from specific genes located in chromosomal or extrachromosomal DNA, forming the genetic foundation of AMR. The proteins produced by these genes drive the development of resistance phenotypes, constituting the phenotypic component of AMR [105–107]. To gain meaningful insights into AMR dissemination, it is essential to characterize both its genetic and phenotypic components (Fig. 3).

Characterization of genetic component involves detecting the presence of AMR genes and analyse the genetic information. AMR genes can move across the community though genetic elements like plasmids, transposons, bacteriophages and genomic islands acquire and disperse AMR genes in their habitat. They act as genetic vectors in dissemination [108, 109]. Therefore, genetic characterization is not limited to identification of the AMR genes, but extended in recognition of the vectors involved in the dissemination. It also unravels the basis behind the development, maintenance, and transfer of resistance genes in bacterial population. Traditional techniques used for genotypic characterization includes PCR, blotting techniques and sequencing analysis. Phenotypic characterization in contrast, involves assessing the proteins, involved in the resistance formation. As discussed previously, resistance proteins are diverse and changes with mechanisms. Even for one antibiotic, bacteria have multiple mechanism involving different proteins to alleviate the toxicity. Thus, phenotypic characterization deals with product of AMR genes. Antibiotic susceptibility test, western blotting analysis and phenotypic screening is used for physical characterization. To evaluate the nature of AMR in s bacteria, simultaneous characterization of both genotypic and phenotypic is essential. Bacteria have stringent regulatory mechanisms to express/repress the production of a resistance protein as they are expensive and compromise bacterial fitness [110, 111]. Thus, a bacteria having genetic determinants of resistance may not produce resistance proteins and remain sensitive to

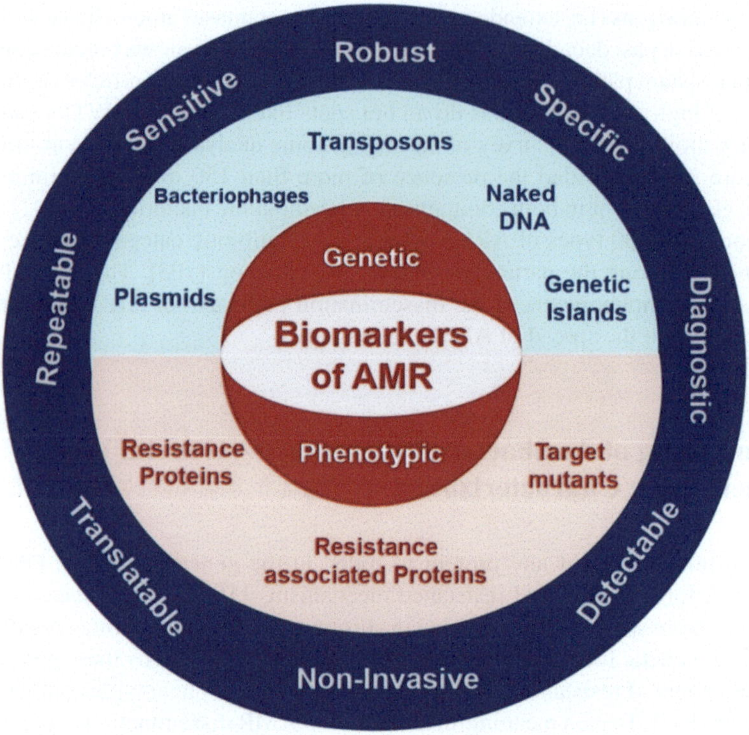

Fig. 3 Antimicrobial resistance characterization encompasses both phenotypic and genotypic approaches. Genotypic characterization involves identifying the genetic elements responsible for resistance development and dissemination, including various horizontal gene transfer mechanisms. Phenotypic characterization focuses on the proteins that confer resistance, such as resistance proteins and mutant variants of target proteins, as well as resistance-associated proteins, where modifications to these genes lead to the development of resistance. Biomarkers selected for the detection are crucial for surveillance. The biomarker should be specific, sensitive, reproducible, translatable and preferentially non-invasive for effectual surveillance

the antibiotic. Therefore, clear understanding of the emergence and dissemination of AMR demands in detail analysis. Nanobiosensors are an ideal candidate for AMR characterization. Their remarkable elasticity in biosensing modules, made genetic and phenotypic characterization possible.

5 Traditional Approaches in AMR Surveillance—Challenges and Limitations

Various methodologies were developed over years to identify and characterize the antimicrobial resistance. They are broadly classified into culture dependent and culture independent methods. In culture dependent methodology the detection was primarily culture-based which requires cultured isolates. They employ systematic procedure including isolation of the target bacteria, and characterizing their antimicrobial resistance by assessing their growth in presence of the antibiotic. This gives a detailed information about the phenotype of antimicrobial resistance [112, 113]. The advent of biotechnology introduces novel techniques in characterization of antimicrobial resistance, which can be culture independent [11]. Few notable techniques which are currently used includes polymerase chain reaction (PCR), Enzyme-Linked Immunosorbent Assay (ELISA), Microarray based methodologies, Sequencing of the DNA and so on [11].

PCR is an efficient tool used for detecting the presence of genes responsible for AMR. Detection is based on the amplification of DNA sequence (definite regions in AMR genes) using primers. Amplification of the DNA indicates the presence of specific AMR genes [114, 115]. ELISA can detect antimicrobial resistance by identifying specific proteins or enzymes produced by the resistant bacteria. Antibodies specific for antimicrobial proteins, are used in this technique. When antibodies bind to resistance-associated proteins, an enzyme-linked secondary antibody is used against it to generate a colour change, indicating the presence of resistance markers [116–118]. Microarray technology enables the high throughput screening of large number of samples in a single chip. Microarrays are chips embedded with thousands of probes (DNA) that bind to specific DNA sequences associated with resistance genes. When a sample DNA is applied to the microarray, complementary binding occurs between the probes and any matching resistance genes. Binding of probe and sample DNA can be detected using fluorescent labelling techniques. This approach is not only rapid but allows for multiplex profiling of resistance markers across multiple pathogens [119–122]. DNA sequencing technologies are greatly employed to detect and characterize antimicrobial resistance. Sequencing technologies directly read the nucleotide sequence from the sample and provide the sequence information, which can be used to identify known resistance genes or mutations associated with resistance. Sequencing methods, like whole-genome sequencing (WGS) or targeted sequencing, allow for detailed analysis of bacterial genomes, revealing the presence and types of resistance genes or gene mutations [58, 123, 124]. Culture independent methodologies are very advantages in identifying resistance related DNA even in low levels of sample. Culture dependent methodologies are sometime use to validate the results which gave new insights in the understanding antimicrobial resistance.

Despite being more efficient in detecting and characterizing AMR, these technologies couldn't meet the growing demand in AMR surveillance due to the following disadvantages; (i) ***Demand instruments and infrastructure***: Culture independent methods employ specialized instrumentation setup which is often bulky and expensive. This dependency limits its feasibility to perform the test in low-resource or field settings without reliable access to power or computing resources. (ii) ***Elaborate Sample processing***: The discussed methodologies require extensive sample preparation which can be require lab-grade equipment and expensive chemicals. In some cases (like mass spectrometry, DNA sequencing and microarray) the sample preparation itself may be time consuming which makes it complicated to perform in field conditions. (iii) ***Required Skilled labour***: Most of the detection methodology required skilled personnel, having enough technical skills capable of performing extensive laboratory protocols, operate instruments, along with trouble shooting and interpreting skills for result analysis. (iv) ***Time consuming***: Even though culture independent techniques are relatively fast, still the time taken for these methods to generate the results are still longer. Even a conventional PCR takes 2–4 h which is very slow for field analysis, where critical timely response is required. (v) ***Expensive reagents***: Although operational costs are decreasing, still modern methodologies use expensive reagents and may require specialized operational and storage conditions which are not available under low-resource settings. (vi) ***Limited multiplexing abilities***: Bacteria becomes multi-drug resistance, by subsequently acquiring more than one pathway which can require multiplex detection abilities. Many of the discussed techniques exhibit limited multiplexing abilities, which renders multiple tests to characterize multi-drug resistance. (vii) ***Limited real-time monitoring***: For real-time monitoring of resistance evolution within healthcare setting or in a community, frequent sampling and testing is required, which is impractical with many of the detection techniques. These disadvantages make them less compatible and unsuitable for continuous AMR surveillance, especially in dynamic clinical settings.

6 Nanobiosensors for the Detection of Antimicrobial Resistant Bacteria (ARB)

Nanobiosensors are state to the art analytical devices, can perform rapid detection of antimicrobial resistant bacteria from different samples. These sensors combine both nanotechnology and biological components, enabling them precise detection of the target molecule even at extremely low concentrations. Components of nanotechnology used in the biosensors include nanoparticles, nanotubes, and nanowires. Having unique physiochemical properties, they provide an excellent platform for biomolecular interactions. Biological components are antibodies, enzymes, and nucleic acids, interact with the AMR specific biomarker which increases the specificity of the biosensors to several folds. When the target bacteria bind to the biological elements of the sensor, they trigger a signal from the nanomaterial (in the form of electricity or optical changes, or thermal changes or mechanical changes) which can be converted into quantitative or qualitative data (Fig. 4). This allows nanobiosensors to detect the antibiotic-resistant bacteria with highest precision in real time, making

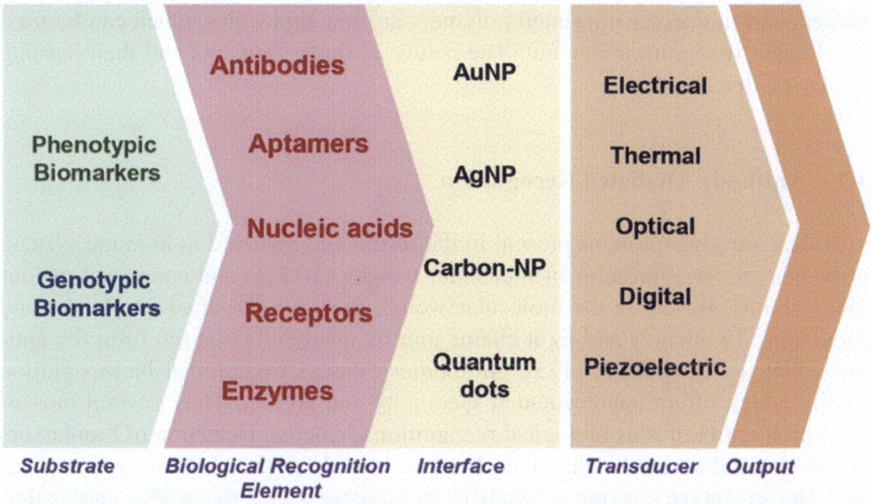

Fig. 4 Structure of a nanobiosensors for AMR surveillance. Configurations include substrate/analyte, biological recognition element, interface, transducer and output. Substrates are biomarkers of genotypic and phenotypic components of resistance. Biological recognition elements are the recognition components of the biomarkers. Substrate binding to recognition element happens in the nanoparticle interface which enhances the interaction. Transducer element, convert the signals to output which is read as quantitative or qualitative data

them highly valuable for surveillance of dissemination of ARB. Since, the structure and function of the nano sensors are described in detail throughout this book, the adoption of biological technologies required for AMR recognition is discussed.

6.1 Biological Recognition Elements

Molecular recognition is the working principle of biological recognition elements. Biological molecules interact with each other to carry out specific activities. The interaction between the molecules inside the cell is highly specific and coordinated. Therefore, all the biological process (from macromolecular biosynthesis to an enzyme catalysed reaction), it requires a systemic, coordinated and specific interaction between two molecules. This property of the biomolecules is exploited for the development of biosensors, by integrating several biological elements which confers specificity. Providing specificity towards analyte is the primary function of a biological recognition element in a biosensor. Biological recognition element achieves this by exhibiting strong affinity towards the analyte. Therefore, the selection of biological recognition element determines the performance of biosensors. Though several classes of biomolecules are utilized for biosensing, for detection of AMR products, only few class of biorecognition molecules can be used. Antibodies, nucleic acids, aptamers and molecular imprinted polymers are few molecules which can be used as biological recognition element. The nature of those elements and their sensing abilities are discussed.

6.1.1 Antibody Mediated Recognition

Antibodies are glycoproteins present in the serum and involved in immune system. Antibodies are large proteins of molecular weight 150 kDa and composed of four peptide chains. Based on the molecular weight they are classified into heavy and light chains. Two heavy and light chains join by disulphide bond to form the antibody which is Y in shape [125]. Antibodies possess exceptional biorecognition capacity which offers unprecedented specificity and affinity. Thus, several biosensors exploit antibodies as biological recognition elements. Detection of biomarkers through specific antibodies has long been practiced for decades for disease diagnosis. The advantage of using antibody is, it can sense a wide range of biomolecules, including, carbohydrates [126, 127], proteins [128–130], lipids [131] and nucleic acids [132–134]. Thus, antibody mediated biological recognition is used for both genetic and physical characterization of AMR. Though antibody mediated recognition is wide used for medical diagnosis, in context of antimicrobial few notable attempts were made. Usage of hetero-specific antibodies were efficient in detection of methicillin resistance in *S. aureus* [135]. Antibodies generated from birds and aquatic animals offers excellent specificity in detection of bacterial epitopes [136]. Single domain antibodies (also called nanobodies) are capable of accessing even the hard-to-reach epitopes in bacteria offering excellent sensitivity [136–138]. Antibodies targeting P-glycoprotein serve as an excellent diagnostic reagent for detecting multidrug resistant cell lines [139]. Culture free detection of antimicrobial resistance through specific antibodies were proven to be successful in detection of body fluids [140]. Lateral flow based immunodetection methodologies employing hetero-specific antibodies identified carbapenem resistance from multi-drug-resistant gram-negative organisms [141]. Hetero-specific antibodies are found to be rapid and reliable in the multiplex detection of toxins and resistance specific markers produced by clinically relevant *Staphylococcus aureus* [142]. Thus, antibody-based recognition is versatile, rapid and enable multiplexing capabilities in AMR recognition.

6.1.2 Aptamer Based Recognition

Aptamers are nucleic acid-based ligands made of ssDNA, RNA, and peptide moiety that can fold into unique three-dimensional structures, allowing them to bind specifically and with high affinity to a variety of target molecules [143]. This recognition and binding ability made them a potential candidate for the diagnostic purposes in health care sector [143]. They are typically 100 base pairs in length and have complementary region at the centre [144] which helps in the formation of three-dimensional structures [145]. Aptamers are generated from synthetic libraries by a systemic procedure called SELEX (**S**ystematic **E**volution of **L**igands by **EX**ponential enriching) [146]. The selection procedure involves generation of a library which contains random aptamers. Aptamers against a specific target is selected by several repeated binding assays. The non-specific aptamers are eliminated during this stem leaving a pool of target specific aptamers. They are then enriched using methods like polymerase chain reaction, peptide biosynthesis and so on [147]. Being more stable than antibodies, aptamers are used in ELISA based techniques as a cheap alternative for antibodies [148, 149]. Aptazymes (also called DNAzyme or RNAzyme) are aptamers with catalytic abilities can perform enhanced biosensing activities [150, 151]. Aptamer based detection systems have been used to monitor antibiotic residues across various environments [54, 152, 153] bacterial contamination [154–156] and specific analyte associated with disease progression [157, 158]. Though, the efficiency of aptamers in detection of antimicrobial resistance genes and proteins remains unexplored. Based on the molecular recognition of aptamers, it is easily understandable that aptamers can bind and detect AMR biomarkers in an efficient manner. However more research is required to address potential challenges and drawbacks.

6.1.3 Molecularly Imprinted Polymers

Molecularly imprinted polymers (MIPs) are synthetic analogues of biological receptors. They work in a similar manner of cellular receptors, bind specifically to a substrate/analyte by molecular recognition [159–161]. MIPs are made up of two kinds of monomers: (i) functional monomers and (ii) crosslinking monomers. Functional monomers are crucial for the substrate interaction and molecular recognition of analyte. They interact with substrate via hydrogen bonds and ionic interactions helps in formation of MIP-substrate complex. Cross-linking monomers are building matrix helps in the structural stability of the MIPs. Monomers like methacrylic acid, acrylic acid, styrene, allylamine are examples of functional monomers whereas tetraethyleneglycol dimethacrylate, divinylbenzene and trimethylopropane trimethacrylate are examples of cross-linking monomers [162, 163]. Synthesis of MIPs involve several methodologies including phase inversion, surface polymerization, soft lithography and so on [164–167]. Despite the difference in the production, the MIPs are designed over the substrate to leaves behind imprinted cavities that are complementary to the target molecule in size, shape, and functional groups. MIPs are good candidates of recognition part for the detection of antimicrobial resistant proteins [159, 168]. Detection of nucleic acids are also achieved with MIPs [160], making suitable for genetic and phenotypic characterization of AMR.

6.1.4 Nucleic Acid-Based Recognition

Nucleic acid-based recognition system exploits the complementarity property for recognition of substrate [169, 170]. DNA is the substrate and binding of the substrate is based on the sequence specificity. Based on the need, an artificial fragment of DNA is synthesized and immobilized to the substrate which act as recognition element. Exposure of the DNA with complementary sequence, result in formation of base pairing with recognition element and subsequent signal transduction to the sensor [171–173]. Modified nucleotides are used instead of normal bases in the recognition element resulted in peptide nucleic acids and locked nucleic acids [174–176]. Though primarily DNA based, nucleic acid-based recognition systems can be extended to in biosensing of proteins by exploiting natural DNA–protein interaction. This system can be used for extensive genotypic characterization of AMR.

6.2 Nanoparticles for Biosensing

Nanoparticles are very critical for the versatility of nanobiosensors. These particles, with size ranging from 1 to 100 nm in size exhibit unique properties with high surface area-to volume, optical, electrical and magnetic properties are valued in the biosensing applications. They by providing a platform the biosensing increases the sensitivity, selectivity and operational flexibility by enabling the detection of biomolecules even at very low levels [177–179]. Notable functions of nanoparticles in biosensors include, immobilization of biomolecules, catalysis of chemical reactions, augmented electron transfer, labelling and reactant. Immobilization of biomolecules often resulted in the denaturation and subsequent loss of bioactivity of the molecule, over the course of time. Nanoparticles have charge and can electrostatically absorb the biomolecule on to the surface without loss of activity [180]. Gold nanoparticles, silver nanoparticles, quantum dots, magnetic nanoparticles, carbon-based nanoparticles, silica nanoparticles and polymeric nanoparticles are used for nanobiosensors for various health care applications. Gold nanoparticles are most widely used for the immobilization of biomolecules. They form covalent and noncovalent interactions with the biomolecules, and immobilizes it. Hetero-specific antibodies conjugated with gold nanoparticles rapidly detected the presence of pathogenic *E. coli* in liquid samples [181]. Silver nanoparticles are used in the biosensors for their remarkable transducing capacity. Their excellent electron transfer capacity made them candidate for detection systems like surface plasmon resonance, Foerster resonance energy transfer and SERS based sensors [182]. Chemoreceptor embedded nanoparticles, detect bacterial metabolites and aids in the identification of bacteria in liquid samples [183]. Apart from gold and silver, titanium, zirconium, and iron nanoparticles are used for the biosensing applications [184, 185]. An attempt to identify pathogenic bacteria using monoclonal antibodies conjugated with magnetic nanoparticles reduced the detection time to few minutes with low level of samples (10 cfu/ml) [181]. DNA probes attached to magnetic nanoparticles are efficient in genotyping of bacterial species. The sensor can recognize and differentiate 13 different species within 2 h. These rapid, and robust detection was accomplished with low species [186]. Silica nanoparticles conjugated antibody can detect the DNA and protein of pathogenic *E. coli* O157:H7 from food samples [187]. Rajendran et al. [188], integrated the fluorescent nanobiosensors capable of detecting food borne pathogens to smartphone allows rapid detection and environmental monitoring. Biosensing by surface-enhanced Raman scattering (SERS), using silver nanoparticles not only detected bacterial species, but also differentiated their subspecies in water samples [189].

Biological samples are often complex and contains components which inhibit the detection and also creates high noise-to-signal ratio. This creates direct sensing from the samples very challenging. Nanoprobes immobilized in iron nanoparticles, can detect Mycobacterium avium directly in blood and milk without sample processing and enrichment [190]. Magnetic nanoparticles with functionalized lipopolysaccharides detect dangerous pathogenic bacteria *Brucella abortus* in animal serum samples [191]. Lee and coworkers [192] integrated a microfluidic chamber and detected the presence of *Mycobacterium bovis* from native biological samples. The sensor can detect the presence of 20 cfu/ml in less than 30 min which offers flexible point-of-care applications. Though the given references it is evident that; by using suitable biological recognition elements, antibiotic resistance bacteria can be detected from various sources.

Once when biological recognition occurs, a signal is produced and transducer convert the signal to measurable output like electrical, optical, thermal or piezoelectrical signals. The produced signals are then amplified by the amplifiers by enhancing the signal-to-noise ratio, enabling accurate detection of even very low concentration of target analytes. Optical output includes calorimetric, fluorescence signals. Calorimetric and fluorometric methodologies use chromophore and fluorophores which produce the output signals which are recorded. Digital output is nowadays used in nanobiosensors. Digital platforms coupled with detection systems provide real-time data that can be visualized in electronic devices like digital screens, computers, and IoT devices. By integrating detection, amplification, and output capabilities, nanobiosensors hold vast potential to transform AMR surveillance.

7 Challenges and Concluding Remarks

Competent AMR surveillance is closely tied with secured human health. Therefore, novel and capable methodologies should always emerge and meet the changing demands. Lack of inexpensive, sensitive devices is a major challenge in detection of AMR. The rapid evolution and adaptation of pathogens to antibiotics underscore the urgency for innovative detection and monitoring strategies in AMR surveillance, and nanobiosensors have emerged as a revolutionary technology in this field. Nevertheless, the adoption of nanobiosensors for AMR surveillance is still primitive due to several challenges. The manufacturing of Nano-biosensors often requires sophisticated fabrication processes, which is expensive and complex to scale and required low-cost alternatives. The nanobiosensors have a limited life span due to the stability of nanoparticles and biological recognition elements, which needs improvements. While some nano-biosensors have multiplexing ability, it still remains challenging to develop sensors that can simultaneously detect multiple resistance genes or proteins in a single sample accurately. Integration of nanobiosensors to the existing diagnostic workflows can be challenging. Additionally, the compatibility of nanobiosensors with current technology ensures standardization, calibration and seamless adoption in high throughput analysis.

The transformative potential of nanobiosensors in antimicrobial resistance (AMR) surveillance needs to be recognized. Nanobiosensors present several significant advantages over conventional methods of AMR detection, including high sensitivity, rapidity in detection, low sample requirements, and possibilities for miniaturization and multiplexing. Emerging trends like surface-enhanced Raman scattering (SERS), electrochemical biosensors, and optical biosensors further blurred the limits of the sensors to sensing of single molecules. The versatility of nanobiosensors also allows them to be adapted for various formats, such as wearable sensors and mobile diagnostic platforms. Additionally, the compatibility of nanobiosensors with artificial intelligence and machine learning will transform, surveillance by enhancing pattern recognition and predicting resistance trends, making AMR surveillance systems more proactive and predictive. These innovations make nanobiosensors not only accessible to healthcare facilities but also to community health workers in low-resource settings, where AMR is often underreported. Such adaptability has the potential to democratize AMR surveillance, promoting more equitable healthcare across regions with differing resource levels.

References

1. Paterson IK, Hoyle A, Ochoa G, Baker-Austin C, Taylor NGH (2016) Optimising antibiotic usage to treat bacterial infections. Sci Rep 6:37853
2. Kolář M (2022) Bacterial infections, antimicrobial resistance and antibiotic therapy. Life (Basel, Switzerland) 12
3. Cheadle WG (1992) Current perspectives on antibiotic use in the treatment of surgical infections. Am J Surg 164:44S-47S
4. Polk HC, Christmas AB (2000) Prophylactic antibiotics in surgery and surgical wound infections. Am Surg 66:105–111
5. Cosgrove SE (2006) The relationship between antimicrobial resistance and patient outcomes: mortality, length of hospital stay, and health care costs. Clin Infect Dis 42(Suppl 2):S82–S89
6. Wozniak TM, Dyda A, Lee X (2022) The increased length of hospital stay and mortality associated with community-associated infections in Australia. Open forum Infect Dis 9:ofac133
7. Christaki E, Marcou M, Tofarides A (2020) Antimicrobial resistance in bacteria: mechanisms, evolution, and persistence. J Mol Evol 88:26–40
8. Lerminiaux NA, Cameron ADS (2019) Horizontal transfer of antibiotic resistance genes in clinical environments. Can J Microbiol 65:34–44
9. Wang Q, Wei S, Silva AF, Madsen JS (2023) Cooperative antibiotic resistance facilitates horizontal gene transfer. ISME J 17:846–854
10. Holmes AH, Moore LSP, Sundsfjord A, Steinbakk M, Regmi S, Karkey A, Guerin PJ, Piddock LJV (2016) Understanding the mechanisms and drivers of antimicrobial resistance. Lancet (London, England) 387:176–187
11. Anjum MF, Zankari E, Hasman H (2017) Molecular methods for detection of antimicrobial resistance. Microbiol Spectr 5

12. Burcham ZM, Schmidt CJ, Pechal JL, Brooks CP, Rosch JW, Benbow ME, Jordan HR (2019) Detection of critical antibiotic resistance genes through routine microbiome surveillance. PLoS ONE 14:e0213280
13. Rajapaksha P, Elbourne A, Gangadoo S, Brown R, Cozzolino D, Chapman J (2019) A review of methods for the detection of pathogenic microorganisms. Analyst 144:396–411
14. Mabey D, Peeling RW, Ustianowski A, Perkins MD (2004) Diagnostics for the developing world. Nat Rev Microbiol 2:231–240
15. Hassan RYA (2022) Advances in electrochemical nano-biosensors for biomedical and environmental applications: from current work to future perspectives. Sensors (Basel) 22
16. Kant K (2023) Nano-biosensors for detection and monitoring (volume 1). Biosensors 13
17. Batool R, Rhouati A, Nawaz MH, Hayat A, Marty JL (2019) A review of the construction of nano-hybrids for electrochemical biosensing of glucose. Biosensors 9
18. Cox G, Wright GD (2013) Intrinsic antibiotic resistance: mechanisms, origins, challenges and solutions. Int J Med Microbiol 303:287–292
19. Li X-Z, Plésiat P, Nikaido H (2015) The challenge of efflux-mediated antibiotic resistance in gram-negative bacteria. Clin Microbiol Rev 28:337–418
20. Arzanlou M, Chai WC, Venter H (2017) Intrinsic, adaptive and acquired antimicrobial resistance in gram-negative bacteria. Essays Biochem 61:49–59
21. Impey RE, Hawkins DA, Sutton JM, Soares da Costa TP (2020) Overcoming intrinsic and acquired resistance mechanisms associated with the cell wall of gram-negative bacteria. Antibiot (Basel, Switzerland) 9
22. Dai J-S, Xu J, Shen H-J, Chen N-P, Zhu B-Q, Xue Z-J, Chen H-H, Ding Z-S, Ding R, Qian C-D (2024) The induced and intrinsic resistance of Escherichia coli to sanguinarine is mediated by AcrB efflux pump. Microbiol Spectr 12:e0323723
23. Choi BJ, Choi U, Ryu D-B, Lee C-R (2024) PhoPQ-mediated lipopolysaccharide modification governs intrinsic resistance to tetracycline and glycylcycline antibiotics in Escherichia coli. mSystems 9:e0096424
24. Maybin M, Ranade AM, Schombel U, Gisch N, Mamat U, Meredith TC (2024) IS1-mediated chromosomal amplification of the arn operon leads to polymyxin B resistance in Escherichia coli B strains. MBio 15:e0063424
25. Chen L, Duan K (2016) A PhoPQ-regulated ABC transporter system exports tetracycline in Pseudomonas aeruginosa. Antimicrob Agents Chemother 60:3016–3024
26. Younis RM, Faisal RM (2024) Plasposon mutagenesis in Pseudomonas aeruginosa isolates illustrates the role of ABC transporter in intrinsic resistance to antibiotics. J Appl Nat Sci 16:1256–1264
27. Çöl B, Kürkçü MS, Di Bek E (2024) Genome-wide screens identify genes responsible for intrinsic boric acid resistance in Escherichia coli. Biol Trace Elem Res 202:5771–5793
28. Wellner SM, Alobaidallah MSA, Fei X, Herrero-Fresno A, Olsen JE (2024) Genome-wide identification of fitness-genes in aminoglycoside-resistant Escherichia coli during antibiotic stress. Sci Rep 14:4163
29. Leshchiner D, Rosconi F, Sundaresh B, Rudmann E, Ramirez LMN, Nishimoto AT, Wood SJ, Jana B, Buján N, Li K, Gao J, Frank M, Reeve SM, Lee RE, Rock CO, Rosch JW, van Opijnen T (2022) A genome-wide atlas of antibiotic susceptibility targets and pathways to tolerance. Nat Commun 13:3165
30. Balachandra Dass S, Jayaraman R (1985) Conditional rifampicin sensitivity of arif mutant of Escherichia coli: rifampicin induced changes in transcription specificity. J Biosci 9:213–221
31. Zhang H, Dai X, Hu P, Tian L, Li C, Ding B, Yang X, He X (2024) Comparison of targeted next-generation sequencing and the Xpert MTB/RIF assay for detection of Mycobacterium tuberculosis in clinical isolates and sputum specimens. Microbiol Spectr 12:e0409823
32. Torokaa PR, Majigo MV, Kileo H, Urio L, Mbwana MR, Monah MC, Ntibabara SS, Kimambo J, Seleman P, Franklin C, Balama R, Kisonga RM, Joachim A (2024) The pattern of rpoB gene mutation of Mycobacterium tuberculosis and predictors of rifampicin resistance detected by GeneXpert MTB/RIF assay in Tanzania. PLoS ONE 19:e0296563

33. Zaw MT, Emran NA, Lin Z (2018) Mutations inside rifampicin-resistance determining region of rpoB gene associated with rifampicin-resistance in Mycobacterium tuberculosis. J Infect Public Health 11:605–610
34. Rezaei S, Tajbakhsh S, Naeimi B, Yousefi F (2024) Investigation of gyrA and parC mutations and the prevalence of plasmid-mediated quinolone resistance genes in Klebsiella pneumoniae clinical isolates. BMC Microbiol 24:265
35. Molodtsov V, Sineva E, Zhang L, Huang X, Cashel M, Ades SE, Murakami KS (2018) Allosteric effector ppGpp potentiates the inhibition of transcript initiation by DksA. Mol Cell 69:828-839.e5
36. Shemediuk AL, Dolia BS, Ochi K, Fedorenko VO, Ostash BO (2022) Properties of spontaneous rpsL mutant of Streptomyces albus KO-1297. Cytol Genet 56:31–36
37. Shafipour M, Shirzad-Aski H, Mohammadzadeh A, Ghazvini K, Zamani S, Koohi PM, Ghaemi S, Ghaemi EA (2022) Evaluation of mutations related to streptomycin resistance in Mycobacterium tuberculosis clinical isolates. Curr Microbiol 79:343
38. Smittipat N, Juthayothin T, Billamas P, Jaitrong S, Rukseree K, Dokladda K, Chaiyasirinroje B, Disratthakit A, Chaiprasert A, Mahasirimongkol S, Yanai H, Yamada N, Tokunaga K, Palittapongarnpim P (2016) Mutations in rrs, rpsL and gidB in streptomycin-resistant Mycobacterium tuberculosis isolates from Thailand. J Glob Antimicrob Resist 4:5–10
39. Masi M, Winterhalter M, Pagès J-M (2019) Outer membrane porins. Subcell Biochem 92:79–123
40. Zhou G, Wang Q, Wang Y, Wen X, Peng H, Peng R, Shi Q, Xie X, Li L (2023) Outer membrane porins contribute to antimicrobial resistance in gram-negative bacteria. Microorganisms 11
41. Choi U, Lee C-R (2019) Distinct roles of outer membrane porins in antibiotic resistance and membrane integrity in Escherichia coli. Front Microbiol 10:953
42. Dé E, Baslé A, Jaquinod M, Saint N, Malléa M, Molle G, Pagès JM (2001) A new mechanism of antibiotic resistance in Enterobacteriaceae induced by a structural modification of the major porin. Mol Microbiol 41:189–198
43. Zheng P, Lun J, Yu F, Huang T, Peng T, Li J, Hu Z (2024) Deletion of ArmPT, a LamB-like protein, increases cell membrane permeability and antibiotic sensitivity in Vibrio alginolyticus. Ecotoxicol Environ Saf 270:115855
44. Kuang S-F, Xiang J, Zeng Y-Y, Peng X-X, Li H (2024) Elevated membrane potential as a tetracycline resistance mechanism in Escherichia coli. ACS Infect Dis 10:2196–2211
45. Du D, Wang-Kan X, Neuberger A, van Veen HW, Pos KM, Piddock LJV, Luisi BF (2018) Multidrug efflux pumps: structure, function and regulation. Nat Rev Microbiol 16:523–539
46. Blair JMA, Richmond GE, Piddock LJV (2014) Multidrug efflux pumps in gram-negative bacteria and their role in antibiotic resistance. Future Microbiol 9:1165–1177
47. Zwama M, Nishino K (2021) Ever-adapting RND efflux pumps in gram-negative multidrug-resistant pathogens: a race against time. Antibiot (Basel, Switzerland) 10
48. Alekshun MN, Levy SB (2007) Molecular mechanisms of antibacterial multidrug resistance. Cell 128:1037–1050
49. Qin Z, Peng K, Feng Y, Wang Y, Huang B, Tian Z, Ouyang P, Huang X, Chen D, Lai W, Geng Y (2024) Transcriptome reveals the role of the htpG gene in mediating antibiotic resistance through cell envelope modulation in Vibrio mimicus SCCF01. Front Microbiol 14
50. Oncel B, Hasdemir U, Aksu B, Pournaras S (2024) Antibiotic resistance in Campylobacter jejuni and Campylobacter coli: significant contribution of an RND type efflux pump in erythromycin resistance. J Chemother 36:110–118
51. De Pascale G, Wright GD (2010) Antibiotic resistance by enzyme inactivation: from mechanisms to solutions. ChemBioChem 11:1325–1334
52. Wright GD (2011) Molecular mechanisms of antibiotic resistance. Chem Commun (Camb) 47:4055–4061
53. Liao J, Qi Q, Kuang L, Zhou Y, Xiao Q, Liu T, Wang X, Guo L, Jiang Y (2024) Chloramphenicol binding sites of Acinetobacter baumannii chloramphenicol acetyltransferase CatB8. ACS Infect Dis 10:870–878

54. Karalewitz AP-A, Miller SI (2018) Multidrug-resistant Acinetobacter baumannii chloramphenicol resistance requires an inner membrane permease. Antimicrob Agents Chemother 62
55. Ramirez MS, Nikolaidis N, Tolmasky ME (2013) Rise and dissemination of aminoglycoside resistance: the aac(6')-Ib paradigm. Front Microbiol 4:121
56. Sarno R, McGillivary G, Sherratt DJ, Actis LA, Tolmasky ME (2002) Complete nucleotide sequence of Klebsiella pneumoniae multiresistance plasmid pJHCMW1. Antimicrob Agents Chemother 46:3422–3427
57. Tolmasky ME, Crosa JH (1987) Tn1331, a novel multiresistance transposon encoding resistance to amikacin and ampicillin in Klebsiella pneumoniae. Antimicrob Agents Chemother 31:1955–1960
58. Koenig C, Kuti JL (2024) Evolving resistance landscape in gram-negative pathogens: an update on β-lactam and β-lactam-inhibitor treatment combinations for carbapenem-resistant organisms. Pharmacotherapy 44:658–674
59. Sawa T, Kooguchi K, Moriyama K (2020) Molecular diversity of extended-spectrum β-lactamases and carbapenemases, and antimicrobial resistance. J Intensive Care 8:13
60. Gutkind GO, Di Conza J, Power P, Radice M (2013) β-lactamase-mediated resistance: a biochemical, epidemiological and genetic overview. Curr Pharm Des 19:164–208
61. Mascellino MT (2022) Multi-drug-resistant gram-negative microorganisms: epidemiology, treatment and alternative approach. Antibiot (Basel, Switzerland) 11
62. Livermore DM (2008) Defining an extended-spectrum beta-lactamase. Clin Microbiol Infect 14(Suppl 1):3–10
63. Darby EM, Trampari E, Siasat P, Gaya MS, Alav I, Webber MA, Blair JMA (2023) Molecular mechanisms of antibiotic resistance revisited. Nat Rev Microbiol 21:280–295
64. Morrill HJ, LaPlante KL (2015) Overconsumption of antibiotics. Lancet Infect Dis 15:377–378
65. Ghafur A (2015) Overconsumption of antibiotics. Lancet Infect Dis 15:377
66. Browne AJ, Chipeta MG, Haines-Woodhouse G, Kumaran EPA, Hamadani BHK, Zaraa S, Henry NJ, Deshpande A, Reiner RC, Day NPJ, Lopez AD, Dunachie S, Moore CE, Stergachis A, Hay SI, Dolecek C (2021) Global antibiotic consumption and usage in humans, 2000–18: a spatial modelling study. Lancet Planet Heal 5:e893–e904
67. Van Boeckel TP, Gandra S, Ashok A, Caudron Q, Grenfell BT, Levin SA, Laxminarayan R (2014) Global antibiotic consumption 2000 to 2010: an analysis of national pharmaceutical sales data. Lancet Infect Dis 14:742–750
68. Ding D, Wang B, Zhang X, Zhang J, Zhang H, Liu X, Gao Z, Yu Z (2023) The spread of antibiotic resistance to humans and potential protection strategies. Ecotoxicol Environ Saf 254:114734
69. Liu X, Shrestha R, Koju P, Maharjan B, Shah P, Thapa P, Li H (2022) The direct medical economic burden of healthcare-associated infections and antimicrobial resistance: a preliminary study in a teaching hospital of Nepal. J Glob Antimicrob Resist 29:299–303
70. Cagle SD, Hutcherson BL, Wiley AT (2022) Health care-associated infections: best practices for prevention. Am Fam Physician 105:262–270
71. He P, Wu Y, Huang W, Wu X, Lv J, Liu P, Bu L, Bai Z, Chen S, Feng W, Yang Z (2020) Characteristics of and variation in airborne ARGs among urban hospitals and adjacent urban and suburban communities: a metagenomic approach. Environ Int 139:105625
72. Tong C, Xiao D, Xie L, Yang J, Zhao R, Hao J, Huo Z, Zeng Z, Xiong W (2022) Swine manure facilitates the spread of antibiotic resistome including tigecycline-resistant tet(X) variants to farm workers and receiving environment. Sci Total Environ 808:152157
73. Liu Y-Y, Wang Y, Walsh TR, Yi L-X, Zhang R, Spencer J, Doi Y, Tian G, Dong B, Huang X, Yu L-F, Gu D, Ren H, Chen X, Lv L, He D, Zhou H, Liang Z, Liu J-H, Shen J (2016) Emergence of plasmid-mediated colistin resistance mechanism MCR-1 in animals and human beings in China: a microbiological and molecular biological study. Lancet Infect Dis 16:161–168
74. Mendelsohn E, Ross N, Zambrana-Torrelio C, Van Boeckel TP, Laxminarayan R, Daszak P (2023) Global patterns and correlates in the emergence of antimicrobial resistance in humans. Proc Biol Sci 290:20231085

75. Smith TC, Gebreyes WA, Abley MJ, Harper AL, Forshey BM, Male MJ, Martin HW, Molla BZ, Sreevatsan S, Thakur S, Thiruvengadam M, Davies PR (2013) Methicillin-resistant Staphylococcus aureus in pigs and farm workers on conventional and antibiotic-free swine farms in the USA. PLoS ONE 8:e63704
76. Vieira AR, Collignon P, Aarestrup FM, McEwen SA, Hendriksen RS, Hald T, Wegener HC (2011) Association between antimicrobial resistance in Escherichia coli isolates from food animals and blood stream isolates from humans in Europe: an ecological study. Foodborne Pathog Dis 8:1295–1301
77. Husain Khan A, Abdul Aziz H, Palaniandy P, Naushad M, Cevik E, Zahmatkesh S (2023) Pharmaceutical residues in the ecosystem: antibiotic resistance, health impacts, and removal techniques. Chemosphere 339:139647
78. Nguyen AQ, Vu HP, Nguyen LN, Wang Q, Djordjevic SP, Donner E, Yin H, Nghiem LD (2021) Monitoring antibiotic resistance genes in wastewater treatment: current strategies and future challenges. Sci Total Environ 783:146964
79. Caioni G, Benedetti E, Perugini M, Amorena M, Merola C (2023) Personal care products as a contributing factor to antimicrobial resistance: current state and novel approach to investigation. Antibiot (Basel, Switzerland) 12
80. Roberts MG, Burgess S, Toombs-Ruane LJ, Benschop J, Marshall JC, French NP (2021) Combining mutation and horizontal gene transfer in a within-host model of antibiotic resistance. Math Biosci 339:108656
81. Evans DR, Griffith MP, Sundermann AJ, Shutt KA, Saul MI, Mustapha MM, Marsh JW, Cooper VS, Harrison LH, Van Tyne D (2020) Systematic detection of horizontal gene transfer across genera among multidrug-resistant bacteria in a single hospital. Elife 9
82. Abe K, Nomura N, Suzuki S (2020) Biofilms: hot spots of horizontal gene transfer (HGT) in aquatic environments, with a focus on a new HGT mechanism. FEMS Microbiol Ecol 96
83. Aminov RI (2011) Horizontal gene exchange in environmental microbiota. Front Microbiol 2:158
84. Fang G-Y, Liu X-Q, Jiang Y-J, Mu X-J, Huang B-W (2024) Horizontal gene transfer in activated sludge enhances microbial antimicrobial resistance and virulence. Sci Total Environ 912:168908
85. 2021. Flies, cockroaches and AMR. Nat Microbiol 6:1213–1214
86. Usui M, Shirakawa T, Fukuda A, Tamura Y (2015) The role of flies in disseminating plasmids with antimicrobial-resistance genes between farms. Microb Drug Resist 21:562–569
87. Onwugamba FC, Fitzgerald JR, Rochon K, Guardabassi L, Alabi A, Kühne S, Grobusch MP, Schaumburg F (2018) The role of "filth flies" in the spread of antimicrobial resistance. Travel Med Infect Dis 22:8–17
88. Nayduch D, Neupane S, Pickens V, Purvis T, Olds C (2023) House flies are underappreciated yet important reservoirs and vectors of microbial threats to animal and human health. Microorganisms 11
89. Gwenzi W, Chaukura N, Muisa-Zikali N, Teta C, Musvuugwa T, Rzymski P, Abia ALK (2021) Insects, rodents, and pets as reservoirs, vectors, and sentinels of antimicrobial resistance. Antibiot (Basel, Switzerland) 10
90. Poudel A, Kang Y, Mandal RK, Kalalah A, Butaye P, Hathcock T, Kelly P, Walz P, Macklin K, Cattley R, Price S, Adekanmbi F, Zhang L, Kitchens S, Kaltenboeck B, Wang C (2020) Comparison of microbiota, antimicrobial resistance genes and mobile genetic elements in flies and the feces of sympatric animals. FEMS Microbiol Ecol 96
91. Mosaheb MU-WFZ, Khan NA, Siddiqui R (2018) Cockroaches, locusts, and envenomating arthropods: a promising source of antimicrobials. Iran J Basic Med Sci 21:873–877
92. Rodovalho CM, Santos AL, Marcolino MT, Bonetti AM, Brandeburgo MAM (2007) Urban ants and transportation of nosocomial bacteria. Neotrop Entomol 36:454–458
93. Literak I, Dolejska M, Rybarikova J, Cizek A, Strejckova P, Vyskocilova M, Friedman M, Klimes J (2009) Highly variable patterns of antimicrobial resistance in commensal Escherichia coli isolates from pigs, sympatric rodents, and flies. Microb Drug Resist 15:229–237

94. Channaiah LH, Subramanyam B, McKinney LJ, Zurek L (2010) Stored-product insects carry antibiotic-resistant and potentially virulent enterococci. FEMS Microbiol Ecol 74:464–471
95. Zhou Z, Lin Z, Shuai X, Achi C, Chen H (2024) Antibiotic resistance genes alterations in murine guts microbiome are associated with different types of drinking water. J Hazard Mater 465:133422
96. Li Y, Fernández R, Durán I, Molina-López RA, Darwich L (2020) Antimicrobial resistance in bacteria isolated from cats and dogs from the Iberian Peninsula. Front Microbiol 11:621597
97. Smoglica C, Evangelisti G, Fani C, Marsilio F, Trotta M, Messina F, Di Francesco CE (2022) Antimicrobial resistance profile of bacterial isolates from urinary tract infections in companion animals in Central Italy. Antibiot (Basel, Switzerland) 11
98. Fernández M, Garcias B, Duran I, Molina-López RA, Darwich L (2023) Current situation of bacterial infections and antimicrobial resistance profiles in pet rabbits in Spain. Vet Sci 10
99. Chen Y, Liu Z, Zhang Y, Zhang Z, Lei L, Xia Z (2019) Increasing prevalence of ESBL-producing multidrug resistance Escherichia coli from diseased pets in Beijing, China from 2012 to 2017. Front Microbiol 10:2852
100. Liu X, Thungrat K, Boothe DM (2016) Occurrence of OXA-48 carbapenemase and other β-lactamase genes in ESBL-producing multidrug resistant Escherichia coli from dogs and cats in the United States, 2009–2013. Front Microbiol 7:1057
101. Becsei Á, Solymosi N, Csabai I, Magyar D (2021) Detection of antimicrobial resistance genes in urban air. Microbiologyopen 10:e1248
102. Maestre-Carballa L, Navarro-López V, Martinez-Garcia M (2024) Metagenomic airborne resistome from urban hot spots through the One Health lens. Environ Microbiol Rep 16:e13306
103. Hu J, Zhao F, Zhang X-X, Li K, Li C, Ye L, Li M (2018) Metagenomic profiling of ARGs in airborne particulate matters during a severe smog event. Sci Total Environ 615:1332–1340
104. Crick F (1970) Central dogma of molecular biology. Nature 227:561–563
105. Pontarotti G, Mossio M, Pocheville A (2022) The genotype-phenotype distinction: from Mendelian genetics to 21st century biology. Genetica 150:223–234
106. Rozwandowicz M, Brouwer MSM, Fischer J, Wagenaar JA, Gonzalez-Zorn B, Guerra B, Mevius DJ, Hordijk J (2018) Plasmids carrying antimicrobial resistance genes in Enterobacteriaceae. J Antimicrob Chemother 73:1121–1137
107. Rekadwad BN, Pramod N, Rao MPN, Hashem A, Avila-Quezada GD, Abd Allah EF (2023) Identification and specificity validation of unique and antimicrobial resistance genes to trace suspected pathogenic AMR bacteria and to monitor the development of AMR in non-AMR strains in the environment and clinical settings. Saudi J Biol Sci 30:103869
108. Tokuda M, Shintani M (2024) Microbial evolution through horizontal gene transfer by mobile genetic elements. Microb Biotechnol 17:e14408
109. Moura de Sousa J, Lourenço M, Gordo I (2023) Horizontal gene transfer among host-associated microbes. Cell Host Microbe 31:513–527
110. Wendling CC, Refardt D, Hall AR (2021) Fitness benefits to bacteria of carrying prophages and prophage-encoded antibiotic-resistance genes peak in different environments. Evolution 75:515–528
111. Hernando-Amado S, Sanz-García F, Blanco P, Martínez JL (2017) Fitness costs associated with the acquisition of antibiotic resistance. Essays Biochem 61:37–48
112. Gajic I, Kabic J, Kekic D, Jovicevic M, Milenkovic M, Mitic Culafic D, Trudic A, Ranin L, Opavski N (2022) Antimicrobial susceptibility testing: a comprehensive review of currently used methods. Antibiot (Basel, Switzerland) 11
113. Salam MA, Al-Amin MY, Pawar JS, Akhter N, Lucy IB (2023) Conventional methods and future trends in antimicrobial susceptibility testing. Saudi J Biol Sci 30:103582
114. Tran CM, Tanaka K, Watanabe K (2013) PCR-based detection of resistance genes in anaerobic bacteria isolated from intra-abdominal infections. J Infect Chemother 19:279–290
115. Volkmann H, Schwartz T, Bischoff P, Kirchen S, Obst U (2004) Detection of clinically relevant antibiotic-resistance genes in municipal wastewater using real-time PCR (TaqMan). J Microbiol Methods 56:277–286

116. Peng J, Cheng G, Huang L, Wang Y, Hao H, Peng D, Liu Z, Yuan Z (2013) Development of a direct ELISA based on carboxy-terminal of penicillin-binding protein BlaR for the detection of β-lactam antibiotics in foods. Anal Bioanal Chem 405:8925–8933
117. Odekerken JCE, Logister DMW, Assabre L, Arts JJC, Walenkamp GHIM, Welting TJM (2015) ELISA-based detection of gentamicin and vancomycin in protein-containing samples. Springerplus 4:614
118. Wang W, Liu L, Xu L, Ma W, Kuang H, Xu C (2013) Detection of β-lactamase residues in milk by sandwich ELISA. Int J Environ Res Public Health 10:2688–2698
119. Jakobsen L, Garneau P, Kurbasic A, Bruant G, Stegger M, Harel J, Jensen KS, Brousseau R, Hammerum AM, Frimodt-Møller N (2011) Microarray-based detection of extended virulence and antimicrobial resistance gene profiles in phylogroup B2 Escherichia coli of human, meat and animal origin. J Med Microbiol 60:1502–1511
120. Vora GJ, Meador CE, Bird MM, Bopp CA, Andreadis JD, Stenger DA (2005) Microarray-based detection of genetic heterogeneity, antimicrobial resistance, and the viable but nonculturable state in human pathogenic Vibrio spp. Proc Natl Acad Sci USA 102:19109–19114
121. Schmitt S, Stephan R, Huebschke E, Schaefle D, Merz A, Johler S (2020) DNA microarray-based characterization and antimicrobial resistance phenotypes of clinical MRSA strains from animal hosts. J Vet Sci 21:e54
122. Perreten V, Vorlet-Fawer L, Slickers P, Ehricht R, Kuhnert P, Frey J (2005) Microarray-based detection of 90 antibiotic resistance genes of gram-positive bacteria. J Clin Microbiol 43:2291–2302
123. Boolchandani M, D'Souza AW, Dantas G (2019) Sequencing-based methods and resources to study antimicrobial resistance. Nat Rev Genet 20:356–370
124. Rooney AM, Raphenya AR, Melano RG, Seah C, Yee NR, MacFadden DR, McArthur AG, Schneeberger PHH, Coburn B (2022) Performance characteristics of next-generation sequencing for the detection of antimicrobial resistance determinants in Escherichia coli genomes and metagenomes. mSystems 7:e0002222
125. Conroy PJ, Hearty S, Leonard P, O'Kennedy RJ (2009) Antibody production, design and use for biosensor-based applications. Semin Cell Dev Biol 20:10–26
126. Haji-Ghassemi O, Blackler RJ, Martin Young N, Evans SV (2015) Antibody recognition of carbohydrate epitopes†. Glycobiology 25:920–952
127. Huflejt ME, Vuskovic M, Vasiliu D, Xu H, Obukhova P, Shilova N, Tuzikov A, Galanina O, Arun B, Lu K, Bovin N (2009) Anti-carbohydrate antibodies of normal sera: findings, surprises and challenges. Mol Immunol 46:3037–3049
128. van Regenmortel MH (2000) The recognition of proteins and peptides by antibodies. J Immunoassay 21:85–108
129. Chersi A, Di Modugno F, Rosano L (1997) Aims and limitations in the use of antipeptide antibodies in molecular biology. Biol Chem 378:635–640
130. Brady RA, Leid JG, Camper AK, Costerton JW, Shirtliff ME (2006) Identification of Staphylococcus aureus proteins recognized by the antibody-mediated immune response to a biofilm infection. Infect Immun 74:3415–3426
131. Spellerberg MB, Chapman CJ, Mockridge CI, Isenberg DA, Stevenson FK (1995) Dual recognition of lipid A and DNA by human antibodies encoded by the VH4-21 gene: a possible link between infection and lupus. Hum Antibodies Hybridomas 6:52–56
132. Radic MZ, Weigert M (1995) Origins of anti-DNA antibodies and their implications for B-cell tolerance. Ann N Y Acad Sci 764:384–396
133. Wang X, Xia Y (2019) Anti-double stranded DNA antibodies: origin, pathogenicity, and targeted therapies. Front Immunol 10:1667
134. Pisetsky DS, Garza Reyna A, Belina ME, Spencer DM (2022) The interaction of anti-DNA antibodies with DNA: evidence for unconventional binding mechanisms. Int J Mol Sci 23
135. O'Hara DM, Reynolds PE (1987) Antibody used to identify penicillin-binding protein 2' in methicillin-resistant strains of Staphylococcus aureus (MRSA). FEBS Lett 212:237–241
136. Seixas AMM, Sousa SA, Leitão JH (2022) Antibody-based immunotherapies as a tool for tackling multidrug-resistant bacterial infections. Vaccines 10

137. Kordus SL, Kroh HK, Rodríguez RC, Shrem RA, Peritore-Galve FC, Shupe JA, Wadzinski BE, Lacy DB, Spiller BW (2023) Nanobodies against C. difficile TcdA and TcdB reveal unexpected neutralizing epitopes and provide a toolkit for toxin quantitation in vivo. PLoS Pathog 19:e1011496
138. Tran L-H, Graulus G-J, Vincke C, Smiejkowska N, Kindt A, Devoogdt N, Muyldermans S, Adriaensens P, Guedens W (2022) Nanobodies for the early detection of ovarian cancer. Int J Mol Sci 23
139. Kartner N, Evernden-Porelle D, Bradley G, Ling V. Detection of P-glycoprotein in multidrug-resistant cell lines by monoclonal antibodies. Nature 316:820–823
140. Shi X, Kadiyala U, VanEpps JS, Yau S-T (2018) Culture-free bacterial detection and identification from blood with rapid, phenotypic, antibiotic susceptibility testing. Sci Rep 8:3416
141. Tamma PD, Simner PJ (2018) Phenotypic detection of carbapenemase-producing organisms from clinical isolates. J Clin Microbiol 56
142. Stieber B, Monecke S, Müller E, Büchler J, Ehricht R (2015) Direct, specific and rapid detection of staphylococcal proteins and exotoxins using a multiplex antibody microarray. PLoS ONE 10:e0143246
143. Zhou W, Huang P-JJ, Ding J, Liu J (2014) Aptamer-based biosensors for biomedical diagnostics. Analyst 139:2627–2640
144. Lönne M, Zhu G, Stahl F, Walter J-G (2014) Aptamer-modified nanoparticles as biosensors. Adv Biochem Eng Biotechnol 140:121–154
145. Rozenblum GT, Lopez VG, Vitullo AD, Radrizzani M (2016) Aptamers: current challenges and future prospects. Expert Opin Drug Discov 11:127–135
146. Bayat P, Nosrati R, Alibolandi M, Rafatpanah H, Abnous K, Khedri M, Ramezani M (2018) SELEX methods on the road to protein targeting with nucleic acid aptamers. Biochimie 154:132–155
147. Darmostuk M, Rimpelova S, Gbelcova H, Ruml T (2015) Current approaches in SELEX: an update to aptamer selection technology. Biotechnol Adv 33:1141–1161
148. Kimoto M, Shermane Lim YW, Hirao I (2019) Molecular affinity rulers: systematic evaluation of DNA aptamers for their applicabilities in ELISA. Nucleic Acids Res 47:8362–8374
149. Toh SY, Citartan M, Gopinath SCB, Tang T-H (2015) Aptamers as a replacement for antibodies in enzyme-linked immunosorbent assay. Biosens Bioelectron 64:392–403
150. Nomura Y, Yokobayashi Y (2021) Aptazyme-based riboswitches and logic gates in mammalian cells. Methods Mol Biol 2323:213–220
151. Yang Y, Huang J, Yang X, Quan K, Wang H, Ying L, Xie N, Ou M, Wang K (2016) Aptazyme-gold nanoparticle sensor for amplified molecular probing in living cells. Anal Chem 88:5981–5987
152. Liang G, Song L, Gao Y, Wu K, Guo R, Chen R, Zhen J, Pan L (2023) Aptamer sensors for the detection of antibiotic residues—a mini-review. Toxics 11
153. Robati RY, Arab A, Ramezani M, Langroodi FA, Abnous K, Taghdisi SM (2016) Aptasensors for quantitative detection of kanamycin. Biosens Bioelectron 82:162–172
154. Zhan L, Li CM, Fu ZF, Zou HY, Huang CZ (2022) Dual-aptamer-based enzyme linked plasmonic assay for pathogenic bacteria detection. Colloids Surf B Biointerfaces 214:112471
155. Li D, Liu L, Huang Q, Tong T, Zhou Y, Li Z, Bai Q, Liang H, Chen L (2021) Recent advances on aptamer-based biosensors for detection of pathogenic bacteria. World J Microbiol Biotechnol 37:45
156. Torres-Chavolla E, Alocilja EC (2009) Aptasensors for detection of microbial and viral pathogens. Biosens Bioelectron 24:3175–3182
157. Li Y, Zhang D, Zeng X, Liu C, Wu Y, Fu C (2024) Advances in aptamer-based biosensors for the detection of foodborne mycotoxins. Molecules 29
158. Calabria D, Zangheri M, Pour SRS, Trozzi I, Pace A, Lazzarini E, Calabretta MM, Mirasoli M, Guardigli M (2022) Luminescent aptamer-based bioassays for sensitive detection of food allergens. Biosensors 12

159. Piletsky S, Canfarotta F, Poma A, Bossi AM, Piletsky S (2020) Molecularly imprinted polymers for cell recognition. Trends Biotechnol 38:368–387
160. Nawaz N, Abu Bakar NK, Muhammad Ekramul Mahmud HN, Jamaludin NS (2021) Molecularly imprinted polymers-based DNA biosensors. Anal Biochem 630:114328
161. El-Schich Z, Zhang Y, Feith M, Beyer S, Sternbæk L, Ohlsson L, Stollenwerk M, Wingren AG (2020) Molecularly imprinted polymers in biological applications. Biotechniques 69:406–419
162. BelBruno JJ (2019) Molecularly imprinted polymers. Chem Rev 119:94–119
163. Zhang H (2020) Molecularly imprinted nanoparticles for biomedical applications. Adv Mater 32:e1806328
164. Wang HY, Xia SL, Sun H, Liu YK, Cao SK, Kobayashi T (2004) Molecularly imprinted copolymer membranes functionalized by phase inversion imprinting for uracil recognition and permselective binding. J Chromatogr B Analyt Technol Biomed Life Sci 804:127–134
165. Lee M-H, Thomas JL, Ho M-H, Yuan C, Lin H-Y (2010) Synthesis of magnetic molecularly imprinted poly(ethylene-co-vinyl alcohol) nanoparticles and their uses in the extraction and sensing of target molecules in urine. ACS Appl Mater Interfaces 2:1729–1736
166. Wolfe DB, Qin D, Whitesides GM (2010) Rapid prototyping of microstructures by soft lithography for biotechnology. Methods Mol Biol 583:81–107
167. Mele E, Pisignano D (2009) Nanobiotechnology: soft lithography. Prog Mol Subcell Biol 47:341–358
168. Pilvenyte G, Ratautaite V, Boguzaite R, Ramanavicius A, Viter R, Ramanavicius S (2023) Molecularly imprinted polymers for the determination of cancer biomarkers. Int J Mol Sci 24
169. Wang J (2000) From DNA biosensors to gene chips. Nucleic Acids Res 28:3011–3016
170. Vercoutere W, Akeson M (2002) Biosensors for DNA sequence detection. Curr Opin Chem Biol 6:816–822
171. Hua Y, Ma J, Li D, Wang R (2022) DNA-based biosensors for the biochemical analysis: a review. Biosensors 12
172. Chao J, Zhu D, Zhang Y, Wang L, Fan C (2016) DNA nanotechnology-enabled biosensors. Biosens Bioelectron 76:68–79
173. Stangherlin S, Liu J (2023) Nanomaterials enabled and enhanced DNA-based biosensors. J Mater Chem B 11:6994–7003
174. Ferapontova EE (2018) DNA electrochemistry and electrochemical sensors for nucleic acids. Annu Rev Anal Chem (Palo Alto Calif) 11:197–218
175. Teengam P, Siangproh W, Tuantranont A, Henry CS, Vilaivan T, Chailapakul O (2017) Electrochemical paper-based peptide nucleic acid biosensor for detecting human papillomavirus. Anal Chim Acta 952:32–40
176. D'Agata R, Giuffrida MC, Spoto G (2017) Peptide nucleic acid-based biosensors for cancer diagnosis. Molecules 22
177. Nie L, Liu F, Ma P, Xiao X (2014) Applications of gold nanoparticles in optical biosensors. J Biomed Nanotechnol 10:2700–2721
178. Eivazzadeh-Keihan R, Saadatidizaji Z, Mahdavi M, Maleki A, Irani M, Zare I (2024) Recent advances in gold nanoparticles-based biosensors for tuberculosis determination. Talanta 275:126099
179. Peng H-I, Miller BL (2011) Recent advancements in optical DNA biosensors: exploiting the plasmonic effects of metal nanoparticles. Analyst 136:436–447
180. Luo X, Morrin A, Killard AJ, Smyth MR (2006) Application of nanoparticles in electrochemical sensors and biosensors. Electroanalysis 18:319–326
181. Wang Y, Alocilja EC (2015) Gold nanoparticle-labeled biosensor for rapid and sensitive detection of bacterial pathogens. J Biol Eng 9:16
182. Beck F, Loessl M, Baeumner AJ (2023) Signaling strategies of silver nanoparticles in optical and electrochemical biosensors: considering their potential for the point-of-care. Mikrochim Acta 190:91
183. Webster MS, Cooper JS, Chow E, Hubble LJ, Sosa-Pintos A, Wieczorek L, Raguse B (2015) Detection of bacterial metabolites for the discrimination of bacteria utilizing gold nanoparticle chemiresistor sensors. Sensors Actuators B Chem 220:895–902

184. Wang M, Jin L, Hang-Mei Leung P, Wang-Ngai Chow F, Zhao X, Chen H, Pan W, Liu H, Li S (2024) Advancements in magnetic nanoparticle-based biosensors for point-of-care testing. Front Bioeng Biotechnol 12:1393789
185. Karnwal A, Kumar Sachan RS, Devgon I, Devgon J, Pant G, Panchpuri M, Ahmad A, Alshammari MB, Hossain K, Kumar G (2024) Gold nanoparticles in nanobiotechnology: from synthesis to biosensing applications. ACS Omega 9:29966–29982
186. Chung HJ, Castro CM, Im H, Lee H, Weissleder R (2013) A magneto-DNA nanoparticle system for rapid detection and phenotyping of bacteria. Nat Nanotechnol 8:369–375
187. Zhao W, Wang L, Tan W (2007) Fluorescent nanoparticle for bacteria and DNA detection. Adv Exp Med Biol 620:129–135
188. Rajendran VK, Bakthavathsalam P, Jaffar Ali BM (2014) Smartphone based bacterial detection using biofunctionalized fluorescent nanoparticles. Microchim Acta 181:1815–1821
189. Zhou H, Yang D, Ivleva NP, Mircescu NE, Niessner R, Haisch C (2014) SERS detection of bacteria in water by in situ coating with Ag nanoparticles. Anal Chem 86:1525–1533
190. Kaittanis C, Naser SA, Perez JM (2007) One-step, nanoparticle-mediated bacterial detection with magnetic relaxation. Nano Lett 7:380–383
191. Fornara A, Johansson P, Petersson K, Gustafsson S, Qin J, Olsson E, Ilver D, Krozer A, Muhammed M, Johansson C (2008) Tailored magnetic nanoparticles for direct and sensitive detection of biomolecules in biological samples. Nano Lett 8:3423–3428
192. Lee H, Yoon T-J, Weissleder R (2009) Ultrasensitive detection of bacteria using core-shell nanoparticles and an NMR-filter system. Angew Chem Int Ed Engl 48:5657–5660

The manufacturer's authorised representative in the EU is Springer Nature Customer Service Centre GmbH, Europaplatz 3, 69115 Heidelberg, Germany. If you have any concerns regarding our products, please contact ProductSafety@springernature.com

Printed and bound by CPI Group (UK) Ltd, Croydon, CR0 4YY
01/10/2026
02225928-0016